AF247475

International Review of Science

Organic Chemistry
Series Two

Consultant Editor
D. H. Hey, F.R.S.

Publisher's Note

The International Review of Science is an important venture in scientific publishing presented by Butterworths. The basic concept of the Review is to provide regular authoritative reviews of entire disciplines. Chemistry was taken first as the problems of literature survey are probably more acute in this subject than in any other. Biochemistry and Physiology followed naturally. As a matter of policy, the authorship of the Review of Science is international and distinguished, the subject coverage is extensive, systematic and critical.

The Review has been conceived within a carefully organised editorial framework. The overall plan was drawn up, and the volume editors appointed by seven consultant editors. In turn, each volume editor planned the coverage of his field and appointed authors to write on subjects which were within the area of their own research experience. No geographical restriction was imposed. Hence the 500 or so contributions to the Review of Science come from many countries of the world and provide an authoritative account of progress. The publication of Organic Chemistry Series One was completed in 1973 with ten text volumes and one index volume; in accordance with the stated policy of issuing regular reviews to keep the series up to date, volumes of Series Two will be published between the middle of 1975 and early 1976; Series Two of Physical Chemistry will be published at the same time, while Inorganic Chemistry Series Two was published during the first half of 1975. Volume titles are the same as in Series One but the articles themselves either cover recent advances in the same subject or deal with a different aspect of the main theme of the volume. In Series Two an index is incorporated in each volume and there is no separate index volume.

Butterworth & Co. (Publishers) Ltd.

BIOCHEMISTRY SERIES ONE

Consultant Editors
H. L. Kornberg, F.R.S.,
*Department of Biochemistry
University of Leicester* and
D. C. Phillips, F.R.S., *Department of
Zoology, University of Oxford*

Volume titles and Editors

PHYSIOLOGY SERIES ONE

Consultant Editors
A. C. Guyton,
*Department of Physiology and
Biophysics, University of Mississippi
Medical Center* and
D. F. Horrobin,
*Department of Physiology, University
of Newcastle upon Tyne*

Volumes titles and Editors

International Review of Science

Organic Chemistry
Series Two

Volume 2
Aliphatic Compounds

Edited by **N. B. Chapman**
University of Hull

BUTTERWORTHS
LONDON - BOSTON
Sydney - Wellington - Durban - Toronto

THE BUTTERWORTH GROUP

ENGLAND
Butterworth & Co (Publishers) Ltd
London: 88 Kingsway, WC2B 6AB

AUSTRALIA
Butterworths Pty Ltd
Sydney: 586 Pacific Highway, NSW 2067
Also at Melbourne, Brisbane, Adelaide and Perth

CANADA
Butterworth & Co (Canada) Ltd
Toronto: 2265 Midland Avenue,
Scarborough, Ontario M1P 4S1

NEW ZEALAND
Butterworths of New Zealand Ltd
Wellington: 26–28 Waring Taylor Street, 1

SOUTH AFRICA
Butterworth & Co (South Africa) (Pty) Ltd
Durban: 152–154 Gale Street

USA
Butterworths (Publishers) Inc
Reading: 161 Ash Street, Boston, Mass. 01867

Library of Congress Cataloging in Publication Data
Main entry under title:

Aliphatic chemistry.

(Organic chemistry, series two; v. 2) (International
review of science)
Includes index.
1. Aliphatic compounds. I. Chapman, Norman
Bellamy, 1916– II. Series. III. Series: International
review of science.
QD245.073 vol. 2 [QP301] 547′.008s [547′.4]
ISBN 0 408 70614 7   75-19382

Typeset by Amos Typesetters, Hockley, Essex
Printed and bound in Great Britain by
REDWOOD BURN LIMITED
Trowbridge and Esher

Consultant Editor's Note

The ten volumes in Organic Chemistry in the Second Series of the biennial reviews in the International Review of Science follow logically from those of the First Series. No major omissions have come to light in the overall coverage of the First Series. The titles of the ten volumes therefore remain unchanged but there are three new Volume Editors. The volume on Structure Determination in Organic Chemistry has been taken over by Professor Lloyd M. Jackman of Pennsylvania State University, that on Alicyclic Compounds by Professor D. Ginsburg of Technion-Israel Institute of Technology, and that on Amino Acids, Peptides and Related Compounds by Professor H. N. Rydon of the University of Exeter. The international character of the Series is thus strengthened with four Volume Editors from the United Kingdom, two each from Canada and the U.S.A., and one each from Israel and Switzerland. An even wider pattern is shown for the authors, who now come from some sixteen countries. The reviews in the Second Series are mainly intended to cover work published in the years 1972 and 1973, although relevant results published in 1974 and 1975 are included in some cases, and earlier work is also covered where applicable.

It is my pleasure once again to thank all the Volume Editors for their helpful cooperation in this venture.

London   D. H. Hey

Preface

This volume follows the same pattern as that adopted in Series One. As a matter of editorial policy no attempt has been made to eliminate small overlaps between the material of the several chapters, in the belief that the reader's interests are served by having the items discussed in more than one context. In Chapter 1 the term 'carbenium ion' has been systematically used, but elsewhere the same chemical species are often referred to as 'carbonium ions'. In the editor's judgment there is much force in Olah's advocacy of the former term, but this term has not yet been so widely accepted as to make its use mandatory.

The richness of the material for review has not diminished; if anything it is greater than for 1970–71, and the authors have been able to confine themselves to the limits set only by the most stringent selection of material. As for Series One, an astonishing range of activity in the area of hydrocarbons is seen. Boron compounds again provide some remarkably fascinating material and there is no lack of vitality in the field of nitrogen, oxygen and sulphur compounds. Phosphorus compounds and halogeno compounds also compete strongly for eminence. In short, despite its long and rich history, aliphatic organic chemistry continues to yield excitement and satisfaction to its devotees.

Hull   N. B. Chapman

Contents

1
Hydrocarbons

D. E. WEBSTER
University of Hull

1.1   INTRODUCTION

This chapter reviews the 1972 and 1973 literature on aliphatic hydrocarbons. As in the corresponding review in Series One[1], heterogeneous catalysis and the patent literature have both been excluded; there is insufficient space to include all the worthwhile work, and selection of a small part would be most arbitrary. The chemistry reviewed is that which is considered to be of interest to organic chemists. Studies of the physical chemistry of aliphatic hydrocarbons, and of their inorganic chemistry (or more accurately organometallic chemistry) where the emphasis of the work is on the inorganic part of the molecule, have also been omitted.

1.2   THEORETICAL STUDIES

1.2.1   Electronic structures and molecular properties

Theoretical chemists continue to apply a wide range of wave-mechanical methods to the treatment of the electronic structures of molecules. The ready access to computers and the easy interchange of computer programs make the study of multi-electron systems easier year by year. It has become extremely difficult to decide which of the plethora of methods is the best for any particular job. However, one thing is certain: the ready ability to 'describe' the theoretical model in pictorial terms no longer exists for many models — the theoretical chemists continue to move into a world all of their own.

The geometries and energies of the isomeric $C_3H_5^+$ cations have been calculated by using various sets of Gaussian fitted Slater-type orbitals, STO-3G, and the extended sets 4-31G and 6-31G*[2]. The planar allyl cation

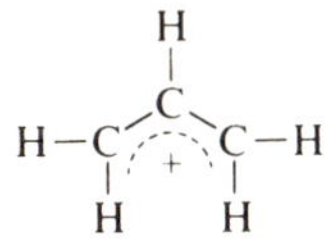

(1) is found to be the most stable $C_3H_5^+$ isomer, the 'perpendicular' form (2) being 35 kcal mol^{-1} higher in energy. The cation (1) is more stable than the ethyl cation by 26 kcal mol^{-1}. The cyclopropyl cation (3) is 39 kcal mol^{-1} less stable than (1). The only other stable minimum on the $C_3H_5^+$ energy surface, at 17 kcal mol^{-1} higher than that for (1), corresponds to the 2-propenyl cation (4). The 1-propenyl cation (5) is a further 16 kcal mol^{-1} higher in energy and should readily rearrange to the 2-isomer (4). Protonation of allene will give (4) preferentially since it is much more stable than the 'perpendicular' allyl cation (2). Barriers to 1,2-shifts are substantially higher in vinylic than in alkane systems, e.g. that for the ethyl cation [equation (1.1)] is 7.3 kcal mol^{-1} but that for the vinyl cation [equation (1.2)] is 19.2

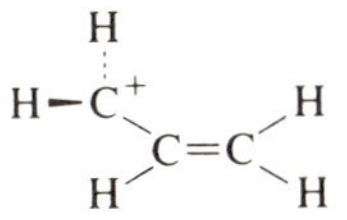

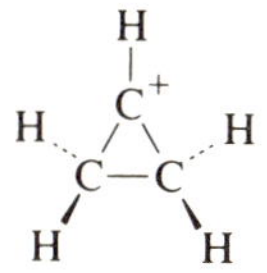

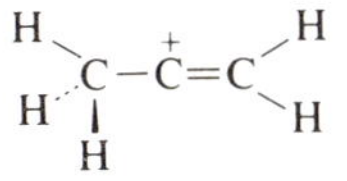

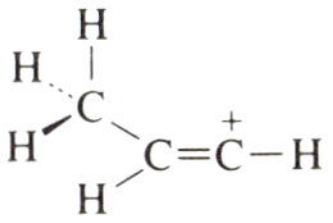

$$CH_3CH_2^+ \longrightarrow H_2C \overset{H}{\underset{+}{=}} CH_2 \tag{1.1}$$

$$H_2C=CH^+ \longrightarrow HC \overset{H}{\underset{+}{=}} CH \tag{1.2}$$

kcal mol^{-1}. A variety of semi-empirical methods have been compared for studying the C-atom hybridisation in hydrocarbons[3]. The results obtained by using extended Hückel (EH), self-consistent charge, complete neglect of differential overlap (CNDO), and the maximum overlap approximation (MOA) have been compared with *ab initio* results. The bond orders for methane, ethane, ethylene, acetylene, propene and buta-1,3-diene have been evaluated. The MOA method gives bond orders that agree most closely with *ab initio* values and those derived from n.m.r. ^{13}C—H coupling constants. Also, the MOA method is the most pictorial and the simplest; whereas the CNDO method requires five parameters and the EH method four, the MOA method needs only two. However, the structures calculated by the different methods are qualitatively the same and it was concluded that 'the choice of a particular method for the investigation of gross molecular properties of hydrocarbons is a matter of personal taste'. The molecular mechanics method for calculating electronic structures has been extended to hydrocarbons with delocalised π-systems, and has been applied to buta-1,3-diene[4].

An *ab initio* generalised self-consistent valence-bond method (GVB) has been developed[5] and applied to methane, ethane, ethylene, propene and acetylene. The energies obtained are lower than the Hartree–Foch energies, and the method has two major conceptual advantages: the orbitals correspond closely to the chemist's 'intuitive' ideas of bonds and lone pairs in molecules, and the process of breaking chemical bonds is correctly described since the GVB orbitals of the molecule change smoothly into the atomic orbitals of the products. An example is ethylene, *cf.* Figure 1.1, which shows

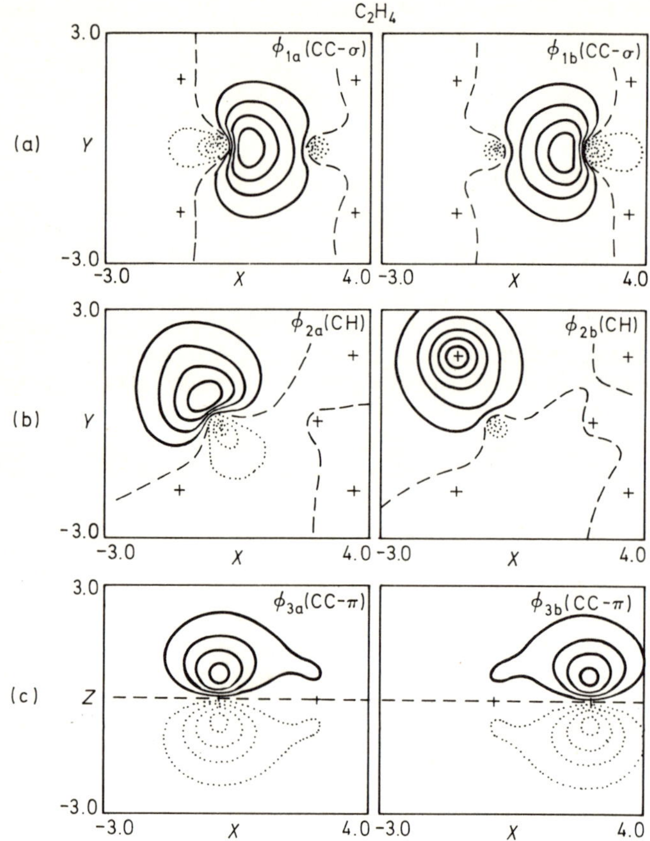

Figure 1.1 The bonding GVB orbitals of ethylene. Orbitals are shown for only one of the C—H bonds (ϕ_{2a} and ϕ_{2b}); the core orbitals are not shown. The contour increments are 0.01 a.u.; the long dashes indicate nodel lines and the solid lines indicate positive contours. (From Hay *et al.*[5], by courtesy of the American Chemical Society)

electron density contours in increments of 0.01 au. The orbitals depicted are (a) the C—C-σ, (b) the C—H-σ (only one) and (c) the C—C-π. It must be realised that each orbital contains only one electron, and so a two-electron bond involves two orbitals.

Another method recently developed is that using Floating Spherical Gaussian Orbitals (FSGO)[6]. These orbitals have the property of not being constrained to be centred on a nucleus. The FSGO model has been used to calculate the geometry of propane, of n-butane and of isobutane, simply by employing FSGO data unchanged from those used for ethane. Bond angles within 0.3° of the experimental values were obtained[7]. ESCA chemical shifts for methane, ethane, ethylene, acetylene, propane, propene, isobutane and neopentane have all been calculated by using the FSGO model[8]. An extended FSGO model has also been used to study the internal rotational barrier in ethane, and a value in the range 3.7–4.0 kcal mol^{-1} was obtained[9]. This compares well with the experimental value of 3.0 kcal mol^{-1}. A modified form of the semi-empirical SCF–MO method, with complete neglect of differential overlap, has been described by Boyd and Whitehead[10]. This CNDO/BW method has been used to calculate the potential functions for internal rotation in several small organic molecules[11]. For ethane, ethylene and propene the geometry of the most stable conformer is predicted by this method, but only for ethylene is the calculated height of the rotational barrier large enough. The force-field method, previously used for alkanes[12], has been extended to alkenes[13]. Ethylene, propene and isobutene are used as standard structures for obtaining the force-field parameters. Structures and strain energies of a range of alkenes from ethylene to 1,2-di-t-butylethylene (2,2,5,5-tetramethylhex-3-ene) were calculated. Correlation diagrams for the molecular orbitals on moving from the staggered ethane (D_{3d} symmetry) to the bridged diborane structure (D_{2h} symmetry) show that the extra two electrons in ethane require the shape to be as found experimentally, as the highest bonding MO in an equivalent bridged structure would have too high an energy[14]. Binding energies for methane, ethane, ethylene and acetylene have been calculated by a semi-empirical MO correlation method[15]. The values obtained are within 0.5 eV of the experimental if accurate Hartree–Foch energies are used, and the method is not sensitive to the molecular wavefunctions used. A multi-configuration SCF method that is strongly dependent on the basis set of atomic orbitals (Slater type) has been used to obtain the energies of methane, ethane and ethylene[16]. For ethylene the σ–π model gives rather better values than the completely localised 'banana-type' orbitals. Methane has also been treated by using a spherical Gaussian orbital model[17], and by a pseudopotential technique which considers only valence electrons[18]. Both methods give good agreement with experiment for the various parameters of the molecule. An *ab initio* study which compares several basis sets of Gaussian-type functions has yielded energies and population numbers for ethylene and acetylene[19]. The CNDO/2 method has been used to calculate the electronic structure of ethylene, of propene and of a series of isomeric butenes and pentenes[20]. The ionisation potentials obtained are all higher than the experimental values, but it must be remembered that such methods give vertical potentials, whereas the experimental potentials are adiabatic. The ionisation potentials obtained are interpreted in terms of charge distribution effects of the alkyl groups. The same method gives good agreement with the experimental value for the C≡C bond length in propyne and but-2-yne[21].

Despite the growth of *ab initio* and SCF methods, which are very time consuming and costly, the simple Hückel method continues to be developed. An extended version containing new parameters, to give better agreement between theory and experiment, has been applied to a range of hydrocarbons[22]. The ionisation energies and dissociation energies of 31 hydro-carbons and hydrocarbon radicals have been calculated. The extended Hückel calculation of the structure of *cis*-but-2-ene by Hoffmann[23] has been reinvestigated[24]. The original study found the staggered conformation

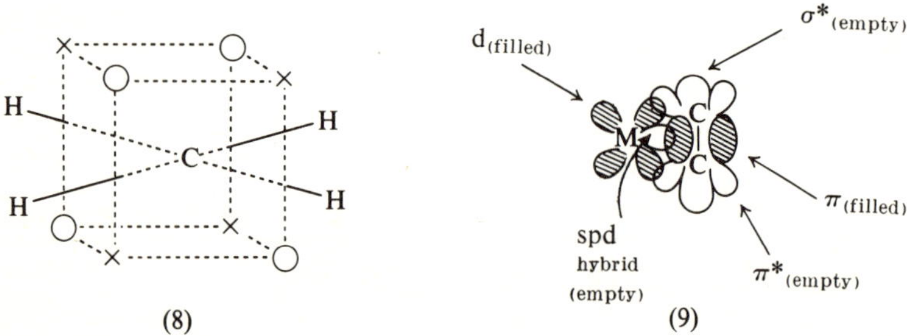

(6) to be 10.6 kcal mol^{-1} more stable than the eclipsed conformation (7), contrary to the experimental fact. This has now been shown to arise because standard bond angles were assumed. If angle relaxation is allowed for, then the eclipsed conformer is calculated to be 1.2 kcal mol^{-1} more stable than the staggered.

Chemists are always trying to prepare compounds with novel structures. Two papers relevant to the intriguing problem of planar 4-coordinate carbon disagree on its stability[25,26]. Murrell *et al.* have calculated that planar methane will have a triplet electronic structure in its lowest energy state[25]. The energy of the planar triplet state is calculated to be 180 kcal mol^{-1} higher than that of the tetrahedral ground state, and the planar singlet state is a further 70 kcal mol^{-1} higher in energy. However, Firestone[26], by using the Linnett model[27,28] for planar methane (8) — ○ and × signify the electrons (in 2 tetrahedra) of the Linnett model — calculates that the energy of the planar structure is only 35–40 kcal mol^{-1} greater than that of the tetrahedral, and it may, therefore, be possible to prepare compounds containing planar carbon.

Two papers have dealt with the nature of the bonding of ethylene to a transition metal[29,30]. These completely confirm the well known Dewar–Chatt model (9). In the silver–ethylene bond in (9; M = Ag$^+$) it is concluded that the σ-donation from the ethylene is significantly larger than the π back-donation from the silver ion[29], but in the titanium–ethylene and the nickel–ethylene bond in (9; M = Ti^{3+} or Ni^{3+}) the π back-donation is extremely important for both metals even though one is an early, and the other a late, transition metal[30]. π Back-donation into the σ^* orbitals of ethylene is also significant.

Theoretical studies on some simple reactions have been reported[31,32]. The energies of hydrogenation of ethane, ethylene and acetylene calculated by using the 6-31G* set of functions are in fair agreement with experimental values (Table 1.1)[31]. A non-empirical LCAO–MO–SCF calculation on the

Table 1.1 Energies of hydrogenation[31]

| | ΔE/kcal mol^{-1} | |
	Calc.	*Expt.*
$C_2H_6 + H_2 \rightarrow 2CH_4$	-23.5	-18.1
$C_2H_4 + 2H_2 \rightarrow 2CH_4$	-65.9	-57.2
$C_2H_2 + 3H_2 \rightarrow 2CH_4$	-117.8	-105.4

approach of an electrophile (simulated by a unit positive charge) to acetylene shows that the lowest-energy path is at right angles to the molecular axis, towards the centre of the C—C bond, and that the energy becomes a minimum when the electrophile is 0.3 nm from the bond[32]. From this point the lowest-energy path is then roughly parallel to the molecular axis. This treatment suggests a likely route for the conversion of a bridged species into a vinyl cation.

Photoelectron spectra have been used to study the orbital sequence in buta-1,3-diene and in *cis*- and *trans*-hexa-1,3,5-triene[33]; these are π, π, σ and π, π, σ, π, respectively[33]. The results of SPINDO calculations agree with these experimental results [the SPINDO method (spectroscopic potentials adjusted for INDO) is the semi-empirical INDO method modified by empirical corrections to give better agreement between observed photoelectron ionisation potentials and calculated values[34]]. The nature of the internal rotation barrier in ethane has been further studied[35]. The closed-shell overlap repulsion model for the barrier proposed[36] in 1968 has been optimised and the barriers for the fixed and the optimised conformer are 2.6 and 3.1 kcal mol^{-1}, respectively. This confirms the conclusion of the earlier paper that the repulsive overlap of the closed-shell bond functions, as a result of the Pauli exclusion principle, is an essential element.

1.2.2 Woodward–Hoffmann orbital symmetry rules

Simple, easy to use, orbital symmetry rules have been described for unimolecular reactions[37]. The symmetry of the bonds broken must be the same as that of the bonds to be made, the symmetry being related only to those elements of symmetry that are conserved. Reactions discussed include the symmetry-forbidden suprafacial loss of hydrogen from ethane [equation (1.3)], and the symmetry-allowed Diels–Alder retrogression of cyclohexene [equation (1.4)]. If the full symmetry of the reactants is used, additional features of the unimolecular reactions are revealed. For example, in the metal-catalysed alkene metathesis [equations (1.5) and (1.6)] the alkene must rotate

$$(1.3)$$

$$(1.4)$$

relative to the metal centre [equation (1.6)] or the symmetry must change from planar to tetrahedral as the reaction proceeds [equation (1.5)]. The importance of preservation of the coordinate bonding in such concerted metal-catalysed reactions has been stressed by Mango[38].

Planar   Tetrahedral

$$(1.5)$$

$$(1.6)$$

What might be considered to be the valence-bond analogue of the Woodward–Hoffmann rules (which apply to molecular orbital functions) has been developed by Goddard[39]. The method, which involves the orbital-phase continuity principle (OPCP), recognises that, for a reaction to have a low activation energy, certain phase relationships must exist between reactants and products. The OPCP does not depend on molecular symmetry and hence can be applied to reaction systems that have no symmetry. The principle is based on a generalised valence bond (GVB) self-consistent field method[5,40]. Figure 1.1 shows the GVB orbitals for ethylene; for typical molecules GVB orbitals are qualitatively similar to hybridised and somewhat delocalised valence-bond orbitals. The OPCP has the following features: (i) bonding pairs of orbitals must remain strongly overlapping during reactions; (ii) when bonding and non-bonding orbitals must be delocalised over the same region, they will tend to become orthogonal in the transition state, and hence they will show opposite phase-change characteristics; (iii) the phase relations among the orbitals must change continuously during a reaction. When these three conditions are satisfied, a reaction will have a low activation energy, i.e. it will be favoured. Reactions which cannot retain the appropriate orbital-phase continuity will generally require relatively high activation energies and will not be favoured. The OPCP has been applied to cycloadditions, electrocyclic reactions, sigmatropic reactions, eliminations and additions of carbene and open-shell molecules to an alkenic linkage. The method is often very simple to apply; the bond shifts which take place can easily be drawn and the phase changes counted.

1.3 ALKANES

1.3.1 Conformational analysis

Two studies agree, within experimental error, on the value for the enthalpy difference between the *anti*- and the *gauche*-isomer of n-butane[41,42]. N.m.r. chemical shifts and coupling constants are used, and in one of the studies, [1,1,1,4,4,4-$^2H_6$]butane was studied[42]. The *anti*-form (10) is more stable than the *gauche*-isomer (11). The enthalpy difference is 840 $\pm$ 200 cal mol^{-1} [41],

(10)   (11)

(12)   (13)

or 681 $\pm$ 35 cal mol^{-1} [42], and the entropy difference is -1.376 cal mol^{-1} K^{-1} [42]. The enthalpy and entropy differences between the two conformational isomers of 2-[$^2H_9$]methylbutane (12) and (13) are found to be 888 $\pm$ 18 cal mol^{-1} and -1.376 cal mol^{-1} K^{-1}, respectively, by studying n.m.r. coupling constants and chemical shifts of 2-[$^2H_9$]methylbutane[43].

1.3.2 Reactions

1.3.2.1 Electrophilic substitutions

The activation of alkanes by super-acids has been developed further during the past two years. Three very similar papers[44-46] review the nomenclature that is suggested for positively charged hydrocarbon ions, *carbocations,* and the subdivision into trivalent classical *carbenium* ions, and penta- or tetra-coordinate (non-classical) *carbonium* ions. A naming scheme, in accordance with IUPAC rules of nomenclature has been given[44]. Chlorinations[47,48] and brominations[49] of alkanes have been reported. The chlorinations involve a super-acid system, but for the bromination the solvent is dichloromethane. If SbF_5 with Cl_2 in SO_2ClF is used at $-78\ °C$, both chlorination by substitution and chlorinolysis by bond cleavage occur[48]. The final products from methane, ethane and propane are the corresponding dialkylchlorinium ions[47]. The reaction for ethane [equation (1.7)] gives the dimethylchlorinium and the

$$\text{(1.7)}$$

diethylchlorinium ions in the ratio of $7 : 3$, by insertion of Cl^+ into the C—C and into a C—H bond, respectively. The selectivity of the chlorinating agent is changed by changing the Friedel–Crafts catalyst. SbF_5 is the strongest catalyst: if $AlCl_3$ is used, it coordinates with the Cl_2 to produce a weaker electrophile which gives substitution reactions only, and with $AgSbF_6$ as catalyst the electrophile is so weak that it reacts only with cycloalkanes and strained systems[48]. Isobutane and isopentane are brominated by Br_2–$AgSbF_6$ in dichloromethane[49]. The reaction is as shown in equations (1.8) and (1.9).

$$Br_2 + AgSbF_6 \rightleftharpoons Br^{\delta+}-Br^{\delta-} \rightarrow Ag^+[SbF_6]^-$$

$$\text{(1.8)}$$

$$R_3C-H + Br^{\delta+}-Br^{\delta-} \rightarrow Ag[SbF_6]^- \longrightarrow \left[R_3C \cdots Br \atop H \right]^+ + [SbF_6]^- + AgBr$$

$$\text{(1.9)}$$

$$\longrightarrow R_3CBr + HSbF_6$$

Alkylcarbenium hexafluoroantimonates exchange hydrogen with alkanes, unless the alkane and the carbenium ion are sterically hindered[50]. Propane and $Me_2\overset{+}{C}H$ interchange a hydrogen atom on the secondary carbon, as do other pairs such as $Me_2CHCHMe_2$ and $Me_2\overset{+}{C}CHMe_2$. But, because of steric factors, Bu^tCHMe_2 and $Bu^t\overset{+}{C}Me_2$ do not interchange hydrogen. Methyl and ethyl hexafluoroantimonates in SO_2ClF are hydrogen-abstracting agents, as well as alkylating agents, for methane, ethane, propane, n- and iso-butane and neopentane[51]. At 60 °C MeF–SbF$_5$ reacts extremely slowly with methane, CD$_3$F–SbF$_5$ reacts much faster to give a deuterated ethane and higher condensation products, as well as the hydrogen-transfer product

$$CH_4 + CD_3F \rightarrow SbF_5 \rightleftharpoons \left[\begin{array}{c} H \\ | \\ H-C \cdots CD_3 \\ | \quad H \\ H \end{array}\right]^+ \quad \begin{array}{l} \rightleftharpoons CD_3H + MeF \rightarrow SbF_5 \\[2mm] \xrightarrow{-H^+} CH_3CD_3 \\ \downarrow \\ C_4 \text{ and } C_6 \text{ compounds} \end{array} \quad (1.10)$$

[equation (1.10)]. Methylation of ethane proceeds via the alkonium-ion transition states (14) and (15), formed by insertion of CH$_3{}^+$ into a C—H or

$$\left[CH_3CH_2 \cdots \begin{array}{c} H \\ CH_3 \end{array}\right]^+ \qquad \left[CH_3 \cdots \begin{array}{c} CH_3 \\ CH_3 \end{array}\right]^+$$

$$(14) \hspace{6cm} (15)$$

C—C bond. The complex (14) gives methane by a hydride transfer, or propane by elimination of a proton (alkylation). The ethyl cation formed by C—H bond cleavage will react with more ethane to give the alkonium ions (16)

$$\left[CH_3CH_2 \cdots \begin{array}{c} H \\ CH_2CH_3 \end{array}\right]^+ \quad \left[CH_3 \cdots \begin{array}{c} CH_3 \\ CH_2CH_3 \end{array}\right]^+ \quad \left[R \cdots \begin{array}{c} H \\ H \end{array}\right]^+$$

$$(16) \hspace{3.5cm} (17) \hspace{3.5cm} (18)$$

and (17) and thence the C$_3$ and C$_4$ hydrocarbons. Other hydrocarbons react by similar but increasingly complex reaction sequences. Hydrogen–deuterium exchange has been studied in 21 alkanes from methane to neopentane[52]. The key to full understanding of the extensive data is the realisation that the protolytic attack takes place on the main lobes of the C—H and C—C bond orbitals, where the major part of the electron density resides, and not at the hydrogen or carbon atoms themselves. The suggested structure of the intermediate (18) involved in the formation of stable carbenium ions by oxidising saturated hydrocarbons with SbF$_5$ has been questioned[53]. The abstraction of H$^-$ from ethane and from neopentane is faster when SbF$_5$ alone is used than when SbF$_5$ with HF or with FSO$_3$H is used. The reaction, it is suggested, is as set out in equations (1.11), (1.12) and (1.13), the overall reaction

$$2\,RH + 2\,SbF_5 \rightleftharpoons 2\,R^+[SbF_5H]^- \tag{1.11}$$

$$SbF_5H^- + 2\,SbF_5 \rightarrow SbF_6{}^- + SbF_3 + HSbF_6 \tag{1.12}$$

$$SbF_5H^- + HSbF_6 \rightarrow SbF_6{}^- + H_2 + SbF_5 \tag{1.13}$$

$$2\,RH + 3\,SbF_5 \rightarrow 2\,R^+SbF_6{}^- + SbF_3 + H_2 \qquad (1.14)$$

being represented by equation (1.14). Ethane and neopentane have equal specific reaction rates. This suggests that abstraction of hydride by SbF_5 involves more than the simple transfer of a hydride ion.

Alkanes are chlorinated by sulphuryl chloride in sulpholan as solvent[54]. Good yields of chloroalkanes are obtained at 60 °C from hexane or iso-octane. For hexane, with a hexane : sulphuryl chloride ratio of 1 : 3, the yield of chlorohexanes after 18 h is 40%. The products are 2-chloro- (56%), 3-chloro- (24%) and 1-chloro-hexane (20%). For iso-octane with an iso-octane : sulphuryl chloride ratio of 1 : 3.7 the yield of chlorinated products after 4 h is 30%. The products are 1-chloro-2,2,4-trimethylpentane (43%) and 3-chloro-2,2,4-trimethylpentane (28%) and 1-chloro-2,4,4-trimethyl-pentane (29%). The chlorination proceeds by an ionic mechanism but this has not been described. Sulpholan itself is chlorinated in the absence of the

$$\underset{SO_2}{\bigsqcup} + SO_2Cl_2 \xrightarrow{60\ C} \underset{SO_2}{\overset{Cl}{\bigsqcup}} \qquad (1.15)$$

alkane [equation (1.15)] and some interaction of the solvent and sulphuryl chloride clearly takes place during this reaction.

1.3.2.2 Electrochemical reactions

There is increasing interest in electrochemical reactions, although alkanes have not been much studied so far. In solution in fluorosulphonic acid containing carboxylic acids, alkanes undergo anodic oxidation to form $\alpha\beta$-unsaturated ketones[55]. Earlier work[56] had shown that in molar acetic acid in fluorosulphonic acid (AcOH is a strong base in this solvent), by using a smooth platinum electrode, alkanes containing more than four carbon atoms are oxidised to give $\alpha\beta$-unsaturated ketones containing two more carbon atoms than the starting material, the extra carbon atoms coming from the acetic acid. It has now been found[55] that with less base the smaller alkanes, ethane, propane and butane, can be oxidised. The study is complicated by the fact that if acetyl cations are present, the alkanes undergo chemical reactions to give the same products as do the electrode reactions. The electrochemical method for producing carbenium ions which react with the acetic acid has advantages over the method using SbF_5 (previous section), since this is both expensive and unpleasant to use. A typical reaction, at 60 °C and giving 95% yield, is that of isobutane [equation (1.16)]. If the anodic oxidation

$$Me_3CH \xrightarrow{\text{2M AcOH in FSO}_3\text{H}} Me_2C{=}CHCOMe \qquad (1.16)$$

takes place in solution in acetonitrile or in trifluoroacetic acid containing a tetra-alkylammonium tetrafluoroborate, the secondary carbenium ions formed at the anode react rapidly with the solvent to give isomeric N-s-alkyl-acetamides and isomeric s-alkyl trifluoroacetates[57]. Pentane, hexane, heptane and octane have been used in acetonitrile, and, in addition, isobutane has been used with trifluoroacetic acid as the solvent. The reactions are as shown in equation (1.17).

$$RH \xrightarrow{-2e^-, -H^+} R^+ \begin{array}{c} \xrightarrow{MeCN} MeC^+=NR \xrightarrow[-H^+]{+H_2O} MeCONHR \\ \xrightarrow{CF_3CO_2H} CF_3CO_2R \end{array} \tag{1.17}$$

1.3.2.3 Homogeneous catalysis

It was suggested in the previous review[1] that the extension of transition-metal catalysis to alkane reactions other than hydrogen–deuterium exchange might occur and hence begin to open up a new area of alkane chemistry. This has indeed happened as this section will show. The extensive hydrogen–deuterium exchange studies on alkanes reported in the previous review have now been published[58,59]. The involvement of alkylplatinum complexes in this exchange has received experimental support from studies in which D-containing platinum complexes were alkylated with organomercury compounds, and the deuteriated alkanes formed were identical with those obtained from 'light' alkanes[60]. A multiple-exchange analysis[61], similar to that used by others[62], has been developed, and a study of the influence of halogen substituents in the alkane on H–D exchange rates has enabled a linear free energy relationship to be derived[63]. H–D exchange studies on decane and more complex hydrocarbons did not reveal any products formed by oxidation of the Pt–alkyl intermediates; decyl acetate, for example, could not be detected[64].

An important extension of this work is the observation that, in these platinum-salt-alkane systems, the alkane can be converted into a chloroalkane by Pt^{IV} [65]. It is also found that Pt^{IV} (added H_2PtCl_6) converts acetic acid into chloroacetic acid[66], and so aqueous trifluoroacetic acid is the preferred solvent. In this solvent at 100–120 °C, small alkanes are converted, in low yield, into chloroalkanes[65]. Table 1.2 gives the results of Shilov and

Table 1.2 Chloroalkanes formed by oxidation of alkanes[65] with Pt^{IV}

Hydrocarbon	Reaction time* /h	Products
Methane	4	MeCl
Ethane	4	EtCl
Propane	4	Pr^nCl (75%), Pr^iCl (25%)
Butane	3	Bu^nCl (78%), Bu^sCl (22%)
Isobutane	3	Bu^iCl (82%), Bu^tCl (18%)
Hexane	10	$n\text{-}C_6H_{13}Cl$

*T = 120 °C, with 0.1–0.7 M aqueous H_2PtCl_6 containing 5% of K_2PtCl_4 in a sealed ampoule or in an autoclave.

his co-workers[65]. The chloro products are mainly due to substitution at a primary carbon atom, as was found for H–D exchange. Pt^{II}, the H–D exchange catalyst, is a catalyst for this reaction. It is, of course, a reaction product also [equation (1.18)] so the reaction is autocatalytic. In fact H–D

$$RH + PtCl_6{}^{2-} \rightarrow RCl + PtCl_4{}^{2-} + HCl \tag{1.18}$$

exchange and this reaction are intimately related, having the same intermediates. This is shown in Scheme 1.1[60]. During this oxidative halogenation,

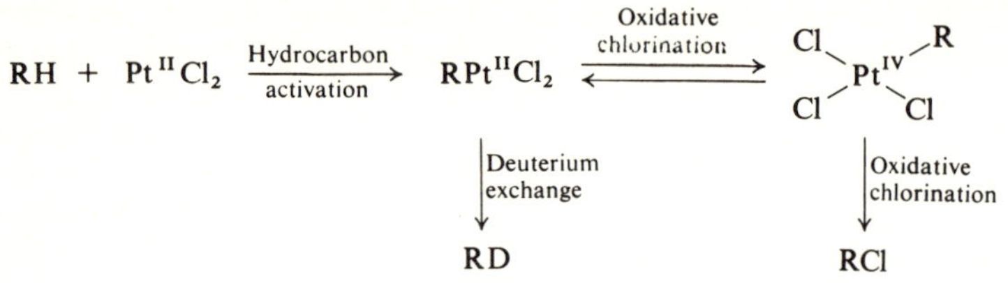

$$RH + Pt^{II}Cl_2 \xrightarrow[\text{activation}]{\text{Hydrocarbon}} RPt^{II}Cl_2 \underset{\xrightarrow{\hspace{1cm}}}{\xleftarrow{\text{Oxidative chlorination}}}$$

Deuterium exchange → RD

Oxidative chlorination → RCl

Scheme 1.1

both stable platinum–alkyl and platinum–chloroalkyl complexes can be detected, and the amount of platinum(II) formed is much more than that of the chloroalkane produced[67]. This must mean that further oxidation of the chloroalkane takes place. The other reaction products have not yet been identified.

A very short intriguing report is that C_1 to C_4 alkanes are oxidised to alcohols at room temperature in acetonitrile solution by oxygen in the presence of stannous chloride[68]. At 20 °C the relative reactivities are methane (0.01), ethane (1), propane (2.85), butane (5.0) and isobutane (3.7). The mechanism of this reaction is not clear; further details are awaited with interest.

The use of cobalt(III) salts to catalyse the oxidation of butane to acetic acid by air is of great industrial interest[69-71]. The activity of cobalt(III) acetate in acetic acid is enhanced by strong organic or inorganic acids, and n-alkanes can be significantly oxidised even at low temperatures[72]. Thus n-heptane, n-decane and n-dodecane were studied at 40 °C. The reaction, which is of the second order in Co^{III} and inhibited by Co^{II}, gives esters if oxygen is absent and ketones when oxygen is present. The predominant substitution is at the second carbon atom of the alkane chain. The mechanism

$$RH + Co^{III} \rightarrow \dot{R} + Co^{II} \tag{1.19}$$

$$\dot{R} + Co^{III} \rightarrow R^+ + Co^{II} \tag{1.20}$$

$$R^+ + AcOH \rightarrow AcOR + H^+ \tag{1.21}$$

$$\dot{R} + O_2 \rightarrow R\dot{O}O \tag{1.22}$$

[equations (1.19) and (1.20) followed by (1.21) or (1.22)] involves a reaction of Co^{III} with the alkane to give Co^{II} and an alkyl radical, which gives a carbenium ion or a peroxyl radical which can react further. If n-butane is the alkane used then the major product of oxidation with oxygen at 110 °C and *ca.* 20 atm for 4 h is acetic acid (83.5%)[73]. Small quantities of propanoic acid (5.4%), butanoic acid (3.5%) and butanone (4.4%) are also produced. Oxidation of n-pentane at 104 °C and 17 atm for 4 h gives acetic acid (48%), propanoic acid (27%), and small quantities of butanoic acid, pentanoic acid and pentan-2- and pentan-3-one. Oxidation of isobutane at 110 °C and 24 atm for 4 h gives a little acetone, t-butyl alcohol and methanol (10% conversion). For this low-temperature reaction Mn^{III} is not a catalyst, and it is suggested that the reaction involves electron transfer from the C—H σ-bond of the alkane to the Co^{III}, the Mn^{III} ion being insufficiently strong an oxidant to abstract electrons from the C—H σ-bond. The interplay between free-radical and electron-transfer reactions in these systems is very subtle; it is temperature dependent, and probably depends also on many other parameters.

Alkylation of aromatic hydrocarbons with alkanes by using copper(II) chloride and aluminium chloride at room temperature has been reported[74]. Alkylcarbenium ions are produced and these react with the benzene. For example, pentylbenzenes, together with smaller amounts of ethylbenzene, isopropylbenzene and 1,1-diphenylethane, are formed when a mixture of benzene and 2-methylbutane is treated with a mixture of copper(II) chloride and aluminium chloride at 17–21 °C. The reaction is shown as an oxidative process involving copper(II) chloride [equations (1.23), (1.24) and (1.25)].

$$Me_2CHEt + CuCl_2 + AlCl_3 \rightarrow EtMe_2\overset{+}{C} + HCl + CuCl + AlCl_3 \qquad (1.23)$$

$$C_6H_6 + EtMe_2\overset{+}{C} \rightarrow PhCMe_2Et + H^+ \qquad (1.24)$$

$$H^+ + CuCl_2 \rightarrow HCl + CuCl \qquad (1.25)$$

$$Me_2CHEt + C_6H_6 + 2\,CuCl_2 \xrightarrow{AlCl_3} PhCMe_2Et + 2\,CuCl + 2\,HCl \qquad (1.26)$$

The overall reaction is given by equation (1.26). If ethane is used, 1,1-diphenylethane is formed in 40–45 mol % yield based on the copper(II) chloride present. If the reaction mixture is exposed to air the yield of alkylbenzene increases since the copper(II) chloride is regenerated from the copper(I) chloride [equation (1.27)].

$$4\,CuCl + 4\,HCl + O_2 \rightarrow 4\,CuCl_2 + 2\,H_2O \qquad (1.27)$$

Copper(I) carbonyl is a catalyst for the carbonylation of alkanes[75]. In concentrated sulphuric acid at room temperature and atmospheric pressure the alkanes given in Table 1.3 react as indicated. The mechanism involves

Table 1,3   Products from the Cu^I carbonyl catalysed carbonylation of alkanes[75]

Alkane	Alkene	Products	Yield/%
2-Methylpentane	Oct-1-ene	2,2-Dimethylpentanoic acid	40
		2-Ethyl-2-methylbutanoic acid	20
		t-C_9 acids	30
2-Methylbutane	Oct-1-ene	2,2-Dimethylbutanoic acid	30
		t-C_9 acids	60
		2,2-Dimethylpropanoic acid	5
Octane	Hex-1-ene	t-C_9 acids	0
		t-C_7 acids	64
2,2,4-Trimethylpentane	Hex-1-ene	t-C_9 acids	1
		t-C_7 acids	85

hydride abstraction from the alkane by a carbenium ion generated from an alkene [equations (1.28) and (1.29)], and saturated hydrocarbons with tertiary carbons alone undergo this reaction.

$$\overset{+}{R^1} + R^2{}_2H \rightarrow R^1H + \overset{+}{R^2} \tag{1.28}$$

$$\overset{+}{R^2} + CO \xrightarrow{Cu(CO)_3{}^+} R^2\overset{+}{C}O \xrightarrow{H_2O} R^2CO_2H \tag{1.29}$$

$$(R^1 = \text{n-, s- or t-alkyl}, \quad R^2 = Me_2Pr^nC \text{ or } EtMe_2C)$$

1.4 ALKENES

1.4.1 Preparation

The synthesis of alkenes from β-ketosilanes has been reported[76]. Trimethylsilylacetone reacts with a Grignard reagent, or an organo-lithium compound, to give a β-hydroxyalkylsilane, which in acetic acid saturated with sodium acetate gives the alkene [equation (1.30)]. Terminal alkenes with long-chain

$$Me_3SiCH_2COMe \xrightarrow[Bu^nLi \text{ in } Et_2O]{Bu^nMgBr \text{ or }} Me_3SiCH_2C(OH)Bu^nMe \xrightarrow[NaOAc]{HOAc} CH_2{=}CBu^nMe \tag{1.30}$$

alkyl groups are formed by removal of the phenylmercapto group from a hydroxyalkyl phenyl sulphide, prepared from an aliphatic aldehyde, R′CHO, by using a trivalent phosphorus compound [equation (1.31)][77]. The reaction

$$PhSCH_2CH(OH)R^1 \xrightarrow[\substack{(ii)\ MeLi \\ 100\,°C,\ 5\,h}]{(i)\ (R^2O)_2PCl} \begin{array}{c} PhS{-}CH_2 \\ \diagdown \\ \overset{..}{(R^2O)_2}\overset{..}{P}{-}O \end{array}\!\!CHR^1 \longrightarrow CH_2{=}CHR^1$$

$$+$$

$$(R^1 = \text{n-}C_9H_{19} \text{ or n-}C_{11}H_{23}) \qquad PhSP(O)(OR^2)_2 \tag{1.31}$$

involves nucleophilic attack of the phosphorus on sulphur and elimination of the *OO*-dialkyl *S*-phenyl phosphorothiolate. *N*-Methylphenylsulphonimidoylmethyl-lithium (19), prepared as in equation (1.32), is known to add

$$PhMeS({:}O) \xrightarrow[\substack{H_2SO_4 \\ \text{in } CHCl_3}]{NaN_3} PhMeS({:}O){=}NH \xrightarrow[HCO_2H]{CH_2O} PhMeS({:}O){=}NMe$$

$$\xrightarrow[THF]{Bu^nLi} LiCH_2SPh({:}O){=}NMe \quad (19) \tag{1.32}$$

readily to aldehydes and ketones to give β-hydroxysulphoximines (20). The C—S bond in these can be cleaved by aluminium amalgam in aqueous THF to give alcohols [equation (1.33)]. By making a very simple modification to

$$(19) \xrightarrow{R^1R^2CO} R^1R^2C(OH)CH_2\overset{\overset{\textstyle O}{\|}}{\underset{\underset{\textstyle NMe}{\|}}{S}}Ph \xrightarrow[THF-H_2O]{Al/Hg} R^1R^2C(OH)Me + PhS({:}O)NHMe$$

$$(20) \tag{1.33}$$

the conditions at this reduction stage, *viz.* adding acetic acid, an alkene is formed instead of the alcohol [equation (1.34)][78]. The carbonyl compounds (21), (23) and (25) are converted into alkenes (22), (24) and (26) in 90%, 60%

$$(20) \xrightarrow[\text{THF-H}_2\text{O-AcOH}]{\text{Al/Hg}} \left[R^1R^2C(\overset{+}{O}H_2)CH_2\overset{\overset{\textstyle O}{\|}}{\underset{\overset{\textstyle \|}{NMe}}{S}}Ph \right] \xrightarrow{2e^-} R^1R^2C{=}CH_2 \quad (1.34)$$

n-C$_{15}$H$_{31}$COMe   n-C$_{15}$H$_{31}$CMe=CH$_2$ n-C$_7$H$_{15}$CHO   n-C$_7$H$_{15}$CH=CH$_2$
(21)   (22)   (23)   (24)

Bu$^n_2$CO   Bu$^n_2$CH=CH$_2$
(25)   (26)

and 80% yield, respectively. The role of the acid is unclear; it is thought that it protonates the hydroxy group and makes it a better leaving group [equation (1.34)]. Alkenes are formed if bis(phenylthio)ethanes are thermolysed at 350 °C [79]. 1,2-Bis(phenylthio)ethane (27; $R^1 = R^2 = R^3 = R^4 = H$) gives ethylene [equation (1.35)], and the diastereoisomeric 2,3-bis-

$$\underset{(27)}{\overset{\text{PhS}}{\underset{R^4}{\diagdown}}\underset{\underset{R^1}{|}}{\overset{}{C}}-\underset{\underset{R^2}{|}}{\overset{}{C}}\overset{\text{SPh}}{\underset{R^3}{\diagup}}} \xrightarrow{350\,°C} R^1R^4C{=}CR^2R^3 \ + \ PhSSPh \ + \ PhSH \quad (1.35)$$

(phenylthio)butanes (27; $R^1 = R^3 = H$; $R^2 = R^4 = Me$; and $R^1 = R^2 = H$, $R^3 = R^4 = Me$) give *cis*- and *trans*-but-2-ene [equation (1.35)]. The reaction involves a free-radical intermediate such as (28). Tri-t-butylethylene

$$\underset{(28)}{\overset{\text{PhS}}{\underset{H}{\diagdown}}\underset{\underset{Me}{|}}{\overset{}{C}}-\underset{\underset{H}{|}}{\overset{\bullet}{C}}\underset{Me}{\diagup}} \qquad\qquad \underset{(29)}{CHBu^t{=}CBu^t_2} \qquad\qquad \underset{(30)}{Bu^tCH_2C\underset{\underset{Bu^t}{|}}{\overset{\diagup CMe_2}{\diagdown}}CH_2}$$

(2,2,5,5-tetramethyl-4-t-butylhex-3-ene) (29) has been prepared by thermolysis of neopentyl-di-t-butylcarbinyl *p*-nitrobenzoate[80]. In addition to tri-t-butylethylene, the cyclopropane derivative (30), the rearranged alkene (31)

Bu^{t}CH$_2$CMeBu^{t}CMe=CH$_2$ Me$_2$C=CMeCH$_2$Bu^t
(31)   (32)

and the fragmentation product (32) are also formed. A reaction that is formally the reverse of permanganate or osmium tetroxide oxidation of alkenes has been used to prepare docos-1-ene[81]. Vicinal dialkoxides undergo reductive deoxygenation with tungsten(IV) halide derivatives in THF. The reaction, e.g. equation (1.36), also gives docos-1-ene (44%) from docosane-1,2-diol after two days. Alkenes can also be formed from aldehydes, ketones

$$Me_2C(OH)C(OH)Me_2 \xrightarrow{MeLi} 2Li^+ \left[\begin{array}{c} O^- \\ \quad CMe_2 \\ | \\ \quad CMe_2 \\ O^- \end{array} \right] \xrightarrow[\text{in THF}]{K_2WCl_6} \left[\begin{array}{c} Cl \\ Cl\ \ | \ O\ \ CMe_2 \\ W \\ Cl\ \ | \ O\ \ CMe_2 \\ Cl \end{array} \right]^{2-}$$

$$(1.36)$$

$$\longrightarrow \quad \begin{array}{c} Cl \\ Cl\ \ | \ O \\ W \\ Cl\ \ | \ O \\ Cl \end{array} \quad + \quad Me_2C=CMe_2$$

and epoxides[82] by similar reductive deoxygenations. In this way but-2-ene is formed from butanone and oct-4-ene, in high yield, from *cis-* and from *trans*-4,5-epoxyoctane. Reductive deoxygenation of aldehydes and ketones can also be accomplished by using low-valent titanium complexes[83]. If the THF complex of $TiCl_3$ is treated with magnesium under argon, highly reactive low-valent titanium complexes are obtained. With such a complex acetone is reductively coupled to give 2,3-dimethylbut-2-ene (98%), and propanal gives the alkene (33) in 60% yield (30% of the *cis* and 30% of the *trans* isomer). Thionocarbonates of vicinal diols react with bis(cyclo-octa-1,5-diene)nickel(0) under mild conditions to give alkenes in high yield[84]. The

$$EtCH=CHEt$$

$$(33)$$

$$(34)$$

$$(35)$$

thionocarbonates of *erythro-* and *threo*-4-methylpentan-2,3-diol (34; $R^1 = H$, $R^2 = Me$, and $R^1 = Me$, $R^2 = H$, respectively) react with $Ni(cod)_2$ in DMF at 65 °C during 46 h, to give *cis*-4-methylpent-2-ene (35; $R^1 = H$, $R^2 = Me$) and *trans*-4-methylpent-2-ene (35; $R^1 = Me$, $R^2 = H$) in 79% and 99% yield, respectively. Addition of an α-methylallyl group and of an $AlEt_2$ group across the double bond occurs when Et_3Al and triscrotyl-boron react with a terminal alkene [equation (1.37)][85]. Reductive removal of

$$R^1CH=CH_2 + Et_3Al + B(CH_2CH=CHMe)_3 \xrightarrow{-R^2_2BEt}$$

$$CH_2=CHCHMeCHR^1CH_2AlEt_2 \qquad (1.37)$$

the $AlEt_2$ gives a higher alkene. In this way but-1-ene is converted into 3,4-dimethylhex-1-ene and 3-methylbut-1-ene into 3,4,5-trimethylhex-1-ene. Treatment of allylic alcohols with an excess of a Grignard reagent in the presence of catalytic amounts of $(Ph_3P)_2NiCl_2$ gives a mixture of alkenes [equation (1.38)][86].

$$R^1CH=CHCH(OH)R^2 + MeMgX + (Ph_3P)_2NiCl_2 \rightarrow R^1CH=CHCHMeR^2 +$$

$$R^1CHMeCH=CHR^2 \qquad (1.38)$$

1.4.2 Reactions

1.4.2.1 Isomerisation

Double bond migrations in alkenes catalysed by solutions of transition-metal complexes continue to be much studied. Details of the mechanisms of the reactions have been obtained by using isotopically labelled alkenes[87-90]. Recent papers discuss isomerisation by $Fe_3(CO)_{12}$[90], $RuHCl(PPh_3)_3$[87,88], $RuCl_2(PPh_3)_3$[87], $OsHCl(CO)(PPh_3)_3$[89], $CoHN_2(PPh_3)_3$[87], $RhH(CO)(PPh_3)_3$[87,91,92], $RhCl(CO)(PPh_3)_2$[91], $IrH(CO)(PPh_3)_3$[91], $IrCl(CO)(PPh_3)_2$[91-93], 'reduced' $(DMSO)_3RhCl_3$[94], 'reduced' $(Ph_3P)_2NiX_2$[95], $Ni\{P(OEt)_3\}_4$ with CF_3CO_2H[89] or H_2SO_4[96], $Ni\{Ph_2P(CH_2)_4PPh_2\}_2$ with an acid[97], a 'reduced' Pd–DMSO complex[98] and $PtClH(PMe_2Ph)_2$ with $MeSO_3F$[99]. The $Fe_3(CO)_{12}$-catalysed isomerisation of 3-ethyl[3-^{2}H]pent-1-ene (36) to

$$CH_2{=}CHCDEt_2 \qquad\qquad MeCH{=}CEt_2 \text{ (monodeuteriated)}$$
$$(36) \qquad\qquad\qquad\qquad\qquad (37)$$

the pent-2-ene derivative (37) is accompanied by deuterium becoming scrambled into the methyl groups[90]. The results can be interpreted unambiguously in terms of a π-allylic hydride intermediate. A comparative study of the isomerisation and hydrogenation of hept-1-ene catalysed by $RhH(CO)(PPh_3)_3$, $IrH(CO)(PPh_3)_3$, $RhCl(CO)(PPh_3)_2$ and $IrCl(CO)(PPh_3)_2$ shows the rhodium hydride complex to be the most effective catalyst[91]. This is to be expected from the results of earlier studies. The observation that u.v. light catalyses these isomerisations[92] is unexpected, and further work must undoubtedly be done to see whether this phenomenon is more general. Does it occur in other transition-metal complex catalysed reactions? It has also been reported that the capacity for isomerisation of $IrX(CO)(PPh_3)_2$ (X = halogen) is increased under hydrogen[93]. This is not a surprising observation, for metal hydrides are known to be necessary catalysts. A catalyst formed by the 'reduction' of the DMSO complex of rhodium trichloride [$(DMSO)_3RhCl_3$] in an unspecified manner is found to catalyse the isomerisation and hydrogenation of linear alkenes more than those of branched alkenes[94], in agreement with all other related observations. Alkenes with a substituent on one of the unsaturated carbon atoms, e.g. 2-methylbut-1-ene or 2-methylbut-2-ene, are neither isomerised nor hydrogenated. The *cis–trans* ratio of the internal alkene formed when a terminal alkene is isomerised has not, as yet, received a satisfactory explanation. In many cases the amount of the thermodynamically less favoured *cis* isomer far exceeds that of the *trans* alkene in the early stages of the reaction. A good example is the isomerisation of but-1-ene by $(PPh_3)_2NiX_2$ (X = Cl, Br, I) with added $SnCl_2$. The Ni^I complex produces but-2-ene with a *cis : trans* ratio of 49 if the nickel iodide complex is used[95]. An explanation of this large ratio has now been proposed based on competitive steric interactions within the hydrocarbon (as a ligand coordinated to the transition metal) and between the hydrocarbon and other ligands on the metal[100]. When there are many ligands, the *cis* isomer is favoured. When the metal is 'bare', the *trans* isomer is favoured. An interesting survey, applying the 16- and 18-electron rule to reactions catalysed by transition-metal complexes[101], postulates that 'organometallic reactions, including catalytic ones, proceed by elementary steps involving only intermediates with 16 or 18 metal valence-electrons'. The application of this postulate to the isomerisation of but-1-ene by $Ni\{P(OEt)_3\}_4 + H_2SO_4$ in methanol, the mechanism of which involves hydride addition and elimination involving a coordinatively unsaturated hydride complex, $HNiL_3^+$, is shown in Figure 1.2[96]. Here the symbols in brackets (e.g. d^8, 4) show the number

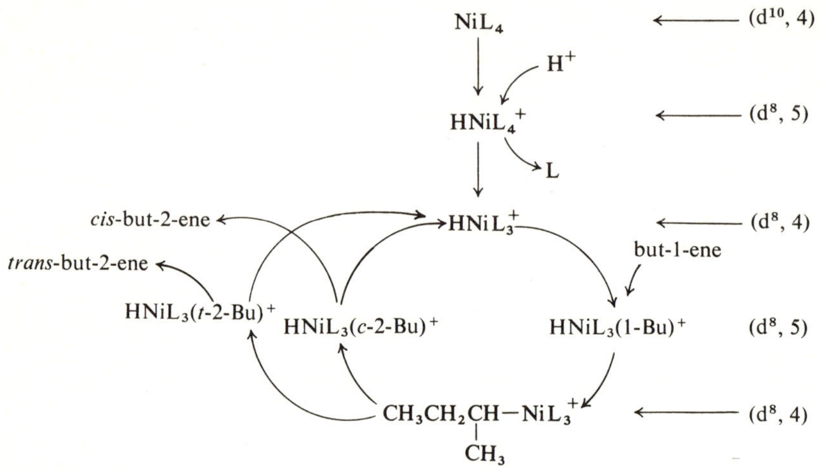

Figure 1.2 Mechanism of isomerisation[96] of but-1-ene catalysed by $HNiL_3^+$

of d-electrons and ligands on the nickel atom at each stage of the reaction. Both σ-alkyl and π-allyl intermediates are reported to occur in the isomerisation of pent-1-ene by $Ni\{Ph_2P(CH_2)_4PPh_2\}_2$ with added acid[97]. With trifluoroacetic acid or sulphuric acid the σ-alkyl route is followed, and with hydrogen cyanide the π-allyl route is preferred. A 'reduced' palladium–DMSO complex is an isomerisation catalyst only if hydrogen is present[98]. A potentially very interesting method for producing catalytic complexes is to use methyl fluorosulphonate ('magic methyl') to abstract halide from a non-catalytic complex to give a catalytic cationic complex[99]. 'Magic methyl' (and trimethyloxonium tetrafluoroborate) are among the most powerful methylating agents used in organic chemistry. They react with transition-metal complexes in four different ways[102]: halide abstraction, oxidative addition, oxidation of the metal atom or by alkylating a coordinated ligand. Of particular importance here is the ability to give a coordinatively unsaturated cation by halide ion abstraction. Oct-1-ene and hex-1-ene are rapidly isomerised by $trans$-$[PtClH(PMe_2Ph)_2]$–$MeSO_3F$ in dichloromethane. The catalyst is the cation $[PtH(PMe_2Ph)_2]^+$. Pent-1-ene is also isomerised as expected, but here the triphenylphosphine analogue of the platinum complex is more active than the dimethylphenylphosphine complex[103], in contrast to its behaviour in the oct-1-ene isomerisation above[99]. Wilkinson's complex, $RhCl(PPh_3)_3$, is also much activated as an isomerisation catalyst by 'magic methyl'[103].

A study relevant to the activation of alkanes (Refs. 58–67) involves hydrogen–deuterium exchange in alkenes[104]. The alkenes $RCMe_2CH=CH_2$

(R = Et, Pr^n, Bu^n), undergo H–D exchange almost exclusively at C-5 under the conditions used for alkanes by Hodges *et al.*[58]. The reaction probably takes place with the alkene coordinated to the platinum, when hydrogen abstraction takes place at C-5, to give the intermediate (38) (shown for 3,3-dimethylpent-1-ene).

(38)

1.4.2.2 *Disproportionation and cycloaddition*

Two reviews of alkene disproportionation (alkene metathesis) [equation (1.39)] have recently been published[105,106]. The reaction is catalysed by

$$R^1CH{=}CHR^2 + R^3CH{=}CHR^4 \rightarrow R^1CH{\parallel}R^3CH + CHR^2{\parallel}CHR^4 \tag{1.39}$$

transition-metal compounds, and both heterogeneous systems (oxides and carbonyls) and homogeneous systems are discussed. The most common catalyst is WCl_6 with a reducing agent. A homogeneous system that involves butylmagnesium iodide as the reducing agent has been used to prepare but-2-ene and dec-5-ene from hept-2-ene at room temperature [equation (1.39); $R^1 = R^3 = Bu^n$, $R^2 = R^4 = Me$][107]. Both $NaBH_4$[108] and $LiAlH_4$[108,109] are also very effective reducing agents for use with WCl_6 in alkene metathesis, although they are not just reducing agents — WCl_4 is not the active species. The appearance of two almost identical papers[108,109] in the same journal within a few months of each other calls for comment. This is provided by an editorial footnote to the second paper[109] saying that the Chemical Society accepts full responsibility for the long delay between receipt and publication. And so they should — the communication (for rapid publication) was received more than three years before publication! One can only speculate on the reasons for such a long delay (rejection until an identical paper appeared, followed by protests from the authors?).

Cycloadditions are closely related to disproportionations, in that disproportionations probably proceed through intermediates that are stable reaction products in cycloadditions. The cycloaddition of tetrafluoroethylene to *cis*- and to *trans*-but-2-ene[110] and to ethylene[111] have been studied. The products

(39) (40) (41)

(39) and (40) are both formed from *cis*- and from *trans*-but-2-ene, in the ratio (39) : (40) of 58 : 42, and (39) : (40) of 72 : 28, respectively. The reaction occurs via a biradical intermediate since two geometrically isomeric products are formed[110]. The addition of tetrafluoroethylene to [1,2-^{2}H$_2$]ethylene is used to show that even an alkene as unhindered as ethylene reacts by a radical mechanism and not in a concerted manner[111]. The photosensitised cycloaddition of *cis*- and of *trans*-but-2-ene to cyclopentadiene gives seven products; three stereoisomeric 5,6-dimethylnorbornenes (41)–(43) and four

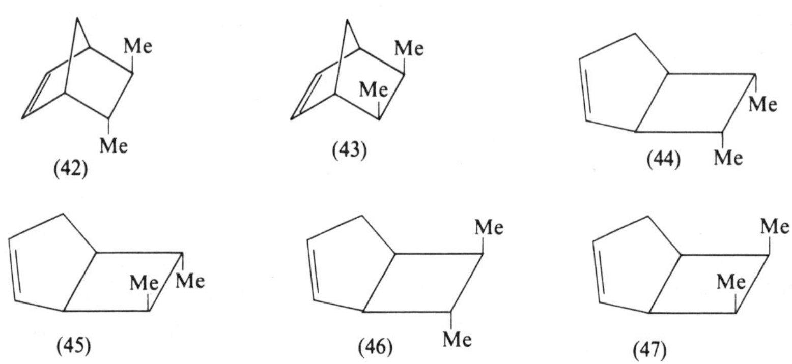

stereoisomeric 6,7-dimethylbicyclo[3,2,0]heptanes (44)–(47)[112]. The reaction products are explained by the reaction being made up of the following steps. (*a*) Photoexcitation of the sensitiser (β-acetonaphthone). (*b*) Intersystem crossing of excited singlet sensitiser to triplet. (*c*) Transfer of triplet energy from excited sensitiser to cyclopentadiene. (*d*) Formation of a triplet biradical from excited cyclopentadiene and but-2-ene. (*e*) Stereochemical equilibration of the biradical by internal rotation. (*f*) Spin inversion of one of the odd electrons in the biradical. (*g*) Ring closure of the product.

1.4.2.3   *Reduction*

It is well known that alkenes can be reduced to alkanes with gaseous hydrogen by using a transition-metal complex as a soluble catalyst. Work has continued on this deceptively simple reaction. Complexes which have been used as catalysts include [(π-cyclopentadienyl)$_2$Ti]$_{1-2}$[113], Mn$_2$(CO)$_{10}$[114], RuCl$_2$-(PPh$_3$)$_3$[115], RuCl$_2$(CO)$_2$(PPh$_3$)$_2$[116], (π-C$_6$H$_6$)RuCl$_2$[117,118], (π-mesitylene)-RuCl$_2$[118], (π-Ph$_3$C$_6$H$_3$)RuCl$_2$[118], (PhCN)$_2$PtCl$_2$[119], (H$_2$PPH$_2$)PdCl$_2$[119] and (Ph$_2$PCH$_2$PPh$_2$)$_3$Pd$_2$[120]. A ruthenium–ethylene complex (48) has been

$$RuH_2(PPh_3)_4 + 2\,C_2H_4 \rightarrow RuC_2H_4(PPh_3)_3 + PPh_3 + C_2H_6 \qquad (1.40)$$
$$(48)$$

isolated as a white crystalline compound from the reaction of RuH$_2$(PPh$_3$)$_4$ with ethylene [equation (1.40)][121]. This lends support to the accepted theory that such a complex must be formed during any reaction of an alkene at a transition-metal centre. An active titanocene catalyst is prepared by adding sodium or lithium naphthalenide in THF to $[\pi\text{-}Cp_2TiCl]_2$ or $\pi\text{-}Cp_2TiCl_2$ and the alkene in a hydrogen atmosphere[113]. Dec-1-ene reacts very rapidly to give decane in high yield. However, $Mn_2(CO)_{10}$ is not a very efficient catalyst. Terminal alkenes are hydrogenated faster than internal alkenes, but temperatures of 160 °C and hydrogen pressures above atmospheric must be used[114]. $RuCl_2(CO)_2(PPh_3)_2$ is first converted into a hydride, and this is the actual catalytic species[116], and a hydride is also formed from the π-arene–ruthenium complexes[117]. Formic acid has been used as the hydrogen source for the reduction of alkenes[122,123]. The reaction [equation (1.41)] is catalysed by

$$RCH{=}CH_2 + HCO_2H \rightarrow RCH_2CH_3 + CO_2 \qquad (1.41)$$

a range of transition-metal complexes[122], the most effective being $RuHCl(PPh_3)_3$[123], and the best solvent is DMF. This ruthenium complex has been the most extensively studied. It is a good catalyst for the decomposition of formic acid to hydrogen and carbon dioxide, but it is reported that this decomposition is not the source of the hydrogen for reduction, as under identical conditions hydrogen does not hydrogenate the terminal alkenes studied. More evidence is needed to clarify ideas about this system.

The development of a new group of compounds with considerable potential for catalysing alkene reactions may now be mentioned. These are the supported homogeneous catalysts[124,125], in which the transition-metal complex is chemically bonded to a polymeric support; although the catalyst is then heterogeneous, its chemistry closely parallels that of the related soluble complex. These heterogeneous catalysts are of particular interest to the industrial chemist because they can be readily separated from reaction products and recycled. Two types of polymer have been most widely studied: that formed from chloromethylated polystyrene in which an organophosphine group has been substituted for the chlorine [equation (1.42)][126–128], and a

$$\text{polystyrene}\text{—}\langle\text{C}_6\text{H}_4\rangle \xrightarrow[\text{SnCl}_4]{\text{EtOCH}_2\text{Cl}} \text{polystyrene}\text{—}\langle\text{C}_6\text{H}_4\rangle\text{—CH}_2\text{Cl}$$

$$\xrightarrow{\text{LiPPh}_2} \text{polystyrene}\text{—}\langle\text{C}_6\text{H}_4\rangle\text{—CH}_2\text{PPh}_2 \qquad (1.42)$$

support based on silica with an organophosphine attached to the silica through silicon–oxygen bonds [equation (1.43)][129]. A soluble complex may

$$Ph_2PH + CH_2{=}CHSi(OEt)_3 \rightarrow Ph_2PCH_2CH_2Si(OEt)_3 \xrightarrow{\text{Silica}}$$

$$Ph_2PCH_2CH_2Si{-}(O)_3{-}\text{silica} + 3EtOH \qquad (1.43)$$

be equilibrated with either of these supports when phosphine–ligand exchange takes place and the complex becomes attached to the support, e.g. (49)[130]. Reduction, hydroformylation, hydrosilylation, acetoxylation, polymerisation and oligomerisation of many alkenes have been carried out by using these

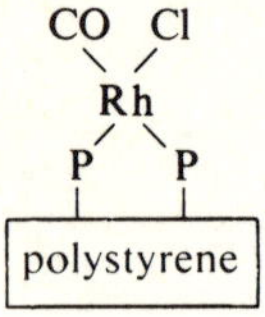

(49)

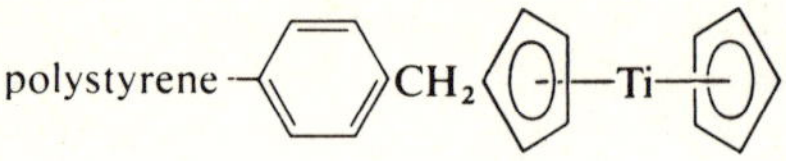

(50)

supported catalysts[125]. An interesting supported titanocene catalyst (50) has been prepared and used as a catalyst for reduction of alkenes. Its use may be compared with that of titanocene (Ref. 113).

1.4.2.4 Dimerisation

The codimerisation of ethylene with propene and with certain pentenes has been studied by using a variety of alkali-metal catalysts[131]. The catalysts, dispersions of lithium hydride or of sodium or potassium on Group IA or IIA metal carbonates, differ in the same selectivity. Catalysts containing lithium and those containing the Group IIA metal carbonates show high selectivity for linear heptanes and nonenes, compared with catalysts involving Group IA metal carbonates which give branched-chain alkenes. The differences in selectivity are explained by the nature of the bonding between the metal and the organic anion intermediate [equation (1.44)]; there is

$$MeCH\!=\!CH_2 \xrightarrow{\;K\;} [CH_2\text{---}CH\text{---}CH_2]^- K^+ \xrightarrow{\;C_3H_6\;} CH_2\!=\!CHCH_2CHMe\bar{C}H_2\,K^+$$

$$\xrightarrow{\;C_3H_6\;} CH_2\!=\!CHCH_2CHMe_2 + [CH_2\text{---}CH\text{---}CH_2]^-\,K^+ \qquad (1.44)$$

greater covalent binding between lithium and Group IIA metals and the anion, thus keeping the anion in one mesomeric form. The reaction, e.g. equation (1.44), gives allylic anions which can react at either end, e.g. equation (1.45). The products are given in Table 1.4, in which the total yields and

$$
\begin{array}{ccc}
Et\bar{C}HCH\!=\!CH_2 & \longleftrightarrow & EtCH\!=\!CH\bar{C}H_2 \\
\big\downarrow\, CH_2\!=\!CH_2 & & \big\downarrow\, CH_2\!=\!CH_2 \\
EtCHCHCH\!=\!CH_2 & & EtCH\!=\!CHCH_2CH_2\bar{C}H_2 \\
\big\downarrow\, CH_2CH_2{}^- & & \big\downarrow \\
\text{3-ethylpent-1-ene} & & \text{hept-3-ene}
\end{array}
\qquad (1.45)
$$

proportions of linear alkenes are given. A number of titanium compounds (tetrabenzyltitanium, π-cyclopentadienyltribenzyltitanium and two-component Zeigler-type systems) are catalysts for the codimerisation of ethylene and buta-1,3-diene to vinylcyclobutane[132], in contrast to other catalysts which give hexa-1,4-diene. The mechanism of the reaction is given in equation

Table 1.4   Co-Oligomerisation of 3 : 1 molar ethylene–propene mixtures[131]

Catalyst	Temp. /°C	Pentenes /% (w/w)	Heptenes /% (w/w)	Heptenes /% Linear	Nonenes /% (w/w)	Nonenes /% Linear
Li/K_2CO_3	150	49	32	64	11	35
Li/Rb_2CO_3	150	60	23	63	7	48
Li/Cs_2CO_3	150	44	32	54	11	30
K/Li_2CO_3	150	58	24	54	< 5	
$K/CaCO_3$	180	52	27	58	5	48
$K/SrCO_3$	180	71	24	58	1	
$K/BaCO_3$	180	50	30	51	8	34
$K/BaCO_3$	150	48	35	54	5	
Na/K_2CO_3	150	14	57	20	12	7
K/K_2CO_3	150	33	53	20	7	1
Na/Cs_2CO_3	150	45	46	32	4	< 1
K/Cs_2CO_3	150	51	32	28	8	4
K/Cs_2CO_3	180	22	57	24	12	3

(1.46). In addition to vinylcyclobutane, some hexa-1,4-diene, 4-vinylcyclohex-1-ene, open-chain octatrienes, cyclohexene and alkene isomerisation products

$$\equiv Ti-H \; + \; CH_2{=}CHCH{=}CH_2 \; + \; C_2H_4 \; \longrightarrow \qquad\qquad (1.46)$$

$$\longrightarrow \left[\underset{Ti}{\overset{CHMe}{\square}} \right] \longrightarrow \square^{CH{=}CH_2} \; + \; \equiv Ti-H$$

are also formed. Rhodium trichloride trihydrate is also a catalyst for the ethylene–buta-1,3-diene codimerisation to give hexa-1,4-diene[133], but the catalyst becomes deactivated as the Rh^{III} is converted into Rh^{I}. The activity is restored if HCl is added; this reacts oxidatively with the Rh^{I} to give an active Rh^{III} complex[134,135]. Organic chlorides can be used instead of hydrogen chloride[136]; allylic chlorides, benzylic chlorides, α-chloroethers, and compounds with a C—Cl bond adjacent to a carbonyl group are the best. Ethylene and buta-1,3-diene also give hexa-1,4-diene with a $CoCl_2$–$(Ph_2PCH_2)_2$–Et_3Al catalyst in 1,2-dichloroethane[137], and with a similar catalyst and a Grignard reagent as the reducing agent[138]. Nickel–phosphine complexes have also been used with boron trifluoride etherate[139], or ethyl-aluminium sesquichloride as the co-catalyst[140], for alkene dimerisation. The mechanism of hexadiene formation is shown in equation (1.47)[137]. From the codimerisation of propene with buta-1,3-diene, the branched-chain 2-methyl-hexa-1,4-diene is usually the product, but by using a mixture of di-μ-chloro-di-π-allyldipalladium with aluminium chloride and a tertiary phosphine as the catalyst, the straight-chain heptadienes are also formed, and the ratio of methylhexadiene to heptadiene is changed by changing the structure of the

$$(1.47)$$

phosphorus ligands[141]. With $(C_6H_{11})_3P$ as the phosphine the ratio of straight-chain to total diene is 0.22, but with $(PhO)_3P$ this ratio is 0.80. The ratios for the compounds Bu^n_3P, Ph_3P and $(EtO)_3P$ are between these two extremes.

1.4.2.5 Hydroboration

Work on the hydroboration of alkenes, particularly by H. C. Brown and his co-workers, continues[142–151]. This topic is discussed in Chapter 9.

1.4.2.6 Reactions with organic compounds

Phenylation and acylation of alkenes continue to be studied[152–155]. Dodecenes, decenes and octenes alkylate benzene with HF as the catalyst, and at 0 °C and at 55 °C the alkylbenzenes produced have different isomer distributions[152]. However, if hex-1-ene or *trans*-hex-3-ene are used, similar amounts of 2- and 3-phenylhexane are produced at the two temperatures. The alkene and the catalyst react to give a carbenium ion, and the final product composition depends on the rate of isomerisation relative to the rate of attack of the carbenium ion on benzene. For all the alkenes the isomerisation rate is increased by adding BF_3. At 35 °C or above, isomerisation gives the equilibrium distribution of carbenium ions and hence the product distribution is the same regardless of the position of the double bond in the reactant alkene. At 0 °C isomerisation is suppressed and the isomer distribution of the products depends on the position of the double bond in the initial alkene. The effect of the BF_3 is to replace F^- ions in the ion-pairs formed during reaction by BF_4^- ions. The competition between isomerisation and alkylation has been further studied by using 8-methylnon-1-ene [see (51)], as this

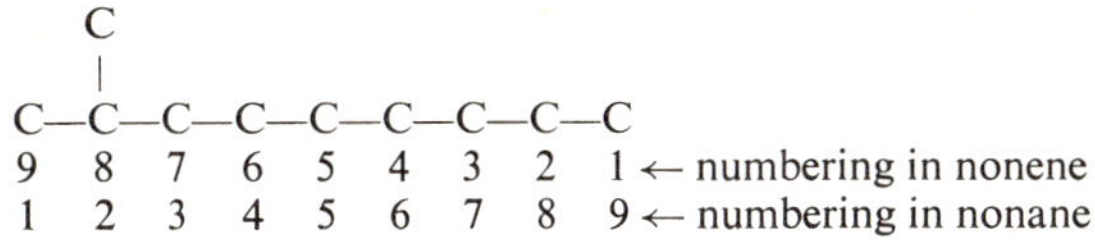

$$
\begin{array}{ccccccccc}
 & & \text{C} & & & & & & \\
 & & | & & & & & & \\
\text{C}-&\text{C}-&\text{C}-&\text{C}-&\text{C}-&\text{C}-&\text{C}-&\text{C}-&\text{C} \\
9 & 8 & 7 & 6 & 5 & 4 & 3 & 2 & 1 \leftarrow \text{numbering in nonene} \\
1 & 2 & 3 & 4 & 5 & 6 & 7 & 8 & 9 \leftarrow \text{numbering in nonane}
\end{array}
$$

$$(51)$$

enables the precursor for each phenylalkane to be determined[153]. With HF as the catalyst the product is 2-methyl-2-phenylnonane [see (51)]; the secondary carbenium ion initially formed has isomerised to the more stable tertiary carbenium ion before alkylation. If $Al(HSO_4)Cl_2$ is used as the catalyst there is much less of the 2-isomer and much more of the 8-isomer [see (51)]; apparently the presence of the large anion in the ion-pair slows down the isomerisation and allows the alkylation to compete more effectively. $AlCl_3 \cdot HCl$ gives the smallest amount of the tertiary isomer, but because of the vulnerability of the t-alkylbenzene to attack by the strong acid $AlCl_3 \cdot HCl$, dealkylation occurs to give substantial amounts of iso-alkanes. A review of acylation of double bonds has appeared[154], as has a report of acylation by using acyl tetrafluoroborates and acyl hexafluoroanti-monates[155]. For the latter compounds, alkene polymerisation is suppressed if the conditions are such that there is no excess of alkene in the system. Arylation of alkenes catalysed by palladium salts has been reviewed[156], and the effect of using triarylphosphines as the source of the aryl group has been studied[157]. By using the complex $Pd(OAc)_2(PAr_3)_2$, good yields of arylated alkenes are obtained [equation (1.48)]; the mechanism involves a phenyl-

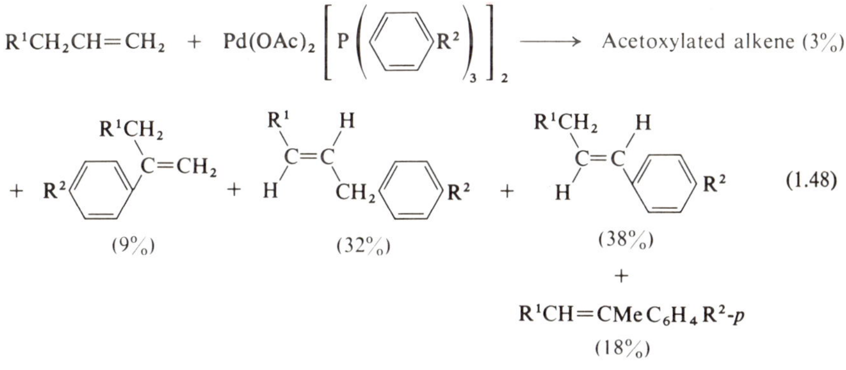

$$
R^1CH_2CH{=}CH_2 \;+\; Pd(OAc)_2\left[P\!\left(\!\bigcirc\!\!-\!R^2\right)_3\right]_2 \;\longrightarrow\; \text{Acetoxylated alkene (3\%)}
$$

$$
+\; R^2\!\bigcirc\!\!\begin{array}{c}R^1CH_2 \\ \diagdown \\ C{=}CH_2\end{array} \quad (9\%)
\;+\; \begin{array}{c}R^1 \quad H \\ \diagdown\;/ \\ C{=}C \\ /\;\diagdown \\ H \quad CH_2\end{array}\!\!-\!\bigcirc\!\!-\!R^2 \quad (32\%)
\;+\; \begin{array}{c}R^1CH_2 \quad H \\ \diagdown\;/ \\ C{=}C \\ / \\ H\end{array}\!\!-\!\bigcirc\!\!-\!R^2 \quad (38\%)
\tag{1.48}
$$

$$
+
$$

$$
R^1CH{=}CMe\,C_6H_4R^2\text{-}p
$$

$$(18\%)$$

palladium intermediate [equation (1.49)]. Ferrocene is an aromatic compound and so can be used in place of benzene, when ferrocenylation of an alkene

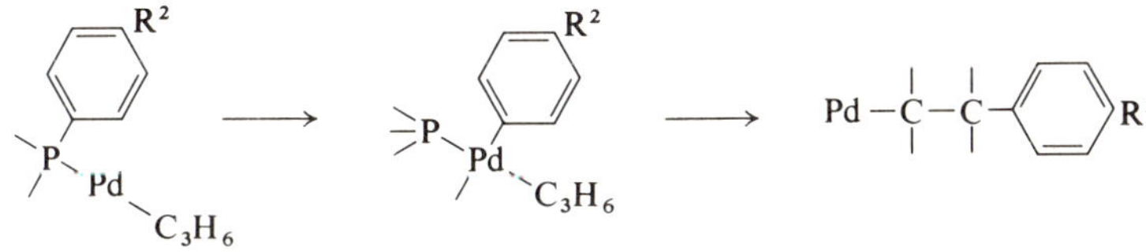

$$
\begin{array}{c} \diagup\!\!\bigcirc\!\!-\!R^2 \\ \diagup P \\ Pd \\ \diagdown C_3H_6 \end{array}
\;\longrightarrow\;
\begin{array}{c} \diagup\!\!\bigcirc\!\!-\!R^2 \\ {\geq}P \\ Pd \\ \diagdown C_3H_6 \end{array}
\;\longrightarrow\;
Pd\!-\!\underset{|}{\overset{|}{C}}\!-\!\underset{|}{\overset{|}{C}}\!\!-\!\bigcirc\!\!-\!R^2
\tag{1.49}
$$

occurs[158]. A ferrocenylpalladium salt is used, and with ethylene, vinylferrocene (45%) and biferrocenyl (4%) are formed [equation (1.50)].

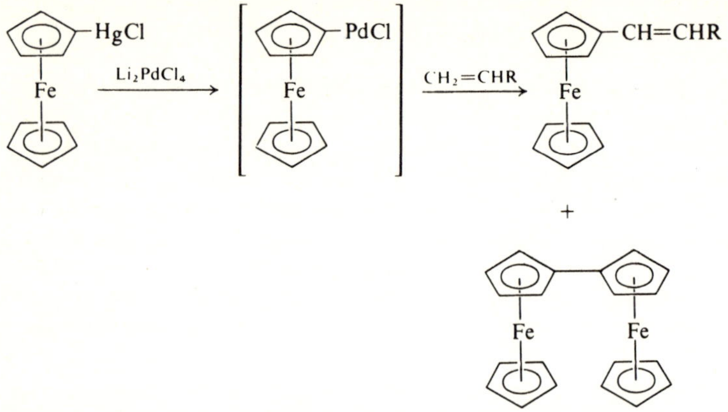

$$(1.50)$$

Under the influence of radiation from a ^{60}Co source, triethyl methanetricarboxylate adds to alkenes [equation (1.51)][159]. The yield of the adduct

$$RCH{=}CH_2 + HC(CO_2Et)_3 \xrightarrow[\text{radiation}]{^{60}Co} RCH_2CH_2C(CO_2Et)_3 \qquad (1.51)$$

increases as the chain length of the alkene increases (hex-1-ene to dodec-1-ene were studied). A convenient method for making chloro(formyloxy)alkanes at room temperature in high yield is by reaction with NCS in DMF[160]. The addendum reacts by *trans* addition. Hexafluoroacetone azine (52) with *cis*-

$$(CF_3)_2C{=}N{-}N{=}C(CF_3)_2$$
(52)

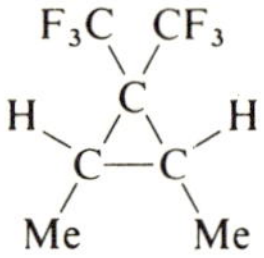

(53)

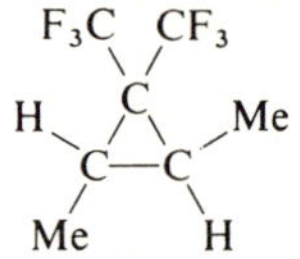

(54)

and with *trans*-but-2-ene under thermal and photochemical conditions gives the products (53) and (54), formally derived from bis(trifluoromethyl)-carbene[161]. 5,6-Dihydro-4H-1,3-thiazines are formed by a polar 1,4-cycloaddition of thioamidoalkyl ions to alkenes[162,163]. The addition is a one-step interaction of a thioamide, an aldehyde and an alkene [equation (1.52)];

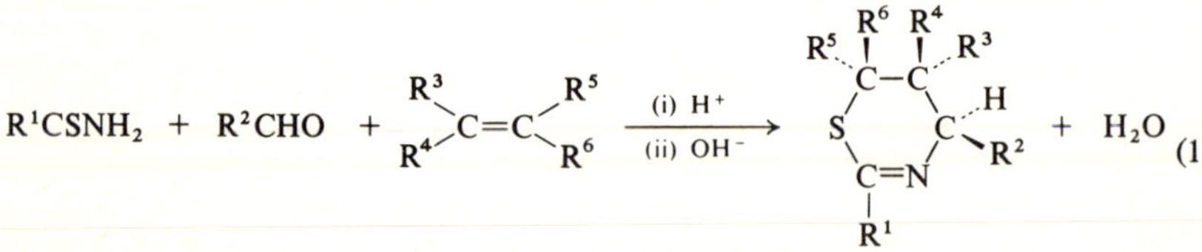

$$(1.52)$$

it is stereospecific and regiospecific. Propene, but-1-ene, but-2-enes and dodec-1-ene have been studied by one of two experimental methods. Either the aldehyde, the thioamide and concentrated sulphuric acid or HCl gas are mixed in acetic acid and added to the alkene in acetic acid, or boron trifluoride etherate is added to the aldehyde, the thioamide and the alkene in chloroform.

Amines such as NN'-tetramethylethylenediamine react with ethylene and with propene in the presence of catalytic amounts of an alkyl-lithium to give dimethylvinylamine and ethyldimethyl- or isopropyldimethyl-amine[164]. The advantages of this method over the more conventional sodium-catalysed reaction is that the pressures and temperatures required are much lower (70 compared with 1000 atm, and 140 compared with 225 °C). The amines used must be carefully selected, e.g. dimethylamine adds to propene but diethylamine does not. Both O-alkylation and C-alkylation occur when alkenes react with halogenotrinitromethanes or tetranitromethane in ether [equation (1.53)][165]. O-Alkylation gives the unstable ether of the aci-trinitro-

$$Me_2C{=}CH_2 \xrightarrow[\substack{X\ =\ NO_2,\\ Br\ or\ I}]{XC(NO_2)_3} CH_2X\overset{+}{C}Me_2 + (O_2N)_3C^-$$

$$\underset{(56)}{\overset{\displaystyle CH_2{-}C(NO_2)_2}{Me_2C\underset{O}{\diagdown}\diagup NOCMe_2CH_2X}} \xleftarrow{Me_2C{=}CH_2} \underset{(55)}{XCH_2CMe_2ON({\rightarrow}O){=}C(NO_2)_2} \qquad (1.53)$$

$$+$$

$$XCH_2CMe_2C(NO_2)_3$$

$$(57)$$

methane (55), and this undergoes a 1,3-dipolar addition to give the substituted 3,3-dinitroisoxazolidine (56). C-Alkylation gives derivatives of 3-halogeno-1,1,1-trinitropropane or 1,1,1,3-tetranitropropane (57). Stereospecific addition of a singlet arylnitrene to an alkene occurs when pentafluoronitroso-benzene is deoxygenated with triethyl phosphite in the presence of 2,3-dimethylbut-2-ene[166]. The aziridines (58) and (59) are formed from cis- and

$$C_6F_5N\diagup\diagdown\overset{R}{\underset{R}{\diagdown}}\diagup H \qquad\qquad C_6F_5N\diagup\diagdown\overset{R}{\underset{H}{\diagdown}}\diagup H$$

$$(58) \qquad\qquad\qquad (59)$$

trans-but-2-ene, respectively. Here the arylnitrene cannot be prepared from an azide, as is usual, as azides add to alkenes at lower temperatures than those required to generate the nitrene. Cyanogen azide adds to alkenes to give N-cyanoaziridines (60) and 1-alkylalkylidenecyanamides (61) [equation (1.54)][167]. The reaction is a 1,3-dipolar concerted addition of the azide, followed by opening of the triazoline ring to a diazonium zwitterion, and loss of nitrogen from this labile species to give the

$$Me_2C=CH_2 + N_3CN \xrightarrow[\text{35 C, 20h}]{\text{MeCN}} \left[\begin{array}{c} CH_2-CMe_2 \\ | \quad\quad | \\ N \quad\quad NCN \\ \diagdown N \diagup \end{array} \right] \longrightarrow Me_2C-CH_2 \quad (34\%)$$

$$(60)$$

$$+ \quad\quad (1.54)$$

$$EtMeC=NCN \quad (48\%)$$

$$(61)$$

products. The 1-alkylalkylidenecyanamides formed are readily hydrolysed to ketones, so the reaction provides a method of making ketones from alkenes. The relative amounts of the two products (60) and (61) vary greatly from alkene to alkene. Linear alkenes give high yields of compounds of the type (61), often to the exclusion of the cyanoaziridine; ethylene, propene, but-1-ene, but-2-ene, hex-1-ene, dec-1-ene and dodec-1-ene all fall into this class. More highly branched alkenes, particularly those with no hydrogen on the carbon to which the cyano-bearing nitrogen becomes attached, such as isobutene, 2-methylbut-1-ene, 3-methylbut-1-ene and 3,3-dimethylbut-1-ene, give significant yields of compounds of the type of both (60) and (61). Steric factors may partly determine the product composition, e.g. *cis*-3-methylpent-2-ene gives 40% of (62) and 60% of (63), whereas the *trans* alkene gives 13%

$$EtPr^iC=NCN \quad\quad\quad EtMeCHCMe=NCN$$

$$(62) \quad\quad\quad\quad\quad (63)$$

and 87%, respectively. Hydrocyanation of alkenes is catalysed by nickel(0) phosphite complexes activated with a Lewis acid such as zinc chloride or aluminium chloride, the catalyst being the $HNiL_3{}^+$ cation[168]. The reaction is as shown in equations (1.55)–(1.58). The catalyst formed in equation (1.55)

$$NiL_4 + HCN + ZnCl_2 \rightarrow (HNiL_3)^+ (ZnCl_2CN)^- + L \quad\quad (1.55)$$

$$HNiL_3{}^+ + RCH=CH_2 \rightarrow RCH_2CH_2NiL_3{}^+ + RCHMeNiL_3{}^+ \quad\quad (1.56)$$

$$RCH_2CH_2NiL_3{}^+ + ZnCl_2CN^- \rightarrow RCH_2CH_2CN + NiL_3 + ZnCl_2 \quad (1.57)$$

$$RCHMeNiL_3{}^+ + ZnCl_2CN^- \rightarrow RCHMeCN + NiL_3 + ZnCl_2 \quad\quad (1.58)$$

reacts with the alkene [equation (1.56)] to give both the terminal and the internal cationic nickel complex; these lead to the terminal nitrile [equation (1.57)] and the internal nitrile [equation (1.58)], respectively. The amounts of terminal and of internal nitrile depend on the alkene; linear alkenes such as propene and hex-1-ene give significant amounts of internal products, but branched-chain alkenes give the terminal nitrile as the major product. Transition-metal complexes also catalyse the addition to alkenes of urea[169] and of methyl trichloroacetate[170]. Urea and ethylene give acetylurea in the presence of oxygen, a nitrile, sodium tetrachloropalladate and copper(II) chloride in acidic aqueous solution[169]. The reaction [equation (1.59)] is

$$CH_2=CH_2 + CO(NH_2)_2 \xrightarrow{Pd^{2+}} [CH_2=CHNHCONH_2] + 2H^+ + Pd^0$$

$$\xrightarrow[Pd^{2+}-H_2O]{} MeCONHCONH_2 + 2H^+ + Pd^0 \quad\quad (1.59)$$

catalysed by palladium(II), the Pd^0 formed being reoxidised by the oxygen, with the Cu^{II} as the oxygen-transfer catalyst (*cf.* the Wacker process for preparing acetaldehyde from ethylene and oxygen). The addition of methyl trichloroacetate to alkenes has been investigated by using dicobalt octacarbonyl, cyclopentadienylmolybdenum tricarbonyl dimer, and cyclopentadienyliron dicarbonyl dimer as catalyst[170]. The reaction product with the

$$R^1R^2C{=}CH_2 + Cl_3CCO_2Me \xrightarrow{Co_2(CO)_8} R^1R^2CClCH_2CCl_2CO_2Me \qquad (1.60)$$

cobalt catalyst [equation (1.60)] is different from that with the Mo or Fe catalyst [equation (1.61)]. Whether the final lactonisation takes place

$$R^1R^2C{=}CH_2 + Cl_3CCO_2Me \xrightarrow[\text{or } [CpFe(CO)_2]_2]{[CpMo(CO)_3]_2} \begin{array}{c} H_2C\!\!-\!\!-\!\!CR^1R^2 \\ | \qquad | \\ Cl_2C \diagdown \quad O \\ CO \end{array} \qquad (1.61)$$

[equation (1.61)] depends on the interaction of the methoxycarbonyl group with the metal in the intermediate (64) which precedes the products. If the

$$R^1R^2CCH_2CCl_2CO_2Me$$
$$|$$
$$M(CO)_xL_y$$
$$(64)$$

interaction is strong, as it must be with the Mo and the Fe catalyst, the reaction is as in equation (1.61); if the interaction is weak, as it must be with the Co catalyst, the reaction is as in equation (1.60).

The products and relative rates of cycloaddition of butenes, substituted butenes and pentenes to dimethylketene have been further studied[171]; the steric and electronic factors accord with a concerted reaction in which the

$$Me_2C{=}C{=}O \; + \; {\Large{>}}C{=}C{\Large{<}} \; \longrightarrow \; \begin{array}{c} Me \quad O \\ | \quad \diagup\!\!\diagup \\ Me{-}C{-}C \\ | \quad | \\ {-}C{-}C{-} \\ | \quad | \end{array} \qquad (1.62)$$

ketene adds antarafacially [equation (1.62)]. An interesting cycloaddition, that of acrylonitrile to ethylene, has also been studied [equation (1.63)][172].

$$C_2H_4 + CH_2{=}CHCN \; \longrightarrow \; \begin{array}{c} H_2C{-}CHCN \\ | \qquad | \\ H_2C{-}CH_2 \end{array} \qquad (1.63)$$

The cyclodimerisation of the ketene and the nitrile are known, but this cocyclisation has not been reported before. It is not a very efficient reaction; the yields are only about 10% under very forcing conditions, *viz.* 1000 atm of ethylene and 300–345 °C. Oxygen and free-radical initiators must be carefully removed from the reactants or polymers and tars are formed. Nickelocene is found to be a weak catalyst.

A number of radical reactions have been studied[173–184]. Diphenylcarbene

(Ph_2C:) reacts with alkenes by addition and by hydrogen abstraction; the reaction pathway traversed is determined mainly by steric factors[173]. The reaction of diphenylcarbene with but-2-enes has been studied before [equa-

$$Ph_2C: \quad + \quad \underset{(65)}{\overset{Me}{\underset{H}{>}}C=C\overset{H}{\underset{Me}{<}}} \quad \longrightarrow \quad \underset{\underset{(66)\quad(22\%)}{(cis\ and\ trans)}}{MeHC\!\!-\!\!CHMe \atop CPh_2} \quad + \quad \underset{(67)\ (28\%)}{Ph_2CHCHMeCH=CH_2}$$

$$\hspace{4cm} (1.64)$$

$$+ \quad Ph_2CHCH_2CH=CHMe$$

$$(cis\ and\ trans)$$

$$(68)\ (50\%)$$

tion (1.64)]; more highly branched alkenes give higher percentages of abstraction–recombination products; the yields of cyclopropane, e.g. (66), and abstraction–recombination products, e.g. (67)–(68), for various alkenes are given in Table 1.5. The non-stereospecific reaction of cis-but-2-ene with

Table 1.5 The mode of reaction of diphenylcarbene with alkenes[173]

Alkene	Cyclopropane/%	Abstraction–recombination product/%
Propene	100	0
Isobutene	100	0
But-1-ene	74	26
3-Methylbut-1-ene	52	48
trans-But-2-ene	22	78
2-Methylbut-2-ene	8	92
2,3-Dimethylbut-2-ene	0	100

nitrophenylcarbenes, produced by photolysis of diazo compounds, has been studied [equation (1.65)][174]. Isobutene reacts with n-butyl-lithium followed

$$ArCH:N_2 \quad + \quad \underset{H}{\overset{Me}{>}}C=C\underset{H}{\overset{Me}{<}}$$

$$\Big\downarrow hv$$

$$(1.65)$$

by n-butyl bromide in hexane containing tetramethylethylenediamine to give (69) and (70)[175]. The reacting species are the methylallyl anion (71) and the tri-

$$\underset{(69)}{CH_2=CMe(CH_2)_4Me} \qquad \underset{(70)}{CH_2=C[(CH_2)_4Me]_2}$$

$$\text{CH}_2\text{—}\overset{\displaystyle\cdots\cdots\diagup\text{CH}_2}{\text{CMe}}\qquad\qquad\left[\text{CH}_2\text{==}\overset{\diagup\text{CH}_2}{\underset{\diagdown\text{CH}_2}{\text{C}}}\right]^{2-}$$

$$(71)\qquad\qquad\qquad\qquad(72)$$

methylenemethane dianion (72). Acetic acid adds to alkenes, when di-t-butyl peroxide is used as a radical initiator, to give normal-chain aliphatic acids, such as butanoic, hexanoic, octanoic, decanoic, lauric, myristic and palmitic acid, by radical addition[176]. Some α-branched-chain acids are also formed. The best yield is obtained with ethylene as the alkene, but this is only 7% after 6 h at 150 °C. Alcohols may be added to alkenes and alkynes by a radical mechanism by using radiation from a quartz mercury lamp as an initiator[177], e.g. propan-2-ol adds to hex-1-ene to give 2-methyloctan-2-ol in 45% yield. A wide range of products is produced if alkenes are irradiated with u.v. light in hydroxylic solvents[178]. The products from 2-methylbut-2-ene and 2,3-dimethylbut-2-ene are given in equations (1.66) and (1.67). The

$$\text{Me}_2\text{C}\text{=CHMe}\xrightarrow[\text{MeOH}]{h\nu}\text{Me}_2\text{CHCHMeOMe} + \text{CH}_2\text{=CMeCHMeOMe} + \text{EtMe}_2\text{COMe} +$$
$$\text{CH}_2\text{=CHCMe}_2\text{OMe}$$
$$+ \text{Me}_2\text{CHEt} + \text{Me}_2\text{CHCH}\text{=CH}_2 + \text{CH}_2\text{=CMeEt} \qquad (1.66)$$

$$\text{Me}_2\text{C}\text{=CMe}_2 \rightarrow \text{CH}_2\text{=CMePr}^i + \text{Me}_2\text{CHCHMe}_2 + \text{CH}_2\text{=CMeCMe}_2\text{OMe}$$
$$+ \text{Me}_2\text{CHCMe}_2\text{OMe} \qquad (1.67)$$

products from the latter compound in an aprotic medium are given in equation (1.68). γδ-Unsaturated acids and γ-lactones are conveniently

$$\text{Me}_2\text{C}\text{=CMe}_2 \rightarrow \text{CH}_2\text{=CMePr}^i + \text{CH}_2\text{=CHBu}^t + \overset{\displaystyle\text{CH}_2}{\underset{\displaystyle\text{MeHC—CMe}_2}{\diagup\diagdown}} \qquad (1.68)$$

prepared by an oxidative radical addition to alkenes[179]. Diethyl malonate and diethyl alkylmalonates react with alkenes and Mn^{III} or Co^{III} acetate in the presence of Cu^{II} acetate to give diethyl alk-2-enylmalonates, which on hydrolysis give γδ-unsaturated acids or γ-lactones. The method is a modification of the free-radical addition of malonic esters to alkenes. If diethyl malonate (0.5 mol), Mn(OAc)$_3\cdot$2H$_2$O (0.025 mol) and hept-1-ene (0.05 mol) are heated at 90 °C with catalytic amounts of Cu(OAc)$_2$, diethyl hept-2-enyl-malonate is the only product. The reaction [equation (1.69); $R^1 = H$ or Bu^n, $R^2 = H$, Bu^n or n-C$_5$H$_{11}$, $R^3 = H$ or Me; $M^{III} = Mn^{III}$ or Co^{III}] involves oxidation of the radical formed from the reactants to give the diester, which on hydrolysis gives a γδ-unsaturated acid. If the product is an alkenyl ester [equation (1.69); $R^2 = H$, $R^3 = Me$] then in acidic solution

$$R^1\text{CH}(\text{CO}_2\text{Et})_2 + \text{Mn}^{III}(\text{OAc})_3 \rightarrow \bullet\text{CR}^1(\text{CO}_2\text{Et})_2 + \text{Mn}^{II}(\text{OAc})_2 + \text{AcOH}$$

$$R^2\text{CH}_2\text{CR}^3\text{=CH}_2 + \bullet\text{CR}^1(\text{CO}_2\text{Et})_2 \rightarrow R^2\text{CH}_2\overset{\bullet}{\text{C}}\text{R}_3\text{CH}_2\text{CR}^1(\text{CO}_2\text{Et})_2$$

$$\xrightarrow[-\text{H}\bullet]{\text{Cu}^{II}(\text{OAc})_2} R^2\text{CH}\text{=CR}^3\text{CH}_2\text{CR}^1(\text{CO}_2\text{Et})_2 \xrightarrow[-\text{CO}_2]{\text{H}_2\text{O, OH}^-} R^2\text{CH}\text{=CR}^3\text{CH}_2\text{CHR}^1\text{CO}_2\text{H}$$

$$\underset{-\text{CO}_2}{\searrow}\;\overset{\text{H}_2\text{O, H}^+}{}$$

$$\overset{\displaystyle R^1\text{HC}\text{——}\text{CH}_2}{\underset{\displaystyle\underset{\text{O}}{\text{OC}\diagdown\;\;\diagup\text{CR}^3\text{CH}_2\text{R}^2}}{\big|\qquad\big|}} \qquad (1.69)$$

the γ-lactone is formed. The addition of carbon tetrachloride and chloroform to alkenes is catalysed by ruthenium(II) phosphine complexes[180], and by copper and iron salts[181]. With $RuCl_2(PPh_3)_3$ as the catalyst, CCl_4 adds faster than $CHCl_3$[180]. The reaction, which involves $Cl_3C\cdot$ radicals, is given in equation (1.70). With copper or iron salts in acetonitrile the reaction takes

$$\equiv\overset{|}{\underset{|}{Ru}}- \ + \ CCl_4 \ \longrightarrow \ \equiv\overset{|}{\underset{|}{Ru}}-Cl \ + \ \cdot CCl_3$$

$$(1.70)$$

$$\cdot CCl_3 \ + \ RCH{=}CH_2 \ \longrightarrow \ R\overset{\bullet}{C}HCH_2CCl_3 \ \xrightarrow{\ \equiv\overset{|}{\underset{|}{Ru}}-Cl\ } \ RCHClCH_2CCl_3$$

a different course and two alkene molecules react with each CCl_4 molecule[181], the products being a mixture of the 2 : 1 telomer (73) and the two-

$$2\,CH_2{=}CHR \ + \ CCl_4 \rightarrow CCl_3CH_2CHRCH_2CHRCl$$
$$(73)$$
$$+ \ RCHClCH_2CCl_2CH_2CHRCl \qquad (1.71)$$
$$(74)$$

fold addition product (74) [equation (1.71)]. Carbonyl compounds add to alkenes by a free-radical mechanism initiated by $Mn^{III}(OAc)_3$[182–184]. Aldehydes give free radicals such as (75) and (76) in which the radical centre

$$RCH_2CHO \ \xrightarrow{\ Mn(OAc)_3\ } \ RCH_2\overset{\bullet}{C}{=}O \ + \ R\overset{\bullet}{C}HCHO \qquad (1.72)$$
$$\hspace{5.5cm}(75) \hspace{2.5cm} (76)$$

is at the carbonyl C atom or the α C atom[182]. In the absence of a solvent or in a non-polar solvent, ketones are the preferred product; in acetic acid solution α-alkylaldehydes and 4-acetoxy derivatives are favoured [equations

$$R^1CH_2\overset{\bullet}{C}{=}O \ + \ CH_2{=}CHR^2 \rightarrow R^1CH_2COCH_2\overset{\bullet}{C}HR^2 \ + \ R^1CH_2COCHR^2\overset{\bullet}{C}H_2$$
$$(75)$$
$$\Big\downarrow R^1CH_2CHO \hspace{3cm} \Big\downarrow \qquad (1.73)$$
$$R^1CH_2COCH_2CH_2R^2 \quad R^1CH_2COCHMeR^2$$

$$R^1\overset{\bullet}{C}HCHO + CH_2{=}CHR^2 \rightarrow R^1CH(CHO)CH_2\overset{\bullet}{C}HR^2 \xrightarrow{\ R^1CH_2CHO\ } R^1CH(CHO)CH_2CH_2R^2$$
$$(76) \hspace{2cm} \Big| Mn(OAc)_3$$
$$\hspace{4cm} \longrightarrow R^1CH(CHO)CH_2CHR^2OAc$$
$$(1.74)$$

(1.72), (1.73) and (1.74)][183]. These systems are heterogeneous because of the low solubility of the $Mn^{III}(OAc)_3$. If the concentration of $Mn(OAc)_3$ is low then ketones are the preferred products [equation (1.73)]; if its concentration is high then the α-alkylalkanals and their γ-acetoxy derivatives are preferred [equation (1.74)]. With acetone and $Cu(OAc)_2$ added to the $Mn(OAc)_3$, unsaturated ketones are formed in high yield [equations (1.75) and (1.76)][184].

$$Me_2CO + Mn(OAc)_3 \rightarrow MeCO\dot{C}H_2 + Mn(OAc)_2 + AcOH \qquad (1.75)$$

$$MeCO\dot{C}H_2 + CH_2{=}CH(CH_2)_nMe \rightarrow MeCO(CH_2)_2\dot{C}H(CH_2)_nMe \rightarrow$$

$$\underset{\text{β,γ-unsaturated ketone}}{MeCOCH_2CH{=}CH(CH_2)_nMe} \xrightarrow[\text{with Cu(OAc)}_2 \text{ present}]{\text{Further oxidation}} \underset{\text{γ-acetoxy-α,β-unsaturated ketone}}{MeCOCH{=}CHCH(OAc)(CH_2)_{n-1}Me}$$

$$(1.76)$$

1.4.2.7 Carbonylation and hydroformylation

Alkenes and carbon monoxide are inserted into nickel carbonyl complexes: carbonylated products, *viz.* acids and ketones, are thus formed[185,186]. A typical reaction for ethylene is given in equation (1.77), where the product

$$MeCH{=}CHCH_2Cl + CH_2{=}CH_2 + CO + H_2O \xrightarrow[\substack{\text{aq. acetone}\\ \text{30 atm.}\\ \text{45 °C}}]{Ni(CO)_4} \begin{array}{l} MeCH{=}CH(CH_2)_3CO_2H \\ \qquad\qquad + HCl \end{array}$$

$$(1.77)$$

is an acid. Higher alkenes give much lower yields of acid, the cyclic ketone being the favoured product [equation (1.78)]. Alkenes can be converted into

$$MeCH{=}CHCH_2Cl + CH_2{=}CH(CH_2)_3Me + 2CO + H_2O$$

$$\downarrow Ni(CO)_4$$

$$\begin{array}{c} CH_2{-}CH_2 \\ | \qquad\quad | \\ Bu^nHC \qquad CHCHMeCO_2H \\ \diagdown_{CO}\diagup \end{array} \quad (1.78)$$

dialkyl succinates, and substituted dialkyl succinates by using palladium salts as catalysts[187,188]. Thus 'methoxycarbonylpalladium chloride' reacts first with an alkene, then with carbon monoxide, and then with methanol; diesters are formed in *ca.* 60% yield [equations (1.79) and (1.80)][187]. Alkenes

$$ClHgCO_2Me + PdCl_2 \rightleftharpoons [ClPdCO_2Me] + HgCl_2 \qquad (1.79)$$

$$[ClPdCO_2Me] + RCH{=}CH_2 \rightarrow RCH(PdCl)CH_2CO_2Me \xrightarrow{CO}$$

$$[RCH(COPdCl)CH_2CO_2Me] \xrightarrow{MeOH} RCH(CO_2Me)CH_2CO_2Me + [HPdCl]$$

$$(1.80)$$

studied include hex-1-ene[187], *cis-* and *trans*-hex-3-ene[187], *trans*-hex-2-ene[187], propene[188] and oct-1-ene[188]. The synthesis of saturated carboxylic acids from alkenes, CO and H_2O by using palladium–phosphine catalysts has been reported earlier[189]. The effects on the reaction of catalyst concentration, phosphine substituents and additional reagents, such as lithium chloride and iron carbonyl, have been studied[190]. When terminal alkenes are used both normal-chain and branched-chain acids are formed; the proportion of normal-chain acid decreases with increase in CO pressure. Water in the system is found to have a most pronounced effect. The yield of products is a maximum with 5–15% water, and the proportion of normal-chain acid is a minimum. A complex picture emerges from this work, for the catalyst undergoes several changes as the reaction proceeds. Tertiary carboxylic acids are also produced by carboxylation of alkenes if a Cu^I carbonyl catalyst is used[191]. The reaction involves carbenium ions and takes place at room temperature and atmospheric pressure. The Cu^I carbonyl is produced *in situ* by adding a Cu^I compound together with CO to concentrated sulphuric acid. The tertiary acid is formed because the carbenium ion forms the more stable tertiary

$$RCH\!=\!CH_2 \xrightarrow{H^+} R\overset{+}{C}HMe \xrightarrow{conc.\ H_2SO_4} R^1R^2\overset{+}{C}Me \xrightarrow{CO,\ Cu(CO)_3^+} R^1R^2CMe\overset{+}{C}O$$

$$\xrightarrow{H_2O} R^1R^2CMeCO_2H \qquad (1.81)$$

isomer before hydrolysis to the acid [equation (1.81); R^1 = Me, Et, Pr^n, Bu^n, $n\text{-}C_5H_{11}$, $n\text{-}C_6H_{13}$, $n\text{-}C_7H_{15}$, R^2 = Me, Et, Pr^n, Bu^n].

Hydroformylation is catalysed by $Mn_2(CO)_{10}$[114], $Co_2(CO)_8$[192], $Rh_4(CO)_{12}$[193] and a chiral bisphosphine rhodium complex[194]. $Rh_4(CO)_{12}$ has been used previously only at high pressure (50–100 atm) and 75 °C. In toluene with propene and hydrogen, it gives aldehydes in high yield at atmospheric pressure and room temperature [equation (1.82)][193]. Asymmetric

$$3Rh_4(CO)_{12} + 4MeCH\!=\!CH_2 + 4H_2 \rightarrow 2Pr^nCHO +$$
$$2Pr^iCHO + 2Rh_6(CO)_{16} \qquad (1.82)$$

hydroformylation, using a catalyst prepared *in situ* from (−)-2,3-*O*-iso-propylidene-2,3-dihydroxy-1,4-bis(diphenylphosphino)butane and $RhH(CO)(PPh_3)_3$ has been reported[194]. Table 1.6 lists the compounds studied and the 'optical purity' of the products. Such asymmetric syntheses with chiral phosphines as ligands for soluble transition-metal complexes is one of the current growth points of this area.

Table 1.6   Asymmetric hydroformylation of alkenes[194]

Alkene	Product	$p_{H_2} + p_{CO}$ /atm	T/°C	Optical purity/%
But-1-ene	2-Methylbutan-1-ol	0.67	25	18.8
Pent-1-ene	2-Methylpentanoic acid	1	25	19.7
3-Methylbut-1-ene	2,3-Dimethylbutan-1-ol	1	25	15.2
Oct-1-ene	2-Methyloctan-1-ol	1	25	16.5
cis-But-2-ene	2-Methylbutan-1-ol	0.67	20	8.1
trans-But-2-ene	2-Methylbutan-1-ol	84	98	3.2
cis-Hex-2-ene	2-Methylhexanoic acid	82	95	7.6
	2-Ethylpentanoic acid			5.8
trans-Hex-2-ene	2-Methylhexanoic acid	82	95	1.4
	2-Ethylpentanoic acid			2.9

1.4.2.8   Oxidation

The molybdenum-catalysed epoxidation of alkenes with alkyl hydroperoxides

$$Me_2C{=}CMe_2 \ + \ RO_2H \ \xrightarrow{\text{catalyst}} \ Me_2C\overset{O}{\overbrace{}}CMe_2 \ + \ ROH \tag{1.83}$$

[equation (1.83)] is a well-known reaction. The way in which the catalyst works has now been discovered[195,196]. The epoxidation involves oxygen transfer from a Mo^{VI} hydroperoxide complex to the alkene [equation

$$\tag{1.84}$$

(1.84)][195]. A molybdenum complex (77) has been isolated by using several alkenes; for example, if the alkene is propene, R in (77) is Me, if oct-1-ene, R is n-C_6H_{13}[196]. In a useful review of some oxidations of alkenes with transition-metal salts as catalysts, palladium-, rhodium- and ruthenium-

(77)

$$Me_2C\overset{O}{\overbrace{}}CMe_2$$

(78)

catalysed systems are discussed[197]. The oxidation of 2,3-dimethylbut-2-ene by using $MCl(CO)(PPj_3)_2$ ($M = Rh^I$ or Ir^I) as catalyst is rapid and selective under mild conditions[198]; the major products are 2,3-dimethyl-2,3-epoxy-butane (78) and 2,3-dimethyl-3-hydroxybut-1-ene (79), with about 10% of

$$Me_2C(OH)CMe{=}CH_2 \qquad\qquad Me_2C(OOH)CMe{=}CH_2$$
$$\text{(79)} \qquad\qquad\qquad\qquad \text{(80)}$$

the minor product, 2,3-dimethyl-3-hydroperoxybut-1-ene (80). The reaction is a free-radical initiated autoxidation; it is more rapid for 2,3-dimethylbut-2-ene than for the less branched alkenes, which suggests that coordination to the metal is not an important factor. The role of the metal is unclear. A reaction similar to the palladium-catalysed Wacker process requires rhodium (III) halides as the catalysts[199,200]. Aqueous solutions of $RhCl_5H_2O^{2-}$ in the presence of iron(III) salts, and related chlororhodium ions, are the catalysts used[199]; the most effective is $RhCl_4(H_2O)_2^{-}$, but both $RhCl_6^{3-}$ and $RhCl_3 \cdot 3H_2O$ are inactive[200]. The palladium-catalysed oxidations have been further studied[201-203]. Palladium(II) acetate-catalysed oxidation of hex-1-ene or hex-2-ene takes place by oxypalladation of the alkene to give the complex (81) (or a similar complex) followed by elimination of $PdH(OAc)$

$$\overset{\displaystyle CH_2}{\underset{\displaystyle AcOCBu^n}{\|}}\!\!\!\xrightarrow{} PdH(OAc)$$

(81)

Table 1.7 Products of the Pd(OAc)$_2$-catalysed oxidation of hex-1-ene[201]

Product	Mole fraction
Hex-1-en-2-yl acetate	0.71
cis-Hex-2-en-2-yl acetate	0.1
Hex-2-en-1-yl acetate	0.1
Hex-1-en-3-yl acetate	0.02
trans-Hex-2-en-2-yl acetate ⎫	
Hex-3-en-2-yl acetate ⎬ 0.03	
Hex-4-en-3-yl acetate ⎭	
Hexan-2-one	0.01
Hexan-2-ols and hexen-2-ones	0.02

to give the products (Table 1.7)[201]. Palladium(II) salts in acetic acid oxidise ethylene to vinyl acetate. Addition of copper(II) chloride or nitrate increases the reaction rate and changes the products to 1,2-disubstituted ethanes, such as mono- and di-acetoxyethane and 2-chloroethyl acetate. Some new oxidants, *viz.* K$_2$Cr$_2$O$_7$, NaNO$_3$, CuBr$_2$, MnO$_2$, Pb(OAc)$_4$, Tl(OAc)$_3$, TlCl$_3$ and HAuCl$_4$, have been added to the Pd^{II} salts[202]. CuBr$_2$ gives 2-bromoethyl acetate; Tl(OAc)$_3$, TlCl$_3$ and HAuCl$_4$ give appreciable quantities of ethylidene diacetate. Mercuric acetate may be used with ethylene glycol to solvomercuriate an alkene [equation (1.85)]. The compound (82) so formed then

$$Hg(OAc)_2 + RCH{=}CH_2 + HOCH_2CH_2OH \overset{p\text{-}MeC_6H_4SO_3H}{\rightleftharpoons}$$
$$HO(CH_2)_2OCHRCH_2HgOAc + HOAc$$
$$(82) \tag{1.85}$$

reacts with palladium(II) chloride to give a σ-bonded palladium intermediate which decomposes to an ethylene acetal [equation (1.86)][203].

$$(82) + PdCl_2 \longrightarrow HO(CH_2)OCHRCH_2PdCl + Hg(OAc)Cl$$

$$\tag{1.86}$$

$$\begin{array}{c} H_2C{-}CH_2 \\ O \quad O \\ \diagdown X \diagup \quad + Pd^0 + HCl \\ R \ \ CH_3 \end{array}$$

Ozonolysis of alkenes has been studied by using a number of techniques[80,204–206]. Infrared methods have been used[204] to examine the products of the reaction of ozone with ethylene, propene, 2-methylpropene, *cis*- and *trans*-but-2-ene, 2-methylbut-2-ene and 2,3-dimethylbut-2-ene at −175 to −80 °C. At −175 °C two species are formed, a π-complex and a primary ozonide, presumably the 1,2,3-trioxolane (83). On being warmed to −80 °C the π-complex reverts to the reactants and the primary ozonide gives the stable products. These include secondary ozonides (84), ketones, aldehydes, acids and polymers. The major product from ethylene is the secondary ozonide; formaldehyde and formic acid are also formed. The secondary ozonide is also the major product from *trans*-but-2-ene; acetaldehyde, acetic acid and polymer are also formed. The results support the mechanism of Criegee and Werner [equation (1.87)][207]. Further support for the Criegee

$$Me_2C{=}CMe_2 \; + \; O_3 \; \longrightarrow \; (83) \; \longrightarrow \; Me_2\overset{+}{C}O\bar{O} \; + \; Me_2CO \tag{1.87}$$

$$\longrightarrow \; (84) \; + \; (Me_2COO)_{\overline{n}}$$

mechanism is provided by a study of the ozonolysis of ethylene in the presence of ^{18}O formaldehyde[205]. If the mechanism involves aldehyde interchange [equation (1.88) routes (a) and (b)][208], then all three ozonides (85),

$$CH_2{=}CH_2 \; + \; O_3 \; \xrightarrow{H_2CO^*} \; \xrightarrow{(b)} \; H_2\overset{+}{C}O\bar{O} \; + \; H_2CO \tag{1.88}$$

(85) and (86) (87) $[O^* \equiv {}^{18}O]$

(86) and (87) should be formed. If the Criegee mechanism is correct [equation (1.88) route (b)][207] only (86) and (87) should be formed, and this is observed. The structure of the secondary ozonide of ethylene has been determined from its microwave spectrum[206]. Ozonolysis of tri-t-butylethylene gives the expected t-butyl compounds [equation (1.89)][80].

$$Bu^t_2C{=}CHBu^t \; + \; O_3 \; \rightarrow \; Bu^t_2CO \; + \; Bu^tCHO \tag{1.89}$$

Fluoro-sulphonic acid[209] and -sulphonyl chloride[210] both add to alkenes. Fluorosulphonic acid addition is a convenient way to prepare alkyl fluorosulphonates [e.g. equation (1.90)][209]. Sulphuryl chloride is well known to

$$MeCH{-}CH_2 \; + \; HOSO_2F \; \xrightarrow{-40\,^{\circ}C} \; Pr^iOSO_2F \tag{1.90}$$

add to alkenes, with copper salts as catalysts, to give bis-(2-chloroalkyl) sulphones and their dehydrohalogenation products[211]. This reaction has now been extended to substituted sulphonyl chlorides ($XC_6H_4SO_2Cl$ in which for example, $X = p\text{-Br}, p\text{-Me}, p\text{-CH}_2Cl, m\text{-SO}_2Cl$ or $m\text{-CO}_2H$)[210].

The selenium dioxide oxidation of alkenes at 25 °C gives the products by hydrolysis of a Se—O bond of a Se^{II} ester [equation (1.91)][212].

$$Bu^nCH\!-\!CMe_2 \xrightarrow{SeO_2} Bu^nCH[\overset{+}{Se}(OH)\overset{-}{O}]CMe\!=\!CH_2 \xrightarrow{2,3\ shift} Bu^nCH\!=\!CHMeCH_2OSe(OH)$$

$$\xrightarrow{H_2O} Bu^nCH\!=\!CMeCH_2OH + Se^{II}(OH)_2 \qquad (1.91)$$

1.4.2.9 Halogenation

Hydrofluorination[213], halogenofluorination[214] and nitrofluorination[215] of alkenes can all be conveniently carried out by using nitrogen bases as the solvent. In the past, hydrofluorination has often required high-pressure equipment, and always the handling of hydrogen fluoride, and the alkene has often polymerised also. Alkenes can now be hydrofluorinated at atmospheric pressure by using stable highly concentrated solutions of HF in pyridine or in a trialkylamine [equation (1.92); X = H][213]. Propene, but-2-ene

$$R^1R^2C\!=\!CR^3R^4 \xrightarrow{\text{(see text)}} R^1R^2CFCR^3R^4X \qquad (1.92)$$

and 2-methylpropene give 35, 20 and 60% yield, respectively. In a much simpler halogenofluorination, the alkene is added to NIS, or iodine, or iodine and silver nitrate, together with anhydrous hydrogen fluoride in pyridine[214]. For bromofluorination, NBS, or bromine and silver nitrate, is used; for chlorofluorination, NCS. The reactions are given in equation (1.92) (X = I, Br or Cl) and typical products and yields are given in Table 1.8[214].

Table 1.8　Products of halogenofluorination and nitrofluorination of alkenes[214,215]

Alkene	Halogen	Product	Yield/%
Ethylene	I	2-Fluoro-1-iodoethane	25
,,	Br	1-Bromo-2-fluoroethane	30
,,	(NO_2)	1-Fluoro-2-nitroethane	60
Propene	I	2-Fluoro-1-iodopropane	32
,,	Br	1-Bromo-2-fluoropropane	40
,,	Cl	1-Chloro-2-fluoropropane	35
,,	(NO_2)	2-Fluoro-1-nitropropane	65
2-Methylpropene	I	2-Fluoro-1-iodo-2-methylpropane	60
,,	Br	1-Bromo-2-fluoro-2-methylpropane	85
,,	Cl	1-Chloro-2-fluoro-2-methylpropane	60
But-2-ene	(NO_2)	2-Fluoro-3-nitrobutane	60
Hex-1-ene	I	2-Fluoro-1-iodohexane	70
,,	Br	1-Bromo-2-fluorohexane	90
,,	Cl	1-Chloro-2-fluorohexane	40
,,	(NO_2)	2-Fluoro-1-nitrohexane	65
Hex-3-ene	I	4-Fluoro-3-iodohexane	65
,,	Br	3-Bromo-4-fluorohexane	85
,,	Cl	3-Chloro-4-fluorohexane	80

In a similar manner, if $\overset{+}{N}O_2 \ \overset{-}{B}F_4$ in 70% HF–pyridine is used, nitrofluorination of the alkene [equation (1.92); X = NO$_2$] occurs (Table 1.8)[215]. If branched-chain alkenes are treated with a cyanogen halide and a Lewis acid (AlCl$_3$ or ZnCl$_2$) in nitromethane, unsaturated nitriles are formed by substitution, e.g. (88), and halogenonitriles by addition, e.g. (89) [equation

$$Me_2C=CMe_2 \xrightarrow[\substack{MeNO_2,\ AlCl_3 \\ 0\ ^\circ C}]{CNCl} \underset{(88)}{CH_2=CMeCMe_2CN} + \underset{(89)}{Me_2CClCMe_2CN} \qquad (1.93)$$

(1.93)][216]. The ratio of substitution to addition products is determined by the structure of the alkene and the nature of the cyanogen halide. For 2-methylbut-2-ene the products are given by equation (1.94). If X = Cl, compounds

$$MeCH=CMe_2 \xrightarrow[\substack{MeNO_2,\ AlCl_3\ or \\ ZnCl_2,\ 0\ ^\circ C}]{XCN} \underset{(90)}{MeCH(CN)CMe=CH_2} + \underset{(91)}{Me_2C=CMeCN}$$

$$+ \underset{(92)}{Me_2CXCHMeCN} + \underset{(93)}{MeCHXCMe_2CN} + \underset{(94)}{CH_2=CHCMe_2CN} \qquad (1.94)$$

(90), (91) and (92) are formed; if X = Br, compounds (90), (91), (93) and (94) are formed; and if X = I, compound (93) alone is formed. The reason for the different products with the different cyanogen halides lies in the type of complex each forms with the Lewis acid: CNCl gives $CN^{\delta+}AlCl_4^{\delta-}$, CNI gives $I^{\delta+}AlCl_3CN^{\delta-}$ and CNBr gives both types of complex. Hence $CN^{\delta+}AlCl_3Br^{\delta-}$ gives (90) and (91) and $Br^{\delta+}AlCl_3CN^{\delta-}$ gives (93). Iodine monochloride with a mixture of epoxyethane and ethylene gives 1-chloro-2-iodoethane (95), 1-chloro-5-iodo-3-oxapentane (96) and the telomerisation

$$\underset{(95)}{ICH_2CH_2Cl} \qquad \underset{(96)}{ICH_2CH_2OCH_2CH_2Cl} \qquad \underset{(97)}{ICH_2CH_2OCH_2CH_2OCH_2CH_2Cl}$$

product, 1-chloro-8-iodo-3,6-dioxaoctane (97)[217]. A detailed mechanism for the addition of iodonium nitrate to alkenes has been reported[218]. TlCl$_3 \cdot$ 4H$_2$O in CCl$_4$ reacts with alkenes or alkadienes to give vicinal dichlorocompounds [equation (1.95)] and monochloro compounds [equations (1.96)

$$>C=C< \ + \ 2TlCl_3 \ \longrightarrow \ \left[\underset{C}{\overset{C}{\underset{\|}{}}} \cdots \overset{\delta+}{Cl} - \overset{\overset{\displaystyle Cl}{|}}{Tl} - Cl \cdots \overset{\delta-}{TlCl_3} \right]$$

$$\Big\downarrow \qquad\qquad (1.95)$$

$$TlCl_3 \ + \ \underset{\underset{\displaystyle Cl}{|}}{\overset{\overset{\displaystyle Cl}{|}}{-C-C-}} \ \xleftarrow{TlCl_4^-} \ \left[\underset{C}{\overset{C}{\overset{|}{}}}\!\!\overset{+}{Cl} \right] \ + \ TlCl \ + \ TlCl_4^-$$

$$CCl_4 \ + \ TlCl_3 \cdot 4H_2O \rightarrow COCl_2 \ + \ 2HCl \ + \ TlCl_3 \cdot 2H_2O \qquad (1.96)$$

$$\text{C=C} + HCl \longrightarrow \underset{\underset{H\quad Cl}{|\quad\;|}}{\text{C-C}} \tag{1.97}$$

and (1.97)][219]. Oct-1-ene, oct-2-ene and buta-1,3-diene have been studied, and the reactive species is thought to be a polarised form of dimeric $TlCl_3$ [equation (1.95)]. Yields of the dichloro compounds are much increased by adding $CuCl_2 \cdot 2H_2O$ or $FeCl_3$ and bubbling oxygen through the solution. $CuCl_2 \cdot 2H_2O$ or $FeCl_3$ alone do not chlorinate the alkene, but they recon-

$$TlCl + 2CuCl_2 \text{ (or } 2FeCl_3) \rightarrow TlCl_3 + 2CuCl \text{ (or } 2FeCl_2) \tag{1.98}$$

$$CuCl \text{ (or } FeCl_2) \xrightarrow{\;O_2,\,aq.\,HCl\;} 2\,CuCl_2 \text{ (or } FeCl_3) \tag{1.99}$$

vert the $TlCl$ formed in equation (1.95) to $TlCl_3$ [equation (1.98)], and are then themselves reoxidised by oxygen [equation (1.99)].

1.4.2.10 Electrochemical reactions

In recent years the electrochemical synthesis of organic compounds has attracted much interest. Very little, so far, has been done on the electrochemical reactions of alkenes. Anodic oxidation of octenes in acetic acid or in methanol gives allylically substituted products [equations (1.100) and

$$CH_2{=}CH(CH_2)_5Me \xrightarrow{\;-\bar{e},\,HOAc\;} CH_2{=}CHCH(OAc)(CH_2)_4Me +$$
$$71\%$$
$$AcOCH_2CH{=}CH(CH_2)_4Me \tag{1.100}$$
$$29\%$$

$$MeCH{=}CH(CH_2)_4Me \xrightarrow{\;-\bar{e},\,HOAc\;} MeCH{=}CHCH(OAc)(CH_2)_3Me$$
$$12\%$$
$$+\ MeCH(OAc)CH{=}CH(CH_2)_3Me + AcOCH_2CH{=}CH(CH_2)_4Me$$
$$50\% \qquad\qquad\qquad 14\% \tag{1.101}$$

(1.101)][220,221]. The detailed mechanism of the reaction has now been published[222]. Yields of acetoxyoctenes from oct-1-ene and oct-2-ene are given in equations (1.100) and (1.101). Electrolysis of a mixture of ethylene and carbon dioxide in a suitable aprotic solvent, e.g. DMF, gives a mixture of oxalic and succinic acids[223]. The reaction is carried out at 20 °C with 0.1M tetrabutylammonium bromide as the base electrolyte, a pressure of *ca.* 20 atm of ethylene and an aluminium anode. If a solution of ethylene alone is electrolysed, a cathodic reaction takes place; the products have not been identified.

1.4.2.11 Organometallic complexes

The molecular orbital and the valence bond descriptions of the bonding in metal–alkene and –alkyne complexes have been examined with regard to their ability to explain the known properties of such complexes[224]. The MO description is preferred, and the complexes are shown to be members of one of two classes: *viz.* those in which there is essentially free rotation about the metal–alkene bond and those in which there is not. The first complex containing a metal and ethylene only, has been reported[225]. It is tris(ethylene)-nickel(0) (98); it is prepared by treating *trans*-1,5,9-cyclododecatrienenickel(0)

$$\text{Ni} \qquad (= = C_2H_4)$$

(98)

in diethyl ether with ethylene at 0 °C. On being cooled to −78 °C, the mixture gives colourless needles of compound (98).

1.5 ALKADIENES

1.5.1 Alka-1,2-dienes

1.5.1.1 Preparation

A new general route for the synthesis of alka-1,2-dienes is the reaction of 1-halogenoallenes or 3-chloroalk-1-ynes with lithium dialkylcuprates at low temperatures [equation (1.102)][226]. The reaction temperature depends on the

$$R^1_2CuLi + R^2R^3C{=}C{=}CHX \rightarrow R^2R^3C{=}C{=}CHR^1 \qquad (1.102)$$

copper compound: for Me_2CuLi it is −5 °C, for Et_2CuLi −30 °C and for Bu^n_2CuLi −60 °C. Reaction times are between 1 and 3 h. Compounds studied and yields are given in Table 1.9. 1,1-Dimethylallene (3-methylbuta-

Table 1.9 Alka-1,2-dienes prepared by reactions of $R^2R^3C{=}C{=}CHX$ with R'_2CuLi

R^1	R^2	R^3	X	Yield/%
Me	Me	Et	Br	85
Me	H	Pr	Br	85
Me	H	Pr	I	68
Et	H	Pr	Br	68
Bu^n	Me	Et	Br	87
Bu^n	Me	Pr	I	64

1,2-diene) is prepared in 6–7% yield, and 1-methyl-1-t-butylallene in 15% yield, when 4,5,5-trimethyl-3-nitroso-oxazolid-2-one (99; R = Me) or

$$\underset{(99)}{\overset{\displaystyle R}{\underset{\displaystyle Me}{\diagdown}}\overset{\displaystyle O{-}CO}{\underset{\displaystyle HC{-}NNO}{\overset{\displaystyle |}{\underset{\displaystyle Me}{\overset{\displaystyle |}{C}}}}}} \xrightarrow[\text{MeOH}]{\text{NaOMe}} MeCR{=}C{=}CH_2 \qquad (1.103)$$

4,5-dimethyl-3-nitroso-5-t-butyloxazolid-2-one (99; R = Bu^t) is treated with methanolic sodium methoxide [equation (1.103)][227]. A review, covering the literature to the end of 1971, on the synthesis of chiral allenes has been published[228].

1.5.1.2   Reactions

A careful systematic study of the thermal oligomerisation of allenes in benzene shows that the reaction products include the dimer, 1,2-bis(methylene)cyclobutane, and various [2 + 2] and [2 + 4] cycloaddition products, as well as products from sigmatropic rearrangements and electrocyclic reactions [equation (1.104)][229]. If 2,4-dimethylpenta-2,3-diene (tetramethyl-

$$n\,CH_2{=}C{=}CH_2 \longrightarrow \quad (1.104)$$

allene) is co-oligomerised with allene, only one molecule of the substituted allene is incorporated into the product molecules; the reaction pathways [equation (1.105)] are similar to those for allene alone. Ketenes give cyclo-

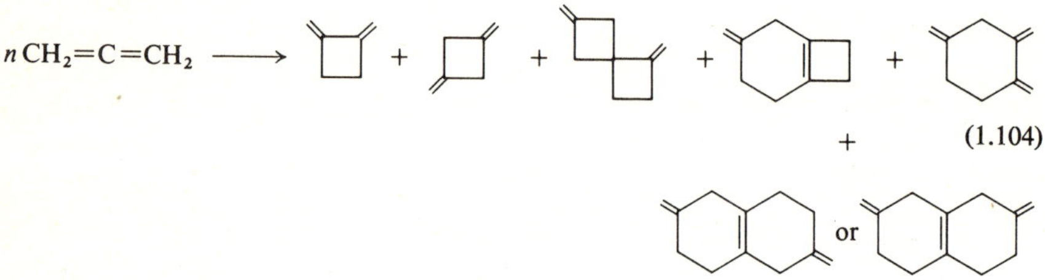

$$CH_2{=}C{=}CH_2 \;+\; Me_2C{=}C{=}CMe_2 \qquad (1.105)$$

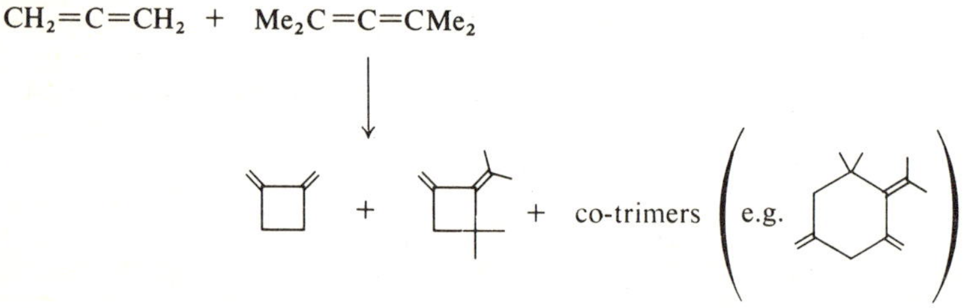

addition products with allenes[230,231], e.g. 2-methylbuta-2,3-diene (1,1-dimethylallene) adds to diphenylketene or to chlorophenylketene to give the cyclobutanones (100) or (101) (X = Ph or Cl)[230]. These four 2-alkylidenecyclobutanones only are formed, i.e. the ketene carbonyl carbon atom always becomes joined to the central allenic atom. The effect of methyl substituents on the reactivity of allenes with diphenylketene is noteworthy. Allene is unreactive, 1,1-dimethylallene is moderately reactive and tetramethylallene is more reactive still, in spite of expected steric effects. Cyano-t-butylketene with S-(+)-penta-2,3-diene gives four products [equation (1.106)][231]. Their optical activity is used to support a non-concerted mechan-

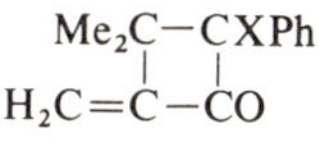

$$\begin{array}{cc} \text{Me}_2\text{C}-\text{CXPh} & \text{H}_2\text{C}-\text{CXPh} \\ | \quad\quad | & | \quad\quad | \\ \text{H}_2\text{C}=\text{C}-\dot{\text{C}}\text{O} & \text{Me}_2\text{C}=\text{C}-\dot{\text{C}}\text{O} \\ (100) & (101) \end{array}$$

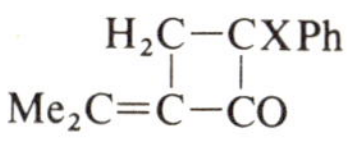

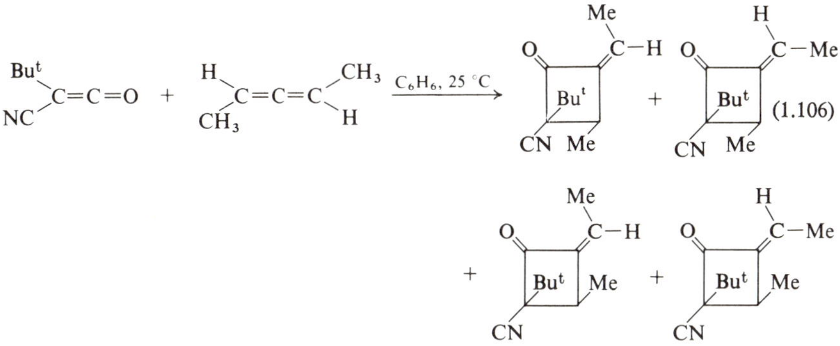

$$\text{Bu}^t\text{(NC)C}=\text{C}=\text{O} + \text{(H)(CH}_3)\text{C}=\text{C}=\text{C(CH}_3)\text{(H)} \xrightarrow{\text{C}_6\text{H}_6,\ 25\ ^\circ\text{C}} \quad (1.106)$$

ism. In the light-induced cycloadditions of allene and benzene, the 1,4-product (102) predominates, about half as much 1,3-adduct (103) being

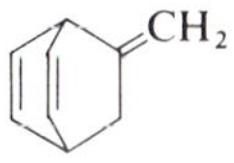

$$(102) \qquad\qquad (103)$$

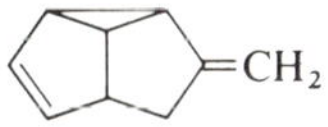

produced[232]. Two other minor unidentified 1 : 1 adducts are also formed. Photo-oxygenation of tetramethylallene in CS_2 with eosin as sensitiser involves singlet oxygen[233]. The reaction [equation (1.107)] is, in fact, a 1,4-

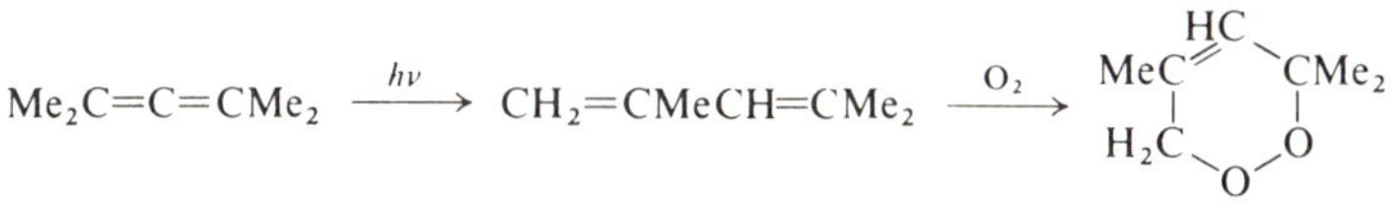

$$\text{Me}_2\text{C}=\text{C}=\text{CMe}_2 \xrightarrow{h\nu} \text{CH}_2=\text{CMeCH}=\text{CMe}_2 \xrightarrow{\text{O}_2} \quad (1.107)$$

addition to 2,4-dimethylpenta-1,3-diene formed by a light-induced isomerisation of the tetramethylallene. Hindered allenes react with ozone to give cyclopropanones (105) and allene epoxides (108) as the initial products[234]. One molar proportion of ozone with 1,1-di-t-butylallene (104; $R^1 = R^2 = Bu^t$, $R^3 = H$) in CH_2Cl_2 at $-78\ ^\circ C$ gives 62 % of di-t-butyl ketone and 27 % of 2,2-di-t-butylcyclopropanone (105; $R^1 = R^2 = Bu^t$, $R^3 = H$). Two molar proportions of ozone with 1,1,3-tri-t-butylallene (104; $R^1 = R^2 = R^3 = Bu^t$) give compound (106; $R^1 = R^2 = R^3 = Bu^t$) which rearranges to (107; $R^1 = R^2 = R^3 = Bu^t$) on being kept at room temperature. With only one molar proportion of ozone, 1,1,3-tri-t-butylallene gives 41 % of (106; $R^1 = R^2 = R^3 = Bu^t$) and 5 % of the allene epoxide (108; $R^1 = R^2$

$$R^1R^2C{=}C{=}CHR^3 \xrightarrow{O_3} \quad (109) \quad \longrightarrow \quad R^1R^2C{=}C{-}CHR^3 \quad (110)$$

$$(104)$$

$$(111) \longrightarrow R^1R^2C{-}CHR^3 + R^1R^2CO \quad (105)$$

$$(108) \longrightarrow (106) \longrightarrow (107)$$

$$(1.108)$$

$=R^3=Bu^t$). 1,3-Di-t-butylallene (104; $R^1 = R^3 = Bu^t$, $R^2 = H$) reacts as a normal unhindered allene and gives only pivalaldehyde, Me_3CCHO [equation

$$Bu^tCH{=}C{=}CHBu^t \xrightarrow{O_3} 2Bu^tCHO + CO \tag{1.109}$$

(1.109)]. For the two hindered allenes the reaction is as in equation (1.108); it involves primary electrophilic attack on the central carbon to give the

$$Me_2C{=}C{=}CMe_2 \longrightarrow \text{further products} \qquad Me_2CHCOCMe_2OAc \quad (112)$$

$$Me_2C{=}C{-}CMe_2 \longrightarrow Me_2C{\cdots}+{\cdots}CMe_2 \longrightarrow Me_2C{\cdots}+{\cdots}CMe_2$$

$$Me_2CHCOC(OMe)Me_2 \quad (115)$$

$$Me_2C{-}CMe_2 \longrightarrow Me_2C{-}CMe_2 \longrightarrow Me_2C{=}CMe_2 + Me_2C{-}CMe_2 \quad (113)$$

$$\xrightarrow{AcOOH}$$

$$(114)$$

$$(1.110)$$

intermediate (109), followed by ring closure to the primary ozonide (110) or loss of oxygen to give the zwitterion (111) as the precursor of (105) and (108). Peroxyacid oxidation of tetramethylallene and 1,1-dimethylallene gives allene epoxides as the initial products; these reactive species then undergo competitive partitioning between valence isomerism to the related cyclopropanone, and further oxidation to dioxaspiropentane derivatives, followed by yet further reactions[235]. Thus with an excess of peroxyacetic acid in buffered methanol, tetramethylallene gives (112), (113) and (114) [equation (1.110)]. Under acidic conditions the only product is the methoxyketone (115). Similarly, 1,1-dimethylallene reacts as in equation (1.111), but in

$$Me_2C{=}C{=}CH_2 \longrightarrow Me_2\overset{O}{\overset{\diagup\diagdown}{C-C}}{=}CH_2 \longrightarrow Me_2\overset{CO}{\overset{\diagup\diagdown}{C-CH_2}} \longrightarrow Me_2\overset{O}{\overset{\diagup\diagdown}{C-CH_2}}$$

$$\begin{array}{c} \overset{O-CO}{\underset{Me_2C-CH_2}{|\quad|}} \end{array} \qquad \tag{1.111}$$

$$\overset{OC-O}{\underset{Me_2C-CH_2}{|\quad|}} \qquad Me_2C(OAc)COMe$$

methanol alone only acetone and methyl acetate are formed. The factors which influence orientation during radical additions to allenes have been discussed with reference to the products formed by addition of toluene-*p*-sulphonyl iodide and of a halogenomethane[236]. With the exception of propadiene, allenes react with arylsulphonyl radicals to give allylic iodides by attack on the central carbon of the allenic system. Halogenomethanes (CF_3I, CH_3I and $CBrCl_3$) give products by terminal attack by CX_3 radicals (X = F, H or Cl), with 41–49% attack at the central carbon by $CBrCl_3$. Thermal cycloaddition of chlorotrifluoroethylene and of 1,1-dichloro-2,2-difluoroethylene to allene gives in each case two 1 : 1 adducts, the 1-chloro-2,2-difluoro-1-halogeno-3-methylenecyclobutanes (116; X = F or Cl) and

$$\begin{array}{cc} \overset{CH_2{=}C-CH_2}{\underset{F_2C-CXCl}{|\quad|}} & \overset{H_2C-C{=}CH_2}{\underset{F_2C-CXCl}{|\quad|}} \\ (116) & (117) \end{array}$$

the 2-chloro-1,1-difluoro-2-halogeno-3-methylenecyclobutanes (117; X = F or Cl)[237]. If, instead of allene, tetramethylallene is used, the reaction takes a quite different course [equation (1.112)][238]. The allene is first isomerised to

$$Me_2C{=}C{=}CMe_2 \xrightarrow{\text{isomerisation}} CH_2{=}CMeCH{=}CMe_2$$

$$\Big\downarrow \begin{array}{l} CF_2{=}CClX \\ (X = F \text{ or } Cl) \\ 120{-}150\,^{\circ}C \end{array}$$

$$\overset{H_2C-CMeCH{=}CMe_2}{\underset{F_2C-CXCl}{|\quad|}} \tag{1.112}$$

2,4-dimethylpenta-1,3-diene and the products are a 47 : 53 mixture of *cis*- and *trans*-2-chloro-2,3,3-trifluoro-1-methyl-1-(2-methylprop-1-enyl)cyclobutane (from chlorotrifluoroethylene) and 2,2-dichloro-3,3-difluoro-1-methyl-1-(2-methylprop-1-enyl)cyclobutane (from 1,1-dichlorodifluoroethene). Acrylonitrile also adds after isomerisation of the allene, the product being 4-cyano-1,3,3-trimethylcyclohexane (118) if the reaction is carried out in a pyrex reactor with wet acrylonitrile[238]. This is in contrast to the substituted cyclobutane (119) obtained when the reaction is carried out in a steel reactor

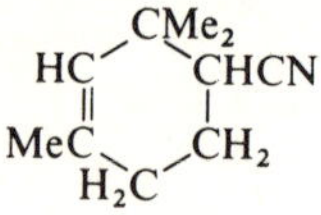

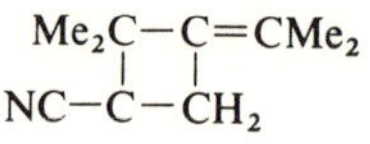

in the presence of alkali[239]. Tetramethylallene also rearranges to 2,4-dimethyl-penta-1,3-diene before it reacts with hexafluoroacetone[240]. The Diels–Alder adduct 3,6-dihydro-4,6,6-trimethyl-2,2-bis(trifluoromethyl)-2H-pyran (120) is formed both thermally and photochemically. 1,1-Dimethylallene gives a 4 : 1 mixture of 1,1,1-trifluoro-4-methyl-3-methylene-2-trifluoromethylpent-4-en-2-ol (121), by direct reaction with the allene, and 3,6-dihydro-4-methyl-

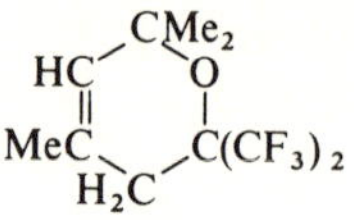

$$(F_3C)_2C(OH)C(:CH_2)CMe=CH_2$$
$$(121)$$

2,2-bis(trifluoromethyl)-2H-pyran (122), by reaction with the isomerised allene. At 175 °C an additional product, 3,6-dihydro-5-methyl-4-(2,2,2-trifluoro-1-hydroxy-1-trifluoromethylethyl)-2,2-bis(trifluoromethyl)-2H-pyran

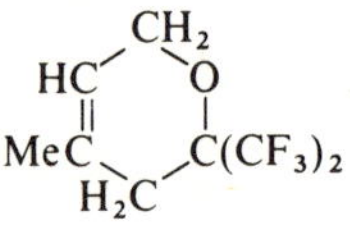

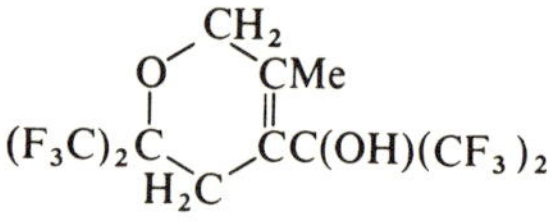

(123), is formed by reaction of $(CF_3)_2CO$ with the alcohol (121). In HF–BF$_3$, HF–PF$_5$ or H$_2$SO$_4$ in dry sulpholan, hexynes and allenic hexadienes isomerise by a series of protonations to vinyl cation intermediates, followed by their deprotonation[241]. All five normal-chain C_6H_{10} isomers exist in equilibrium [equation (1.113)].

$$HC{\equiv}CBu^n + H^+ \rightleftharpoons H_2C{=}\overset{+}{C}Bu^n \rightleftharpoons H_2C{=}C{=}CHPr^n + H^+ \rightleftharpoons Me\overset{+}{C}{=}CHPr^n$$
$$\rightleftharpoons MeC{\equiv}CPr^n + H^+ \rightleftharpoons \text{etc.} \qquad (1.113)$$

Transition-metal catalysed reactions of allenes have been much studied. Rh(CO)$_2$Cl$_2$ with 2 moles of PPh$_3$ in ethanol and one of allene after 6 h at 40 °C gives the tetramer (124), the pentamer (125) and a new hexamer (126)[242]. Similar cycloadditions with various Ni^0–phosphine catalysts give the same pentamer, the trimer (127) and the tetramer (128)[243,244]. Examples

(124)   (125)

(126)

(127)   (128)

are Ni{P(OPh)$_3$}$_4$ in benzene, which gives 72% of (127), 25% of (128) and 3% of (125), and Ni(cyclo-octadiene)$_2$ in benzene, which gives 52% of (125) together with higher oligomers and polymers. Catalytic reduction of allenes has not been previously reported. By using RhCl(PPh$_3$)$_3$, selective hydrogenation of one of the double bonds occurs to give an alkene[245]. For example, nona-1,2-diene is reduced to *cis*-non-2-ene (66%) and 3-ethylpenta-1,2-diene gives 3-ethylpent-2-ene (50%). Allene and buta-1,3-diene react in the presence of a palladium–maleic anhydride–triphenylphosphine catalyst at 120 °C to give a 30% yield of a 3 : 1 mixture of *trans*- and *cis*-2-methyl-3-methyleneocta-1,5,7-triene (129)[246]. Similarly, with bicyclo[2,2,1]hepta-2,5-

$$CH_2=CH-CH=CHCH_2C(:CH_2)CMe=CH_2$$
(129)

(130)

diene a 25% yield of *exo*-3-methyltricyclo[4,2,1,0^{2,5}]nona-3,7-diene (130) is produced. Allene reacts with amines and active methylene compounds with a Ni^0 catalyst to give 3 : 1 adducts[247]. Morpholine, pyrrolidine or n-butylamine and allene, NiBr$_2$ and di-isopropoxyphenylphosphine in ethanol

$$XH+CH_2=C=CH_2 \rightarrow XCH_2CH=CH_2 + XCH_2C(:CH_2)CH_2\dot{C}H=CH_2 +$$
$$\qquad\qquad\qquad\quad (131) \qquad\qquad\qquad\qquad (132)$$

$$XCH_2C(:CH_2)CMe=CH_2 + XCH_2C(:CH_2)C(:CH_2)CH_2CMe=CH_2 +$$
$$\qquad (133) \qquad\qquad\qquad\qquad\qquad (134)$$

$$XCH_2C(:CH_2)CH_2C(:CH_2)CMe=CH_2$$
$$(135)$$

[X is an amine or other residue]   (1.114)

heated in a sealed tube at 100 °C for 1 h, react as in equation (1.114), the yields of (131)–(135) being, for morpholine, a trace, 1, 22, 72 and 5%, respectively; for pyrrolidine, 7, 10, 36, 43 and 4%, respectively; and for n-butylamine, a trace, 21, 33, 33 and 9%, respectively. With nickel acetyl-acetonate, di-isopropoxyphenylphosphine, $NaBH_4$ and NaOPh in ethanol after 16 h at 75 °C, benzyl methyl ketone gives (133; 14%), (134; 63%) and (135; 23%); benzyl cyanide gives (133; 25%), (134; 60%) and (135; 15%); and diethyl malonate gives (133; a trace), (134; 100%) and (135; a trace). The reactions occur by Ni^{II} being reduced to Ni^0 by di-isopropoxyphenyl-phosphine, and a series of nickel–allene complexes are then formed [equation (1.115)]. The major product (134) arises from (136) with preferential reaction

$$NiL_2 + CH_2{=}C{=}CH_2 \longrightarrow L_2Ni \quad (139) \quad \xrightarrow{CH_2{=}C{=}CH_2} \quad LNi \quad (138)$$

$$+ \qquad (1.115)$$

$$LNi \quad (136) \quad \xleftarrow{CH_2{=}C{=}CH_2} \quad LNi \quad (137) \quad + \quad LNi$$

at the π-allyl group adjacent to the *exo*-methylene group, whereas (132) is formed by attack at the allylic carbon adjacent to nickel in (137), and (133) is formed in a similar manner from (138). The small amount of (131) is formed from (139). The stability of (136) is the factor controlling the product composition. The complex (136) occurs in other systems[248,249]; it has been isolated and its structure has been determined by x-ray crystallography[250]. With bis(tri-2-biphenylylphosphite)nickel at 20 °C in toluene, allene gives the orange crystalline complex $C_9H_{12}NiL$ [L = $(o\text{-}PhC_6H_4O)_3P$] (136)[249]. This reacts quantitatively with CO or triphenylphosphine to give 1,2,4-tri-methylenecyclohexane (140). In contrast, if 1,1-dimethylallene is used at

(140)

(141)

$$CH_2{=}CPr^iC(CMe{=}CH_2){=}CH_2$$
(142)

20 °C a pale yellow complex (141) is formed; this with CO or triphenyl-phosphine at 80 °C gives 2-isopropenyl-3-isopropylbuta-1,3-diene (142). The reactions of allene with amines and carbon acids with other Group VIII metal complexes as catalysts give derivatives of 2,3-dialkylbuta-1,3-dienes (133; X = OAc, NR_2 or CR_3) as the major products[251]. 19 Amines (prim-ary, secondary and tertiary) have been studied and the complexes used were $PdCl_2$, $RhCl_3 \cdot 3H_2O$, $(Ph_3P)_4Pd$ and $(Ph_3P)_2Pd \cdot$ alkene. The catalysts are low valent, *viz.* Pd^0 or Rh^I, which are produced from Pd^{II} or Rh^{III} by reduction *in situ* by the amine. Triethylsilane also adds to allene to give allyltriethylsilane[251]. Dodecatrienylnickel reacts with allene to give a new nickel complex which, with more allene, gives two dimethylenecyclohexa-

$$(1.116)$$

(143)

decatrienes, and with CO gives two cyclic hydrocarbons and the carbonyl compound (143) [equation (1.116)][252]; compound (143) on hydrogenation gives ($\pm$)-muscone. A review of carbonylation of dienes with carbon monoxide has been published[253].

1.5.2 Alka-1,3-dienes

1.5.2.1 Preparation and conformational analysis

Dehydrohalogenation of dichloropentanes and dichloroheptanes in aprotic solvents such as HMPT or DMSO with alkoxides as the base give products different from those obtained in protic solvents with sodium or potassium hydroxide as the base[254]. In the aprotic solvents, two molecules of HCl are eliminated and a conjugated diene or alk-2-yne is formed. In protic media only one mole of HCl is lost and the products are chloroalkenes.

The conformations of buta-1,3-diene have been studied by using the temperature dependence of the n.m.r. chemical shifts and coupling constants

(146) skew

(144) *trans*

(145) *cis*

(147) skew

in two deuterium-substituted buta-1,3-dienes, *viz.* 2,3-dideuteriobuta-1,3-diene and 1,1,4,4-tetradeuteriobuta-1,3-diene[255]. It was suggested by Hückel in 1932 that buta-1,3-diene is a mixture of the coplanar *trans* (144) and the *cis* isomer (145). There is considerable experimental evidence that the major conformer is (144), but the evidence concerning the minor component is scanty. It is now found that it is not (145) but a mixture of the skew conformers (146) and (147).

1.5.2.2 *Reactions*

Lithium in THF adds both 1,2 and 1,4 to buta-1,3-diene, and the products, when treated with n-propyl bromide, give those in equation (1.117)[256]. The

$$CH_2{=}CH{-}CH{=}CH_2 + Li \xrightarrow{\ Pr^nBr\ } Bu^nCHPr^nCH{=}CH_2 + Bu^nCH{=}CHBu^n$$

$$\qquad\qquad\qquad\qquad\qquad\qquad\qquad (26\%)\qquad\qquad\qquad (39\%)$$

$$(1.117)$$

photochemical addition of conjugated dienes to aromatic compounds gives a range of products[257–260]. With benzene, 2,3-dimethylbuta-1,3-diene (148) gives three major products, (149), (150) and (151), and two minor products, (152) and (153) [equation (1.118)][257]. Only one product (154), the analogue of

$$CH_2{=}CMeCMe{=}CH_2 + \text{(benzene)} \xrightarrow{\ h\nu\ } (149) + (150)$$

$$(148)$$

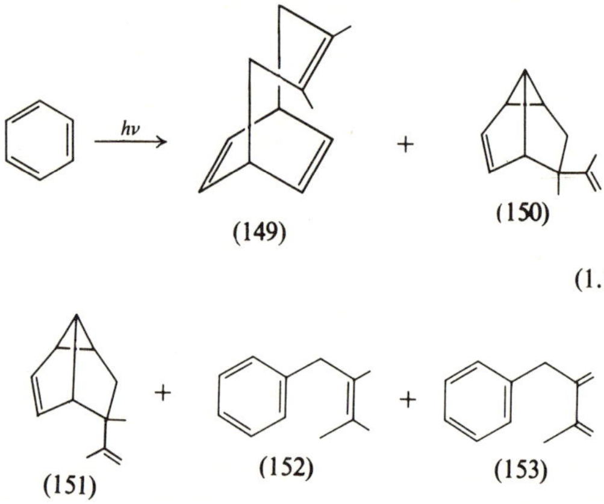

$$(1.118)$$

$$+ (151) + (152) + (153)$$

(149), is obtained if *s-cisoid*-2,4-dimethylpenta-1,3-diene (155) is used [equation (1.119)]. If (155) reacts with toluene or benzonitrile the products

$$s\text{-cisoid-}Me_2C{=}CHCMe{=}CH_2 + \text{(benzene)} \xrightarrow{\ h\nu\ } (154)$$

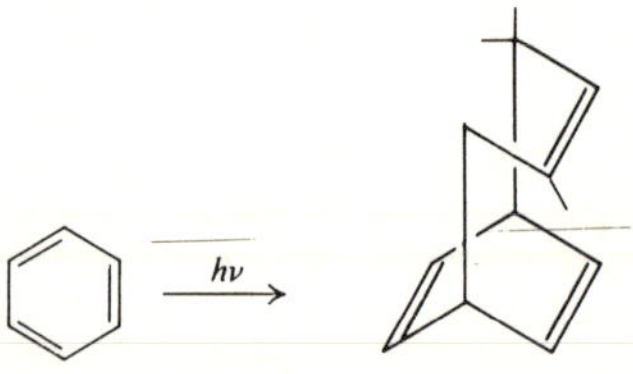

$$(155)$$

$$(1.119)$$

are as in equations (1.120) and (1.121), respectively; electron-donating substituents in the benzene ring direct addition of the diene to the unsubstituted

Me$_2$C=CHCMe=CH$_2$ + [CH$_3$-benzene] $\xrightarrow{hv}$ [structures] + [structure] (1.120)
(155)

Me$_2$C=CHCMe=CH$_2$ + [CN-benzene] $\xrightarrow{hv}$ [structure, CN] + [structure, CN] (1.121)
(155)

positions, whereas electron-withdrawing substituents direct addition to the substituted positions[257]. With naphthalene, 2,4-dimethylpenta-1,3-diene (155) gives (156) and (157) in the ratio 7 : 1 [equation (1.122)], and 2,5-

Me$_2$C=CHCMe=CH$_2$
(155)
+
[naphthalene] $\xrightarrow{hv}$ [structure] + [structure]
(156)
(157) (1.122)

dimethylhexa-2,4-diene (158) gives only one product [equation (1.123)][258]. With anthracene the photochemical addition of 2,5-dimethylhexa-2,4-diene

Me$_2$C=CHCH=CMe$_2$ + [naphthalene] $\xrightarrow{hv}$ [structure] (1.123)
(158)

(158) gives a high yield of the 7,8,9,10-dibenzobicyclo[4,2,2]-9,9-*trans*-3-decatriene (159; R^1 = R^2 = R^3 = R^4 = Me)[259]. This is photoisomerised to the *cis* isomer (160). Also, *trans,trans*-hexa-2,4-diene (161) gives the compound (159; R^1 = R^4 = Me, R^2 = R^3 = H), and this also isomerises to the *cis* isomer (160). If there are electron-withdrawing substituents in the anthracene ring, the reaction takes a different route[260]. With *trans,trans*-

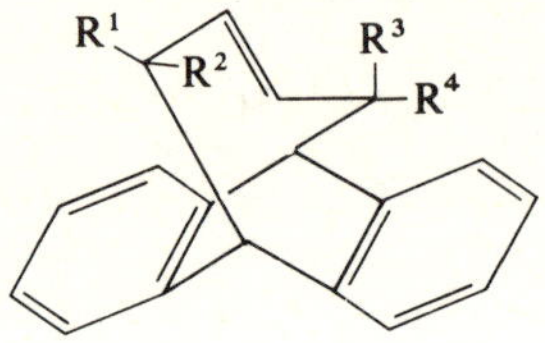

$$\text{(159)} \qquad\qquad \text{(160)}$$

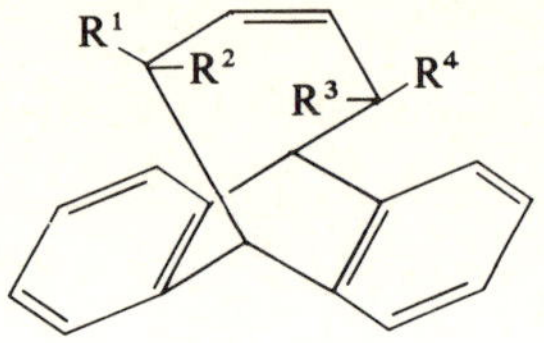

$$MeCH \overset{t}{=} CH - CH \overset{t}{=} CHMe$$

$$(161)$$

hexa-2,4-diene (161), 9-cyanoanthracene (162) gives the product (163; $R^1 = R^4 = Me$, $R^2 = R^3 = H$, $Y = CN$), and *cis,cis*-hexa-2,4-diene gives an isomer (163; $R^1 = R^4 = H$, $R^2 = R^3 = Me$, $Y = CN$). *cis-* and *trans-*

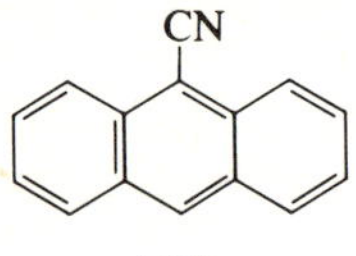

$$\text{(162)} \qquad\qquad \text{(163)}$$

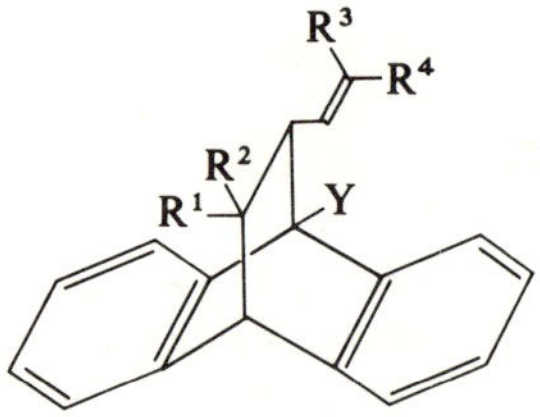

penta-1,3-diene give related products (163; $Y = CHO$), as do all the dienes, in reactions with 9-anthraldehyde[260]. The addition of amines to 1,3-dienes is catalysed by lithium amides[261]. These are generated *in situ* by adding butyl-lithium to a reaction mixture containing an amine and the 1,3-diene.

$$R_2^1N^- \, Li^+ + R^2CH{=}CH_2 \rightarrow R_2^1NCH_2\bar{C}HR^2Li^+ \xrightarrow{\ CH_2=CHR^2\ } \text{polymer}$$

$$\Big\downarrow R^1{}_2NH$$

$$R_2^1N^- \, Li^+ + R_2^1NCH_2CH_2R^2 \qquad\qquad (1.124)$$

The reaction [equation (1.124)] has been studied for a number of amines; it is a convenient route to alkenylalkylamines. Ketene iminium cations (164) undergo 1,2-cycloadditions with 1,3-dienes[262]. The cations are generated from 1-chloro-*N,N,*2-trimethylpropenylamine in the presence of silver

$$Me_2C{-}CClNMe_2 + AgBF_4 \xrightarrow[\;-60\,°C\;]{CH_2Cl_2} Me_2C{=}C{=}\overset{+}{N}Me_2 \; BF^-_4 + AgCl$$

$$(164)$$

$$(1.125)$$

tetrafluoroborate [equation (1.125)]. The reaction [equation (1.126)] gives high yields ($> 85\%$) of products from buta-1,3-diene (165; $R^1 = R^2 = H$) and *cis-* and *trans-*penta-1,3-diene (165; $R^1 = H$, $R^2 = Me$). Tetranitromethane reacts with alkenes by an electrophilic addition leading to an ion

$$(164) + R^1R^2C{=}CHCH{=}CH_2 \longrightarrow R^1R^2C{=}CHCH{-}\overset{\underset{\displaystyle CH_2{-}\overset{+}{C}{=}NMe_2}{|\quad\quad|}}{C}Me_2$$

(165)

$$OH^- \Big|\ \text{hydrolysis}$$

$$R^1R^2C{=}CHCH{-}\overset{\underset{\displaystyle CH_2{-}CO}{|\quad\quad|}}{C}Me_2 \qquad (1.126)$$

pair consisting of a nitrocarbenium ion and the $(NO_2)_3C^-$ anion. With 1,3-dienes the conjugated system allows 1,2 or 1,4 addition [equation (1.127)][263]. 2,5-Dimethylhexa-2,4-diene (166; $R^3 = R^4 = H$, $R^1 = R^2 =$

$$R^1R^2C{=}CR^3CR^4{=}CR^5R^6 + C(NO_2)_4 \rightarrow R^1R^2\overset{\delta+}{C}CR^3{\cdots}CR^4{\cdots}\overset{\delta+}{C}R^5R^6 + C(NO_2)^-_3$$

(166) $| $
                                         NO_2

$$\rightarrow R^1R^2CCR^3{=}CR^4CR^5R^6 \text{ or } R^1R^2C{-}CR^3CR^4{=}CR^5R^6$$

$$\underset{\underset{\text{(167)}}{\text{(1:4 addition)}}}{\underset{\displaystyle NO_2 \qquad C(NO_2)_3}{|\qquad\qquad|}} \qquad \underset{\underset{\text{(168)}}{\text{(1:2 addition)}}}{\underset{\displaystyle O_2N \quad C(NO_2)_3}{|\quad\ |}} \qquad\qquad (1.127)$$

$R^5 = R^6 = Me$), 2,4-dimethylpenta-1,3-diene (166; $R^1 = R^2 = R^4 = H$, $R^3 = R^5 = R^6 = Me$) and 2-methylpenta-2,4-diene (166; $R^1 = R^2 = R^3 = R^4 = H$, $R^5 = R^6 = Me$) give the 1,4-addition product (167), and 2,3-dimethylbuta-1,3-diene (166; $R^1 = R^2 = R^5 = R^6 = H$, $R^3 = R^4 = Me$) gives the 1,2-addition product (168). Other conjugated dienes react in a more complex way. Penta-1,3-diene gives a nitrocarbenium ion with insignificant delocalisation of charge; it O-alkylates the trinitromethane anion to give a nitroalkenyl derivative of aci-trinitromethane. This then enters into a 1,3-dipolar cycloaddition with a second molecule of the diene to give a derivative of 3,3-dinitroisoxazolidine (169). A similar compound (170)

$$MeCH{=}CHCH\underset{O}{\overset{\underset{\displaystyle CH_2{-}C(NO_2)_2}{|\quad\quad\ |}}{\diagdown\ \diagup}}NOCH(CH_2NO_2)CH{=}CHMe$$

(169)

$$CH_2{=}CMeCH\underset{O}{\overset{\underset{\displaystyle CH_2{-}C(NO_2)_2}{|\quad\quad\ |}}{\diagdown\ \diagup}}NOCH_2CH{=}CMeCH_2NO_2$$

(170)

is formed by isoprene. The hexa-1,5-diene ion also O-alkylates the trinitromethane anion, and the product undergoes an intramolecular cyclisation to form 1-aza-2-oxa-3-nitromethyl-8,8-dinitro-1,6-epoxycyclo-octane (171)[263]. The structure of the crystalline adduct of nitrosobenzene and 2,3-dimethylbuta-1,3-diene is given by formula (172)[264]. The major product from the

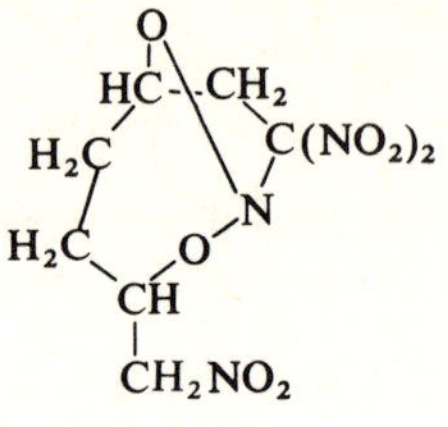

(171)

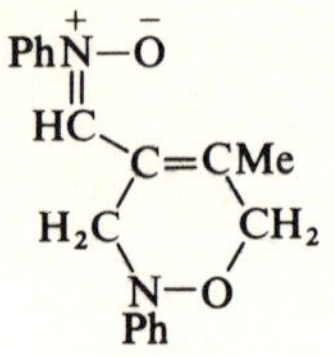

(172)

photosensitised oxygenation of 1,3-dienes is a 3,6-dihydro-o-dioxin (173); for example, isoprene in CCl_2F_2 with 2% of MeOH, when irradiated under oxygen for 7 h at 0 °C with methylene blue or rose bengal as a sensitiser, gives 50% of 4-methyl-3,6-dihydro-o-dioxin [equation (1.128)][265]. The radical

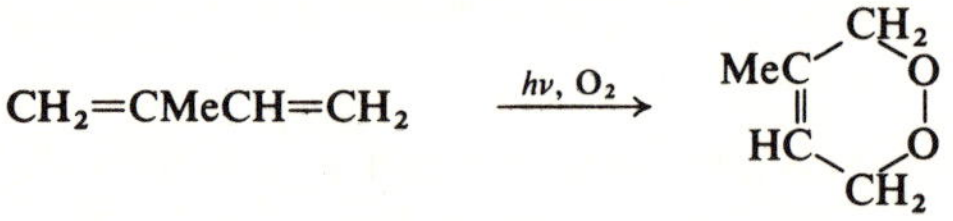

$$CH_2{=}CMeCH{=}CH_2 \xrightarrow{h\nu,\ O_2}$$

(173)   (1.128)

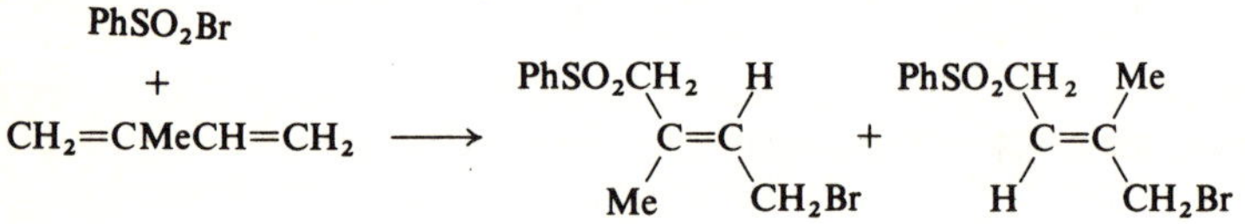

(1.129)

addition of benzenesulphonyl bromide to isoprene [equation (1.129)] has been studied[266]. The photochemical cycloaddition of alka-1,3-dienes to sym-triazolo[4,3-b]pyridazine (174) does not give the expected azetidines, but an unusual 1,3-addition to the 1,8-positions of (174) takes place with

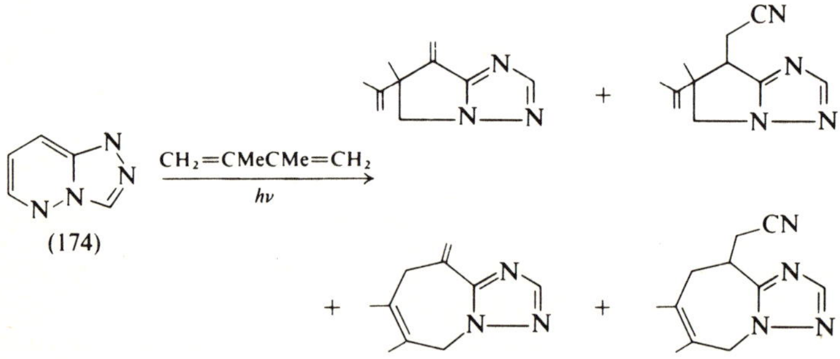

(174)

(1.130)

concurrent opening of the pyridazine ring [equation (1.130)][267]. The addition reactions of buta-1,3-diene with Cl_2, Br_2, ClOAc and BrOAc in a range of solvents have been studied in order to investigate the nature of the bonding in the halogenonium ion intermediates[268,269].

An electrochemical oxidation of aromatic acids and buta-1,3-diene in ethanol or ethanol–DMF gives the two adducts (175; $Ar = o\text{-}C_6H_4CO_2Me$ or $o\text{-}HOC_6H_4$) [equation (1.131)[270].

$$2 \text{ArCO}_2^- + \text{CH}_2\!=\!\text{CH}\!-\!\text{CH}\!=\!\text{CH}_2 \xrightarrow{-2e^-} \text{ArCOOCH}_2\text{CH}\!=\!\text{CHCH}_2\text{OCOAr}$$

(175)   (1.131)

SiF_2 is a structural analogue of carbenes, but it is extraordinarily stable and quite unreactive in the gas phase. Reactions so far observed are attributable to the $(SiF_2)_x$ polymer. If SiF_2 is produced by nuclear recoil it reacts with buta-1,3-diene to give a unique product by 1,4 addition[271]. A mixture of PF_3 and buta-1,3-diene in the vapour phase is irradiated with fast neutrons, and Si atoms are formed by the $^{31}\text{P(n,p)}^{31}\text{Si}$ nuclear transformation. These abstract fluorine from PF_3 and give SiF_2 which reacts to give the silacyclopentene [equations (1.132), (1.133) and (1.134)].

$$^{31}\text{Si} + PF_3 \rightarrow {}^{31}\text{SiF} + PF_2 \tag{1.132}$$

$$^{31}\text{SiF} + PF_3 \rightarrow {}^{31}\text{SiF}_2 + PF_2 \tag{1.133}$$

$$^{31}\text{SiF}_2 + \text{CH}_2\!=\!\text{CH}\!-\!\text{CH}\!=\!\text{CH}_2 \rightarrow \begin{array}{c} {}^{31}\text{SiF}_2 \\ \text{H}_2\text{C} \diagup \diagdown \text{CH}_2 \\ \text{HC}\!=\!\text{CH} \end{array} \tag{1.134}$$

The remainder of this section will be concerned with reactions of alka-1,3-dienes catalysed by transition-metal complexes. There is an extensive literature on the reactions of conjugated dienes activated by coordination to a transition metal. For example, there is much work on, and interest in, the polymerisation of buta-1,3-diene. Only those reactions that give small oligomers will be reviewed here. A number of nickel catalysts are known for the cyclodimerisation of butadiene to 2-methylene-1-vinylcyclopentane (176). The complex $NiX_2(PBu^n{}_3)_2$ (X = Cl, Br, or NO_3) with $NaBH_4$ or sodium alkoxide forms a catalyst for the formation of (176) and octa-1,3,7-triene

(176)   (1.135)

(177) [equation (1.135)] from buta-1,3-diene[272]. A nickel catalyst with one site blocked with, for example, a phosphine ligand gives a mixture of cyclo-

octa-1,5-diene, *cis*-1,2-divinylcyclobutane and 4-vinylcyclohexene[273]. The stereochemistry of the products from methyl-substituted 1,3-dienes is used to show that the mechanism of the oligomerisation is multistep rather than concerted. In contrast, a careful study of the non-catalytic Diels–Alder dimerisation of buta-1,3-diene to vinylcyclohexene[274] favours a concerted two-stage mechanism, as proposed by Woodward and Katz[275]. The oligomerisation of buta-1,3-diene with a Rh^I complex depends critically on the solvent[276]. In acetone or alcohols, cyclo-octa-1,5-diene and 1,4-*trans*-polybutadiene are formed in addition to 4-vinylcyclohexene, but in DMF, 4-vinylcyclohexene is the only reaction product [equation (1.136); L = ben-

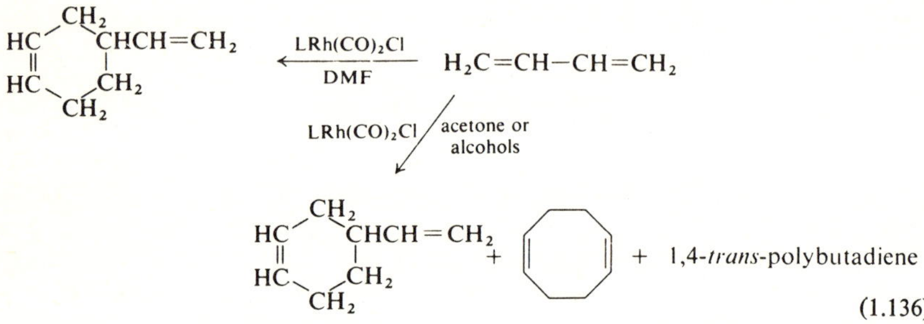

$$\text{(1.136)}$$

zotriazole, acridine, imidazole, benzimidazole or *m*-phenylenediamine]. 5-Vinylcyclohexa-1,3-diene is the major product when buta-1,3-diene and acetylene are co-oligomerised with a Ni^0–tri-n-alkylphosphine complex catalyst[277], and 1-phenylhexa-1,4-diene is produced when buta-1,3-diene and styrene are co-oligomerised with a π-allylpalladium dichloride–boron-trifluoride–triphenylphosphine complex catalyst[278]. The nickel-based catalyst system used above[272] also catalyses the reaction of buta-1,3-diene with strained cycloalkenes, when a compound with a four-membered ring is

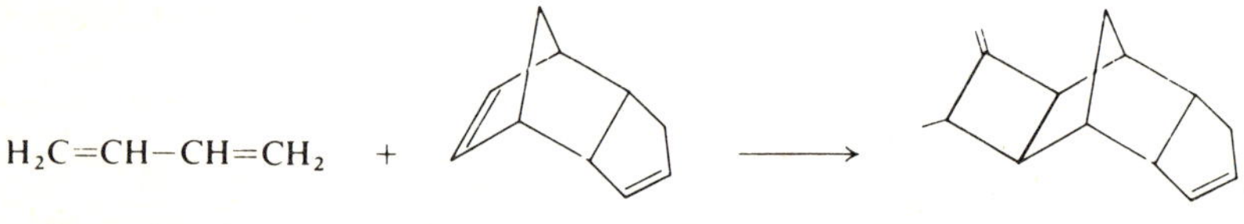

$$\text{(1.137)}$$

formed [equation (1.137)][279]. $(Bu^n_3P)_2NiBr_2$ is the best catalyst; the corresponding chloride has about 20% of the activity, and the corresponding cyanide and nitrate are inactive. The codimerisation of buta-1,3-diene with methyl methacrylate by using iron acetylacetonate with triethylaluminium as the catalyst gives as the products methyl hepta-2,5-dienoate (178) and methyl cyclohex-3-enecarboxylate (179)[280]. The amount of each product is influenced by the presence of an added ligand [L in equation (1.138)]; this may be pyridine, 2,2'-bipyridyl, a phosphite or a phosphine. In the absence of the ligand [equation (1.138)] one of the hydrogens of the hydrocarbon chain becomes activated and this leads to (178), whereas with the ligand present, (179) is formed. The reaction with this iron catalyst contrasts with the oligomerisation using nickel-based catalysts, when high molecular weight acids

$$H_2C{=}CH{-}CH{=}CH_2 \;+\; Fe(acac)_3 \;+\; AlEt_3 \;\longrightarrow$$

(178)

$$MeCH{=}CHCH_2CH{=}CHCO_2Me$$

(179)

(1.138)

are formed. Hydrogenation of buta-1,3-diene with Wilkinson's complex, $RhCl(PPh_3)_3$[281], and pentacyanocobaltate(II)[282] have been reported. With $RhCl(PPh_3)_3$, 1,2- or 1,4-addition of hydrogen can occur. 1,2-Addition is observed for the substituted butadienes 2-methylbuta-1,3-diene and 2,3-dimethylbuta-1,3-diene, whereas 1,4-addition is observed for buta-1,3-diene and the isomeric penta-1,3-dienes[281]. These differ, however, in that with butadiene *cis* addition of hydrogen occurs, and with the pentadienes the addition is *trans*.

Reactions of buta-1,3-diene with active methylene compounds are catalysed by nickel[283] and palladium[284] complexes. With nickel acetylacetonate and di-isopropoxyphenylphosphine–$NaBH_4$–NaOPh, a 90% conversion of buta-1,3-diene and benzyl methyl ketone into two 2 : 1 and two 1 : 1 adducts

$$PhCH_2COMe + H_2C{=}CH{-}CH{=}CH_2 \xrightarrow[16\,h]{25\,^\circ C}$$

$$H_2C{=}CH(CH_2)_3CH(CH{:}CH_2)CHPhCOMe \;+$$

$$CH_2{=}CH(CH_2)_3CH{=}CHCH_2CHPhCOMe \;+$$

$$MeCH{=}CHCH_2CHPhCOMe \;+\; CH_2{=}CHCHMeCHPhCOMe \qquad (1.139)$$

takes place [equation (1.139)][283]. Benzyl methyl ketone also reacts with isoprene and 2,3-dimethylbuta-1,3-diene, but the reactivity decreases as the diene is changed. With $PdBr_2(Ph_2PCH_2CH_2PPh_2)_2$ and NaOPh, buta-1,3-diene, isoprene, penta-1,3-diene or hexa-2,4-diene react with acetylacetone, ethyl acetoacetate or diethyl malonate to give similar derivatives of the active methylene compound [equation (1.140); $R^1 = COMe$ or CO_2Et, $R^2 =$

$$R^1R^2CH_2 + H_2C=CH-CH=CH_2 \rightarrow R^1R^2CHCHMeCH=CH_2$$
$$+ MeCH=CHCH_2CHR^1R^2 + R^1R^2C(C_4H_7)_2 \qquad (1.140)$$

COMe or $CO_2Et]^{284}$. The 1 : 1 adducts are favoured because the bidentate phosphine ligand prevents simultaneous coordination of two diene molecules to the palladium.

The chloroplatinic acid-catalysed addition of silanes to dienes has been known for some time. The effect that systematic changes in the structure of the diene have on the reaction has now been studied[285]. 1,2-Addition, with silicon becoming attached to carbon atom 1 or 2, and 1,4-addition both take place, and the different reaction products are explained in terms of the different types of complexes possible[285]. $NiCl_2(PEt_3)_2$ catalyses the addition of *sym*-tetramethyldisilane to buta-1,3-dienes [equation 1.141)][286]. For buta-1,3-diene [equation (1.141); $R^1 = R^2 = H$] the only product (180) is

$$H_2C=CR^1CR^2=CH_2 \ + \ Me_2SiHSiHMe_2$$

$$\downarrow$$

that in which the Si–Si bond of the disilane has been broken; for isoprene [equation (1.141); $R^1 = H$, $R^2 = Me$] about 20% of the product (181) is that formed by Si–H addition, and for 2,3-dimethylbuta-1,3-diene [equation (1.141); $R^1 = R^2 = Me$] the major product is (181).

A number of catalysed reactions of alka-1,3-dienes with amines have been reported[287-291]. Rhodium(I) complexes catalyse the alkylation of secondary amines and primary aromatic amines by buta-1,3-diene[287]. For example, $RhCl_3 \cdot 3H_2O$ in ethanol catalyses the reaction of buta-1,3-diene with morpholine [equation (1.142); R^1 and $R^2 = -CH_2CH_2OCH_2CH_2-$] to

$$H_2C=CH-CH=CH_2 + R^1R^2NH \rightarrow H_2C=CHCHMeNR^1R^2 +$$
$$(182)$$

$$MeCH=CHCH_2NR^1R^2 + H_2C=CH(CH_2)_3CH(NR^1R^2)CH=CH_2$$
$$(183) \qquad\qquad (184)$$

$$+ H_2C=CH(CH_2)_3CH=CHCH_2NR^1R^2$$
$$(185) \qquad\qquad (1.142)$$

give an 85% yield of (182; 70%) and (183; 30%). Addition of triphenylphosphine (1 : 1 with $RhCl_3$) alters the reaction to give an 85% yield of (182; 26%), (183; 11.5%), (184; 5%) and (185; 57.5%). Similarly, for di-n-propylamine [equation (1.142); $R^1 = R^2 = Pr^n$] and aniline [equation (1.142); $R^1 = H$, $R^2 = Ph$] the yields of the 2 : 1 adducts (184) and (185) are increased when triphenylphosphine is added, and for these latter two amines the efficiency of the process is also increased when PPh_3 is present. The 1 : 1 adducts arise from a π-allyl complex (186) formed from a rhodium hydride which has itself been formed from the Rh^{III} and the

(186)   (187)

amine. When the stabilising phosphine ligand is added, complex (186) reacts with buta-1,3-diene to give (187), and this complex leads to the octadienyl adducts. Similar reactions occur with nickel[288–290] and palladium[291] complexes as catalysts. The yields of amination products are increased by adding protic acids to the nickel catalyst[288]; bis(cyclo-octa-1,5-diene)nickel and tributylphosphine with buta-1,3-diene and morpholine give an 80% yield of octa-1,3-triene, but if 0.2 mole of trifluoroacetic acid is added then octadienylamines (184) and (185) (R^1 and $R^2 = —CH_2CH_2OCH_2CH_2—$) are formed in 60% yield. With an acid to nickel ratio of 10 : 1, butenylamines (182) and (183) (R^1 and $R^2 = —CH_2CH_2OCH_2CH_2—$) are selectively formed. The acid promotes the formation of the nickel hydride. If the catalyst is a mixture of $Ni(acac)_2$, $PPh(OPr^i)_2$ and $NaBH_4$, the $NaBH_4$ is an essential component. However, if the $Ni(acac)_2$ is replaced by $NiBr_2$, $NiCl_2$, $Ni(OAc)_2$ or nickel laurate the $NaBH_4$ is not essential[289]. The borohydride is needed with the acetylacetonate because with this ligand on the nickel the amines form complexes, such as $Ni(acac)_2(morpholine)_2$, and this complex formation competes with the reduction of Ni^{II} to Ni^0 by the alkoxyphosphine; $NaBH_4$ breaks down these complexes, and even when it is not essential it marginally increases reaction rates by assisting the reduction of the nickel salt. A specific effect of particular phosphines has also been noted[290]. Only certain phosphines of the type $R^1P(OR^2)_2$ are effective in the catalyst mixture. These are those in which (*a*) there is an aryl group directly attached to P, and (*b*) there are two alkoxy groups from secondary alcohols attached to P[290]. Thus $PhP(OMe)_2$, $PhP(OEt)_2$ and $MeP(OPr^i)_2$ were ineffective but $PhP(OCHMeEt)_2$, $PhP(OPr^i)_2$ and related phosphines were effective for catalysing the reaction of buta-1,3-diene with diethylamine involving nickel(II) laurate and $NaBH_4$. The reason for this effect may be that there is some interaction of the nickel with the secondary hydrogen of the alkoxy group. A wide range of addition reactions with buta-1,3-diene catalysed by palladium complexes has been reviewed[291,292]. They are of the type shown in the general equation (1.143). With ammonia, for example,

$$(1.143)$$

octadienylamine, di- and tri-octadienylamines are obtained[291]. Reactions with alcohols, phenols, water, enamines, active methylene compounds, hydrosilanes and carbon monoxide also occur. Various ketones have been studied in some detail[292].

Octa-1,6-dienes are the only products from the reaction of 1,3-dienes with formic acid, with palladium(II) acetate in triethylamine as the catalyst

$$2CH_2{=}CH{-}CH{=}CH_2 + HCO_2H \rightarrow CH_2{=}CH(CH_2)_3CH{=}CHMe + CO_2$$

$$(1.144)$$

[equation (1.144)][293]. The mechanism follows the same general pattern as reactions already described, *viz.* Pd–H addition to a C_8 chain derived from two buta-1,3-diene molecules bonded to the palladium by two π-allylic groups. With acetic acid, buta-1,3-diene gives acetoxyoctadienes, butenyl acetates and octa-1,3,7-triene. If *ortho*-substituted triaryl phosphites, e.g. $(o\text{-}PhC_6H_4O)_3P$, are added to the palladium catalyst, quantitative yields of the acetoxyoctadienes are obtained[294]. The effect of the addition of a range of Group V ligands (N, P, As, Sb and Bi compounds) to the nickel acetyl-acetonate–aluminium chloride catalyst on the relative yields of the 2 : 1 and 4 : 1 buta-1,3-diene–methyl acrylate co-oligomerisation products

$$2\,CH_2{=}CHCH{=}CH_2 + CH_2{=}CMeCO_2Me \rightarrow$$

$$CH_2{=}CH(CH_2)_3CH{=}CH(CH_2)_2C(CO_2Me){=}CH_2 \xrightarrow{\;2\,CH_2{=}CH\,CH{=}CH_2\;}$$

$$CH_2{=}CH(CH_2)_3CH{=}CHCH_2CH{=}C(CO_2Me)(CH_2)_2CH{=}CH(CH_2)_3CH{=}CH_2$$

$$(1.145)$$

[equation (1.145)] has been examined[295]. $AlEtCl_2$ is better than either $AlEt_3$ or $AlEt_2Cl$ as a catalyst for the Tischenko reaction [equation (1.146)][296].

$$CH_2{=}CH{-}CH{=}CH_2 \;+\; CH_2{=}CHCHO \;\longrightarrow\; \text{(cyclohexene-CHO)} \qquad (1.146)$$

Phenoxybutenes (188) and (189) are favoured over phenoxyoctadienes (190) and (191) in the nickel(0)–phosphine-catalysed addition of phenol to buta-1,3-diene by electron-donor ligands, an excess of ligand, high phenol concentrations and low conversions [equation (1.147)][297]. π-Allyl complexes of

$$(R_3P)_4Ni \;+\; \text{(phenol, OH)} \;+\; CH_2{=}CH{-}CH{=}CH_2$$

$$\downarrow$$

$$CH_2{=}CHCHMe(OPh) \;+\; MeCH{=}CHOPh$$
$$\text{(188)} \qquad\qquad\qquad \text{(189)}$$

$$+\; CH_2{=}CHCH(OPh)(CH_2)_3CH{=}CH_2 \;+\; CH_2{=}CH(CH_2)_3CH{=}CHCH_2OPh$$
$$\text{(190)} \qquad\qquad\qquad\qquad\qquad \text{(191)} \qquad\qquad\qquad (1.147)$$

buta-1,3-diene and octadiene are again involved, as they are in the addition of CO and an alcohol to buta-1,3-diene by using a palladium–triphenyl-phosphine catalyst [equation (1.148)][298]. By combining the polymerisation of

$$CH_2{=}CH{-}CH{=}CH_2 + CO + ROH \rightarrow MeCH{=}CHCH_2CO_2R \xrightarrow{\;C_4H_6\;}$$

$$CH_2{=}CH(CH_2)_3CH{=}CH\,CH_2\,CO_2R$$

$$(1.148)$$

buta-1,3-diene using π-allylnickel halide catalysts with the formation of esters of unsaturated acids when CO and an alcohol react with acetylene, esters of long-chain unsaturated acids[299] and the free acids[300] have been synthesised.

Buta-1,3-diene reacts with 1,1,1-trichloroethane and 1,1,1,3-tetrachloropropane in the presence of $Fe(CO)_5$ in propan-2-ol according to equation (1.149) ($R = H$ or $ClCH_2$) to give the 1 : 1 and the 2 : 1 adducts[301].

$$RCH_2CCl_3 + H_2C-CHCH=CH_2 \rightarrow RCH_2CCl_2CH_2CH-CHCH_2Cl$$
$$+ RCH_2CCl_2CH_2CH=CH(CH_2)_2CH=CHCH_2Cl \qquad (1.149)$$

Copper and copper salts initiate radical reactions with alka-1,3-dienes[302,303]. If ethyl diazoacetate is decomposed by copper powder in the presence of a hexa-2,4-diene, then ethyl 2-methyl-3-propenylcyclopropanecarboxylates are formed [equation (1.150); R^1, R^2, R^3, and R^4 are two methyl groups

$$(1.150)$$

and two hydrogen atoms][302]. Buta-1,3-diene or isoprene reacts with benzenediazonium chloride in the presence of $CuCl_2$ or $CuCl$. Cu^I is the catalyst; it is produced from added $CuCl_2$ by reduction. Aryl radicals and Cl radicals add to the diene [equation (1.151)][303].

$$ArN_2Cl + CuCl \rightleftharpoons [ArN_2CuCl]^+Cl^- \longrightarrow Ar\cdot + CuCl$$

$$\left[ArN_2Cu\right]^{2+} Cl_2^- \xrightarrow[-CuCl]{-N_2} ArCH_2CH=CHCH_2Cl \qquad (1.151)$$

The mechanism of insertion of 1,3-dienes into allyl–palladium bonds has been elucidated[304]. The diene, acting as a monodentate ligand, is inserted by an electrocyclic mechanism [equation (1.152)][304]. If propene reacts with a

$$(1.152)$$

solution formed by the reduction of $[(\pi\text{-}C_6H_6)Mo(\pi\text{-}C_3H_5)Cl]_2$ by $EtAlCl_2$, propane and an alka-1,3-diene complex (192) are formed[305]. This solution also undergoes other reactions with alkenes, alkynes or alkadienes. Nickel atoms can be condensed into buta-1,3-diene when the dark red volatile

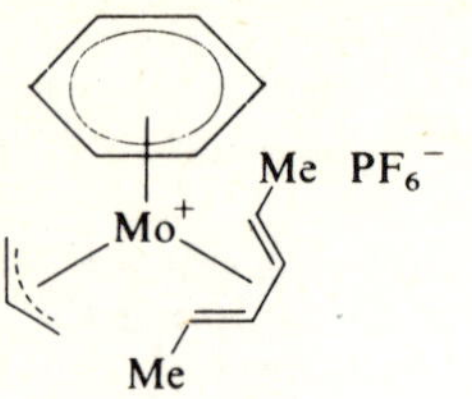

(192)

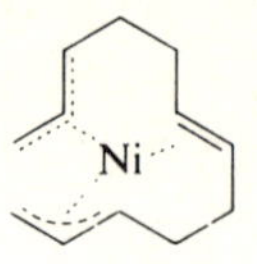

(193)

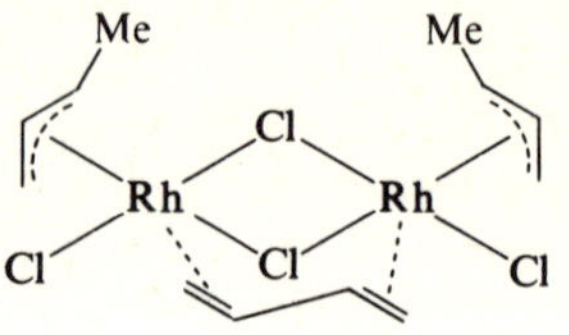

(194)

complex (193) is formed[306], and the alkadiene complex (194) with three butadiene molecules and two rhodium atoms with chloride bridges has been isolated[307] during the reaction of buta-1,3-diene with alcohols catalysed by $RhCl_3 \cdot 3H_2O$.

1.5.3 Other alkadienes

1.5.3.1 Preparation

Penta-1,4-diene is one of the two products formed by the photochemical decomposition of 2,3-diazabicyclo[3,2,0]hept-2-ene [equation (1.153)][308].

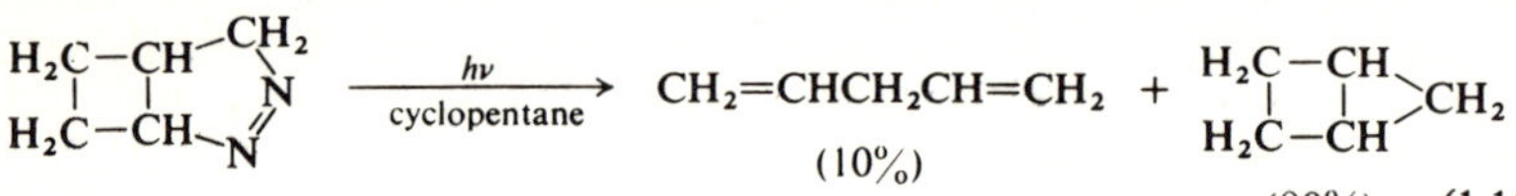

$$CH_2=CHCH_2CH=CH_2 \quad (10\%) \qquad (90\%) \qquad (1.153)$$

1.5.3.2 Reactions

Terminal alkadienes, $H_2C=CH(CH_2)_nCH=CH_2$ ($n = 1$–4), give mono and bis adducts with perfluoroalkyl radicals formed from iodoperfluoroalkanes, except for hepta-1,6-diene which preferentially cyclises [equation (1.154)][309]. 5,7-Dimethylocta-1,6-diene and 3,7-dimethylocta-1,6-diene both cyclise with

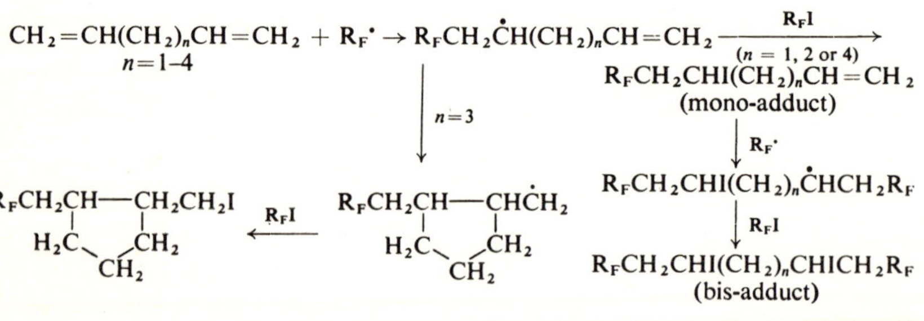

$$(1.154)$$

a HCO_2H–BF_3 or a HCO_2H–strong mineral acid catalyst to give cyclic formates [equations (1.155) and (1.156)][310]. The 3,7-dimethyl compound

(1.155)

(1.156)

gives two products, one by a bond migration in the carbenium ion intermediate [equation (1.156)]. A number of papers concerning the hydroboration of higher dienes have been published[311-315]. These are reviewed in Chapter 9.

1.5.4 Alkatrienes

1.5.4.1 Reactions

Hydrogenation of *cis*-hexa-1,3,5-triene at 160 °C with a hydrogen pressure of 30 atm by using a (methyl benzoate)–Cr(CO)$_3$ catalyst gives cyclohexene as the main product[316]. This is formed by 1,4-addition of hydrogen to cyclohexa-1,3-diene formed by a thermal Cope-type cyclisation of the

(1.157)

hexatriene [equation (1.157)]. Hydrogenation of *trans*-hexa-1,3,5-triene takes place by 1,4-addition and a 1,3-shift of a hydrogen atom as shown in equation (1.157)[316]. The kinetics of the thermal Cope cyclisation of 1- and 3-alkylhexa-1,3,5-trienes have been studied[317]. The results are explained by the donor abilities of the different alkyl groups, steric retardation and differences in ground-state energies of the different molecules.

1.6 ALKYNES

1.6.1 Preparation

The facile β-elimination of β-halogenoalkylsilanes to give alkenes is well known. The equivalent reaction, β-elimination from a β-halogenosilylalkene, has not previously been reported. If KF in DMSO is used, a smooth elimination takes place to give acetylene [equation (1.158)][318]. KF must be used; both

$$\begin{array}{c} H \quad\quad H \\ C = C \quad + \quad F^- \quad\longrightarrow\quad HC\equiv CH \;+\; (CH_3)_3SiF \;+\; Cl^- \\ Me_3Si \quad\quad Cl \end{array}$$

$$(1.158)$$

KCl and KI are ineffective. 1,1-Dibromoalkenes are converted into alkynes when treated with Bu^nLi in THF at $-78\,°C$, or preferably[319] with 1.5% Li amalgam in THF at 25 °C. Bis-hydrazones of α-diketones give alkynes in high yield when oxidised with oxygen in pyridine solution with $CuCl_2$ as a catalyst[364]. For example, the bis-hydrazone of octane-4,5-dione (195) gives

$$\begin{array}{c} Pr^nC - CPr^n \\ \| \quad\; \| \\ H_2N-N \quad N-NH_2 \end{array}$$
$$(195)$$

an 89% yield of oct-4-yne when added to a solution of $CuCl_2$ and O_2 in pyridine. A synthesis of t-butylacetylene (3,3-dimethylbutyne), superior in all respects to previous methods, involves the bromination of t-butylethylene (3,3-dimethylbutene) and double dehydrobromination with potassium t-butoxide in DMSO[320]. Polyynes, $H(C\equiv C)_nH$ ($n = 4$–10, 12), have been prepared by a reaction sequence involving copper-catalysed oxidative coupling of silyl-protected terminal alkynes, partial desilylation of the products by alkali, recoupling, and complete desilylation[321]. A typical reaction sequence is given in equation (1.159). Substituted polyynes can be prepared by the same general method[322].

$$Et_3SiC\equiv CH \rightarrow Et_3Si(C\equiv C)_2SiEt_3 \rightarrow Et_3Si(C\equiv C)_2H \rightarrow$$
$$Et_3Si(C\equiv C)_4SiEt_3 \rightarrow etc.$$
$$(1.159)$$

1.6.2 Reactions

1.6.2.1 Reactions with main-group metals and metalloidal compounds

Diacetylenes are metallated with methyl-lithium or butyl-lithium[323]. The usual products from hexa-2,4-diynes are trilithio derivatives, but prolonged treatment with butyl-lithium gives the tetralithio compound. Octa-2,4-diyne gives a trilithio and octa-3,5-diyne gives a dilithio compound. The two triple bonds in hepta-1,6-diyne are metallated independently to give di-, tri- and tetra-lithio derivatives. Diethylstrontium forms a 1 : 1 adduct with hex-1-en-3-yne in diethyl ether, which on hydrolysis gives octa-3,4-diene in 40–50% yield[324]. In this respect the reaction of the organo-strontium compound is the same as that of organo-lithium, -calcium and -barium compounds.

A number of papers concerning the hydroboration of alkynes are reviewed in Chapter 9[325-329]. The first oxythallation of alkynes has been reported[330]. Thallium(III) acetate in acetic acid gives a mixture of *cis*- and *trans*-alkenes as in equation (1.160). Alkenylmercuric compounds[331] and geminal alkane

$$RC{\equiv}CR \xrightarrow[\text{HOAc}]{\text{Tl(OAc)}_3} \underset{\substack{\text{AcO} \quad\quad \text{Tl(OAc)}_2}}{\overset{\substack{R \quad\quad\quad R}}{C{=}C}} + \underset{\substack{\text{AcO} \quad\quad\quad R}}{\overset{\substack{R \quad\quad \text{Tl(OAc)}_2}}{C{=}C}} \qquad (1.160)$$

dimercurials[332] have been prepared by using hydroboration and cleavage of the C—B bonds by mercury salts [equations (1.161) and (1.162)]. Acetylene[333]

$$(1.161)$$

$$B_2H_6 + RC{\equiv}CH \xrightarrow[]{\text{THF}} RCH_2CH(BH_2)_2 \xrightarrow[]{\text{4MeOH}} RCH_2CH[B(OMe_2)_2]_2$$

$$\xrightarrow[\text{NaOH}]{\text{HgCl}_2} RCH_2CH(HgCl)_2 \qquad (1.162)$$

and its homologues[334] react with SiF_2 to give 1 : 2 and 2 : 2 adducts by a radical mechanism [equation (1.163)]. Organotin and organolead hydrides add to alkynes[335]. With but-1-yne, anti-Markovnikov addition occurs with both the tin and the lead hydrides [equation (1.164); R = Pr^n, M = Sn; R = Ph, M = Sn; R = Et, M = Pb]. With vinylacetylene, organotin hydrides undergo 1,4-addition to give allenic derivatives [equation (1.165); R = Pr^n or Ph, M = Sn], and organolead hydrides give 3,4-addition pro-

$$HC\equiv CH + F_2\dot{Si}\dot{Si}F_2 \longrightarrow H\dot{C}=CHSiF_2\dot{Si}F_2 \xrightarrow{\text{ring closure}} \begin{array}{c} HC-SiF_2 \\ \| \quad | \\ HC-SiF_2 \end{array}$$

$$\Big\downarrow HC\equiv CH$$

$$\text{Polymer} \longleftarrow H\dot{C}=CHSiF_2SiF_2CH=\dot{C}H$$

ring closure   1,5-migration

$$HC\equiv CSiF_2SiF_2CH=CH_2$$

$(SiF_2\,?)$

$$\tag{1.163}$$

ducts [equation (1.166); R = Et, M = Pb]. These latter compounds might
be formed by 1,4-addition followed by rearrangement.

$$R_3MH + HC\equiv CEt \rightarrow R_3MCH=CHEt \tag{1.164}$$

$$R_3MH + HC\equiv C-CH=CH_2 \rightarrow R_3MCH=C=CHMe \tag{1.165}$$

$$R_3MH + HC\equiv C-CH=CH_2 \rightarrow R_3MCH=CH-CH=CH_2 \tag{1.166}$$

1.6.2.2   *Reactions with organic nitrogen compounds*

Substituted pyridines are produced when a β-aminocrotonic ester reacts with
diacetylene with sodium as a catalyst [equation (1.167)][336]. Buta-1,3-diyne
and its homologues react with semicarbazide to give 3-alkylpyrazole-1-

$$MeC(NH_2)=CHCO_2Et + HC\equiv C-C\equiv CH \longrightarrow \tag{1.167}$$

$$HC\equiv C-C\equiv CH + H_2NNHCONH_2 \longrightarrow \quad\longrightarrow \tag{1.168}$$

(197)   (R = H)   (196)

carbonamides (196) [equation (1.168)][337]. For buta-1,3-diyne, but not for
penta- or hexa-1,3-diyne, some of the 5-alkylpyrazole isomer (197) is also
formed. If a strong acid with an anion of low nucleophilicity (e.g. HBF$_4$ or
HSbF$_6$) is added to an alkyne in nitromethane, the product formed is a
ketone derivative which has the molecular formula of a 1 : 1 adduct of the
alkyne and nitromethane[338,339]. The reaction has been examined in detail
and various alternative pathways rejected in favour of that in equation

$$R^1C{\equiv}CR^2 + H^+ \longrightarrow [R^1\overset{+}{C}{=}CHR^2] \xrightarrow{CH_3NO_2}$$

$$R^1COCHR^2ON{=}CH_2 \qquad (1.169)$$

(1.169). The not unexpected fact that nucleophilic addition of secondary amines to diacetylenes takes place at a greater rate in polar aprotic solvents, such as nitromethane or acetonitrile, than in hexane has been reported[340]. Alkynes react with chlorosulphonyl isocyanate to give 6-chloro-1,2,3-oxathiazine 2,2-dioxide cycloadducts [equation (1.170)][341]. But-2-yne, hex-2-yne, hex-3-yne and oct-4-yne all react in this way; however, hex-1-yne

$$R^1C{\equiv}CR^2 + ClSO_2NCO \longrightarrow \qquad (1.170)$$

$$Bu^nC{\equiv}CH + ClSO_2NCO \longrightarrow \longrightarrow Bu^nC{\equiv}CCONHSO_2Cl$$

$$(198)$$

$$\downarrow H_2O$$

$$Bu^n C{\equiv}CCONH_2$$

$$(199) \quad (1.171)$$

gives only hept-2-ynamide derivatives (198) and (199) [equation (1.171)]. With oct-1-en-4-yne only the alkyne part of the molecule reacts with the chlorosulphonyl isocyanate. 2H-Azirines are formed[342] when N-aminoph-

$$\xrightarrow{Pb(OAc)_4} \qquad (1.172)$$

thalimide is oxidised with lead tetra-acetate in the presence of propyne, but-2-yne, pent-1-yne or hex-3-yne in CH_2Cl_2 between -20 and $+20\,°C$. It is presumed that the products are formed by the ready rearrangement of the elusive $1H$-aziridines [equation (1.172)]. A comprehensive two-part review of 1,3-dipolar cycloadditions to alkynes includes additions of cyanoimines ($-\overset{+}{C}=N-\overset{-}{N}-$), cyano-oxides ($-\overset{+}{C}=N-\overset{-}{O}$), diazoalkanes ($\overset{+}{N}=N-\overset{-}{C}=$), azides ($\overset{+}{N}=N-\overset{-}{N}-$), dinitrogen monoxide ($\overset{+}{N}=N-\overset{-}{O}$), carbenes and nitrenes[343,344].

1.6.2.3 Oxidation and reactions with organic oxygen and sulphur compounds

Autoxidation of alkynes by air at 20–$30\,°C$ gives the peroxy compounds as shown in equation (1.173), in which the first peroxy group is formed on the

$$R^1C\equiv CPr^i \xrightarrow{\text{air}} R^1C\equiv C-C(OOH)Me_2 \xrightarrow{\text{air}} R^2C(OOH)C\equiv CC(OOH)Me_2 \quad (1.173)$$

secondary carbon atom adjacent to the alkyne bond, and the second peroxy group is formed on the primary carbon on the other side of the alkyne bond[345]. During the ozonisation of simple alkynes, i.r. examination showed the appearance of a strong carbonyl band in the unstable precursor of the acid anhydride that leads to the final products[346]. It had been suggested earlier[347] that the reaction involves an unstable anhydride; this is illustrated for but-2-yne by equation (1.174). The compound initially formed cannot,

$$MeC\equiv CMe + O_3 \rightarrow MeCOOCOMe \rightarrow CH_2=C=O + MeCO_2H \qquad (1.174)$$

$$\underset{(200)}{\underset{RC\overset{\displaystyle O-O}{\underset{O}{\rule{0pt}{0pt}}}CR}{}} \qquad \underset{(201)}{RC=CR\ (O_2\text{-ozonide})} \qquad \underset{(202)}{RCO-C\cdots C-COR}$$

therefore, be ozonide (200) or (201), as neither of these contain a carbonyl group. It might be a dimer (or polymer) of an acyl carbonyl oxide (202). Propan-2-ol adds to the triple bond of 3,3-dimethylbut-1-yne to give a 16% yield of the tertiary allylic alcohol when irradiated for 17 h at 18–$20\,°C$ with u.v. light [equation (1.175)][348]. Diacetylenes react with carbonyl compounds

$$Bu^tC\equiv CH + Pr^iOH \xrightarrow{h\nu} \underset{H}{\overset{Bu^t}{\diagup}}C=C\underset{C(OH)Me_2}{\overset{H}{\diagdown}} \qquad (1.175)$$

$$HC\equiv C-C\equiv CH + R^1R^2CO \xrightarrow[\text{OH form}]{\text{Resin}} R^1R^2C(OH)C\equiv C-C\equiv CH +$$

$$R^1R^2C(OH)C\equiv C-C\equiv C-C(OH)R^1R^2 \qquad (1.176)$$

with OH ion-exchange resins as catalysts [equation (1.176)][349]; a range of diacetylenes have been studied. Diacetylenes also add o-, m- and p-amino-phenols in dioxan solution with KOH as the catalyst [equation (1.177)][350].

$$HC{\equiv}C{-}C{\equiv}CH \;+\; HO{-}\text{(aryl)}{-}NH_2 \xrightarrow{\text{KOH}} HC{\equiv}C{-}CH{=}CHO{-}\text{(aryl)}{-}NH_2 \quad (1.177)$$

Cyano-t-butylketene, generated *in situ* from 2,5-diazido-3,6-di-t-butyl-1,4-benzoquinone, reacts with alkynes to give good yields (40–80%) of substituted cyclobutanones [equation (1.178)][351]. The addition of nitro-methane to alkynes[338,339] has been reported on p. 68. A corresponding reaction occurs if methoxymethyl cations ($MeOCH_2{}^+$) or acyl cations (RCO^+) are used, the products being the equivalent methoxymethyl or acyl

$$\text{(quinone)} \longrightarrow Bu^tC(CN){=}C{=}O \xrightarrow{\;HC{\equiv}CBu^t\;} \begin{array}{c} HC{-}CO \\ \text{‖ ‖} \\ Bu^t C{-}CBu^tCN \end{array} \quad (1.178)$$

$$R^1C{\equiv}CR^2 \;+\; MeOCH_2{}^+\,BF_4{}^- \;+\; R^3CH_2NO_2 \longrightarrow \left[\begin{array}{c} CH_2OMe \\ | \\ R^1C{=}CR^2 \\ O{\diagup}{\diagdown}O \\ N \\ \text{‖} \\ CHR^3 \end{array} \right]$$

$$\longrightarrow R^1COCR^2(CH_2OMe)ON{=}CHR^3 \;+\; HBF_4 \quad (1.179)$$

derivatives [illustrated for methoxymethyl by equation (1.179); $R^1 = R^2 = Bu^n$, $R^3 = H$; $R^1 = R^2 = Bu^n$, $R^3 = Me$; $R^1 = Bu^n$, $R^2 = R^3 = H$][352]. For acyl cations the reaction is not so clean. A competing reaction for dialkyl- and especially monoalkyl-alkynes is the formation of β-diketones that are devoid of the nitrogen-containing moiety [equation (1.180); $R^1 =$

$$R^1C{\equiv}CR^2 \;+\; R^3CO^+\,Y^- \;+\; MeNO_2 \longrightarrow \left[\begin{array}{ccc} COR^3 & & COR^3 \\ | & & | \\ R^1C{=}CR^2 & & R^1C{=}CR^2 \\ O{\diagup}{\diagdown}O & \rightleftharpoons & O^+{\diagup}{\diagdown}O \\ N^+{=}O \;\; Y^- & & H{-}O\;N \\ | & & \text{‖} \\ Me & & C \\ & & H\;\;H \end{array} \right]$$

$$\longrightarrow R^1C(OH){=}CR^2COR^3 \;+\; HC{\equiv}N{\rightarrow}O \;+\; HY \quad (1.180)$$

Bu^n, $R^2 = H$, $R^3 = Me$ or Bu^t; $R^1 = Bu^t$, $R^2 = H$, $R^3 = Bu^t$; $R^1 = R^2 = Bu^n$, $R^3 = Me$][353]. When heated together above 100 °C, acetylene and dimethyl sulphoxide in the presence of alkali and water give more than ten products, including methyl vinyl sulphide and divinyl sulphide[354]. Organic disulphides react with acetylene according to equation (1.181)[355,356]. At comparatively low temperatures the 1,2-bis(alkylthio)ethylenes (203) and alkyl vinyl sulphides (204) are the major products. At 170–180 °C 1,1- and

$$HC{\equiv}CH + RSSR \rightarrow RSCH{=}CHSR + CH_2{=}CHSR + MeCH(SR)_2$$

$$(203) \qquad\qquad (204) \qquad\qquad (205)$$

$$+ RSCH_2CH_2SR \qquad\qquad\qquad (1.181)$$

$$(206)$$

1,2-bis(alkylthio)ethanes (205) and (206) predominate. Di-n-alkyl disulphides are all reactive, but di-isopropyl disulphide is less so and dimethyl, di-t-butyl, bis(3-chlorobut-2-enyl), diallyl and diphenyl disulphide are virtually unreactive.

1.6.2.4 Halogenation

Hydrofluorination and halogenofluorination of alkenes were discussed in Section 1.4.2.9[213-215]. Alkynes undergo analogous reactions: with HF in pyridine, hex-1-yne and hex-3-yne each react with two moles of HF to give a difluoroalkane[213]; with N-halogenosuccinimides, but-2-yne gives the 3-fluoro-2-halogeno-but-2-enes (halogen = Cl, Br or I) and hex-3-yne gives the analogous fluorohalogeno alkenes[214].

1.6.2.5 Transition-metal catalysed reactions

The catalysed trimerisation of conjugated diacetylenes gives both compound (207) and compound (208) [equation (1.182)]; yields are poor (*ca.* 10%), and the separation of (207) and (208) is difficult. By using $Ni(CO)_2(PPh_3)_2$ as

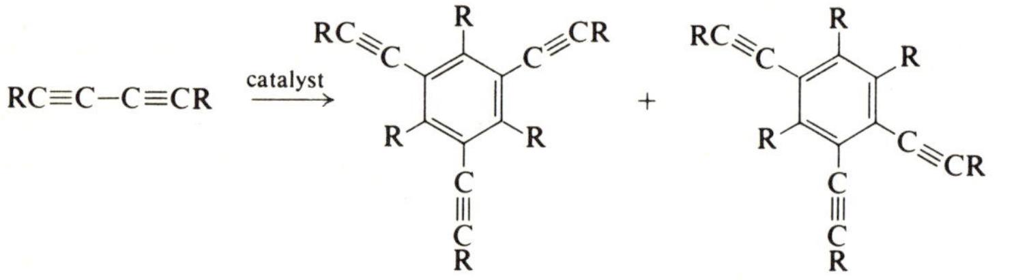

$$RC{\equiv}C{-}C{\equiv}CR \xrightarrow{\text{catalyst}}$$

(207) + (208)   (1.182)

catalyst, high yields (e.g. for R = Me, 77%) of (208) alone are obtained[357]. Higher yields than previously of the mixture are obtained by using $CoCp(CO)_2$ as the catalyst. To obtain good yields the catalyst must be a compound which is not so active that compound (207) or (208) can react further. The hydroformylation of but-1-yne (and but-2-yne) to give a mixture of pentan-1-ol and 2-methylbutan-1-ol is catalysed by $RhH(CO)(PPh_3)_3$ [equation (1.183)][358]. The arylation of acetylene with benzene by using

$$RC{\equiv}CH + CO + H_2 \xrightarrow[\text{180 °C}]{\text{catalyst}} R(CH_2)_3OH + MeCHRCH_2OH \qquad (1.183)$$

normal Friedel–Crafts catalysts, e.g. $AlCl_3$, gives a mixture of styrene, anthracene, $(CH)_n$ polymers and diphenylethane. However, if the bis-arene–cobalt complex $Co[AlCl_4]_2 \cdot 2C_6H_6$ is used as the catalyst, diphenylethane is formed selectively[359]. Dimers of γ-but-2-enolactone are formed in moderate yield when β-bromostyrene reacts with alkynes in DMF with $Ni(CO)_4$ as the catalyst [equation (1.184)][360]. Diacetylenic amines can be prepared by the reaction of diacetylene with dialkylalkoxyamines by using CuCl as the catalyst [equation (1.185); $R^1 = R^2 = $ Me, Et, Bu^n; R^1 and $R^2 = (CH_2)_5$, $CH_2CH_2OCH_2CH_2$; $R^1 = $ Et, $R^2 = $ Ph][361].

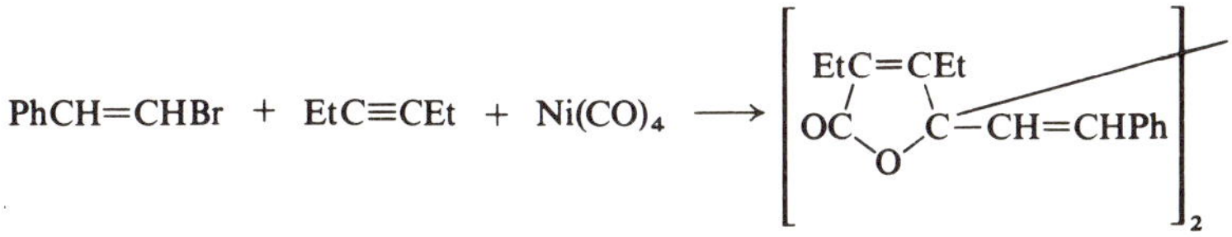

$$PhCH=CHBr \ + \ EtC\equiv CEt \ + \ Ni(CO)_4 \ \longrightarrow \ \left[\begin{array}{c} EtC=CEt \\ OC \qquad C=CH=CHPh \\ O \end{array} \right]_2 \qquad (1.184)$$

$$R^1R^2NH \ + \ CH_2O \ + \ CH_2(OH)R^3 \rightarrow R^1R^2NCH_2OR^4$$

$$\xrightarrow[\ HC\equiv C-C\equiv CH\ ]{\ CuCl\ } R^1R^2NCH_2(C\equiv C)_2CH_2NR^1R^2 \qquad (1.185)$$

The reduction of acetylene to ethylene has been widely used to assay the nitrogen-fixing activity of enzymes[362]. A water-soluble compound, modelled on the nitrogen-fixing enzyme *nitrogenase*, *viz.* *meso*-tetra(*p*-sulphonatophenyl)porphinatocobalt(III) (209), together with $NaBH_4$ reduces acety-

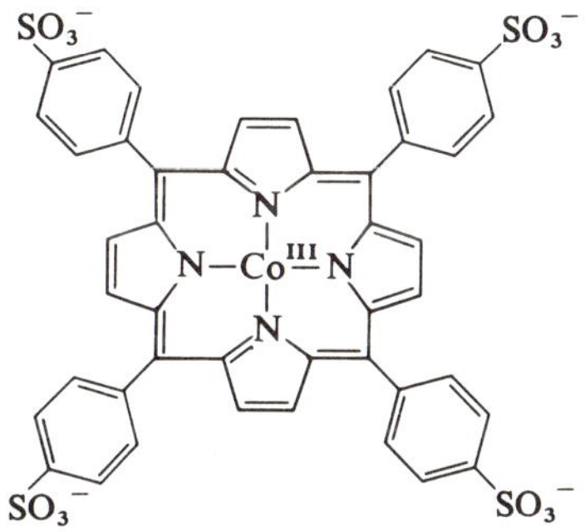

(209)

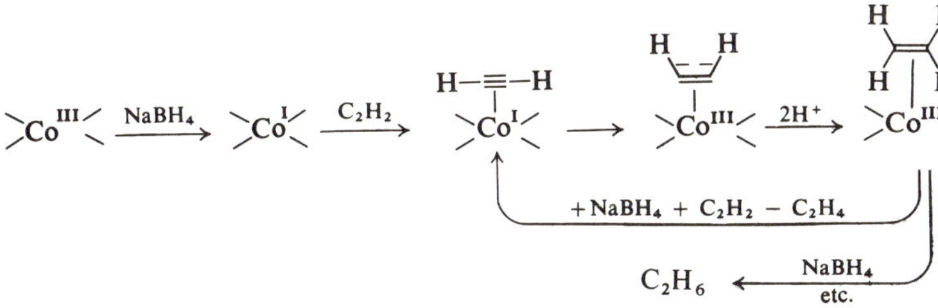

lene to ethylene (with a little ethane) as shown in equation (1.186); a cobalt(I) species is the catalyst[363]. It is particularly interesting, in view of the known composition of nitrogen-fixing enzymes, that the iron(III) and manganese(III) compounds analogous to (209) are not catalysts for this reduction but the molybdenum(III) analogue is.

References

1. Webster, D. E. (1973). *MTP International Review of Science, Organic Chemistry Series One*, Vol. 2, *Aliphatic Compounds* (N. B. Chapman, editor), p. 1 (London: Butterworths)
2. Radom, L., Hariharan, P. C., Pople, J. A. and Schleyer, P. von R. (1973). *J. Amer. Chem. Soc.*, **95**, 6531
3. Maksić, Z. B. and Randic, M. (1973). *J. Amer. Chem. Soc.*, **95**, 6522
4. Allinger, N. L. and Sprague, J. T. (1973). *J. Amer. Chem. Soc.*, **95**, 3893
5. Hay, P. J., Hunt, W. J. and Goddard, W. A. (1972). *J. Amer. Chem. Soc.*, **94**, 8293
6. Chu, S. Y. and Frost, A. A. (1971). *J. Chem. Phys.*, **54**, 764
7. Nelson, J. L. and Frost, A. A. (1972). *J. Amer. Chem. Soc.*, **94**, 3727
8. Nelson, J. L. and Frost, A. A. (1972). *Chem. Phys. Lett.*, **13**, 610
9. Nelson, J. L. and Frost, A. A. (1973). *Theoret. Chim. Acta*, **29**, 75
10. Boyd, J. A. and Whitehead, M. A. (1972). *J. Chem. Soc. Dalton Trans.*, 81 and refs. therein
11. Weber, J. and Gerdil, R. (1973). *Helv. Chim. Acta*, **56**, 1565
12. Allinger, N. L., Tribble, M. T., Miller, M. A. and Wertz, D. H. (1971). *J. Amer. Chem. Soc.*, **93**, 1637
13. Allinger, N. L. and Sprague, J. T. (1972). *J. Amer. Chem. Soc.*, **94**, 5734
14. Gimerc, B. M. (1973). *J. Amer. Chem. Soc.*, **95**, 1417
15. Sinanoglu, O. and Önder Pamuk, H. (1973). *J. Amer. Chem. Soc.*, **95**, 5435
16. Levy, B. (1973). *Chem. Phys. Lett.*, **18**, 59
17. Barthelat, J. C. and Durand, Ph. (1972). *Theoret. Chim. Acta*, **27**, 109
18. Barthelat, J. C. and Durand, Ph. (1972). *Chem. Phys. Lett.*, **16**, 63
19. Fischer-Hjalmars, I. and Siegbahn, P. (1973). *Theoret. Chim. Acta*, **31**, 1
20. Clark, P. A. (1972). *Theoret. Chim. Acta*, **28**, 75
21. Bjorseth, A. (1972). *Acta Chem. Scand.*, **26**, 1278
22. Cherrot, J. L. (1973). *Theoret. Chim. Acta*, **28**, 201
23. Hoffmann, R. (1963). *J. Chem. Phys.*, **39**, 1397
24. Zeelan, F. J. (1973). *Rec. Trav. Chim.*, **92**, 801
25. Durmaz, S., Murrell, J. N. and Pedley, J. B. (1972). *J. Chem. Soc. Chem. Commun.*, 933
26. Firestone, R. A. (1973). *J. Chem. Soc. Chem. Commun.*, 163
27. Linnett, J. W. (1964). *The Electronic Structure of Molecules* (London: Methuen)
28. Firestone, R. A. (1969). *J. Org. Chem.*, **34**, 2621
29. Basch, H. (1972). *J. Chem. Phys.*, **56**, 441
30. Baerends, E. J., Ellis, D. E. and Ros, P. (1972). *Theoret. Chim. Acta*, **27**, 339
31. Pople, J. A. and Hariharan, P. C. (1973). *Theoret. Chim. Acta*, **28**, 213
32. Clark, D. T. and Adams, D. B. (1973). *Tetrahedron*, **29**, 1887
33. Beaz, M., Bieri, G., Bock, H. and Heilbronner, E. (1973). *Helv. Chim. Acta*, **56**, 1028
34. Asbrink, L., Fridh, C. and Lindholm, E. (1972). *J. Amer. Chem. Soc.*, **94**, 5501
35. Stevens, R. M. and Karplus, M. (1972). *J. Amer. Chem. Soc.*, **94**, 5140
36. Sovers, O. J., Kern, C. W., Pitzer, R. M. and Karplus, M. (1968). *J. Chem. Phys.*, **49**, 2592
37. Pearson, R. G. (1972). *J. Amer. Chem. Soc.*, **94**, 8287
38. Mango, F. D. (1973). *Tetrahedron Lett.*, 1509
39. Goddard, W. A. (1972). *J. Amer. Chem. Soc.*, **94**, 793
40. Goddard, W. A. and Ladner, R. C. (1971). *J. Amer. Chem. Soc.*, **93**, 5750
41. Aksnes, D. W. and Albriktsen, P. (1972). *Acta Chem. Scand.*, **26**, 3021
42. Woller, P. B. and Garbisch, E. W. (1972). *J. Amer. Chem. Soc.*, **94**, 5310
43. Lipnick, R. L. and Garbisch, E. W. (1973). *J. Amer. Chem. Soc.*, **95**, 6375
44. Olah, G. A. (1972). *J. Amer. Chem. Soc.*, **94**, 808
45. Olah, G. A. (1972). *Chem. Brit.*, 281
46. Olah, G. A. (1973). *Angew. Chem. Int. Ed. Engl.*, **12**, 173
47. Olah, G. A. and Mo, Y. K. (1972). *J. Amer. Chem. Soc.*, **94**, 6864
48. Olah, G. A., Renner, R., Schilling, P. and Mo, Y. K. (1973). *J. Amer. Chem. Soc.*, **95**, 7686
49. Olah, G. A. and Schilling, P. (1973). *J. Amer. Chem. Soc.*, **95**, 7680

50. Olah, G. A., Mo, Y. K. and Olah, J. A. (1973). *J. Amer. Chem. Soc.*, **95**, 4939
51. Olah, G. A., DeMember, J. R. and Shen, J. (1973). *J. Amer. Chem. Soc.*, **95**, 4952
52. Olah, G. A., Halpern, Y., Shen, J. and Mo, Y. K. (1973). *J. Amer. Chem. Soc.*, **95**, 4960
53. Lukas, J., Kramer, P. A. and Kouwenhaven, A. P. (1973). *Rec. Trav. Chim.*, **92**, 44
54. Tabushi, I., Yoshida, Z. and Tamaru, Y. (1973). *Tetrahedron*, **29**, 81
55. Bertram, J., Coleman, J. P., Fleischmann, M. and Pletcher, D. (1973). *J. Chem. Soc. Perkin Trans. II*, 374
56. Bertram, J., Fleischmann, M. and Pletcher, D. (1971). *Tetrahedron Lett.*, 349
57. Clark, D. B., Fleischmann, M. and Pletcher, D. (1973). *J. Chem. Soc. Perkin Trans. II*, 1578
58. Hodges, R. J., Webster, D. E. and Wells, P. B. (1972). *J. Chem. Soc. Dalton Trans.*, 2571
59. Hodges, R. J., Webster, D. E. and Wells, P. B. (1972). *J. Chem. Soc. Dalton Trans.*, 2577
60. Gol'dshleger, N. F., Moiseev, I. I., Khidekel, M. L. and Shteinman, A. A. (1972). *Dokl. Akad. Nauk SSSR*, **206**, 106
61. Rudakov, E. S. and Shteinman, A. A. (1973). *Kinet. Katal.*, **14**, 1346
62. Anderson, J. R. and Kemball, C. (1954). *Proc. Roy. Soc.*, **A223**, 361
63. Rapka, L. F. and Shteinman, A. A. (1973). Personal communication
64. Littlecott, G. W. and McQuillin, F. J. (1973). *Tetrahedron Lett.*, 5013
65. Gol'dshleger, N. F., Es'kova, V. V., Shilov, A. E. and Shteinman, A. A. (1972). *Zh. Fiz. Khim.*, **46**, 1353
66. Es'kova, V. V., Shilov, A. E. and Shteinman, A. A. (1972). *Kinet. Katal.*, **13**, 534
67. Sanders, J. R., Webster, D. E. and Wells, P. B. (1974). Unpublished observations
68. Muradov, M. Z., Shilov, A. E. and Shteinman, A. A. (1972). *Kinet. Katal.*, **13**, 1357
69. Gulf R. and D. Co. (1969). *Brit. Pat.* 1 266 678
70. Gulf R. and D. Co. (1970). *U.S. Pat.* 3 644 512
71. Labofina, S. A. (1969). *Brit. Pat.* 1 214 417
72. Hanotier, J., Camerman, P., Hanotier-Bridoux, M. and de Radzitzky, P. (1972). *J. Chem. Soc. Perkin Trans. II*, 2247
73. Onopchenko, A. and Schulz, J. G. D. (1973). *J. Org. Chem.*, **38**, 909
74. Schmerling, L. and Vesely, J. A. (1973). *J. Org. Chem.*, **38**, 312
75. Souma, Y. and Sano, H. (1973). *J. Org. Chem.*, **38**, 3633
76. Hudrlik, P. F. and Peterson, D. (1972). *Tetrahedron Lett.*, 1785
77. Kuwajima, I., Sato, S. and Kurata, Y. (1972). *Tetrahedron Lett.*, 737
78. Johnson, C. R., Shanklin, J. R. and Kirchoff, R. A. (1973). *J. Amer. Chem. Soc.*, **95**, 6462
79. Shevlin, P. B. and Greene, J. L. (1972). *J. Amer. Chem. Soc.*, **94**, 8447
80. Abruscato, G. J. and Tidwell, T. T. (1972). *J. Org. Chem.*, **37**, 4151
81. Sharpless, K. B. and Flood, T. C. (1972). *J. Chem. Soc. Chem. Commun.*, 370
82. Sharpless, K. B., Umbreit, M. A., Nieh, M. T. and Flood, T. C. (1972). *J. Amer. Chem. Soc.*, **94**, 6538
83. Tyrlik, S. and Wolochowicz, I. (1973). *Bull. Soc. Chim. Fr.*, 2147
84. Semmelheck, M. F. and Stauffer, R. D. (1973). *Tetrahedron Lett.*, 2667
85. Stefari, A. (1973). *Helv. Chem. Acta*, **56**, 1192
86. Felkin, H. and Swierczewski, G. (1972). *Tetrahedron Lett.*, 1443
87. Hudson, B., Webster, D. E. and Wells, P. B. (1972). *J. Chem. Soc. Dalton Trans.*, 1204
88. Ewing, D. F., Hudson, B., Webster, D. E. and Wells, P. B. (1972). *J. Chem. Soc. Dalton Trans.*, 1287
89. Bingham, D., Webster, D. E. and Wells, P. B. (1972). *J. Chem. Soc. Dalton Trans.*, 1928
90. Casey, C. P. and Cyr, C. R. (1973). *J. Amer. Chem. Soc.*, **95**, 2248
91. Strohmeier, W., Fleischmann, R. and Rehder-Stirnweiss, W. (1973). *J. Organometal. Chem.*, **47**, C37
92. Strohmeier, W. (1973). *J. Organometal. Chem.*, **60**, C60
93. Strohmeier, W. and Fleischmann, R. (1972). *J. Organometal. Chem.*, **42**, 163
94. Freidlin, L. Kh., Kopyttsev, Yu. A., Nazarova, N. M. and Varava, T. I. (1972). *Izv. Akad. Nauk SSSR*, **21**, 1420

95. Kanai, H. (1972). *J. Chem. Soc. Chem. Commun.*, 203
96. Tolman, C. A. (1972). *J. Amer. Chem. Soc.*, **94**, 2994
97. Corain, B. and Puosi, G. (1972). *J. Catal.*, **30**, 403
98. Freidlin, L. Kh., Nazarova, N. M. and Kopytsev, Yu. A. (1972). *Izv. Akad. Nauk SSSR*, **21**, 201
99. Eaborn, C., Farrell, N. and Pidcock, A. (1973). *J. Chem. Soc. Chem. Commun.*, 766
100. Bingham, D., Webster, D. E. and Wells, P. B. (1974). *J. Chem. Soc. Dalton Trans.*, 1519
101. Tolman, C. A. (1972). *Chem. Soc. Rev.*, **1**, 337
102. Eaborn, C., Farrell, N., Murphy, J. L. and Pidcock, A. (1973). *J. Organometal. Chem.*, **55**, C68
103. Legge, G., Sanders, J. R. and Webster, D. E. (1974). Unpublished observations
104. Masters, C. (1972). *J. Chem. Soc. Chem. Commun.*, 1258
105. Hughes, W. B. (1972). *Organometal. Chem. Synth.*, **1**, 341
106. Calderon, N. (1972). *Acc. Chem. Res.*, **5**, 127
107. Takagi, T., Hamaguchi, T., Fukuzumi, K. and Aoyama, M. (1972). *J. Chem. Soc. Chem. Commun.*, 838
108. Chatt, J., Haines, R. J. and Leigh, G. J. (1972). *J. Chem. Soc. Chem. Commun.*, 1202
109. Matlin, S. A. and Sammes, P. G. (1973). *J. Chem. Soc. Chem. Commun.*, 174
110. Bartlett, P. D., Hummer, K., Elliott, S. P. and Minns, R. A. (1972). *J. Amer. Chem. Soc.*, **94**, 2898
111. Bartlett, P. D., Cohen, G. M., Elliott, S. P., Hummel, K., Minns, R. A., Sharts, C. M. and Fuhunoga, J. Y. (1972). *J. Amer. Chem. Soc.*, **94**, 2899
112. Kramer, B. D. and Bartlett, P. D. (1972). *J. Amer. Chem. Soc.*, **94**, 3934
113. Van Tamlen, E. E., Cretney, W., Klaentschi, N. and Miller, J. S. (1972). *J. Chem. Soc. Chem. Commun.*, 481
114. Weil, T. A., Metlin, S. and Wender, I. (1973). *J. Organometal. Chem.*, **49**, 227
115. Litvin, E. F., Freidlin, A. Kh. and Karimov, K. K. (1972). *Izv. Akad. Nauk SSSR*, **21**, 1853
116. Fahey, D. R. (1973). *J. Org. Chem.*, **38**, 3343
117. Hinze, A. G. (1973). *Rec. Trav. Chim.*, **92**, 542
118. Iwato, R. and Ogata, I. (1973). *Tetrahedron*, **29**, 2753
119. Stern, E. W. and Maples, P. K. (1972). *J. Catal.*, **27**, 120
120. Stern, E. W. and Maples, P. K. (1972). *J. Catal.*, **27**, 134
121. Komiya, S., Yamamoto, A. and Ikeda, S. (1972). *J. Organometal. Chem.*, **42**, C65
122. Kolomnikov, I. S., Kukolev, V. P., Chernyshev, V. O. and Vol'pin, M. E. (1972). *Izv. Akad. Nauk SSSR*, **21**, 693
123. Kolomnikov, I. S., Koreshkov, Yu. D, Kukolev, V. P., Mosin, N. A. and Vol'pin, M. E. (1973). *Izv. Akad. Nauk SSSR*, **22**, 175
124. Pittman, C. U. and Evans, G. O. (1973). *Chemtech*, 560
125. Michalska, Z. M. and Webster, D. E. (1974). *Platinum Metals Rev.*, **18**, 65
126. Collman, J. P., Hegedus, L. S., Cooke, M. P., Norton, J. R., Dolcetti, G. and Marquardt, D. N. (1972). *J. Amer. Chem. Soc.*, **94**, 1789
127. Grubbs, R. H., Gibbons, C., Kroll, L. C., Bonds, W. D. and Brubaker, C. H. (1973). *J. Amer. Chem. Soc.*, **95**, 2373
128. Grubbs, R. H., Kroll, L. C. and Sweet, E. M. (1973). *J. Macromol. Sci.*, **A7**, 1047
129. Allum, K. G., Hancock, R. D., McKenzie, S. and Pitkethley, R. C. (1972). *Proc. 5th Internat. Cong. Catalysis*, Palm Beach
130. Bercaw, J. E., Marwick, R. H., Balland, L. G. and Brintzinger, H. H. (1972). *J. Amer. Chem. Soc.*, **94**, 1219
131. Jones, J. R., Chambers, M. R. and Lowther, E. D. (1973). *J. Chem. Soc. Perkin Trans. I*, 248
132. Cannell, L. G. (1972). *J. Amer. Chem. Soc.*, **94**, 6867
133. Alderson, T., Jenner, E. L. and Lindsey, R. V. (1965). *J. Amer. Chem. Soc.*, **87**, 5638
134. Verbanc, J. J. *U.S. Pat.* 3 152 195
135. Cramer, R. (1967). *J. Amer. Chem. Soc.*, **89**, 1633
136. Su, A. C. L. and Collette, J. W. (1972). *J. Organometal. Chem.*, **46**, 369
137. Henrici-Olive, G. and Olive, S. (1972). *J. Organometal. Chem.*, **35**, 381
138. Kagawa, T. and Hashimoto, H. (1972). *Bull. Chem. Soc. Jap.*, **45**, 2586
139. Maruya, K., Mizoroki, T. and Ozaki, A. (1972). *Bull. Chem. Soc. Jap.*, **45**, 2255

140. Eberhardt, G. G. and Myers, H. K. (1972). *J. Catal.*, **26**, 459
141. Ito, T., Kawai, T. and Takami, Y. (1972). *Tetrahedron Lett.*, 4775
142. Negishi, E. and Brown, H. C. (1972). *Synthesis*, 196
143. Negishi, E., Katz, J. J. and Brown, H. C. (1972). *Synthesis*, 555
144. Brown, H. C. and Ravindran, N. (1972). *J. Amer. Chem. Soc.*, **94**, 2112
145. Brown, H. C., Negishi, E. and Katz, J. J. (1972). *J. Amer. Chem. Soc.*, **94**, 5893
146. Pasto, D. J., Lepeska, B. and Cheng, T. C. (1972). *J. Amer. Chem. Soc.*, **94**, 6083
147. Lane, C. F. and Brown, H. C. (1972). *J. Organometal. Chem.*, **34**, C29
148. Mikhailov, B. M., Bubnov, Y. N., Nesmeyanova, O. A., Kiselev, V. G., Rudashev-skaga, T. Y. and Kazansky, B. A. (1972). *Tetrahedron Lett.*, 4627
149. Zweifel, G. and Horng, A. (1973). *Synthesis*, 672
150. Brown, H. C. and Ravindran, N. (1973). *J. Org. Chem.*, **38**, 182
151. Scouten, C. G. and Brown, H. C. (1973). *J. Org. Chem.*, **38**, 4092
152. Alul, H. R. and McEwan, G. J. (1972). *J. Org. Chem.*, **37**, 3323
153. Alul, H. R. and McEwan, G. J. (1972). *J. Org. Chem.*, **37**, 4157
154. Groves, J. K. (1972). *Chem. Soc. Rev.*, **1**, 73
155. Smit, V. A., Semenovskii, A. V., Lyubinskaya, O. V. and Kucherov, V. F. (1972). *Dokl. Akad. Nauk SSSR*, **203**, 604
156. Moritani, I. and Fujiwara, Y. (1973). *Synthesis*, 524
157. Kikukawa, K., Yamane, T., Takagi, M. and Matsuda, T. (1972). *J. Chem. Soc. Chem. Commun.*, 695
158. Kasahara, A., Izumi, T., Saito, G., Yodono, M., Saito, R. and Goto, Y. (1972). *Bull. Chem. Soc. Jap.*, **45**, 895
159. Saus, A. and Dederichs, B. (1972). *Tetrahedron Lett.*, 1291
160. Micev, I., Christova, N., Panajotova, B. and Jortscheff, A. (1973). *Chem. Ber.*, **106**, 606
161. Forshaw, T. P. and Tipping, A. E. (1972). *J. Chem. Soc. Perkin Trans. I*, 1059
162. Giordano, C. (1972). *Synthesis*, 34
163. Abis, L. and Giordano, C. (1973). *J. Chem. Soc. Perkin Trans. I*, 771
164. Lehmkuhl, H. and Reinehr, D. (1973). *J. Organometal. Chem.*, **55**, 215
165. Altukov, K. V., Ratsino, E. V. and Perekalin, V. V. (1973). *Zh. Org. Khim.*, **9**, 269
166. Abramovitch, R. A. and Challand, S. R. (1972). *J. Chem. Soc. Chem. Commun.*, 1160
167. Hermes, M. E. and Marsh, F. D. (1972). *J. Org. Chem.*, **37**, 2969
168. Taylor, B. W. and Swift, H. E. (1972). *J. Catal.*, **26**, 254
169. Inoue, T., Harada, H. and Miyagima, S. (1972). *Bull. Chem. Soc. Jap.*, **45**, 1915
170. Mori, Y. and Tsuji, J. (1972). *Tetrahedron*, **28**, 29
171. Isaacs, N. S. and Stanbury, P. (1973). *J. Chem. Soc. Perkin Trans. II*, 166
172. Hall, H. K., Smith, C. D. and Plorde, D. E. (1973). *J. Org. Chem.*, **38**, 2084
173. Baron, W. J., Hendrick, M. E. and Jones, M. (1973). *J. Amer. Chem. Soc.*, **95**, 6286
174. Goh, S. H. (1972). *J. Chem. Soc. Chem. Commun.*, 512
175. Klein, J. and Medlik, A. (1973). *J. Chem. Soc. Chem. Commun.*, 275
176. Suhara, Y. (1973). *Bull. Chem. Soc. Jap.*, **46**, 990
177. Serebryakov, E. P., Kostachka, L. M. and Kucherov, V. F. (1973). *Zh. Org. Khim.*, **9**, 1606
178. Kropp, P. J., Reardon, E. J., Gaibel, Z. L. F., Willard, K. F. and Haltaway, J. H. (1973). *J. Amer. Chem. Soc.*, **95**, 7058
179. Niksihin, G. I., Vinogradov, M. G. and Fedorova, T. M. (1973). *J. Chem. Soc. Chem. Commun.*, 693
180. Matsumoto, H., Nakano, T. and Nagai, Y. (1973). *Tetrahedron Lett.*, 5147
181. Moore, L. O. (1972). *J. Org. Chem.*, **37**, 2633
182. Nikishin, G. I., Vinogradov, M. G., Verenchikov, S. P., Kostyukov, I. N. and Kereselidze, R. V. (1972). *Zh. Org. Khim.*, **8**, 539
183. Nikishin, G. I., Vinogradov, M. G. and Il'ina, G. P. (1972). *Zh. Org. Khim.*, **8**, 1401
184. Vinogradov, M. G., Verenchikov, S. P. and Nikishin, G. I. (1972). *Zh. Org. Khim.*, **8**, 2467
185. Chiusoli, G. P. and Cometti, G. (1972). *J. Chem. Soc. Chem. Commun.*, 1051
186. Chiusoli, G. P. (1973). *Acc. Chem. Res.*, **6**, 422
187. Heck, R. F. (1972). *J. Amer. Chem. Soc.*, **94**, 2712
188. Fenton, D. M. and Steinwand, P. J. (1972). *J. Org. Chem.*, **37**, 2034
189. von Kutepow, N., Bittler, K. and Neubauer, D. (1969). *U.S. Pat.* 3 437 676

190. Fenton, D. M. (1973). *J. Org. Chem.*, **38**, 3192
191. Souma, Y., Sano, H. and Iyoda, J. (1973). *J. Org. Chem.*, **38**, 2016
192. Casey, C. P. and Cyr, C. R. (1973). *J. Amer. Chem. Soc.*, **95**, 2240
193. Chini, P., Martinengo, S. and Garlaschelli, G. (1972). *J. Chem. Soc. Chem. Commun.*, 709
194. Consiglio, G., Botteghi, C., Salomon, Ch. and Pino, P. (1973). *Angew. Chem. Int. Ed. Engl.*, **12**, 669
195. Sheldon, R. A. (1973). *Rec. Trav. Chim.*, **92**, 253
196. Sheldon, R. A. (1973). *Rec. Trav. Chim.*, **92**, 367
197. Fenton, D. M. and Oliver, K. L. (1972). *Chemtech*, 220
198. Lyons, J. E. and Turner, J. O. (1972). *J. Org. Chem.*, **37**, 2881
199. James, B. R. and Kastner, M. (1972). *Can. J. Chem.*, **50**, 1698
200. James, B. R. and Kastner, M. (1972). *Can. J. Chem.*, **50**, 1708
201. Brown, R. G. and Davidson, J. M. (1972). *J. Chem. Soc. Chem. Commun.*, 642
202. Henry, P. M. (1973). *J. Org. Chem.*, **38**, 1681
203. Hunt, D. F. and Rodeheaver, G. T. (1972). *Tetrahedron Lett.*, 3595
204. Hall, L. A., Hisatsune, I. C. and Heicklen, J. (1972). *J. Amer. Chem. Soc.*, **94**, 4856
205. Gillies, C. W. and Kuczkowski, R. L. (1972). *J. Amer. Chem. Soc.*, **94**, 7609
206. Gilles, C. W. and Kuczkowski, R. L. (1972). *J. Amer. Chem. Soc.*, **94**, 6337
207. Criegee, R. and Werner, G. (1949). *Ann. Chem.*, **546**, 9
208. Murray, R. W., Youssefyeh, R. D. and Story, P. R. (1967). *J. Amer. Chem. Soc.*, **89**, 2429
209. Olah, G. A., Nishimura, J. and Mo, Y. K. (1973). *Synthesis*, 661
210. Sinnreich, J. and Asscher, M. (1972). *J. Chem. Soc. Perkin Trans. I*, 1543
211. Asscher, M. and Vorfsi, D. (1964). *J. Chem. Soc.*, 4962
212. Sharpless, K. B. and Lauer, R. F. (1972). *J. Amer. Chem. Soc.*, **94**, 7154
213. Olah, G. A., Nojima, M. and Kerekes, I. (1973). *Synthesis*, 779
214. Olah, G. A., Nojima, M. and Kerekes, I. (1973). *Synthesis*, 780
215. Olah, G. A. and Nojima, M. (1973). *Synthesis*, 785
216. Bodrikov, I. V. and Danova, B. V. (1972). *Zh. Org. Chim.*, **8**, 2462
217. Shabanov, A. L., Movsumzade, M. M., Khodzhaev, G. Kh. and Gurbanov, P. A. (1972). *Zh. Org. Chim.*, **8**, 2285
218. Lown, J. W. and Joshua, A. V. (1973). *J. Chem. Soc. Perkin Trans. I*, 2680
219. Uemura, S., Sasaki, O. and Okano, M. (1972). *Bull. Chem. Soc. Jap.*, **45**, 1482
220. Shono, T. and Kosaka, T. (1968). *Tetrahedron Lett.*, 6207
221. Shano, T., Ikeda, A. and Kimura, Y. (1971). *Tetrahedron Lett.*, 3599
222. Shono, T. and Ikeda, A. (1972). *J. Amer. Chem. Soc.*, **94**, 7892
223. Gambino, S. and Silvestri, G. (1973). *Tetrahedron Lett.*, 3025
224. Hartley, F. R. (1972). *Angew. Chem. Int. Ed. Engl.*, **11**, 596
225. Fischer, K., Jones, K. and Wilke, G. (1973). *Angew. Chem. Int. Edn.*, **12**, 565
226. Kalli, M., Landor, P. D. and Landor, S. R. (1972). *J. Chem. Soc. Chem. Commun.*, 593
227. Newman, M. S. and Liang, W. C. (1973). *J. Org. Chem.*, **38**, 2435
228. Rossi, R. and Diversi, P. (1973). *Synthesis*, 25
229. Dai, S. H. and Dolbier, W. R. (1972). *J. Org. Chem.*, **37**, 950
230. Brook, P. R., Harrison, J. M. and Hunt, K. (1973). *J. Chem. Soc. Chem Commun.*, 733
231. Duncan, W. G., Weyler, W. and Moore, H. W. (1973). *Tetrahedron Lett.*, 4391
232. Bryce-Smith, D., Foulger, B. E. and Gilbert, A. (1972). *J. Chem. Soc. Chem. Commun.*, 664
233. Greibrokk, T. (1973). *Tetrahedron Lett.*, 1663
234. Crandall, J. K. and Conover, W. W. (1973). *J. Chem. Soc. Chem. Commun.*, 340
235. Crandall, J. K., Machleder, W. H. and Sojka, S. A. (1973). *J. Org. Chem.*, **38**, 1149
236. Byrd, L. R. and Caserio, M. C. (1972). *J. Org. Chem.*, **37**, 3881
237. Taylor, D. R., Warburton, M. R. and Wright, D. B. (1972). *J. Chem. Soc. Perkin Trans. I*, 1365
238. Taylor, D. R. and Wright, D. B. (1973). *J. Chem. Soc. Perkin Trans. I*, 445
239. Martin, J. C., Carter, P. L. and Chitwood, J. L. (1971). *J. Org. Chem.*, **36**, 2225
240. Taylor, D. R. and Wright, D. B. (1973). *J. Chem. Soc. Perkin Trans. I*, 956
241. Barry, B. J., Beale, W. J., Carr, M. D., Hei, S. K. and Reid, I. (1973). *J. Chem. Soc. Chem. Commun.*, 177

242. Scholten, J. P. and van der Ploeg, H. J. (1972). *Tetrahedron Lett.*, 1685
243. Otsuka, S., Tani, K. and Yamagata, T. (1973). *J. Chem. Soc. Dalton Trans.*, 2491
244. Otsuka, S., Nakamura, A., Yamagata, T. and Tani, K. (1972). *J. Amer. Chem. Soc.*, **94**, 1037
245. Bhagwat, M. M. and Devaprabhakara, D. (1972). *Tetrahedron Lett.*, 1391
246. Coulson, D. R. (1972). *J. Org. Chem.*, **37**, 1253
247. Baker, R. and Cook, A. H. (1973). *J. Chem. Soc. Chem. Commun.*, 472
248. Otsuka, S., Nakamura, A., Ueda, S. and Tani, K. (1971). *Chem. Commun.*, 863
249. Englert, M., Jolly, P. W. and Wilke, G. (1972). *Angew. Chem. Int. Ed. Engl.*, **11**, 136
250. Barnett, B. L., Kruger, C. and Hung Tsay, Y. (1972). *Angew. Chem. Int. Ed. Engl.*, **11**, 137
251. Coulson, D. R. (1973). *J. Org. Chem.*, **38**, 1483
252. Baker, R., Blackett, B. N. and Cookson, R. C. (1972). *J. Chem. Soc. Chem. Commun.*, 802
253. Eidus, Ya. T., Lapidus, A. L., Puzitskii, K. V. and Nefedov, B. K. (1973). *Usp. Khim.*, 442
254. Cerceau, C., Laroche, M. and Blouri, B. (1972). *Bull. Soc. Chim. Fr.*, **11**, 737
255. Lipnick, R. L. and Garbisch, E. W. (1973). *J. Amer. Chem. Soc.*, **95**, 6370
256. Davis, A., Morgan, M. H., Richards, D. H. and Scilly, N. F. (1972). *J. Chem. Soc. Perkin Trans. I*, 286
257. Yang, N. C. and Libman, J. (1973). *Tetrahedron Lett.*, 1409
258. Yang, N. C., Libman, J. and Savitzky, M. F. (1972). *J. Amer. Chem. Soc.*, **94**, 9226
259. Yang, N. C. and Libman, J. (1972). *J. Amer. Chem. Soc.*, **94**, 1405
260. Yang, N. C., Libman, J., Barrett, L., Hui, M. H. and Loeschen, R. L. (1972). *J. Amer. Chem. Soc.*, **94**, 1406
261. Schlott, R. J., Falk, J. C. and Narducy, K. W. (1972). *J. Org. Chem.*, **37**, 4243
262. Marchand-Brynaert, J. and Ghosez, L. (1972). *J. Amer. Chem. Soc.*, **94**, 2870
263. Andreeva, L. M., Altukhov, K. V. and Perekalin, V. V. (1972). *Zh. Org. Khim.*, **8**, 1419
264. Oikawa, E. and Tsubaki, S. (1973). *Bull. Chem. Soc. Jap.*, **46**, 1819
265. Kondo, K. and Matsumoto, M. (1972). *J. Chem. Soc. Chem. Commun.*, 1332
266. Zakharkin, L. I. and Zhigareva, G. G. (1973). *Zh. Org. Khim.*, **9**, 891
267. Bradshaw, J. S., Stanovik, B. and Tišler, M. (1973). *Tetrahedron Lett.*, 2199
268. Heasley, V. L., Heasley, G. E., Loghry, R. A. and McConnell, M. R. (1972). *J. Org. Chem.*, **37**, 2228
269. Heasley, G. E., Heasley, V. L., Manatt, S. L., Day, H. A., Hodges, R. V., Kroon, P. A., Redfield, D. A., Rold, T. L. and Williamson, D. E. (1973). *J. Org. Chem.*, **38**, 4109
270. Aniskov, L. V., Dubinin, A. G., Mirkind, L. A. and Fioshin, M. Ya. (1973). *Zh. Org. Khim.*, **9**, 425
271. Tang, Y. N., Gennaro, G. P. and Su, Y. Y. (1972). *J. Amer. Chem. Soc.*, **94**, 4355
272. Kiji, J., Yamamoto, K., Mitani, S., Yoshikawa, S. and Furukawa, J. (1973). *Bull. Chem. Soc. Jap.*, **46**, 1791
273. Barnett, B., Büssemeier, B., Heimbach, P., Jolley, P. W., Krüger, C., Tkatchenko, I. and Wilke, G. (1972). *Tetrahedron Lett.*, 1457
274. Doering, W. von E., Franck-Neumann, M., Hasselmann, D. and Kaye, R. L. (1972). *J. Amer. Chem. Soc.*, **94**, 3833
275. Woodward, R. B. and Katz, T. (1959). *Tetrahedron*, **5**, 70
276. Chekrii, P. S., Khidekel', M. L., Kalechits, I. V., Eremenko, O. N., Karyakine, G. I. and Todozhakova, A. S. (1972). *Isv. Akad. Nauk SSSR*, **21**, 1579
277. Fahey, D. R. (1972). *J. Org. Chem.*, **37**, 4471
278. Ito, T., Takahashi, K. and Takami, Y. (1973). *Tetrahedron Lett.*, 5049
279. Yoshikawa, S., Nishimura, S., Kiji, J. and Furukawa, J. (1973). *Tetrahedron Lett.*, 3071
280. Akhmedov, V. M., Mardanov, M. A. and Zakharkin, L. I. (1972). *Zh. Org. Khim.*, **8**, 1779
281. Litvin, E. F. and Topuridze, L. F. (1972). *Zh. Org. Khim.*, **8**, 669
282. Funabiki, T., Matsumoto, M. and Tarama, K. (1972). *Bull. Chem. Soc. Jap.*, **45**, 2723
283. Baker, R., Halliday, D. E. and Smith, T. N. (1972). *J. Organometal. Chem.*, **35**, C61
284. Takahashi, K., Miyake, A. and Hata, G. (1972). *Bull. Chem. Soc. Jap.*, **45**, 1183

285. Swisher, J. V. and Zullig, C. (1973). *J. Org. Chem.*, **38**, 3353
286. Okinoshima, H., Yamamoto, K. and Kumada, M. (1972). *J. Amer. Chem. Soc.*, **94**, 9263
287. Baker, R. and Halliday, D. E. (1972). *Tetrahedron Lett.*, 2773
288. Kiji, J., Yamamoto, K., Sasakawa, E. and Furukawa, J. (1973). *J. Chem. Soc. Chem. Commun.*, 770
289. Baker, R., Cook, A. H. and Smith, T. N. (1973). *Tetrahedron Lett.*, 503
290. Rose, D. (1972). *Tetrahedron Lett.*, 4197
291. Tsuji, J. (1973). *Acc. Chem. Res.*, **6**, 8
292. Ohno, K., Mitsuyasu, T. and Tsuji, J. (1972). *Tetrahedron*, **28**, 3705
293. Gardner, S. and Wright, D. (1972). *Tetrahedron Lett.*, 163
294. Rose, D. and Lepper, H. (1973). *J. Organometal. Chem.*, **49**, 473
295. Singer, H., Umbach, W. and Dohr, M. (1972). *Synthesis*, 42
296. Miyajima, S. and Inukai, T. (1972). *Bull. Chem. Soc. Jap.*, **45**, 1553
297. Weigert, F. J. and Drinkard, W. C. (1973). *J. Org. Chem.*, **38**, 335
298. Tsuji, J., Mari, Y. and Hara, M. (1972). *Tetrahedron*, **28**, 3721
299. Nechiporenko, V. P., Treboganov, A. D., Chernova, V. P., Myagkova, G. I. and Preobrazhenskii, N. A. (1973). *Zh. Org. Khim.*, **9**, 238
300. Nechiporenko, V. P., Glazkov, A. A., Myagkova, G. I. and Evstigneeva, R. P. (1973). *Zh. Org. Khim.*, **9**, 243
301. Chukovskaya, E. Ts. and Freidlina, R. Kh. (1972). *Isv. Akad. Nauk SSSR*, **21**, 468
302. Mazzocchi, P. H. and Tamburin, H. J. (1973). *J. Org. Chem.*, **38**, 221
303. Ganushchak, N. I., Golik, V. D. and Migaichuk, I. V. (1972). *Zh. Org. Khim.*, **8**, 2356
304. Hughes, R. P. and Powell, J. (1972). *J. Amer. Chem. Soc.*, **94**, 7723
305. Green, M. L. H., Knight, J., Mitchard, L. C., Roberts, G. G. and Silverthorn, W. E. (1972). *J. Chem. Soc. Chem. Commun.*, 987
306. Skell, P. S., Havel, J. J., Williams-Smith, D. L. and McGlinchey, M. J. (1972). *J. Chem. Soc. Chem. Commun.*, 1098
307. Kawazura, H. and Ohmori, T. (1972). *Bull. Chem. Soc. Jap.*, **45**, 2213
308. White, D. H., Condit, P. B. and Bergman, R. G. (1972). *J. Amer. Chem. Soc.*, **94**, 1348
309. Brace, N. O. (1973). *J. Org. Chem.*, **38**, 3167
310. Hall, J. B. and Lala, L. K. (1972). *J. Org. Chem.*, **37**, 920
311. Brown, H. C., Negishi, E. and Burke, P. L. (1972). *J. Amer. Chem. Soc.*, **94**, 3561
312. Brown, H. C. and Negishi, E. (1972). *J. Amer. Chem. Soc.*, **94**, 3567
313. Negishi, E., Burke, P. L. and Brown, H. C. (1972). *J. Amer. Chem. Soc.*, **94**, 7431
314. Burke, P. L., Negishi, E. and Brown, H. C. (1973). *J. Amer. Chem. Soc.*, **95**, 3654
315. Negishi, E. and Brown, H. C. (1973). *J. Amer. Chem. Soc.*, **95**, 6757
316. Frankel, E. N. (1972). *J. Org. Chem.*, **37**, 1549
317. Spangler, C. W., Jondahl, T. P. and Spangler, B. (1973). *J. Org. Chem.*, **38**, 2478
318. Cunico, R. F. and Dexheimer, E. M. (1972). *J. Amer. Chem. Soc.*, **94**, 2868
319. Corey, E. J. and Fuchs, P. L. (1972). *Tetrahedron Lett.*, 3769
320. Collier, W. L. and Macomber, R. S. (1973). *J. Org. Chem.*, **38**, 1367
321. Johnson, T. R. and Walton, D. R. M. (1972). *Tetrahedron*, **28**, 4601
322. Johnson, T. R. and Walton, D. R. M. (1972). *Tetrahedron*, **28**, 5221
323. Klein, J. and Becker, J. Y. (1973). *J. Chem. Soc. Perkin Trans. II*, 599
324. Pis'mennaya, G. I., Bal'yan, Kh. V. and Petrov, A. A. (1973). *Zh. Org. Khim.*, **9**, 845
325. Brown, H. C. and Gupta, S. K. (1972). *J. Amer. Chem. Soc.*, **94**, 4370
326. Zweifel, G., Fisher, R. P., Snow, J. T. and Whitney, C. C. (1972). *J. Amer. Chem. Soc.*, **94**, 6560
327. Zweifel, G., Fisher, R. P. and Horng, A. (1973). *Synthesis*, 37
328. Brown, H. C. and Ravindran, N. (1973). *J. Organometal. Chem.*, **61**, C5
329. Brown, H. C. and Ravindran, N. (1973). *J. Org. Chem.*, **38**, 1617
330. Sharma, R. K. and Fellers, N. H. (1973). *J. Organometal. Chem.*, **49**, C69
331. Larock, R. C., Gupta, S. K. and Brown, H. C. (1972). *J. Amer. Chem. Soc.*, **94**, 4371
332. Larock, R. C. (1973). *J. Organometal. Chem.*, **61**, 27
333. Liu, C. S., Margrave, J. L., Thompson, J. C. and Timms, P. L. (1972). *Can. J. Chem.*, **50**, 459
334. Liu, C. S., Margrave, J. L. and Thompson, J. C. (1972). *Can. J. Chem.*, **50**, 465
335. Juenge, E. C., Hawkes, S. J. and Snider, T. E. (1973). *J. Organometal. Chem.*, **51**, 189

336. Zaichenko, Yu. A., Maretina, I. A. and Petrov, A. A. (1972). *Zh. Org. Khim.*, **8**, 1328
337. Zaichenko, Yu. A., Maretina, I. A. and Petrov, A. A. (1972). *Zh. Org. Khim.*, **8**, 2605
338. Roitburd, G. V., Smit, V. A., Semenovskii, A. V., Kucherov, V. F. and Chizhov, O.S., *Isv, Akad. Nauk SSSR*, **21**, 2225
339. Roitburd, G. V., Smit, W. A., Semenovsky, A. V., Shchegolev, A. A., Kucherov, V. F., Chizhov, O. and Kadentsev, V. I. (1972). *Tetrahedron Lett.*, 4935
340. Ul'yanov, A. A. and Maretina, I. A. (1973). *Zh. Org. Khim.*, **9**, 228
341. Moriconi, E. J. and Shimakawa, Y. (1972). *J. Org. Chem.*, **37**, 196
342. Anderson, D. J., Gilchrist, T. L., Gymer, G. E. and Rees, C. W. (1973). *J. Chem. Soc. Perkin Trans. I*, 550
343. Bastide, J., Hamelin, J., Texier, F. and Quang, Y. Vo. (1973). *Bull. Soc. Chim. Fr.*, 2555
344. Bastide, J., Hamelin, J., Texier, F. and Quang, Y. Vo. (1973). *Bull. Soc. Chim. Fr.*, 2871
345. Chirko, A. I., Ivanov, K. I., Tishchenko, I. G. and Stepin, S. G. (1972). *Zh. Org. Chim.*, **8**, 687
346. DeMore, W. B. and Lin, C. L. (1973). *J. Org. Chem.*, **38**, 985
347. DeMore, W. B. (1971). *Int. J. Chem. Kinet.*, 161
348. Kostochka, L. M., Serebryakov, E. P. and Kucherov, V. F. (1973). *Zh. Org. Khim.*, **9**, 1611
349. Zanina, A. S., Kotlyarevskii, I. L., Shergina, S. I., Sokolov, I. E. and Shishkina, L. I. (1972). *Isv. Akad. Nauk SSSR*, **21**, 690
350. Volkov, A. N., Sokolyanskaya, L. V., Tsetlina, E. O. and Mushii, R. Ya. (1973). *Zh. Org. Khim.*, **9**, 64
351. Gheorghiu, M. D., Drăghici, C., Stănescu, L. and Avram, M. (1973). *Tetrahedron Lett.*, 9
352. Roitburd, G. V., Smit, V. A., Semenovskii, A. V., Kucherov, V. F., Chizhov, O. S. and Kadentsev, V. I. (1972). *Isv. Akad. Nauk SSSR*, **21**, 2232
353. Roitburd, G. V., Smit, V. A., Semenovskii, A. V., Shchegolev, A. A. and Kucherov, V. F. (1972). *Dokl. Akad. Nauk SSSR*, **203**, 1086
354. Trofimov, B. A. and Amosova, S. V. (1972). *Zh. Org. Khim.*, **8**, 2664
355. Gusarova, N. K., Trofimov, B. A. and Amosova, S. V. (1972). *Isv. Akad. Nauk SSSR*, **21**, 1047
356. Trofimov, B. A., Gusarova, W. K., Atavin, A. S., Amosova, S. V., Gusarov, A. V., Kazantseva, N. I. and Kalabin, G. A. (1973). *Zh. Org. Khim.*, **9**, 8
357. Chalk, A. J. and Jerussi, R. A. (1972). *Tetrahedron Lett.*, 61
358. Fell, B. and Beutler, M. (1972). *Tetrahedron Lett.*, 3455
359. Chukhadzhyan, G. A., Abramyan, Zh. I. and Grigoryan, V. G. (1973). *Zh. Org. Khim.*, **9**, 632
360. Ryang, M., Sawa, Y., Somasundaram, S. N., Murai, S. and Tsutsumi, S. (1972). *J. Org. Chem.*, **46**, 375
361. Zanina, A. S. Khabibibulina, G. N., Legkoderya, V. V. and Kotlyarevskii, I. L. (1972). *Zh. Org. Khim.*, **8**, 1527
362. Dilworth, M. J. (1966). *Biochim. Biophys. Acta*, **127**, 285
363. Fleischer, E. B. and Krishnamurthy, M. (1972). *J. Amer. Chem. Soc.*, **94**, 1382
364. Tsuji, J., Takahashi, H. and Kajimoto, T. (1973). *Tetrahedron Lett.*, 4573

2
Halogeno Compounds

G. M. BROOKE
Durham University

2.1 INTRODUCTION

This chapter is organised in the same way as the corresponding chapter in Series One, except that the very large number of reactions of halogeno compounds are now classified in terms of the new bonds which are formed.

Interest in the chemistry of organic fluoro compounds has reached such a level that, in 1971, the Chemical Society published the first volume of a Specialist Periodical Report, 'Fluorocarbon and Related Chemistry', and the *Journal of Fluorine Chemistry* was born. A new book *Fluorine in Organic Chemistry* is notable for the attention given to reaction mechanisms[1].

2.2 PREPARATION OF ALKYL HALIDES

2.2.1 Electrophilic additions to alkenes

The effectiveness of CF_3OF as a mild, selective electrophilic fluorinating agent has been demonstrated in its reaction with griseofulvin[2]: the 5-fluoro derivative is the major product, accompanied by the 3'-fluoro and the 3',5-difluoro compounds. A particularly attractive alternative to CF_3OF is bis(fluoroxy)difluoromethane, $CF_2(OF)_2$, readily available from CO_2–F_2–CsF, both fluoro-oxy groups of which are used for fluorination[3].

The photochemically induced α-chlorination of long-chain acyl chlorides with $SOCl_2$ ($\xrightarrow{h\nu} Cl_2$) is very efficient compared with the chlorine analogue of the Hell–Volhard–Zelinsky reaction, though small amounts of α,α-dichloro compounds are also formed[4]. Treatment of the enol borinates of

$$\underset{|}{R^1CX}-CO- \xleftarrow{\text{NXS}} \underset{|\quad |}{R^1C}=CO\,BR^2_2 \xleftarrow{\overset{+}{N_2}-\overset{-}{\underset{|}{C}}-CO-} R^2_3B$$

$$\xrightarrow[h\nu]{} {\Large >}C=\underset{|}{C}-CO- \qquad (2.1)$$

$$\underset{|\quad\;\;|}{R^3C}-\underset{}{CX}-CO- \xleftarrow{\text{NXS}} \underset{|\quad\;\;|\;\;\,|}{R^3C}-C=CO\,BR^2_2$$

ketones and esters, available by two different routes, with N-halogeno-succinimides (NXS; X = Cl or Br), affords the corresponding *regiospecific, mono-*α-halogenocarbonyl derivative [equation (2.1)][5].

α-Bromoesters can be prepared in high yield and purity from α-alkylaceto-acetates by utilising the following sequence of reagents[6]: (i) NaH; (ii) Br_2; (iii) $Ba(OH)_2$. Copper(II) catalyses the γ-chlorination of $\alpha\beta$-unsaturated aldehydes and ketones, but the mechanism is not fully understood[7]. Selective α'-bromination of, for example, mesityl oxide to give $Me_2C{=}CHCOCH_2Br$ is now possible by using 2,4,4,6-tetrabromocyclohexa-2,5-dienone[8]. The use of silver trifluoroacetate in the Prévost reaction (I_2–RCO_2^-Ag–alkene), followed by mild hydrolysis of the ester, represents a new route to 1,2-iodo-hydrins[9]. Recently, thallium(I) carboxylates have been shown to be attractive alternatives to the silver salts in the preparation of α-iodocarboxylates[10].

Olah has studied electrophilic addition of HF to fluoroalkenes in super-acid media (SbF_5–HF–SO_2ClF)[11]. The ease of initial protonation decreases with increasing fluorine substitution, and the reaction gives regiospecific products derived from intermediates in which the cationic centre is stabilised by back donation from a lone pair of electrons on a fluorine atom. With $CF_2{=}CFX$ [$\to \overset{+}{C}F_2CHFX \to CF_3CHFX$ (X = Cl, Br or I)], neighbouring group participation by X is involved in stabilising the carbenium ion since tetrafluoroethylene (X = F) and hexafluoropropene (X = CF_3) are inert to the reagents.

The stereoselectivity and regiospecificity of ionic addition of compounds of the type RSCl (R = CN^{12}; p-$ClC_6H_4{}^{13}$; 2,4-$(NO_2)_2C_6H_3{}^{13,14}$) to alkenes has been discussed in terms of bridged episulphonium ions and open carbocations.

Evidence has been presented in support of a six-membered transition state for 1,4-molecular addition of HI to conjugated dienes[15].

2.2.2 Nucleophilic additions to alkenes and to carbonyl compounds

The preparation of polyfluoroalkyl compounds based on reactions of carbanions formed by the addition of fluoride ion (usually as the potassium or the caesium salt) to polyfluoroalkenes continues to be widely explored. Fluoride ion is displaced by $(CF_3)_2\bar{C}F$ from perfluoro-pyridine[16], -quino-line[17] and -tetrahydroquinoline[18], while closely related reactions involving $CF_3\bar{C}F_2{}^{19}$, $CF_3\bar{C}FX^{20}$ (X = Cl, Br or I), $(CF_3)_3C^{-21}$, cyclo-$C_6F_{11}^-$, -$C_5F_9^-$ and -$C_4F_7^-{}^{22}$ have been described. The perfluoro-t-butyl carbanion, $(CF_3)_3\bar{C}$, displaces halogen from a saturated carbon atom in alkyl, allyl and benzyl halides[23] and in α-halogenoacetates[24], and gives Michael addition products, $(CF_3)_3CCH_2CH_2X$, with $CH_2{=}CHX$ (X = CN, CHO or CO_2R)[23]. With trityl chloride, however, an electron-transfer reaction takes place [equation (2.2)][25].

$$(CF_3)_3\bar{C} + Ph_3CCl \to Cl^- + [(CF_3)_3\overset{\bullet}{C}\, Ph_3\overset{\bullet}{C}] \to p\text{-}(CF_3)_3CC_6H_4CHPh_2 \quad (2.2)$$

Other reactions of $R_F{}^-$ involve treatment with benzenediazonium chloride[26] ($\to R_FN{=}NPh$) and with sulphur[27]. Perfluoroalkylsilver compounds (from a perfluoroalkene and AgF, or better, CF_3CO_2Ag and KF or CsF) couple with allyl halides[28], form allyl thioethers with S followed by an allyl halide[29] and give perfluoroalkyl iodides with I_2[29].

The addition of caesium fluoride to hexafluoroacetone followed by reaction of the resulting anion, $(CF_3)_2CFO^-$ with perfluoroacyl fluorides, R_FCOF, at $-108\,^\circ C$ gives perfluoroesters, $R_FCOOCF(CF_3)_2$, obtained pure for the first time[30] when isolated from the reaction vessel by sublimation at $<-78\,^\circ C$. They are stable at $25\,^\circ C$ for long periods when prepared in this way. The combination of external fluoride ion and the α-fluorine of the alcohol moiety is responsible for the instability of these esters[31].

Another important synthetic procedure has been described which, by using MoF_6, makes aryl trifluoromethyl ethers readily accessible [equation (2.3)][32].

$$ArONa \xrightarrow[CSCl_2]{} ArOCSCl \xrightarrow{MoF_6} ArOCF_3 \qquad (2.3)$$

2.2.3 Radical additions to alkenes

Quantitative studies of the addition of fluorine atoms to alkenes have been limited by the experimental difficulties associated with the handling of macroscopic quantities of F_2, HF, etc. The production of tracer levels of thermal or near-thermal ^{18}F atoms by the $^{19}F(n,2n)^{18}F$ nuclear reaction in gaseous SF_6 has largely overcome these problems, the identification of radical adducts after abstraction of H· from HI being made by radio–gas chromatography[33]. The reaction of ^{18}F with C_2H_4 is faster than with $CHF{=}CF_2$ by a factor of 3–4, and in the latter compound the less fluorinated carbon atom is the preferred site of attack. The measured ratio of terminal to central carbon atom attack of 1·35 for propene characterises atomic fluorine as an indiscriminate reactive species (*cf.* a ratio of 15 for H addition)[34].

The redox-catalysed addition of organic polyhalides to alkenes (Scheme 2.1) can give two different 1,2-addition products, depending on the conditions (path A[35], $CuCl_2$, benzoin, $Et_2\overset{+}{N}H_2Cl^-$ in MeCN; path B[36], $CuCl_2$, LiCl in MeCN).

$$CCl_2(CH_2CHClCO_2Me)_2 \xleftarrow{\ A\ } CCl_4 + 2CH_2{=}CHCO_2Me$$

$$\text{Two-fold addition} \qquad\qquad \downarrow B$$

$$CCl_3CH_2CH(CO_2Me)CH_2CHClCO_2Me$$

$$2:1 \text{ Telomer}$$

Scheme 2.1

The addition of perfluoroalkyl iodides R_FI to alkenes in photochemical[37] and thermal reactions[37,38] has been studied in detail and a copper-catalysed reaction has been reported[39]. The photochemical decomposition of *N*-halogenoacetamides in the presence of alkenes gives both 1,2-adducts (favoured at $-70\,^\circ C$) and allylic halogenation[40], and a number of additions of polyfluoro compounds containing the $>$N—X group (X = F, Cl or Br) to alkenes have been described[41−43]. CF_3SCl reacts with methyl acrylate when irradiated to give $ClCH_2CH(SCF_3)CO_2Me$ as the major product[44].

2.2.4 Nucleophilic substitution at carbon

Various new methods for the overall conversion of ROH into RHal have been published. Alkyl fluorides are formed with varying degrees of success in the reactions of $PhPF_4$ with trimethylsilyl ethers, and a measure of selectivity has been noted with compounds containing other functional groups (e.g. $C=C$; $C—Cl$)[45]. Chlorodimethylsulphonium chloride selectively converts allylic —OH into —Cl even in the presence of primary or secondary alcoholic groups[46], e.g. equation (2.4).

$$HO(CH_2)_2CMe=CHCH_2OH \xrightarrow{\overset{+}{Me_2}\overset{-}{SCl}\,Cl} HO(CH_2)_2CMe—CHCH_2Cl \quad (2.4)$$

In contrast, the corresponding bromo compound, $Me_2\overset{+}{S}Br\,Br^-$, gives bromo compounds from non-allylic alcohols, mainly through an inversion mechanism[47]. Allylic alcohols can also be converted into chlorides without rearrangement by using $Ph_3P–CCl_4$[48], and the same reagent gives 1,2-dichloroalkanes with oxiranes[49]. Other systems for reaction with alcohols include $(PhO)_3P$ or Ph_3P with N-halogenosuccinimides[50], and Ph_2PCl followed by reaction with HX or X_2 (X = Cl, Br or I)[51].

Magnesium iodide reacts with alkyl tosylates under very mild conditions to give the corresponding iodide (especially good for cyclic alcohols)[52], and methyldicyclohexylcarbodi-imidium iodide will even react with neopentyl alcohol[53].

The conversion of RSH into RCl requires[54] reaction of the thiol with ClSCOCl, and treatment of the resulting product RSSCOCl, with Ph_3P at $-35\,°C$.

2.2.5 Radical substitution at carbon

The controlled direct fluorination process $\geqslant C—H \rightarrow \geqslant C—F$, was originally described by Lagow and Margrave[55]. Recent syntheses involve the conversion of Me_3CCMe_3[56], $MeO(CH_2)_2OMe$ ('glyme')[57] and $[MeO(CH_2)_2]_2O$ ('diglyme')[57] into the corresponding perfluoro compounds.

The reaction of alkyl radicals, produced by the decomposition of diacyl peroxides, with copper(II) halides gives alkyl halides by two simultaneous processes[58]: (a) an atom transfer occurring primarily via a homolytic transition state $[R \cdots X \cdots CuX]^‡$; (b) a competing oxidative substitution proceeding via cationic intermediates, $RCu^{II}X^+\,X^- \rightleftharpoons R^+Cu^I\,X^-_2 \rightarrow RX + Cu^IX$.

The modified Hunsdiecker reaction (degradation of carboxylic acids to alkyl halides by using HgO and halogens in CCl_4) involves the initial formation of $(RCO_2)_2Hg$ and subsequent conversion into RCO_2Br[59]. Moreover, the system $Br_2–HgO$ is also an effective brominating agent for alkenes; in this reaction Br_2O is believed to be a reactive intermediate[60].

2.2.6 Miscellaneous methods

Potassium tetrafluorocobaltate(III) fluorinates and cleaves ketones to give polyfluoroacyl fluorides as the major products [e.g. $Et_2CO \xrightarrow{240\,°C} CH_3CHFCOF$ (18%) + CH_3CF_2COF (45%)][61]. Several 1,1-diphenylalkenes react with XeF_2–HF or XeF_2–CF_3CO_2H to give products from the overall addition of F_2[62].

Interest continues in developing methods of electrophilic halogenation of alkanes in solution under mild conditions. The degree of selectivity of chlorination by chlorine depends on the Friedel–Crafts catalyst used[63]: with SbF_5, substitution of hydrogen and chlorolysis of C—C bonds occur; with $AlCl_3$ there is only substitution of hydrogen; and with $AgSbF_6$ only tertiary C—H groups and strained (and therefore more reactive) systems are substituted. Bromination with Br_2–$AgSbF_6$ can be conducted[64] in dichloromethane at −45 to +25 °C. Sulphuryl chloride in sulpholan is also a mild electrophilic chlorinating agent[65].

2.3 PREPARATION OF VINYLIC HALIDES

2.3.1 Electrophilic additions to alkynes

The polarity of the solvent has a dramatic effect on the relative specific rates of addition of Cl_2 and Br_2 to analogously substituted alkenes and alkynes[66]. Thus, bromination of $PhCH{=}CH_2$ and $PhC{\equiv}CH$ in acetic acid gives $k_{alkene}/k_{alkyne} = 2590$, but in water this factor is reduced to 0.67, i.e. the alkyne is more reactive, contrary to general experience.

Lewis acids catalyse the addition of alkyl halides to alkynes[67]. The reaction proceeds through a linear vinyl cation in which the vacant p-orbital lies in the plane of the molecule. Attack at the positive centre takes place at the least hindered position and is determined by the relative size of the β-groups.

$$Bu^tCl + PhC{\equiv}CH \text{ or } HCl + Bu^tC{\equiv}CPh \longrightarrow \underset{H}{\overset{Bu^t}{\diagdown}}C{=}\overset{+}{C}Ph + Cl^- \qquad (2.5)$$

$$\downarrow$$

$$\underset{H}{\overset{Bu^t}{\diagdown}}C{=}C\overset{Ph}{\underset{Cl}{\diagup}}$$

Both S—Cl bonds in S_2Cl_2 give overall *trans* addition to dimethyl acetylenedicarboxylate to form a disulphide[68].

2.3.2 Nucleophilic additions to alkynes

Miller has found a method for promoting the telomerisation of $CF_3C{\equiv}CCF_3$ by CsF[69]. In the presence of $CF_3CF{=}CBrCF_3$, fluorovinylcaesium intermediates undergo metal–halogen exchange faster than they add to $CF_3C\equiv C\text{-}$ CF_3, and *trans,trans*-$F(CCF_3=CCF_3)_2Br$ has been isolated in 80% yield.

The formation of 1-bromoallenes from prop-1-yn-3-ols by the action of HBr–CuBr proceeds by a stereospecific S_Ni' reaction of the intermediate π-complex[70].

2.3.3 Radical additions to alkynes

Exposure of acetylene to ^{18}F atoms in SF_6 with HI present gives $CH_2=CH^{18}F$ in a controlled reaction[71] (see Section 2.2.3).

2.3.4 Miscellaneous methods

Copper(II) chloride catalyses the addition of CX_3Y (X = Cl or Br; Y = Cl, Br, CN or CO_2Et) to trimethylsilyl ethers, $CH_2=CROSiMe_3$, and the products, $YCX_2CH_2CRXOSiMe_3$, readily eliminate Me_3SiX and HX to form $RCOCH=CXY$[72].

The two geometrical isomers of $RCH=CHX$ (X = Hal) can be synthesised in high stereochemical purity by reactions involving the vinylboronic acid obtained by hydroboration (with catechol–borane), followed by hydrolysis, of the terminal alkyne $RC\equiv CH$ [equation (2.6)][73,74].

$$\underset{H}{\overset{R}{\diagdown}}C=C\underset{I}{\overset{H}{\diagup}} \xleftarrow[\text{Et}_2O,\ 0\ ^\circ C]{\text{NaOH–I}_2} \underset{H}{\overset{R}{\diagdown}}C=C\underset{B(OH)_2}{\overset{H}{\diagup}} \xrightarrow[\text{(ii) MeONa–MeOH}]{\text{(i) Br}_2,\ \text{CH}_2\text{Cl}_2,\ \text{Et}_2\text{O}} \underset{H}{\overset{R}{\diagdown}}C=C\underset{H}{\overset{Br}{\diagup}} \quad (2.6)$$

Halogenoboration of phenylacetylene with BX_3 (X = Cl, Br or I) under kinetic control gives Z- or E-$PhCX=CHBX_2$ and < 10% Z- or E-$PhC(BX_2)=CHX$; E-products are formed in solvents which more readily support a more polar transition state[75].

2.4 PREPARATION OF 1-HALOGENOALKYNES

As well as the addition of ^{18}F atoms to acetylene to give $\overset{\bullet}{C}H=CH^{18}F(\xrightarrow{HI}$ $CH_2=CH^{18}F)$ (Section 2.3.3), Rowland has also shown that 'hot' atoms actually replace hydrogen to give $HC\equiv C^{18}F$ in 2–3% yield[71].

2.5 NEIGHBOURING GROUP INTERACTIONS OF HALOGENS

2.5.1 Interactions with cations, $X(C)_nC^+$

2.5.1.1 *Halogenocarbenium ions (n = 0)*

Olah has measured ^{13}C magnetic resonance shift *differences*, $\Delta\delta^{13}C$, between $CH_2=\overset{+*}{C}XR$ (R = H or Me) and $Me_2\overset{+*}{C}X$ (which eliminates the influence of the inductive effect of the halogen X) and has compared the values with $\Delta\delta^{13}C$ values for $CH_2=\overset{*}{C}RMe$ and $Me_2\overset{+*}{C}R$ (R = H or Me) in order to estimate the degree of back donation of a lone pair of electrons on the halogen atom to the adjacent carbocationic centre[76]. The back donation order is F > Cl $\gg$ Br, and this clearly shows in a very direct manner that, as the size of the halogen is increased, the interaction between the halogen lone-pair and the vacant 2p orbital decreases. This back donation effect is also related to the stability of the halogenocarbenium ion since $Me_2\overset{+}{C}X$ is converted by SbF_5–SO_2 at $-70\,°C$ into $Me_2\overset{+}{C}F$ at different rates: very rapidly for X = Br, but much more slowly for X = Cl. The ion $Me\overset{+}{C}F_2$ has now been prepared[11], as well as the first stable secondary arylfluorocarbenium ion[77] and the first stable fluorinated allyl cation[78], $C_6F_5\overset{+}{C}HF$ and p-$MeOC_6H_4\overset{+}{C}F=CFCF_2$, respectively.

2.5.1.2   1,2-Halogenonium ions (n = 1)

In Series One, dimethylhalogenonium hexafluoroantimonates, $Me_2\overset{+}{X}\,SbF_6{}^-$, were considered as the acyclic counterparts of the three-membered ring 1,2-halogenonium ions. Further non-bridged ions have now been obtained by alkylating a wide variety of halides with RF–SbF_5 (R = Me or Et). In view of the ready preparation and decomposition of arylmethylhalogenonium ions, the possibility that related compounds might be involved in Friedel–Crafts alkylations of halogenobenzenes must now be considered[79]:

$$PhBr + MeF\text{–}SbF_5 \xrightarrow[-78\,°C]{SO_2} Ph\overset{+}{B}rMe\,SbF_6{}^- \xrightarrow{0\,°C} C_6H_3Me_2Br \text{ (mixture)} \qquad (2.7)$$

A number of dipositive dihalogenonium ions have been prepared[80]. The great ability of iodine to stabilise a positive charge is shown in the formation of $MeICH_2IMe$ from CH_2I_2 and the tripositive ion $2,4,6\text{-}(MeI)_3C_6Me_3$ from 2,4,6-tri-iodomesitylene. (The simplest dicarbenium ion requires at least two carbon atoms between the cationic centres.) The simplest dipositive dibrominium ion so far prepared is $Me_2CH\overset{+}{B}rCH_2\overset{+}{B}rMe$, but *no* dipositive chlorinium ion has yet been obtained.

Calculations show that 1,2-bridged ions $[C_2H_4X]^+$ are more stable than the corresponding 2-substituted ethyl cations by 3.58 and 15.81 kcal mol^{-1} for X = F and Cl, respectively[81]. The order of reactivity for the bromination of $CH_2=CXCH_2Cl$, *viz.* X = H $\gg$ F > Cl $\approx$ Br, is interpreted in terms of a bridged brominium ion with a significant contribution to stabilisation by a 1-halogenocarbenium-type interaction[82].

Anchimeric assistance by β-iodine has been found in the solvolysis of *trans*-β-iodovinyl arylsulphonates[83].

2.5.1.3   1,3-Halogenonium ions (n = 2)

The attempted preparation of stable four-membered ring halogenonium ions by the ionisation of various simple 1,3-dihalogeno compounds in strong-acid media gave exclusively either the three- or five-membered ring halogenonium ion, formed by ring contraction or expansion, which indicates the greater thermodynamic stability of these systems[84]. However, one stable substituted 1,3-brominium ion has now been prepared[85]:

$$
\begin{array}{c}
FH_2C \quad CH_2Br \\
C \\
FH_2C \quad CH_2F
\end{array}
\xrightarrow[-55\,^\circ C]{SbF_5-SO_2ClF}
\begin{array}{c}
FH_2C \quad CH_2F \\
C \\
H_2C \overset{+}{\underset{Br}{\quad}} CH_2
\end{array}
\qquad (2.8)
$$

2.5.1.4 *1,4-Halogenonium ions (n = 3)*

Five-membered ring systems are formed not only by rearrangements during the attempted formation of trimethylenehalogenonium ions[84], but also in the attempted formation of *seven*-membered ring systems[86]. The ionisation of 1,6-dihalogenohexanes in strong-acid media gives as the first observable product a 1-ethyltetramethylenehalogenonium ion which can be converted into the thermodynamically more stable 1,1-dimethyltetramethylenehalogenonium ion (for chlorine and bromine). ^{13}C n.m.r. studies of the chlorinium ion have revealed an interesting series of equilibria in which a tertiary carbenium ion is involved [equation (2.9)]. The reactions of some tetramethylenechlorinium ions with relatively weak nucleophiles have also been described[87,88].

$$
Cl(CH_2)_6Cl \xrightarrow[-65\,^\circ C]{SbF_5-SO_2}
Et\overset{+}{\underset{Cl}{\bigtriangleup}}
\xrightleftharpoons[-65\,^\circ C]{-48\,^\circ C}
\underset{Me}{\overset{Me}{}}\overset{+}{\underset{Cl}{\bigtriangleup}}
\longleftrightarrow
Me_2C\overset{+}{\underset{Cl}{\bigtriangleup}}
\qquad (2.9)
$$

2.5.1.5 *1,5-Halogenonium ions (n = 4)*

The ionisation of 1,5-di-iodopentane in SbF_5-SO_2 at low temperatures does yield some pentamethyleneiodinium ion (along with some 1-methyltetramethyleneiodinium ion), but a better method involves the methylation of 1,5-dihalides (Br or I) with $MeF-SbF_5$ in SO_2[89].

2.5.2 Interactions with radicals

The abstraction of hydrogen β to bromine in the photochemical bromination of bromoalkanes (Br_2[90], Br_2-NBS[91] or CCl_3Br[92]) is definitely faster than that from an analogous site in an unsubstituted alkane, and is consistent with bridging by bromine through an anti-periplanar transition state. *cis*-4-t-Butylcyclohexyl bromide has the ideal geometry for anchimeric assistance by bromine for axial 2-H abstraction, and the relative rate constant is 1045 times the value for unassisted abstraction after allowing for the inductive effect of the β-bromine substituent[91]. CIDNP studies show that in $BrCH_2\overset{\cdot}{C}H_2$ the two CH_2 groups are non-equivalent: the radical exists in a preferred conformation which is considered to allow hyperconjugative stabilisation by the halogen[93].

Chlorine and fluorine exert no anchimeric assistance towards hydrogen abstraction[91], but calculations[94], e.s.r. measurements[95] and the chemistry of the products of halogenation of 2-chlorobutanes[96] indicate that chlorine forms unsymmetrical bridges with an adjacent radical centre. A facile 1,2-chlorine migration at $-130\,^{\circ}C$ has been observed ($Me_2CCl\dot{C}H_2 \rightarrow Me_2\dot{C}$-$CH_2Cl$)[95].

No 1,3-iodine transfer occurs in the decomposition of isotopically labelled 4-iodobutanoyl peroxide, so symmetrically bridged four-membered ring species could not possibly be involved as intermediates[97].

2.5.3 Sigmatropic halogen migrations

The base-catalysed conversion of 1,1,1-trichloroalk-2-en-4-ones, $RCH_2COCH=CHCCl_3$, into salts of 5-chlorodienoic acids, $RCCl=CHCH-=CHCO_2H$, involves a 1,5-chlorine shift during the reaction[98]. However, this migration is somewhat different from the thermally allowed suprafacial 1,5-sigmatropic chlorine shift invoked in the isomerisation of $RCH_2COCH=CHCCl_3$ into $RCHClCOCH_2CH=CCl_2$ [via $RCH=C(O^-)CH=CHCCl_3$] (Series One, p. 78).

Cristol has described some photosensitised reactions of a number of allylic halides giving allylic rearrangement products (1,3-sigmatropic reactions) and halogenocyclopropanes (1,2-sigmatropic photocyclisations)[99]:

$$CHCl=CHCH_2Cl \ (cis \text{ and } trans) \xrightarrow{\ h\nu\ } CHCl_2CH=CH_2 + \underset{\triangledown}{\overset{Cl\ Cl}{\bigsqcup}} + \underset{\triangledown_{Cl}}{\overset{Cl}{\bigsqcup}} \quad (2.10)$$

2.6 REACTIONS OF ALKYL HALIDES

2.6.1 Formation of carbon–hydrogen bonds

The controlled-potential electroreduction of substituted benzotrifluorides in methanol gives complete conversion of all C—F into C—H bonds[100].

Attention has been drawn to the use of $LiAlH_4$ to replace halogen by hydrogen in what had previously been considered inert systems (bridgehead and cyclopropyl halides)[101]. $LiAlH_4$ does not give simple products with polyfluoroalkyl iodides[102]; initially a complex of the type $LiAlR_FH_2I$ is formed, which decomposes to fluoro-alkanes and -alkenes. Lithium triethylhydridoborate is the most powerful nucleophile known for effecting an S_N2 replacement; neopentyl bromide is converted by this reagent into neopentane in 96% yield in 3 h in boiling THF.

Halogen abstraction from $PhSO_2CHXAr$ by Ph_3P in aqueous DMF to give $PhSO_2CH_2Ar$ shows the order of reactivity $Br > I$, consistent with the formation of the stronger P—Br bond, which more than compensates for the stronger C—Br bond being broken[104,105].

The single deuterium atom in R^1R^2CDX (X = CO_2R^3, CN or $CONR_2^4$) can be introduced in high isotopic purity in a one-step process by treatment of $R^1R^2C(Hal)X$ with $Zn-D_2O$[106].

Polyhalogenomethyl groups are converted into di- and mono-halogenomethyl groups by $Ni(CO)_4$–THF under mild conditions[107] and by photolysis in THF[108], while cis-$HMn(CO)_4PPh_3$ at room temperature reduces both poly- and mono-halogenoalkanes[109]. Halogen abstraction by $Me_3\overset{\bullet}{S}n$ is involved in the photochemical conversion of alkyl halides into alkanes by Me_3SnH[110].

2.6.2 Formation of carbon–carbon bonds

The heterogeneous reaction of alkyl halides with aqueous NaCN is strongly catalysed by the addition of a small amount of an organic-phase soluble tetra-alkyl-ammonium- or -phosphonium salt ('phase-transfer' catalysts)[111]. 1-Chloro-octane is unchanged after being heated with aqueous NaCN at 105 °C for 3 h, but after the addition of 1.5% of tributylhexadecylphosphonium bromide, 99% conversion into 1-cyano-octane occurs in 1.8 h.

Further important reactions of alkyl halides with copper-containing compounds have been reported. Lithium dimethylcuprate will dimethylate benzal chloride[112] and $\alpha\alpha$-dichloroesters[113], and two new reagents, $CuR(SPh)Li$[114] and $CuR(OBu^t)Li$[115], are particularly useful for the alkylation of primary halides (R = secondary or tertiary alkyl). The t-butoxy compound specifically *mono*-alkylates $\alpha\alpha'$-dibromoketones [equation (2.11)][116].

$$Pr^nCHBrCOCHBrPr^n \xrightarrow{CuBu^t(OBu^t)Li} \left[\begin{array}{c} O \\ \triangle \\ Pr^n \quad Pr^n \end{array} \right] \qquad (2.11)$$

$$\Big\downarrow CuBu^t(OBu^t)Li$$

$$Bu^nCOCHPr^nBu^t$$

Allylic halides couple smoothly with $CuCH_2CN$[117] [from MeCN with (i) Bu^nLi, (ii) CuI] and with $CuCH_2CO_2Et$[118] [from $MeCO_2Et$ with (i) $LiNPr^i_2$, (ii) CuI] without rearrangement, but with $Pr^iSCH=CHCH_2Cu$[119] [from the allyl-lithium and CuI] reaction occurs exclusively by an S_N2' process.

Improved methods for the alkylation of potentially carbanionic species prepared by proton abstraction include reactions of halides (*a*) with enolates from $\alpha\beta$-unsaturated esters, $R^1R^2CHCR^3=CR^4CO_2Et$, and lithium N-isopropylcyclohexylamide[120], and from lactones and lithium di-isopropylamide[121], and (*b*) with organo-allyl compounds, prepared from $>C=\overset{|}{C}-CH<$ and $NaNH_2$–liq. NH_3[122] or a t-amino complex of Bu^nLi[123]. Stork has out-

$$\text{(2.12)}$$

lined a general method for the synthesis of *cis*-annelated bicyclic ketones by using halogenoacetals as starting materials [equation (2.12)][124]. Use of the corresponding *lithium* salt in these reactions leads to a dramatic reversal of the stereochemical course, the *trans*-annelated products being formed[125].

Attention has been drawn to the role of impurities in the magnesium in the reaction of alkyl halides with it and the possibility of their promoting cross-coupling reactions[126]. Efficient reactions take place between allyl halides and Grignard reagents in the presence of transition-metal halides[127] and with alkyl-lithium reagents[128] (prepared *in situ* from the alkyl mesitoate and Li), while $\alpha\alpha'$-disubstituted dimethyl ethers are formed from bis(chloromethyl) ether and $R^1R^2C(ZnBr)CO_2Et$[129]. The phenylation of certain alkyl halides with PhCdCl involves free-radical intermediates[130].

All three alkyl groups of trialkylboranes are transferred to carbon in $CHCl_2OMe$ under mild conditions (15–25 °C in THF) by using the hindered base Et_3COLi. All the alkyl groups may be secondary[131] or one may be tertiary[132]; oxidation of the intermediate $R_3CB(OMe)Cl$ with H_2O_2–HO^- provides a useful route to hindered carbinols. Two syntheses of ketones are based on borane chemistry[133,134]:

$$R_2^1B(OR^2) + CHCl_2OMe \xrightarrow{\ Et_3COLi\ } \xrightarrow{\ H_2O_2-HO^-\ } R_2^1CO \qquad (2.13)$$

$$R_3^1B + LiC{\equiv}CR^2 \rightarrow R_3^1\bar{B}C{\equiv}CR^2 \xrightarrow{\ R^3Br\ } R_2^1BCR^1{-}CR^2R^3 \xrightarrow{\ H_2O_2-HO^-\ } R^1COCHR^2R^3 \qquad (2.14)$$

Nucleophilic addition to a double bond gives a carbanionic species which may displace a halogen by an inter- or an intra-molecular process. Examples of the former reaction are the formation of the dialkyl adducts $PhCXRCRYZ$, and both $RCH_2CH{=}CHCH_2R$ and $RCH_2CHRCH{=}CH_2$ by the reactions of styrene derivatives and buta-1,3-diene respectively with RBr and Li, where R is used both as a nucleophile (RLi) and as an electrophile RBr[135]. Conjugate addition of R_2^1CuLi to activated vinylcyclopropanes followed by R^2Hal also gives dialkylated products[136]:

$$CH_2{=}CH{-}\triangle(CO_2Et)_2 \quad \rightarrow \quad R^1CH_2CH{=}CHCH_2CR^2(CO_2Et)_2$$

$$\triangle CH{=}CH(CO_2Et)_2 \quad \rightarrow \quad \triangle\overset{R^1\ \ R^2}{\diagdown}(CO_2Et)_2$$

An intramolecular alkylation takes place in the addition of NaOMe or KCN to $Me_2CBrCH{=}C(CO_2Me)_2$, which gives cyclopropanes[137]. Closely related cyclopropane ring syntheses are the electrochemical reductions of 2,4-dibromopentanes (which give $Me\bar{C}HCH_2CHBrMe$)[138].

Various useful homologation reactions have been described. Iodomethylations[139] ($RX \rightarrow RCH_2I$) and iodopropenylations[139] ($RX \rightarrow RCH= CHCH_2I$) are carried out as in equation (2.15):

$$RX + Z-SCH_2Li \text{ or } Z-SCHLi-CH=CH_2 \xrightarrow[\text{MeI}]{\text{excess}} RCH_2I \text{ or } \begin{array}{c} R \\ \diagdown \\ H \end{array} C=C \begin{array}{c} H \\ \diagup \\ CH_2I \end{array} \quad (2.15)$$

$$\left(Z = H_2C \underset{S}{\overset{H_2C-N}{\diagdown \diagup}} C- \right)$$

In a new aldehyde synthesis ($RX \rightarrow RCHO$), $EtSCH_2S(O)Et$ is treated with (i) Bu^nLi, (ii) RX and the product $EtSCHRS(O)Et$ is hydrolysed with $HCl-HgCl_2$[140]. However, the anion of the alkylated thioacetal monosulphoxide, $Et\bar{S}CRS(O)Et$ will *also* undergo conjugate addition to $\alpha\beta$-unsaturated carbonyl compounds, and this provides an efficient route to 1,4-dicarbonyl compounds[140a].

The overall reaction $RX \rightarrow RCH_2CHO$ may be accomplished by two different procedures. The first uses 2-methyl-2-thiazoline and gives aldehydes in 50–60% overall yields[141]:

$$\underset{N}{\overset{S}{\diagdown}} Me \xrightarrow[\text{(ii) RX}]{\text{(i) Bu}^n\text{Li}} \underset{N}{\overset{S}{\diagdown}} CH_2R \xrightarrow[\text{(ii) HgCl}_2]{\text{(i) Al–Hg}} RCH_2CHO \quad (2.16)$$

The second method can be performed in a 'one-pot' reaction under very mild conditions and includes the following reactions[142]:

$$Me_2 \overset{Me}{\underset{N}{\diagdown}} Me \xrightarrow[\substack{\text{(ii) NaH–DMF} \\ \text{(iii) RX}}]{\text{(i) MeI}} Me_2 \overset{Me}{\underset{\underset{Me}{N^+}}{\diagdown}} CH_2R \xrightarrow{[H^-]} Me_2 \overset{Me}{\underset{\underset{Me}{N}}{\diagdown}} CHR \quad (2.17)$$

$$\xrightarrow[\text{(i) BH}_4^- \; \text{(ii) H}_2\text{O}]{} RCH_2CHO$$

The use of alkyl halides for electrophilic alkylation is well established in Friedel–Crafts reactions with aromatic compounds. ω-Halogenoalkyl cyanides–$AlCl_3$ alkylate benzene *without* the rearrangements expected from primary halides[143] [e.g. $Cl(CH_2)_3CN \rightarrow Ph(CH_2)_3CN$ (98%)]. The complexes $MeF-SbF_5$ and $EtF-SbF_5$ prepared at low temperature in liquid SO_2 are among the most powerful alkylating agents known, reacting with n-donor bases (e.g. $MeF-SbF_5$ with $CO \rightarrow Me\overset{+}{C}O\ SbF_6^- \xrightarrow{H_2O} MeCO_2H$); with π-bases [e.g. $MeF-SbF_5 + CH_2=CMe_2 \rightarrow Et\overset{+}{C}Me_2$]; and with σ-bases in two ways [equation (2.18)].

The cleavage of carbon–halogen bonds electrochemically at a mercury cathode can take place either by a two-electron transfer or by a one-electron

$$-\overset{\vert}{\underset{\vert}{C}}-H \;+\; MeF\text{-}SbF_5$$

$$\downarrow$$

$$\left[\; -\overset{\vert}{\underset{\vert}{C}}\overset{H}{\underset{Me}{\cdots\!\!\diagdown}} \;\right]^{+} \longrightarrow -\overset{\vert}{\underset{\vert}{C}}-Me \;+\; H^{+} \;;\; -\overset{\vert}{\underset{\vert}{C}}-\overset{\vert}{\underset{\vert}{C}}- \;+\; MeF\text{-}SbF_5$$

$$\downarrow$$

$$-\overset{\vert}{\underset{\vert}{C}}-Me \;+\; \overset{+}{\underset{\diagup\diagdown}{C}} \;\longleftarrow\; \left[\; -\overset{\vert}{\underset{\vert}{C}}\cdots\overset{\diagup}{\underset{Me}{\diagdown C \diagdown}} \;\right]^{+} \qquad (2.18)$$

transfer, depending on the applied potential. In the former, carbanions are formed which have been added to activated alkenes and to carbon dioxide[145]:

$$CCl_4 \xrightarrow{2e^-} Cl^- + CCl_3^- \xrightarrow[\text{(ii) } H_2O]{\text{(i) } CH_2=CHX \;(X=CN \text{ or } CO_2Et)} CCl_3CH_2CH_2X \quad (2.19)$$

$$CH_2=CHCH_2Cl \xrightarrow{2e^-} Cl^- + CH_2=CH\bar{C}H_2 \xrightarrow[\text{(ii) } CH_2=CHCH_2Cl]{\text{(i) } CO_2}$$

$$CH_2=CHCH_2CO_2CH_2CH=CH_2 \qquad (2.20)$$

The one-electron reduction of 6-bromo-1-phenylhex-1-yne at -2.45 V gave among the products both five- and six-membered carbocycles via a radical intermediate[146]:

$$PhC{\equiv}C(CH_2)_4Br$$

$$\downarrow e^-$$

$$Br^- + PhC{\equiv}C(CH_2)_4{}^{\bullet} \longrightarrow \longrightarrow \quad (2.21)$$

Radical intermediates have been implicated in a number of C—C bond-forming reactions from halides. These include the photochemical formation of $PhCH_2CO_2Et$ from benzene and XCH_2CO_2Et–Lewis acid[147]; the zinc-induced homolytic cleavage of phenacyl bromide in DMSO to give $PhCO\overset{\bullet}{C}H_2$, which may dimerise or add to alkenes[148]; and the conversion of aralkyl halides (RX) into R—R almost quantitatively by $VCl_2(py)$[149]. Sodium dibutylboron reacts with alkyl halides by an electron transfer process to give a radical which forms alkanes by hydrogen abstraction and by dimerisation[150].

Transition metals and some of their derivatives have been used in various syntheses. A further important application of $Na_2Fe(CO)_4$ is to prepare ketones R^1COR^2 by reaction with R^1Hal followed by reaction of the intermediate $[R^1Fe(CO)_4]^-$ with another alkyl halide R^2Hal[151]. A particularly important feature of the $[Fe(CO)_4]^{2-}$ reagent is that it will tolerate the groups CN and CO_2Et in the group R^1 [152]. The reaction of $\alpha\alpha'$-dibromo-ketones with $Fe_2(CO)_9$ in the presence of alkenes or 1,3-dienes provides a useful intermolecular route to cyclopentenones[153], cyclopentanones[154] or cycloheptenones, via an oxyallyl zwitterion[154]:

$$R^1{}_2CBrCOC\,BrR^1{}_2$$

$$\downarrow {\scriptstyle Fe_2(CO)_9}$$

$$[R^1{}_2C \!=\! \overset{\overset{\textstyle O^-}{|}}{\underset{+}{C}} \!=\! CR^1{}_2]\,Fe(II)L_n \xrightarrow{\ R^2CH=CH_2\ } \qquad\qquad (2.22)$$

Iron pentacarbonyl also reacts with various α-halogenoketones to give coupled 1,4-diketones and reduced monoketones[155]. The ω-iodoalkynes $Bu^nC\equiv C(CH_2)_nI$ and $Ni(CO)_4-KOBu^t$ form the esters $Bu^nC\equiv C(CH_2)_n$-CO_2Bu^t for $n = 3$ or 5, but cyclo-pentenyl and -pentylidene esters are obtained for $n = 4$ [156]. The intramolecular cross-coupling of allylic dihalides with $Ni(CO)_4$ has been utilised for macrocyclic lactone formation, compounds which are important in the antibiotic field[157]. Palladium catalyses the overall substitution of vinylic hydrogen by benzyl in the reaction of benzyl halides with alkenes in the presence of $Bu^n{}_3N$ [158] (e.g. $PhCH_2Cl + CH_2 = CHCO_2\,Me \rightarrow PhCH_2CH=CHCO_2Me$).

The 2-methoxyallyl cation may be prepared from $CH_2=C(OMe)CH_2Br$ under special conditions and undergoes a cycloaddition reaction with furan[159].

2.6.3   Formation of carbon–nitrogen bonds

Cyanogen chloride is alkylated by alkyl chlorides in the presence of ferric chloride to give novel complexes $(RN=CCl_2)_n(FeCl_3)_m$, which may be converted into RNH_2, $RNHCO_2Et$ or $RN=C=O$ by reaction with water, EtOH or ZnO, respectively[160]. 2-Iodoalkyl azides of known stereochemistry react stereospecifically with alkyl- and aryl-dichloroboranes to give N-alkyl- and N-aryl-aziridines[161].

2.6.4   Formation of carbon–oxygen and carbon–sulphur bonds

The hydrolysis of the strong aliphatic C—F bond is accelerated by the salts of Th(IV) and Al(III) and this has been observed in reactions with $PhCH_2F$, $PhCF_3$ and related HO- and NH_2-substituted compounds, and with $PhOCOCH_2F$ [162].

The formation of orientationally specific α-methoxyketones as well as the parent ketone occurs during the debromination of unsymmetrical $\alpha\alpha'$-dibromoketones with Zn–Cu in methanol. The intervention of a selective 2-oxyallyl cation is proposed for this reaction [equation (2.23)][163].

$$\text{MeCHBrCOCBrMe}_2 \xrightarrow[\text{L = solvent}]{\text{Zn}} \begin{array}{c} \text{Me} \\ \diagdown \\ \text{H} \end{array} \mathrm{C}{=}\mathrm{C} \begin{array}{c} \text{O}-\text{Zn(L)Br} \\ \diagup \\ \diagdown \text{CBrMe}_2 \end{array} \longrightarrow \begin{array}{c} \text{Me} \\ \diagdown \\ \text{H} \end{array} \mathrm{C}{=}\mathrm{C} \begin{array}{c} \text{O}^- \\ \diagup \\ \diagdown \overset{+}{\text{C}}\text{Me}_2 \end{array}$$

$$\text{EtCOCMe}_2\text{OMe} \;+\; \text{EtCOPr}^\text{i}$$
$$76\% \qquad\qquad 24\%$$

(2.23)

The scope of the Darzens' glycidic ester synthesis (involving the formation of the oxirane ring system) has been extended to include acetaldehyde and monosubstituted acetaldehyde derivatives by *prior* formation of the enolates of the α-halogenoesters [$\text{R}^1\text{R}^2\text{CHXCO}_2\text{R}^3$] with $(\text{Me}_3\text{Si})_2\text{N}{-}\text{Na}$ [164] (or $-\text{Li}$)[165], for use in the reaction. Related preparations of oxiranes include the reactions of aldehydes and ketones with $\text{BrCH}_2\text{CH}{=}\text{CHCO}_2\text{Et--KOBu}^\text{t}$–$\text{Bu}^\text{t}\text{OH}$ [166]; with $\text{R}^1\text{R}^2\text{CBrLi}$ [167] (from $\text{R}^1\text{R}^2\text{CBr}_2$ and $\text{Bu}^\text{n}\text{Li}$); with RCBr_2Li [168] (from RCHBr_2 and $\text{Pr}^\text{i}_2\text{NLi}$); and with $\text{ArSCH}_2\text{Cl--KOBu}^\text{t}$–$\text{Bu}^\text{t}\text{OH}$ [169]. The thermolysis of compounds of the type $\text{Bu}^\text{n}_3\text{SnO(CH}_2)_n\text{Br}$ gives cyclic ethers for $n = 2$, 3 and 4 [170,171,172].

Copper(I) carboxylates form esters with alkyl halides[173] and the same overall reaction can be achieved by using the alkyl halide, the carboxylic acid and the $\text{Cu}_2\text{O--Bu}^\text{t}\text{NC}$ complex, in which a key intermediate is $\text{RCO}_2\text{Cu(I)}$–$(\text{Bu}^\text{t}\text{NC})_n$ [174]. The thermal rearrangement of $(\text{ClCH}_2)_2\text{CHOAc}$ to $\text{ClCH}_2\text{CHClCH}_2\text{OAc}$ (63% present after equilibration at 180 °C) proceeds via neighbouring group participation by the acetate group[175]. Monoalkyl hydrogen phosphites ROPH(O)(OH) are formed conveniently by the alkylation of tetramethylammonium t-butyl hydrogen phosphite with an alkyl halide and hydrolysis of the ester $\text{Bu}^\text{t}\text{OPH(O)(OR)}$ with aqueous $\text{CF}_3\text{CO}_2\text{H}$ [176].

The photochemical reaction of perfluoroalkyl iodides, $\text{R}_\text{F}\text{I}$, with dialkyl sulphides[177,178] or disulphides[179,180] provides a convenient route to alkyl polyfluoroalkyl sulphides [e.g. $\text{R}_\text{F}\text{SMe}$ or $\text{R}_\text{F}\text{SEt}$] and thence to sulphonic acids[181] [$\text{R}_\text{F}\text{SMe} \xrightarrow{\text{KMnO}_4} \text{R}_\text{F}\text{SO}_2\text{Me} \xrightarrow{\text{NaOCl}} \text{R}_\text{F}\text{SO}_2\text{CCl}_3 \xrightarrow{\text{HO}^-} \text{R}_\text{F}\text{SO}^-_3$]. Mono- and bis-methyl sulphides have been obtained by similar reactions with $\alpha\omega$-di-iodoperfluoroalkanes[182].

Thi-iranes may be prepared directly from cyclic (but not acyclic) alkenes by using $\text{I}_2 + (\text{SCN})_2$ and preferentially hydrolysing the thiocyanate moiety of the intermediate *trans*-1,2-iodoalkyl thiocyanate[183]. Removal of a thioacetal protecting group under mild conditions is now possible by using MeI–moist acetone[184].

2.6.5 Formation of carbon–halogen bonds

7,7-Dichloro-2,5-diphenylbenzocyclopropene undergoes halogen exchange with AgF–MeCN under very mild conditions and a substituted benzocyclopropenium cation is postulated as an intermediate[185].

Ion cyclotron resonance spectroscopy of alkyl halides shows that the chemistry initiated by the parent ion depends on the halide, and the differences in behaviour can be rationalised on thermochemical grounds[186] as follows:

$RX^+ + RX \rightarrow RXH^+ + RX$; and then $RXH^+ + RX \rightarrow R\overset{+}{X}R + HX$ for $X = F$ or Cl, but $RX^+ + RX \rightarrow R\overset{+}{X}R + X$ for $X = Br$ or I.

With a mixture of methyl chloride and ethyl fluoride under these conditions the ethyl methyl chlorinium ion is formed exclusively by a nucleophilic displacement reaction consistent with the formation of an intermediate complex in which the labile proton is always transferred to the more basic site (ethyl fluoride) before the formation of the dialkylhalogenonium ion produced.

$$Me\overset{+}{Cl}H + EtF \quad\atop\quad Et\overset{+}{F}H + MeCl \longrightarrow \left[\begin{array}{c} MeH_2C\text{---}F \\ \\ Cl\text{----}H^+ \\ | \\ Me \end{array} \right] \longrightarrow Et\overset{+}{Cl}Me + HF \qquad (2.24)$$

Evidence has been presented that the photochemical rearrangement of α-allenic halides [e.g. $CH_2{=}C{=}CRCH_2Cl \rightarrow CH_2{=}CClCR{=}CH_2$] proceeds by the initial homolysis of the carbon–halogen bond[187].

2.6.6 Eliminations

The mechanism of the dehydrohalogenation of alkyl halides continues to be widely investigated[188].

The elimination of an alkyl fluoride from substituted alkoxyperfluoroisobutenes in the presence of Lewis acids to give perfluoro-α-methylacrylate derivatives is a general reaction[189]:

$$(CF_3)_2C{-}C(OR)X + SbF_5 \rightarrow SbF_6^- + [CF_2{\cdots}C(CF_3){\cdots}CX{\cdots}OR]^+$$
$$\rightarrow CF_2{=}C(CF_3)COX + RF + SbF_5 \qquad (2.25)$$

A new protecting group for ketones, the bromoethylethylene acetal group, is readily removed under neutral conditions by activated zinc in boiling methanol[190]. Stilbenes are formed from α-chlorosulphides, $ArCH_2SCHClAr$, by treatment *first* with Ph_3P, then with base[191].

The stereospecificity of dehalogenation of 1,2-dihalides by sodium naphthalenide[192] and by the photochemically induced reaction with hexa-n-butylditin have been reported[193].

2.6.7 Metallations

The first perlithiated alkanes, CLi_4 and C_2Li_6, have been formed by the reaction of an excess of 'atomic' lithium with $\cdot CCl_4$ and C_2Cl_6 respectively at 800–1000 °C [194]. An alternative and generally more efficient procedure for the synthesis of alcohols now involves the simple treatment of the alkyl halide and carbonyl compound (aldehyde, ketone or ester) in THF with lithium[195]. The generation of CH_2BrLi *in situ* from CH_2Br_2 and Li–Hg in THF and reaction with *o*-acylanilines provides a useful route to indoles

$$\text{(2.26)}$$

[equation (2.26)][196]. Metal–halogen exchange with $CCl_3P(O)(OEt)_2$ and Bu^nLi at -80 °C gives $LiCCl_2P(O)(OEt)_2$ which possesses the usual activity of an organolithium compound[197].

An improved synthesis of organocalcium halides in THF demands high-purity calcium for reaction with alkyl halides[198]. In reactions with carbonyl compounds, calcium derivatives possess no special advantages over the more commonly used Grignard reagents[199]. Calcium atoms defluorinate $CF_3CF{=}CFCF_3$ at liquid-nitrogen temperature to give $CF_3C{\equiv}CCF_3$ [200].

Non-solvated fluoro-organozinc compounds, prepared by reaction of zinc atoms with fluoroalkyl iodides on a liquid-nitrogen-cooled surface, are much more reactive and less stable than those formed in solution[201]. Whereas $(CF_3)_2CFZnI$ prepared in solution is very stable and not easily hydrolysed, reaction of zinc atoms with $(CF_3)_2CFI$ saturated with water vapour gives $(CF_3)_2CFH$ in high yield compared with $CF_3CF{=}CF_2$ (by decomposition of non-solvated $(CF_3)_2CFZnI$), which shows that hydrolysis of the organo-metallic compound is efficient even at low temperatures. α-Bromonitriles form organozinc derivatives in THF for reaction with aldehydes and ketones, and for conjugate addition to αβ-unsaturated ketones and to alkylidene diethyl malonates[202].

Pre-formation of $n\text{-}C_7F_{15}Cu$ from $n\text{-}C_7F_{15}I$ and copper in DMSO at 110 °C and reaction with an alkene, $RCH_2CH{=}CH_2$, gives $n\text{-}C_7F_{15}(CH_2)_3R$ and $n\text{-}C_7F_{15}CH_2CH{=}CHR$, possibly by a radical mechanism involving $n\text{-}C_7F_{15}\cdot$. However, treatment of 1-perfluoroalkylethylenes, $R_FCF_2CH{=}CH_2$, and perfluoroalkyliodides, R'_FCF_2I, with copper in DMF at 150 °C produces three materials: $R_FCF_2(CH_2)_2CF_2R'_F$, $R_FCF{=}CFCH_2CF_2R'_F$ and $R_FCF{=}CHCH{=}CFR'_F$ [203]. Other experiments suggest that these products are probably formed by the initial addition of R'_FCF_2Cu across the double bond.

2.7 REACTIONS OF VINYL HALIDES

2.7.1 Formation of carbon–hydrogen bonds

Treatment of β-chlorovinyl aldehydes $Cl\overset{|}{C}{=}\overset{|}{C}CHO$ with zinc powder permits their selective conversion into $H\overset{|}{C}{=}\overset{|}{C}CHO$ [204]. $LiAlH_4$ will conveniently hydrogenolyse vinylic halogen[101] (e.g. 2-bromo-2-methylstyrene → 2-methyl-styrene by reaction in boiling ether for 18 h).

2.7.2 Formation of carbon–carbon and carbon–silicon bonds

The cross-coupling of perfluoroalkyl iodides with halogenoalkenes can be brought about conveniently by reaction in the presence of copper in DMF at 120 °C[205] (e.g. n-$C_7F_{15}I$ + *trans*-CHCl=CHI $\xrightarrow{Cu}$ *trans* -n-C_7F_{15}CH=CHCl). However, when compounds of the type R_FCBr=CH_2 (R_F = C_6F_{13} or C_8F_{17}) are used with copper and perfluoroalkyl iodides R'_FI (R'_F=C_4F_9, C_6F_{13} or C_8F_{17}), the expected products $R_FCR'_F$=CH_2 are *not* formed[206]; surprisingly, the products are R_FCH=CHR$'_F$, obtained in 60–70% yields. The nucleophilic substitution of vinylic bromine by Me_2CuLi takes place with retention of configuration and is compatible with an addition–elimination reaction sequence[207].

The selective cross-coupling of a Grignard reagent and a vinylic halide is catalysed by nickel(II) acetylacetonate[208] and by dichloro-[1,2-bis(diphenyl-phosphine)ethane]nickel(II)[209]. With *trans*-CHCl=CHCl and PhMgBr, the former reaction is stereospecific (→ *trans*-PhCH=CHPh), while the latter gives *cis*- and *trans*-stilbene in the ratio of 43 : 57.

Bis-(1,5-cyclo-octadiene)nickel(0) promotes the self-coupling of vinylic halides in the presence of functional groups (e.g. CO_2Me) under very mild conditions (25 °C), but with variable stereospecificity[210]. The reaction of β-bromostyrene with $Ni(CO)_4$ in the presence of alkynes in DMF at 60 °C yields dimers of γ-but-2-enolactone derivatives[211].

$$PhCH=CHBr + EtC{\equiv}CEt + Ni(CO)_4 \longrightarrow \left[\begin{array}{c} EtC=CEt \\ \diagup C \quad C \diagdown \\ O{=}C \diagdown_O\diagup \quad CH \\ \| \\ CHPh \end{array} \right]_2$$

The replacement of vinylic fluorine by Me_3Si, by the photochemical reaction of $(Me_3Si)_2Hg$ with chlorotrifluoroethylene to give *cis*- and *trans*-CFCl=CFSiMe$_3$, involves the prior formation of the adduct Me_3-$SiCF_2CFClHgSiMe_3$[212].

The halogen in 1-halogenoallenes is directly replaceable by an alkyl group R, without rearrangement, by using R_2CuLi[213], or by $(C{\equiv}C)_2SiMe_3$[214] by using $H(C{\equiv}C)_2SiMe_3$–$CuBr$–Bu^n_3N–DMF.

2.7.3 Formation of carbon–nitrogen and carbon–phosphorus bonds

Potassium phthalimide reacts readily with vinylic bromides in boiling *NN*-dimethylacetamide with copper(I) iodide as catalyst to give the corresponding phthalimido compounds[215]. The mechanism of the replacement of vinylic halogen by amines in the absence of a catalyst has been studied[216].

Treatment of 2-substituted 1-bromo-1-nitroethylenes, $RCH=C(NO_2)Br$, with Ph_3P gives good yields of $Ph_3\overset{+}{P}CHRCN$, but the reaction pathway has not yet been identified[217].

Alkoxyallenic halides ($ROCH_2CBr=C=CHCH_2NMe_2$) undergo allene–acetylene rearrangements on reaction with dimethylamine to form $ROCH_2C\equiv CCH(NMe_2)CH_2NMe_2$[218], but the mechanism must be quite different from that with terminal allenyl bromides, $R^1R^2C=C=CHBr$, which initially lose HBr and form products from further reaction with ($R^1R^2\overset{+}{C}-C\equiv\overset{-}{C}\leftrightarrow R^1R^2C=C=C$:).

2.7.4   Formation of carbon–oxygen and carbon–sulphur bonds

The mechanism of the solvolysis of vinylic halides in the presence of oxygen- and sulphur-containing nucleophiles has been extensively studied by Rapoport[219]. Evidence is given that the reaction of vinyl halides $R^1R^2C=CR^3X$ with mercuric salts in water to give the carbonyl compounds $R^1R^2CHCOR^3$ involves an oxymercuriation–dehalogenomercuriation sequence[220].

Copper(I) carboxylates in refluxing pyridine form esters with vinyl halides[221], while phenyl vinyl sulphides are readily formed with PhSCu[222].

The solvolysis of trisubstituted halogenoallenes involves the formation of a resonance-stabilised cation $[R^1R^2C_3{=\!=}C_2{=\!=}C_1R^3]^+$, which in aqueous acetone gives mainly a propargyl alcohol $R^1R^2C(OH)C\equiv CR^3$ though t-butyl groups at C-3 encourage reaction at C-1, when an αβ-unsaturated carbonyl compound is formed[223]. Steric effects are important in influencing the reactivity of these systems, an aromatic ring being most effective in stabilising the intermediate cation, when constrained to be coplanar with the vacant π-orbital[224].

The nucleophilic substitution of chlorine in $ArCCl=NOMe$ by methoxide gives a stereoselective inversion of configuration at the sp^2-hybridised carbon[225]. Viehe has reviewed the chemistry of the highly reactive compound $CCl_2=\overset{+}{N}Me_2\ Cl^-$, which includes reactions with carbon, nitrogen, oxygen and sulphur nucleophiles[226]. Mechanistic and synthetic studies have been reported for the reactions of 1-chloro- and 1,4-dichloro-1,4-diaryl-2,3-diazabuta-1,3-dienes[227]; azocarbocations are involved in the conversion of the former, $Ar^1CCl=NN=CHAr^2$, with aqueous dioxan into $Ar^1CONHN=CHAr^2$.

2.7.5   Formation of carbon–halogen bonds

The first examples of vinylic fluorine exchange have been reported[228]. Tetrachlorodiborane, B_2Cl_4, replaces one fluorine by chlorine in $CH_2=CF_2$ and in $MeCF=CH_2$; with $CF_2=CHF$, $CF_2=CHCl$ is formed. The reaction is sensitive to structural factors since $CF_3CF=CH_2$ is inert to the reagent.

2.8   REACTIONS OF 1-HALOGENOALKYNES

The mechanisms of the reactions of 1-halogenoalkynes with nitrogen[229], phosphorus[230] and sulphur[231] nucleophiles have been studied further. 2-Phenyl- and 2H-1-halogenoalkynes (but *not* the 2-alkyl derivatives) react with triphenyl- and tributyl-phosphine to give a convenient route to ethynylphosphonium salts. Halogenoalkynes cleave one bond at an amine bridgehead to give ynamines or amides.

The Me_3Si group is a useful protecting group in the cross-coupling of $Me_3SiC{\equiv}CI$ with arylcopper reagents at or below room temperature[232]. Treatment of the product with alkali readily gives $ArC{\equiv}CH$, which on reaction with $Me_3SiC{\equiv}CBr{-}Cu^I$ followed by alkali forms $Ar(C{\equiv}C)_2H$[233]; this in turn has been used to synthesise $Ar(C{\equiv}C)_3H$ [233].

2.9 MISCELLANEOUS REACTIONS INVOLVING CARBENES[234]

Further syntheses of phenylhalogenoalkylmercury compounds, $PhHgR^1$, and their thermal- and sodium iodide-induced decomposition to carbenes, have been described. These include CF_2: ($R^1 = CF_3$)[235-237]; $CFCl$: ($R^1 = CFCl_2$)[238]; CF_3CF: ($R^1 = CF_3CFBr$)[239]; $CFBr$: ($R^1 = CFBr_2$)[240]; $CF(CO_2R^2)$: ($R^1 = CFClCO_2R^2$ or $CFBrCO_2R^2$)[241]; and CCl_2: ($R^1 = CCl_3$ or $CBrCl_2$)[242-244].

Difluorocarbene, CF_2:, is produced in the thermal decomposition of many of polyhalogenocyclopropanes containing the CF_2 group, and this is attributed to the high stability of this species[245,246]. Reaction of CF_2Br_2 with R_3P ($R = Ph$ or Me_2N) and CsF or KF also provides a very convenient route to CF_2: via $[R_3\overset{+}{P}CF_2Br]Br^-$ [247].

Dichlorocarbene is formed in the reactions of Grignard reagents (compared with the usual organolithium reagents) with $CHCl_3$ and with CCl_4[248]. Specific α-elimination from certain β-alkoxides followed by hydrolysis yields α-chloroketones regiospecifically[249]:

$$LiCCl_2CR^1R^2OLi \rightarrow Cl\overset{..}{C}CR^1R^2OLi \xrightarrow{\ H_2O\ } R^1COCHClR^2 \; \dashv \; R^2COCHClR^1$$

Skell has given an account of some intriguing experiments with carbon atoms from a carbon arc. The preferred insertion of carbon atoms into a carbon–halogen bond gives a carbene, the fate of which depends on the neighbouring structural features[250] [equation (2.27)].

$$MeCCl_3$$
$$\downarrow {\scriptstyle C_1}$$
$$[MeCCl_2\overset{..}{C}Cl] \longrightarrow MeCCl{=}CCl_2; \quad Bu^tCl \xrightarrow{\ C_1\ } \begin{array}{c} H_2C{-}H \\ | \\ Me_2C{-}\overset{..}{C}Cl \end{array}$$
$$\downarrow \qquad (2.27)$$
$$Me_2{-}\triangle{-}Cl$$

Carbene radical anions, $R_2C^{\overline{\cdot}}$, previously undocumented species, are proposed as reactive intermediates in the reaction of *gem*-dihalides, R_2CX_2, with sodium naphthalenide[251]. Three products are obtained with 2,2-dichloro-3,3-dimethylbutane:

$$Bu^tCCl_2Me \xrightarrow{Na\ C_{10}H_8^-} \underset{Me_2}{\overset{Me}{\triangle}} + Bu^tCH{=}CH_2 + Bu^tCH_2Me \qquad (2.28)$$

The cyclopropane and the alkene are products expected from intramolecular insertion of carbene into C—H bonds, while the proportion of 2,2-dimethylbutane formed as a function of sodium naphthalenide concentration can only be accommodated by invoking a competing reaction of the carbene, *viz.* its further reduction to $R_2C^{\cdot-}$ ($\xrightarrow{SH} R_2\dot{C}H \xrightarrow{e^-} R_2\bar{C}H \xrightarrow{SH} R_2CH_2$; SH = dimethoxyethane).

References

1. Chambers, R. D. (1973). *Fluorine in Organic Chemistry* (London and New York: Wiley-Interscience)
2. Barton, D. H. R., Hesse, R. H., Ogunkoya, L., Westcott, N. D. and Pechet, M. M. (1972). *J. Chem. Soc. Perkin Trans. I*, 2889
3. Barton, D. H. R., Hesse, R. H., Pechet, M. M., Tarzia, G., Toh, H. T. and Westcott, N. D. (1972). *J. Chem. Soc. Chem. Commun.*, 122
4. Rodin, R. L. and Gershon, H. (1973). *J. Org. Chem.*, **38**, 3919
5. Hooz, J. and Bridson, J. N. (1972). *Can. J. Chem.*, **50**, 2387
6. Slotter, P. L. and Hill, K. A. (1972). *Tetrahedron Lett.*, 4067
7. Dietl, H. K., Normark, J. R., Payne, D. A., Thweatt, J. G. and Young, D. A. (1973). *Tetrahedron Lett.*, 1719
8. Calo, V., Lopez, L., Pesce, G. and Todesco, P. E. (1973). *Tetrahedron*, **29**, 1625
9. Ritchie, R. G. S. and Szarek, W. A. (1972). *Can. J. Chem.*, **50**, 507
10. Cambie, R. C., Hayward, R. C., Roberts, J. L. and Rutledge, P. S. (1973). *J. Chem. Soc. Chem. Commun.*, 359
11. Olah, G. A. and Mo, Y. K. (1972). *J. Org. Chem.*, **37**, 1028
12. Guy, R. G. and Pearson, I. (1973). *J. Chem. Soc. Perkin Trans. I*, 281; (1973). *J. Chem. Soc. Perkin Trans II*, 1359
13. Schmid, G. H., Csizmadia, V. M., Nowlan, V. J. and Garratt, D. G. (1972). *Can. J. Chem.*, **50**, 2457
14. Schmid, G. H. and Nowlan, V. J. (1972). *J. Org. Chem.*, **37**, 3086
15. Gorton, P. J. and Walsh, R. (1972). *J. Chem. Soc. Chem. Commun.*, 782
16. Chambers, R. D., Corbally, R. P. and Musgrave, W. K. R. (1972). *J. Chem. Soc. Perkin Trans. I*, 1281
17. Chambers, R. D., Corbally, R. P., Musgrave, W. K. R., Jackson, J. A. and Matthews, R. S. (1972). *J. Chem. Soc. Perkin Trans. I*, 1286
18. Banks, R. E., Mullen, K., Nickolson, W. J., Oppenheim, C. and Prakash, A. (1972). *J. Chem. Soc. Perkin Trans. I*, 1098
19. Chambers, R. D. and Gribble, M. Y. (1973). *J. Chem. Soc. Perkin Trans. I*, 1405
20. Chambers, R. D. and Gribble, M. Y. (1973). *J. Chem. Soc. Perkin Trans. I*, 1411
21. Bell, S. L., Chambers, R. D., Gribble, M. Y. and Maslakiewicz, J. M. (1973). *J. Chem. Soc. Perkin Trans. I*, 1716
22. Chambers, R. D., Gribble, M. Y. and Marper, E. (1973). *J. Chem. Soc. Perkin Trans. I*, 1710
23. Delyagina, N. I., Pervova, E. Ya. and Knunyants, I. L. (1972). *Izv. Akad. Nauk SSSR, Ser. Khim.*, 376
24. Knunyants, I. L., Pervova, E. Ya. and Delyagina, N. I. (1972). *Otkrytiya Izobret. Prom. Obraztsy Tovarnyl Znaki*, **49** (18), 80
25. Delyagina, N. I., Dyatkin, B. L., Knunyants, I. L., Bubnov, N. N. and Medvedev, B. Ya. (1973). *J. Chem. Soc. Chem. Commun.*, 456

26. Dyatkin, B. L., Zhuravkova, L. G., Martynov, B. I., Sterlin, S. R. and Knunyants, I. L. (1972). *J. Chem. Soc. Chem. Commun.*, 618
27. Dyatkin, B. L., Sterlin, S. R., Zhuravkova, L. G., Martynov, B. I., Mysov, E. I. and Knunyants, I. L. (1973). *Tetrahedron*, **29**, 2759
28. Dubot, G., Mansuy, D., Lecolier, S. and Normant, J. F. (1972). *J. Organometal. Chem.*, **42**, C105
29. Dyatkin, B. L., Martynov, B. I., Martynova, L. G., Kizim, N. G., Sterlin, S. R., Strumbrevichute, Z. A. and Federov, L. A. (1973). *J. Organometal. Chem.*, **57**, 423
30. De Marco, R. A., Couch, D. A. and Shreeve, J. M. (1972). *J. Org. Chem.*, **37**, 3332
31. Majid, A. and Shreeve, J. M. (1973). *J. Org. Chem.*, **38**, 4028
32. Mathey, F. and Bensoam, J. (1973). *Tetrahedron Lett.*, 2253
33. Smail, T., Iyer, R. S. and Rowland, F. S. (1972). *J. Amer. Chem. Soc.*, **94**, 1041
34. Williams, R. L., Iyer, R. S. and Rowland, F. S. (1972). *J. Amer. Chem. Soc.*, **94**, 7192
35. Moore, L. O. (1972). *J. Org. Chem.*, **37**, 2633
36. Mori, Y. and Tsuji, J. (1973). *Tetrahedron*, **29**, 827
37. Fleming, G. L., Haszeldine, R. N. and Tipping, A. E. (1973). *J. Chem. Soc. Perkin Trans. I*, 574
38. Ashton, D. S., Mackay, A. F., Tedder, J. M., Tipney, D. C. and Walton, J. C. (1973). *J. Chem. Soc. Chem. Commun.*, 496
39. Coe, P. L. and Milner, N. E. (1972). *J. Organometal. Chem.*, **39**, 395
40. Touchard, D. and Lessard, J. (1973). *Tetrahedron Lett.*, 3827
41. Banks, R. E., Parker, A. J., Sharp, M. J. and Smith, G. F. (1973). *J. Chem. Soc. Perkin Trans. I*, 5
42. Fleming, G. L., Haszeldine, R. N. and Tipping, A. E. (1972). *J. Chem. Soc. Perkin Trans. I*, 1877
43. Coy, D. H., Fleming, G. L., Haszeldine, R. N., Newlands, M. J. and Tipping, A. E. (1972). *J. Chem. Soc. Perkin Trans. I*, 1886
44. Harris, J. F., Jr. (1972). *J. Org. Chem.*, **37**, 1340
45. Robert, D. U. and Riess, J. G. (1972). *Tetrahedron Lett.*, 847
46. Corey, E. J., Kim, C. U. and Takeda, M. (1972). *Tetrahedron Lett.*, 4339
47. Furukawa, N., Inoue, T., Aida, T. and Oae, S. (1973). *J. Chem. Soc. Chem. Commun.*, 212
48. Snyder, E. I. (1972). *J. Org. Chem.*, **37**, 1466
49. Isaacs, N. S. and Kirkpatrick, D. (1972). *Tetrahedron Lett.*, 3869
50. Bose, A. K. and Lal, B. (1973). *Tetrahedron Lett.*, 3937
51. Hudson, H. R , Qureshi, A. R. and Ragoonanan (1972). *J. Chem. Soc. Perkin Trans. I*, 1595
52. Gore, J., Place, P. and Roumestant, M. L. (1973). *J. Chem. Soc. Chem. Commun.*, 821
53. Scheffold, R. and Saladin, E. (1972). *Angew. Chem. Int. Ed. Engl.*, **11**, 229
54. Clive, D. L. J. and Denyer, C. V. (1972). *J. Chem. Soc. Chem. Commun.*, 773
55. Lagow, R. J. and Margrave, J. L. (1970). *Chem. Eng. News*, **48**, 40
56. Maraschin, N. J. and Lagow, R. L. (1972). *J. Amer. Chem. Soc.*, **94**, 8601
57. Adcock, J. L. and Lagow, R. J. (1973). *J. Org. Chem.*, **38**, 3617
58. Jenkins, C. L. and Kochi, J. K. (1972). *J. Amer. Chem. Soc.*, **94**, 856
59. Cason, J. and Walba, D. M. (1972). *J. Org. Chem.*, **37**, 669
60. Bunce, N. J. (1972). *Can. J. Chem.*, **50**, 3109
61. Bagnall, R. D., Coe, P. L. and Tatlow, J. C. (1972). *J. Chem. Soc. Perkin Trans. I*, 2277
62. Zupan, M. and Pollak, A. (1973). *J. Chem. Soc. Chem. Commun.*, 845
63. Olah, G. A., Renner, R., Schilling, P. and Mo, Y. K. (1973). *J. Amer. Chem. Soc.*, **95**, 7686
64. Olah, G. A. and Schilling, P. (1973). *J. Amer. Chem. Soc.*, **95**, 7680
65. Tabushi, I., Yoshida, Z. and Tamaru, Y. (1973). *Tetrahedron*, **29**, 81
66. Yates, K., Schmid, G. H., Regulski, T. W., Garratt, D. G., Leung, H.-W. and McDonald, R. (1973). *J. Amer. Chem. Soc.*, **95**, 160
67. Moroni, R., Melloni, G. and Modena, G. (1973). *J. Chem. Soc. Perkin Trans. I*, 2491
68. Ried, W. and Ochs, W. (1972). *Chem. Ber.*, **105**, 1093
69. Miller, W. T., Hummel, R. J. and Pelosi, L. F. (1973). *J. Amer. Chem. Soc.*, **95**, 6850
70. Landor, S. R., Demetriou, B., Evans, R. J., Grzeskowiak, R. and Davey, P. (1972). *J. Chem. Soc. Perkin Trans. II*, 1995

71. Williams, R. L. and Rowland, F. S. (1972). *J. Amer. Chem. Soc.*, **94**, 1047
72. Murai, S., Kuroki, Y., Aya, T., Sonoda, N. and Tsutsumi, S. (1972). *J. Chem. Soc. Chem. Commun.*, 741
73. Brown, H. C., Hamaoka, T. and Ravindran, N. (1973). *J. Amer. Chem. Soc.*, **95**, 5786
74. Brown, H. C., Hamaoka, T. and Ravindran, N. (1973). *J. Amer. Chem. Soc.*, **95**, 6456
75. Blackborow, J. R. (1973). *J. Chem. Soc. Perkin Trans. II*, 1989
76. Olah, G. A., Mo, Y. K. and Halpern, Y. (1972). *J. Amer. Chem. Soc.*, **94**, 3551
77. Olah, G. A. and Mo, Y. K. (1973). *J. Org. Chem.*, **38**, 2682; 2686
78. Chambers, R. D., Matthews, R. S. and Parkin, A. (1973). *J. Chem. Soc. Chem. Commun.*, 509
79. Olah, G. A. and Melby, E. G. (1972). *J. Amer. Chem. Soc.*, **94**, 6220
80. Olah, G. A., Mo, Y. K., Melby, E. G. and Lin, H. C. (1973). *J. Org. Chem.*, **38**, 367
81. Clark, D. T. and Lilley, D. M. J. (1973). *Tetrahedron*, **29**, 845
82. Hooley, S. R. and Williams, D. L. H. (1973). *J. Chem. Soc. Perkin Trans. II*, 1053
83. Bassi, P. and Tonellato, U. (1972). *Gazz. Chim. Ital.*, **102**, 387
84. Olah, G. A., Bollinger, J. M., Mo, Y. K. and Brinich, J. M. (1972). *J. Amer. Chem. Soc.*, **94**, 1164
85. Exner, J. H., Kershner, L. D. and Evans, T. E. (1973). *J. Chem. Soc. Chem. Commun.*, 361
86. Henrichs, P. M. and Peterson, P. E. (1973). *J. Amer. Chem. Soc.*, **95**, 7449
87. Peterson, P. E. and Bonazza, B. R. (1972). *J. Amer. Chem. Soc.*, **94**, 5017
88. Peterson, P. E. and Waller, F. J. (1972). *J. Amer. Chem. Soc.*, **94**, 5024
89. Peterson, P. E., Bonazza, B. R. and Henrichs, P. M. (1973). *J. Amer. Chem. Soc.*, **95**, 2222
90. Traynham, J. G., Green, E. E., Lee, Y.-S., Schweinsberg, F. and Low, C.-E. (1972). *J. Amer. Chem. Soc.*, **94**, 6552
91. Shea, K. J., Lewis, D. C. and Skell, P. S. (1973). *J. Amer. Chem. Soc.*, **95**, 7768
92. Hargis, J. H. (1973). *J. Org. Chem.*, **38**, 346
93. Hargis, J. H. and Shevlin, P. B. (1973). *J. Chem. Soc. Chem. Commun.*, 179
94. Biddles, I. and Hudson, A. (1973). *Chem. Phys. Lett.*, **18**, 45
95. Chen, K. S., Elson, I. H. and Kochi, J. K. (1973). *J. Amer. Chem. Soc.*, **95**, 5341
96. Skell, P. S., Pavlis, R. R., Lewis, D. C. and Shea, K. J. (1973). *J. Amer. Chem. Soc.*, **95**, 6735
97. Drury, R. F. and Kaplan, L. (1972). *J. Amer. Chem. Soc.*, **94**, 3982
98. Kiehlmann, E., Menon, B. C. and Wells, J. I. (1972). *Can. J. Chem.*, **50**, 2561
99. Cristol, S. J., Lee, G. A. and Noreen, A. L. (1973). *J. Amer. Chem. Soc.*, **95**, 7067
100. Coleman, J. P., N-u-din Gilde, H. G., Utley, J. H. P., Weedon, B. C. L. and Eberson, L. (1973). *J. Chem. Soc. Perkin Trans. II*, 1903
101. Jefford, C. W., Kirkpatrick, D. and Delay, F. (1972). *J. Amer. Chem. Soc.*, **95**, 8905
102. Dickson, R. S. and Sutcliffe, G. D. (1972). *Aust. J. Chem.*, **25**, 761
103. Brown, H. C. and Krishnamurthy, S. (1973). *J. Amer. Chem. Soc.*, **95**, 1669
104. Jarvis, B. B. and Saukaitis, J. C. (1973). *Tetrahedron Lett.*, 709
105. Jarvis, B. B. and Saukaitis, J. C. (1973). *J. Amer. Chem. Soc.*, **95**, 7708
106. Trzupek, L. S., Stedronsky, E. R. and Whitesides, G. M. (1972). *J. Org. Chem.*, **37**, 3300
107. Kunieda, T., Tamura, T. and Takizawa, T. (1972). *J. Chem. Soc. D.*, 885
108. Mitsuo, N., Kunieda, T. and Takizawa, T. (1973). *J. Org. Chem.*, **38**, 2255
109. Booth, B. L. and Shaw, B. L. (1972). *J. Organometal. Chem.*, **43**, 369
110. Coates, D. A. and Tedder, J. M. (1973). *J. Chem. Soc. Perkin Trans. II*, 1570
111. Starks, C. M. (1973). *J. Amer. Chem. Soc.*, **95**, 3613
112. Pooner, G. H. and Brunelle, D. J. (1972). *Tetrahedron Lett.*, **4**, 293
113. Villieras, J., Disnar, J.-R., Masure, D. and Normant, J. F. (1973). *J. Organometal. Chem.*, **57**, C95
114. Posner, G. H., Whitten, C. E. and Sterling, J. J. (1973). *J. Amer. Chem. Soc.*, **95**, 7788
115. Posner, G. H. and Whitten, C. E. (1973). *Tetrahedron Lett.*, **21**, 1815
116. Posner, G. H. and Sterling, J. J. (1973). *J. Amer. Chem. Soc.*, **95**, 3076
117. Corey, E. J. and Kuwajima, I. (1972). *Tetrahedron Lett.*, **6**, 487
118. Kuwajima, I. and Doi, Y. (1972). *Tetrahedron Lett.*, **12**, 1163
119. Oshima, K., Yamamoto, H. and Nozaki, H. (1973). *J. Amer. Chem. Soc.*, **95**, 7926

120. Rathke, M. W. and Sullivan, D. (1972). *Tetrahedron Lett.*, **41**, 4249
121. Herrmann, J. L. and Schlessinger, R. H. (1973). *J. Chem. Soc. D*, 711
122. Boyce, R., Murphy, W. S. and Klein, K. P. (1972). *J. Chem. Soc. Perkin Trans. I*, 1292
123. Akujana, S. and Hooz, J. (1973). *Tetrahedron Lett.*, **42**, 4115
124. Stork, G., Gardner, J. O., Boeckman, Jr., R. K. and Parker, K. A. (1973). *J. Amer. Chem. Soc.*, **95**, 2014
125. Stork, G. and Boeckman, Jr., R. K. (1973). *J. Amer. Chem. Soc.*, **95**, 2016
126. Allen, R. B., Lawler, R. G. and Ward, H. R. (1973). *Tetrahedron Lett.*, 3303
127. Ohbe, Y. and Matsuda, T. (1973). *Tetrahedron*, **29**, 2989
128. Katzenellenbogen, J. A. and Lenox, R. S. (1972). *Tetrahedron Lett.*, 1471
129. Johnson, P. Y. and Zitsman, J. (1973). *J. Org. Chem.*, **38**, 2346
130. Jones, P. R. and Costanzo, S. J. (1973). *J. Org. Chem.*, **38**, 3189
131. Brown, H. C. and Carlson, B. A. (1973). *J. Org. Chem.*, **38**, 2422
132. Brown, H. C., Katz, J.-J. and Carlson, B. A. (1973). *J. Org. Chem.*, **38**, 3968
133. Carlson, B. A. and Brown, H. C. (1973). *J. Amer. Chem. Soc.*, **95**, 6876
134. Pelter, A., Harrison, C. R. and Kirkpatrick, D. (1973). *J. Chem. Soc. Chem. Commun.*, 544
135. Davis, A., Morgan, M. H., Richards, D. H. and Scilly, N. F. (1972). *J. Chem. Soc. Perkin Trans. I*, 286
136. Grieco, P. A. and Finkelhor, R. (1973). *J. Org. Chem.*, **38**, 2100
137. Kolsaker, P. and Storesund, H. J. (1972). *J. Chem. Soc. Chem. Commun.*, 375
138. Fry, A. J. and Britton, W. E. (1973). *J. Org. Chem.*, **38**, 4016
139. Hirai, K. and Kishida, Y. (1972). *Tetrahedron Lett.*, 2743
140. Richman, J. E., Herrmann, J. L. and Schlessinger, R. H. (1973). *Tetrahedron Lett.*, 3267
140a. Herrmann, J. L., Richman, J. E. and Schlessinger, R. H. (1973). *Tetrahedron Lett.*, 3271
141. Meyers, A. I., Munavu, R. and Durandetta, J. (1972). *Tetrahedron Lett.*, 3929
142. Meyers, A. I. and Nazarenko, N. (1972). *J. Amer. Chem. Soc.*, **94**, 3243
143. Butler, D. E. (1972). *Tetrahedron Lett.*, 1929
144. Olah, G. A., DeMember, J. R., Schlosberg, R. H. and Halpern, Y. (1972). *J. Amer. Chem. Soc.*, **94**, 156
145. Baizer, M. M. and Chruma, J. L. (1972). *J. Org. Chem.*, **37**, 1951
146. Moore, W. M. and Peters, D. G. (1972). *Tetrahedron Lett.*, 453
147. Izawa, Y., Ishihara, T. and Ogata, Y. (1972). *Tetrahedron*, **28**, 211
148. Ghera, E., Perry, D. H. and Shoua, S. (1973). *J. Chem. Soc. Chem. Commun.*, 858
149. Cooper, T. A. (1973). *J. Amer. Chem. Soc.*, **95**, 4158
150. Pasto, D. J. and Wojtkowski, P. W. (1972). *J. Organometal. Chem.*, **34**, 251
151. Collman, J. P., Winter, S. R. and Clark, D. R. (1972). *J. Amer. Chem. Soc.*, **94**, 1788
152. Siege, W. O. and Collman, J. P. (1972). *J. Amer. Chem. Soc.*, **94**, 2516
153. Noyori, R., Yokoyama, K., Makino, S. and Hayakawa, Y. (1972). *J. Amer. Chem. Soc.*, **94**, 1772
154. Noyori, R., Hayakawa, Y., Funakura, M., Takaya, M., Murai, S., Kobayashi, R. and Tsutsumi, S. (1972). *J. Amer. Chem. Soc.*, **94**, 7202
155. Alper, H. and Keung, E. C. H. (1972). *J. Org. Chem.*, **37**, 2566
156. Crandall, J. K. and Michaely, W. J. (1973). *J. Organometal. Chem.*, **51**, 375
157. Corey, E. J. and Kirst, M. A. (1972). *J. Amer. Chem. Soc.*, **94**, 667
158. Heck, R. F. and Nalley, J. P. (1972). *J. Org. Chem.*, **37**, 2320
159. Green, A. E., Greenwood, G. and Hoffmann, H. M. R. (1973). *J. Amer. Chem. Soc.*, **95**, 1338
160. Fuks, R. and Hartemink, M. (1973). *Tetrahedron*, **29**, 297
161. Levy, A. B. and Brown, H. C. (1973). *J. Amer. Chem. Soc.*, **95**, 4067
162. Clark, H. R. and Jones, M. M. (1972). *J. Catalysis*, **24**, 472
163. Hoffmann, H. M. R., Nour, T. A. and Smithers, R. H. (1972). *J. Chem. Soc. Chem. Commun.*, 963
164. Villieras, J., Payan, D., Anguelova, Y. and Normant, J.-F. (1972). *J. Organometal. Chem.*, **42**, C5
165. Borch, R. F. (1972). *Tetrahedron Lett.*, 3761
166. Koppel, G. A. (1972). *Tetrahedron Lett.*, 1507

167. Cainelli, G., Tangari, N. and Ronchi, A. U. (1972). *Tetrahedron*, **28**, 3009
168. Villieras, J., Bacquet, C., Masure, D. and Normant, J.-F. (1973). *J. Organometal. Chem.*, **50**, C7
169. Tavares, D. F. and Estep, R. E. (1973). *Tetrahedron Lett.*, 1229
170. Delmond, B., Pommier, J.-C. and Valade, J. (1927). *J. Organometal. Chem.*, **35**, 91
171. Delmond, B., Pommier, J.-C. and Valade, J. (1973). *J. Organometal. Chem.*, **47**, 337
172. Delmond, B., Pommier, J.-C. and Valade, J. (1973). *J. Organometal. Chem.*, **50**, 121
173. Lewin, A. H. and Goldberg, N. L. (1972). *Tetrahedron Lett.*, 491
174. Saegusa, T., Murase, I. and Itoh, Y. (1973). *J. Org. Chem.*, **38**, 1753
175. Pews, R. G. and Davis, R. A. (1973). *J. Chem. Soc. Chem. Commun.*, 269
176. Zwierzak, A. and Kluba, M. (1973). *Tetrahedron*, **29**, 1089
177. Haszeldine, R. N., Higginbottom, B., Rigby, R. B. and Tipping, A. E. (1972). *J. Chem. Soc. Perkin Trans. I*, 155
178. Haszeldine, R. N., Rigby, R. B. and Tipping, A. E. (1972). *J. Chem. Soc. Perkin Trans. I*, 1506
179. Haszeldine, R. N., Rigby, R. B. and Tipping, A. E. (1972). *J. Chem. Soc. Perkin Trans I*, 159
180. Haszeldine, R. N., Rigby, R. B. and Tipping, A. E. (1972). *J. Chem. Soc. Perkin Trans. I*, 2180
181. Haszeldine, R. N., Hewitson, B. Higginbottom, B., Rigby, R. B. and Tipping, A. E. (1972). *J. Chem. Soc. Chem. Commun.*, 249
182. Haszeldine, R. N., Rigby, R. B. and Tipping, A. E. (1972). *J. Chem. Soc. Perkin Trans. I*, 2438
183. Hinshaw, J. C. (1972). *Tetrahedron Lett.*, 3567
184. Fetizon, M. and Jurion, M. (1972). *J. Chem. Soc. Chem. Commun.*, 382
185. Müller, P. (1973). *J. Chem. Soc. Chem. Commun.*, 895
186. Beauchamp, J. L., Holtz, D., Woodgate, S. D. and Patt, S. L. (1972). *J. Amer. Chem. Soc.*, **94**, 2798
187. Michel, E., Raffi, J. and Troyanowski, C. (1973). *Tetrahedron Lett.*, 825
188. Bordwell, F. G. (1972). *Accounts Chem. Res.*, **5**, 374; Bunnett, J. F. and Eck, D. L. (1973). *J. Amer. Chem. Soc.*, **95**, 1900; Koch, H. F., Dahlberg, D. B., Toczko, A. G. and Solsky, R. L. (1973). *ibid.*, **95**, 2029; Jackson, O. R., McLennan, D. J., Short, S. A. and Wong, R. J. (1972). *J. Chem. Soc. Perkin Trans. II*, 2308
189. Knunyants, I. L., Abduganiev, Yo. G., Rokhlin, E. M., Okulevich, P. O. and Karpushina, N. I. (1973). *Tetrahedron*, **29**, 595
190. Corey, E. J. and Ruden, R. A. (1973). *J. Org. Chem.*, **38**, 834
191. Mitchell, R. H. (1973). *Tetrahedron Lett.*, 4395
192. Adam, W. and Arce, J. (1972). *J. Org. Chem.*, **37**, 507
193. Kuivila, H. G. and Pian, C. H.-C. (1973). *Tetrahedron Lett.*, 2561
194. Chung, C. and Lagow, R. J. (1972). *J. Chem. Soc. Chem. Commun.*, 1078
195. Pearce, P. J., Richards, D. H. and Scilly, N. F. (1972). *J. Chem. Soc. Perkin Trans. I*, 1655
196. Panunzio, M., Tangar, N. and Ronchi, A. U. (1972). *J. Chem. Soc. Chem. Commun.*, 415
197. Seyferth, D. and Marmor, R. S. (1973). *J. Organometal. Chem.*, **59**, 237
198. Kawabata, N.. Matsumura, A. and Yamashita, S. (1973). *Tetrahedron*, **29**, 1069
199. Kawabata, N., Nakamura, H. and Yamashita, S. (1973). *J. Org. Chem.*, **38**, 3403
200. Klabunde, K. J., Low, J. Y. F. and Key, M. S. (1972–73). *J. Fluorine Chem.*, **2**, 207
201. Klabunde, K. J., Key, M. S. and Low, J. Y. F. (1972). *J. Amer. Chem. Soc.*, **94**, 999
202. Goasdoue, N. and Gaudemar, M. (1972). *J. Organometal. Chem.*, **39**, 17
203. LeBlanc, M., Santini, G., Guion, J. and Riess, J. G. (1973). *Tetrahedron*, **29**, 3195
204. Hara, A. and Sekuja, M. (1972). *Chem. Pharm. Bull.*, **20**, 309
205. Burdon, J., Coe, P. L., March, C. R. and Tatlow, J. C. (1972). *J. Chem. Soc. Perkin Trans. I*, 639
206. Santini, G., LeBlanc, M. and Riess, J. G. (1973). *Tetrahedron*, **29**, 2411
207. Klein, J. and Levene, R. (1972). *J. Amer. Chem. Soc.*, **94**, 2520
208. Corriu, R. J. P. and Masse, J. P. (1972). *J. Chem. Soc. Chem. Commun.*, 144
209. Tamao, K., Sumitani, K. and Kumada, M. (1972). *J. Amer. Chem. Soc.*, **94**, 4374
210. Semmelhack, M. F., Helquist, P. M. and Gorzynski, J. D. (1972). *J. Amer. Chem. Soc.*, **94**, 9234

211. Ryang, M., Sawa, Y., Somasundaram, S. N., Murai, S. and Tsutsumi, S. (1972). *J. Organometal. Chem.*, **46**, 375
212. Fields, R., Haszeldine, R. N. and Hubbard, A. F. (1972). *J. Chem. Soc. Perkin Trans. I*, 847
213. Kalli, M., Landor, P. D. and Landor, S. R. (1972). *J. Chem. Soc. Chem. Commun.*, 593
214. Landor, P. D., Landor, S. R. and Leighton, J. P. (1973). *Tetrahedron Lett.*, 1019
215. Bacon, R. G. R. and Karim, A. (1973). *J. Chem. Soc. Perkin Trans. I*, 278
216. Rappoport, Z. and Peled, P. (1973). *J. Chem. Soc. Perkin Trans. II*, 616
217. Devlin, C. J. and Walker, B. J. (1973). *J. Chem. Soc. Perkin Trans. I*, 1428
218. Mavrov, M. V., Rodionov, A. P. and Kucherov, V. F. (1973). *Tetrahedron Lett.*, 759
219. Rappoport, Z., Pross, A. and Apeliog, Y. (1973). *Tetrahedron Lett.*, 2015, and earlier papers in this series.
220. Arzoumanian, H. and Metzger, J. (1973). *J. Organometal. Chem.*, **57**, C1
221. Lewin, A. H. and Goldberg, N. L. (1972). *Tetrahedron Lett.*, 491
222. Burdon, J., Coe, P. L., Marsh, C. R. and Tatlow, J. C. (1972). *J. Chem. Soc. Perkin Trans. I*, 638
223. Schiavelli, M. D., Gilbert, R. P., Boynton, W. A. and Boswell, C. J. (1972). *J. Amer. Chem. Soc.*, **94**, 5061
224. Schiavelli, M. D., Timpanaro, P. L. and Brewer, R. (1973). *J. Org. Chem.*, **38**, 3054
225. Johnson, J. E., Nalley, E. A. and Weidig, C. (1973). *J. Amer. Chem. Soc.*, **95**, 2051
226. Viehe, H. J. and Janousek, Z. (1973). *Angew. Chem. Int. Ed. Engl.*, **12**, 806
227. Cronin, J., Hegarty, A. F., Cashell, P. A. and Scott, F. L. (1973). *J. Chem. Soc. Perkin Trans. II*, 1708
228. Ritter, J. J., Coyle, T. D. and Bellama, J. M. (1972). *J. Organometal. Chem.*, **42**, 25
229. Dickstein, J. I. and Miller, S. I. (1972). *J. Org. Chem.*, **37**, 2175
230. Dickstein, J. I. and Miller, S. I. (1972). *J. Org. Chem.*, **37**, 2168
231. Beltrame, P. L., Cattania, M. G. and Simonetta, M. (1973). *J. Chem. Soc. Perkin Trans. II*, 63
232. Oliver, R. and Walton, D. R. M. (1972). *Tetrahedron Lett.*, 5209
233. Eastmond, R. and Walton, D. R. M. (1972). *Tetrahedron*, **28**, 4591
234. Köbrich, G. (1972). *Angew. Chem. Int. Ed. Engl.*, **11**, 473
235. Seyferth, D. and Hopper, S. P. (1972). *J. Organometal. Chem.*, **44**, 97
236. Seyferth, D., Hopper, S. P. and Murphy, G. J. (1972). *J. Organometal. Chem.*, **46**, 201
237. Seyferth, D. and Hopper, S. P. (1972). *J. Org. Chem.*, **37**, 4070
238. Seyferth, D. and Murphy, G. J. (1973). *J. Organometal. Chem.*, **49**, 117
239. Seyferth, D. and Murphy, G. J. (1973). *J. Organometal. Chem.*, **52**, C1
240. Seyferth, D. and Hopper, S. P. (1973). *J. Organometal Chem.*, **51**, 77
241. Wolovsky, R. and Maoz, N. (1973). *J. Org. Chem.*, **38**, 4031
242. Fedorynski, M. and Makosza, M. (1973). *J. Organometal. Chem.*, **51**, 89
243. Seyferth, D., Tronich, W., Marmor, R. S. and Smith, W. E. (1972). *J. Org. Chem.*, **37**, 1537
244. Seyferth, D. and Shih, H.-m. (1973). *J. Amer. Chem. Soc.*, **95**, 8464.
245. Birchall, J. M., Haszeldine, R. N. and Roberts, D. W. (1973). *J. Chem. Soc. Perkin Trans. I*, 1071
246. Birchall, J. M., Fields, R., Haszeldine, R. N. and Kendall, N. T. (1973). *J. Chem. Soc. Perkin Trans. I*, 1773
247. Burton, D. J. and Naae, D. G. (1973). *J. Amer. Chem. Soc.*, **95**, 8467
248. Davis, M., Deady, L. W., Finch, A. J. and Smith, J. F. (1973). *Tetrahedron*, **29**, 349
249. Villieras, J., Bacquet, C. and Normant, J.-F. (1972). *J. Organometal. Chem.*, **40**, C1
250. Skell, P. S., Havel, J. J. and McGlinchey, M. J. (1973). *Accounts Chem. Res.*, 97
251. Sargent, G. D., Tatum, C. M. Jr. and Kastner, S. M. (1972). *J. Amer. Chem. Soc.*, **94**, 7174

3
Alcohols, Ethers and Related Compounds

S. G. WILKINSON
University of Hull

3.1 INTRODUCTION

In its subject matter and treatment of topics, this chapter resembles its predecessor in Series One. The senescence of many of the topics has not inhibited the urge to innovate, improve and investigate, and significant advances are generally apparent. Even the demise of certain sections included in the first review is less a reflection of quiescence than of increased competition for the available space. Thus, it has again been necessary to select functional groups and topics as well as individual papers: less than 20% of the primary references survived the selection process. Because of their wider interest, priority has usually been given to preparative studies, while least justice has been done to the numerous kinetic investigations. Similarly, greater attention has been paid to hydroxy and epoxy compounds than to peroxy compounds, but deficiencies in this latter area relating to free-radical reactions are made good in Volume 10 of this series.

3.2 MONOHYDRIC ALCOHOLS

3.2.1 Preparation

3.2.1.1 General methods

Most of the important developments have concerned either novel applications of organoboranes (reviewed in Chapter 9) or the reduction or addition reactions of carbonyl compounds.

The pre-eminence of hydride reducing agents for laboratory transformations has been maintained. The versatility of alkoxyhydridoaluminates has been reviewed[1], and this class of reagents is still being actively studied, especially by Czech chemists. The special virtues (solubility, safety and convenience in use) of sodium bis-(2-methoxyethoxy)dihydridoaluminate, $NaAlH_2$-$(OCH_2CH_2OMe)_2$ (1), have prompted the development[2-4] of analogous reagents such as $NaAlH(OCH_2CH_2NMe_2)_3$ (2) containing a tertiary amino group. The reagent (2) is soluble in aliphatic as well as in aromatic hydrocarbons, but is less efficient than (1) for the reduction of esters[2].

Aldehydes can be reduced rapidly and selectively by using tetrabutyl-ammonium cyanotrihydridoborate ($Bu^n_4\overset{+}{N}\ \overset{-}{BH_3}CN$) in HMPT at 25 °C in the presence of mineral acid (0.1–0.15 N); at higher acid concentration (1.5 N), ketones are also reduced[5]. New complex borohydrides, useful for the stereoselective reduction of cyclic ketones, have been described. Lithium hydrido-tri-s-butylborate is readily prepared[6] from lithium hydridotri-

$$LiAlH(OMe)_3 + Bu^s_3B \xrightarrow[15\ min,\ 25\ °C]{THF} LiBHBu^s_3 + Al(OMe)_3 \qquad (3.1)$$
$$(100\%)$$

methoxyaluminate, equation (3.1). An alternative procedure [equation (3.2)] is available[7] for the preparation of the corresponding potassium derivative.

$$KH + Bu^s_3B \xrightarrow[1\ h,\ 20-22\ °C]{THF} KBHBu^s_3 \qquad (3.2)$$

Both derivatives, which are commercially available, reduce ketones rapidly and quantitatively to alcohols of exceptionally high stereochemical purity (Table 3.1). By using lithium hydrido-tri-s-butylborate at −78 °C, even

Table 3.1 Reduction of various cyclic and bicyclic ketones with lithium hydrido-tri-s-butylborate in THF[6]

Ketone[a]	Temp.[a] /°C	Time /h	Major isomer[c]/%	Temp.[b] /°C	Time /h	Major isomer[c]/%
2-Methylcyclopentanone	0	1.0	98, cis	−78	1.0	99.3, cis
2-Methylcyclohexanone	0	0.5	99.3, cis			
3-Methylcyclohexanone	0	1.0	85, trans	−78	2.0	94.5, trans
4-Methylcyclohexanone	0	1.0	80.5, cis	−78	1.0	90, cis
4-t-Butylcyclohexanone	0	3.0	93, cis	−78	3.0	96.5, cis
3,3,5-Trimethylcyclohexanone	0	3.0	99.8, trans			
Norcamphor	0	0.5	99.6, endo			
Camphor	0	1.0	99.6, exo			

[a] Ratio of hydride to ketone was 1.25.
[b] Ratio of hydride to ketone was 2.0.
[c] Analysis by g.l.c. In all cases, total yield was essentially quantitative.

greater purity can be achieved[6]. Potassium hydrido-tri-isopropoxyborate[7,8] should also be a valuable addition to the range of hindered alkoxyhydrido-borates. Compared with other mild reductants (e.g. $NaBH_4$), it is very selective: 92% of the less stable *cis*-alcohol was obtained[8] in the reduction of 2-methylcyclohexanone at 0 °C.

The characteristics of thexylborane ($Me_2CHCMe_2BH_2$) as a selective reducing agent have been systematically evaluated[9] and compared with those previously determined for diborane and disiamylborane[$(Me_2CHCHMe)_2BH$]. Each of these reagents rapidly reduces aliphatic aldehydes and ketones, but the substituted boranes show useful variations in selectivity with functional groups other than carbonyl. The efficient and selective reduction of carboxylic acids to alcohols by diborane in THF has also been examined in more detail[10]. The conversion of monoethyl adipate into ethyl 6-hydroxyhexanoate (88%) illustrates the selectivity which can be achieved. The same transformation (71%) is accomplished[11] by $NaBH_4$ reduction of the reactive enol ester (3), prepared by standard treatment of the acid with *N*-ethyl-5-phenyl-isoxazolium 3′-sulphonate (Woodward's reagent K) [equation (3.3)]. Such

$$\bar{Ar}\underset{O}{\overset{+}{\diagup}}NEt \xrightarrow[MeCN]{RCO_2H,\ Et_3N} EtNHCOCH=\bar{C}\bar{Ar}O_2CR \qquad (3.3)$$

$$(3)$$

$$\bar{Ar} = \quad \text{(benzene ring with } SO_3^- \text{)}$$

reductions, carried out in aqueous solution, may be especially useful in protein and carbohydrate chemistry.

Other novel reducing agents for carbonyl compounds include amino-iminomethanesulphinic acid [formamidinesulphinic acid, thiourea *SS*-dioxide (4)] and polymethylhydrosiloxane (5). The former[12,13] reduces alkali-stable ketones efficiently [equation (3.4)], and has been applied to

$$R_2CO + NH_2C(:\overset{+}{N}H_2)SO_2^- + 2NaOH \xrightarrow[2-7\ h]{90\ °C} R_2CHOH + CO(NH_2)_2 + Na_2SO_3$$

$$(4) \qquad\qquad\qquad\qquad\qquad\qquad\qquad\qquad (3.4)$$

steroidal ketones[14] and in the preparation of 1-deuterio secondary alcohols[15]. In the presence of bis(acetoxydibutyltin) oxide as catalyst, the reagent (5), which is commercially available, selectively reduces aldehydes and ketones[16].

$$Me_3SiO(SiHMeO)_nSiMe_3 \quad (n \approx 35)$$

$$(5)$$

Mild neutral conditions are maintained during the reaction and work-up, and both air and water are tolerated. The reduction of carbonyl compounds by alkali metals (preferably potassium) in HMPT and a protic co-solvent (preferably t-butyl alcohol) has been described[17]: aliphatic tertiary amides are reduced to alcohols under similar conditions[18].

Reduction via hydrosilylation has become important with the discovery of catalysts, notably chlorotris(triphenylphosphine)rhodium, $(Ph_3P)_3RhCl$, for efficient homogeneous reactions[19-21]. Hydrolysis of the first-formed silyl

$$ (3.5) $$

ethers gives the alcohols in high overall yield [equation (3.5)]. The stereo-selectivity in the reduction of alicyclic ketones has been studied[22], and the reaction has been used in the selective reduction of the carbon–carbon double bond in αβ-unsaturated carbonyl compounds via 1,4-hydrosilyla-tion[20,23]. The use of related catalysts in homogeneous asymmetric reduction is described in Section 3.2.1.3.

A more complete account of the one-step preparation of alcohols from carbonyl compounds, alkyl halides and lithium has been given[24]. The organo-lithium compound (6) is a convenient vehicle for the introduction of a hydroxypropyl group [equation (3.6)][25]. A general study of the reaction of

$$ (3.6) $$

lithium alkyls with carbonyl compounds has been made[26], and a useful investigation[27] of the reactions of methylcalcium iodide has confirmed the resemblance between organo-calcium and -lithium reagents.

3.2.1.2 *Unsaturated alcohols*

The potential of oxymercuriation–demercuriation for the synthesis of alkenols by monohydration of dienes has been explored[28]. With symmetrical non-conjugated dienes, yields of the *monohydration* products close to the statistical value of 50% are realised, but much higher yields of specific alkenols are obtained with unsymmetrical dienes in which the double bonds differ significantly in reactivity[29], particularly when mercury(II) trifluoroace-tate is used for oxymercuriation [equation (3.7)].

$$ Me_2C=CH(CH_2)_7CH=CH_2 \rightarrow Me_2C=CH(CH_2)_7CH(OH)Me + $$
$$ 83\% $$
$$ Me_2C(OH)(CH_2)_8CH=CH_2 \qquad (3.7) $$
$$ 1\% $$

A novel route to allylic alcohols, in which an allylic sulphoxide anion is used as a synthon for a vinyl anion, has been developed[30,31]. The derivative (7) produced by α-alkylation (usually the major process) of the sulphoxide is cleaved by treatment with a thiophile such as trimethyl phosphite, via the sulphenate (8) formed by a [2,3]-sigmatropic rearrangement of the allylic

$$PhS(O)CH_2CH=CH_2 \xrightarrow{\text{Base}} PhS(O)CH \cdots \bar{C}H \cdots CH_2 \xrightarrow{\text{RX}} PhS(O)CHRCH=CH_2$$
$$(7)$$

$$(3.8)$$

$$RCH=CHCH_2OH \xleftarrow{(MeO)_3P} RCH=CHCH_2OSPh$$
$$(8)$$

sulphoxide [equation (3.8)]. Yields of alcohols from acyclic sulphoxides are typically *ca.* 75%.

The synthetic range of organocuprates is expanding rapidly[32], and the value of these reagents for joining functionalised residues is much enhanced by the development of mixed cuprates[33] which permit the selective transfer of only one group. For example, the reagent (9) can be used successfully for the attachment of an *O*-protected allylic alcohol residue (10), by conjugate addition, to cyclohex-2-enone in pilot experiments related to prostaglandin

$$R^1R^2CuLi \quad R^1 = \text{pent-1-ynyl}, \ R^2 = \text{group (10)}$$
$$(9)$$

$$R^3 = \text{dimethyl-t-butylsilyl}, \ R^4 = \text{n-pentyl}$$

$$(10)$$

synthesis[34]. Organocopper(I) reagents have been utilised[35] in a stereospecific synthesis of substituted allylic alcohols [equation (3.9)].

$$R^1Cu, \ MgBr_2 \xrightarrow[\text{(ii) } R^3CHXOCH_2CH_2Cl]{\text{(i) } R^2C\equiv CH} \quad \begin{array}{c} R^2 \\ \diagdown \\ R^1 \end{array} C=C \begin{array}{c} H \\ \diagup \\ CHR^3OCH_2CH_2Cl \end{array}$$
$$(X = Br \text{ or } Cl)$$

$$\downarrow Bu^nLi$$

$$(3.9)$$

$$\begin{array}{c} R^2 \\ \diagdown \\ R^1 \end{array} C=C \begin{array}{c} H \\ \diagup \\ CH(OH)R^3 \end{array}$$

Further routes to allylic alcohols from epoxides are described in Sections 3.5.3, 3.5.4.3, 3.5.5.1 and 3.5.5.3.

Allylic organolithium reagents, generated *in situ* by treating allylic mesitoates with lithium in THF, react with aldehydes and ketones to give homo-allylic alcohols[36]. The yields from hindered carbonyl compounds or from those lacking α-hydrogen atoms are comparable with those obtained in reactions of the Reformatsky type. The stereoselective conversion of δ-alkoxyallylic alcohols into *trans*-homoallylic alcohols by treatment with $LiAlH_4$ in dioxan [equation (3.10)] has been reported[37]. Yields are generally good

$$R^2R^3C(OR^1)CH=CHCR^4R^5OH \xrightarrow[\text{in dioxan}]{LiAlH_4} R^2R^3C=CHCH_2CR^4R^5OH \quad (3.10)$$

(60–70%) and the stereoselectivity is high when the alkoxy group is displaced from a secondary carbon atom.

Analogous rearrangements are involved in new syntheses[38,39] of α-allenic

$$R^2R^3C(OR^1)C\equiv CC(OH)R^4R^5 \xrightarrow[\text{Ether}]{LiAlH_4} R^2R^3C=C=CHC(OH)R^4R^5 \quad (3.11)$$

alcohols [equation (3.11)]. In a related synthesis[40], alcohols of this class are prepared by conjugate addition of organocuprates to α-alkynylepoxides

$$R^1-C\equiv C-\underset{\underset{R^2}{|}}{C}\overset{O}{-}\underset{\underset{R^3}{|}}{C}-R^4 \xrightarrow{R^5_2CuLi} R^1R^5C=C=CR^2C(OH)R^3R^4 \quad (3.12)$$

[equation (3.12)]. The base-induced fragmentation of tosylhydrazones of cyclic keto-ethers has been used[41] to prepare various classes of allenic alcohols in fair yields (48–58%). The stoichiometric and structural requirements for these reactions point to the following mechanism [equation (3.13)].

$$Me_2C(OH)C=C=CMe_2 + N_2$$

$$\text{(3.13)}$$

(Ts = tosyl)

New preparations of allenic alcohols which follow the *gem*-dibromocyclopropane route for the generation of allenes have been devised by Bertrand and his co-workers. Almost quantitative yields of α-allenic alcohols are obtained by hydrolysis of methylenebromocyclopropanes (11) [equation (3.14)], themselves prepared by the addition of dibromocarbene to the

$$\xrightarrow[\text{CaCO}_3,\,100\,°C]{H_2O,\,\text{dioxan}} CH_2=C=CHC(OH)R^1R^2 \quad (3.14)$$

(11)

corresponding allenes, followed by partial reduction of the dibromide[42]. Allylic alcohols are used as starting materials in an alternative route[43] [equation (3.15)], and comparable reactions were used[44] for the preparation

$$R^1R^2C{=}CR^3C(OH)R^4R^5 \quad \xrightarrow{:CBr_2} \quad$$

(3.15)

$$R^1R^2C{=}C{=}CR^3C(OH)R^4R^5$$

of hydroxyallenes with the functional groups more widely separated (protection of the hydroxy group as the trimethylsilyl ether before the alkyl-lithium treatment gave improved yields). Other syntheses of β-allenic alcohols have also been described[45]; an application of organoboranes to the synthesis of α-allenic alcohols is outlined in Chapter 9, Section 9.7.

3.2.1.3   Asymmetric synthesis

Methods for the asymmetric reduction of ketones remain under intensive study, and a thorough account of the earlier literature has been published[46].

Important developments have been made in the use of chiral catalysts for the homogeneous hydrogenation of prochiral ketones. The cationic rhodium(I) catalyst (12) first reported[47] gives alcohols enriched in the (R)-enantiomers, but with only low optical purity (1-phenylethanol, 8.6%; butan-2-ol, 1.9%). Similar optical purities are obtained in other studies[48,49]

$$[Rh(nbd)L_2]^+ \quad ClO_4^- \qquad\qquad nbd = norborna\text{-}2,5\text{-}diene$$
$$(12) \qquad\qquad\qquad L = (+)\text{-}(R)\text{-}benzylmethylphenylphosphine$$

in which related catalysts incorporating chiral phosphines are used. Considerable improvement in optical purity is observed when reduction is achieved via hydrosilylation catalysed by non-cationic rhodium(I) complexes[50]. Thus, (+)-(R)-1-phenylethanol (optical purity 43%) is obtained in 92% yield from acetophenone by using dimethylphenylsilane and a catalyst containing

(3.16)

Table 3.2 Asymmetric heterogeneous catalytic hydrosilylation[52] of acetophenone and isobutyrophenone by dihydrosilane[a]

No.[b]	Ketone	Silane	$\dfrac{Ketone}{Rh}$[c]	Time /h	Isolated alcohol; chemical yield/%	Optical yield /%	Optical yield with the soluble Rh-(+) catalyst[e]/%
1	PhCOMe	$H_2SiPhMe$	25	42	(−)-(S)-Phenylmethylcarbinol; 71	12	13
1′	PhCOMe	$H_2SiPhMe$	25	26	(−)-(S)-Phenylmethylcarbinol; 33	8	
2	PhCOMe	H_2SiPh_2	35	24	(−)-(S)-Phenylmethylcarbinol; 82	29	28
2′	PhCOMe	H_2SiPh_2	35	18	(−)-(S)-Phenylmethylcarbinol; 52	22.5	
3	PhCOMe	$H_2SiPh(C_{10}H_7)$[d]	33	45	(−)-(S)-Phenylmethylcarbinol; 100	58.5	58
3′	PhCOMe	$H_2SiPh(C_{10}H_7)$[d]	20	24	(−)-(S)-Phenylmethylcarbinol; 95	55	58
3″	PhCOMe	$H_2SiPh(C_{10}H_7)$[d]	25	30	(−)-(S)-Phenylmethylcarbinol; 84.5	52	
4	PhCOPr[i]	$H_2SiPhMe$	25	48	(−)-(S)-Phenylisopropylcarbinol; 55	6.5	20
5	PhCOPr[i]	H_2SiPh_2	25	48	(−)-(S)-Phenylisopropylcarbinol; 100	28	35

[a] 0.12 mequiv of catalytic units fixed on the resin which was prepared from resin containing 0.5 mequiv of diphosphine unit per gram; silane/ketone =2; solvent, C_6H_6; room temperature.
[b] In each series (1–1′). (2–2′), (3–3′–3″), the insoluble complex was prepared for the first experiment of the series and then reused for other experiment(s) of the series.
[c] Ratio between the number of mmoles of ketone and the total number of catalytic units (0.12).
[d] $H_2SiPh(C_{10}H_7)$, 1-naphthylphenyldihydrosilane.
[e] Results corresponding to experiments where silane/ketone = 2.

$(-)$-(S)-benzylmethylphenylphosphine. Other studies of asymmetric hydrosilylation have involved soluble platinum catalysts[51] and an insoluble polymer-based rhodium(I) complex[52]. The latter catalyst is prepared from a Merrifield resin (chloromethylated polystyrene) as indicated in equation (3.16). With appropriate silanes, useful optical yields of alcohols (Table 3.2) are obtained both with the heterogeneous catalyst and its soluble counterpart based on the diphosphine (13). The advantages of the insoluble catalyst are

its ready separation from the reaction products and that it can be re-used. A soluble complex of rhodium(I) with a chiral phosphine has been assessed[53] for the asymmetric hydroformylation of prochiral alkenes, but reduction

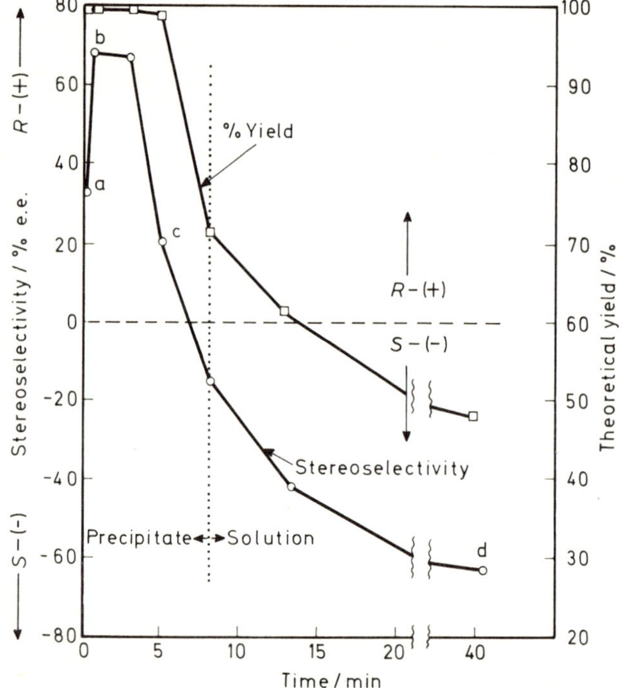

Figure 3.1   Change in stereoselectivity (○—○ left ordinate) and in percent yield (□—□ right ordinate) with age of reagent (LiAlH$_4$; R*OH ratio 1.0 : 2.3) in ether solvent at room temperature. (a) Zero time corresponds to results when acetophenone and RO*H in ether were added to LiAlH$_4$. (b) Acetophenone added to reagent 30 s after mixing. (c) After 8 min of shaking the initial precipitate had just dissolved, at which point acetophenone was added. (d) This sample was refluxed in ether for 3 min as well as standing for 40 min before acetophenone was added. e.e., excess of one enantiomer over the racemate; R*OH = (14) (From Yamaguchi and Mosher[55], by courtesy of the American Chemical Society)

of the first-formed aldehydes gives alcohols with optical purity of only
0.3–1.2%.

The reduction of ketones with chiral reagents derived from $LiAlH_4$ has
been studied further. Good stereoselectivities (40–75% excess of one enantiomer over the racemate) are obtained[54,55] on reducing acetophenone with
complexes produced by treating $LiAlH_4$ with limited amounts of (+)-
(2S,3R)-4-N,N-dimethylamino-3-methyl-1,2-diphenylbutan-2-ol (14) in diethyl ether. However, the nature and degree of the stereoselectivity varied
remarkably with the age of the reagent (Figure 3.1). With each of the carbonyl
compounds tested, the freshly prepared reagent gives high yields of alcohol
with the (−)-(R)-enantiomer in excess. However, the use of aged or previously
heated reagent leads to inferior yields with reversed stereoselectivity (except
for trifluoromethylphenyl ketone). Although some correlation between the
stereoselectivity of the reagent and its solubility is apparent (Figure 3.1), the
course of the reaction is also influenced by other factors (including the
composition of the reagent, and the solvent and temperature used)[55]. High
optical purities are also obtained[56] in the reduction of β-N,N-dialkylamino-
propiophenones ($PhCOCH_2CH_2NR_2$) with complexes derived from $LiAlH_4$
and (−)-menthol. The (+)-(R)-enantiomers predominate and the maximum
optical yield (73%) is obtained with R = Me and by using $LiAlH_4$ and
(−)-menthol in the molar ratio 1 : 3. As in reductions with chiral amino-
alcohols, the nitrogen of the substrate is considered to play an important
part in the asymmetric induction. Preliminary studies of two new chiral
aluminium hydrides (15 and 16), readily soluble in various aprotic solvents

$$CHMePhNMeAlH_2 \qquad\qquad [CHMePhNAlH]_n$$
$$(15) \qquad\qquad\qquad (16) \quad n \approx 6$$

even at −70 °C, have been reported[57]. By using reagent (15) derived from
the (−)-amine in diethyl ether at −71 °C, acetophenone is reduced to
(−)-(S)-1-phenylethanol in 84.5% optical yield, but with only 51% conversion. Although the use of higher reaction temperatures to obtain higher
conversions is detrimental to stereoselectivity, this can partly be offset by
the choice of a less basic solvent.

The numerous other studies of the asymmetric reduction of carbonyl
compounds include reports[58–60] on the use of aluminium alkyls (17) with
the chiral centre in either the β- or the γ-position relative to the aluminium

$$[CHEtMe(CH_2)_n]_3Al \qquad\qquad CHMePhNH_2{-}BH_3$$
$$(17) \quad n = 1 \text{ or } 2 \qquad\qquad (18)$$

atom, and of chiral amine–boranes (18)[61]. Optically active methyl p-tolyl
sulphoxide can be used to prepare alcohols of high optical purity[62]: both

$$CH_3S(O)Ar \xrightarrow[Et_2NH]{Bu^nLi} \bar{C}H_2S(O)Ar \xrightarrow{PhCHO} PhCH(OH)CH_2S(O)Ar$$
$$(19)$$

$$(Ar = p\text{-tolyl})$$

$$\Big\downarrow H_2 \Big| Ni(R) \qquad\qquad (3.17)$$

$$PhCH(OH)Me$$

enantiomers of 1-phenylethanol have been obtained [equation (3.17)]. The diastereoisomers (19) were separated and the individual enantiomeric alcohols were released by reductive desulphurisation.

3.2.1.4 Resolution of racemates

The methods of resolution applicable to alcohols have recently been reviewed[63]. The conversion of racemic alcohols into hydrogen oxalates, resolvable by classical methods, has been recommended[64]; the alcohols are readily regenerated by oxidative cleavage. The coupling of partially resolved alcohol residues by using an achiral reagent (e.g. carbonyl chloride, dichloro-dimethylsilane) gives a *threo* product enriched in the predominant enantiomer. By isolating this product, regenerating the alcohol and repeating the sequence of operations, the optical purity of the alcohol can be increased rapidly[65].

3.2.1.5 Isotopically labelled alcohols

Methods for the preparation of O-deuteriated alcohols[66] and of perdeuteri-ated methanol[67] have been reviewed, and the preparation of all eleven C-deuteriated ethanols has been described[68]. Primary alcohols labelled at C-2 can be obtained by reduction of the corresponding acids in which exchange of α-hydrogens has been effected[69] by heating them at 120 °C with 2 M DCl in acetic acid. Both enantiomers of 1-deuteriopropanol have been prepared by an exchange reaction catalysed by the enzyme diaphorase[70], and the enantiomers of 1-deuterio-1-phenylethanol (70% enantiomeric purity) have been obtained by asymmetric reduction of acetophenone[55]. Simple high-yield procedures for the preparation of ^{18}O-labelled alcohols have been developed[71].

3.2.2 Analytical methods

3.2.2.1 General methods

Progress in the detection, location, determination and classification of hydroxy groups has again been dominated by the application of physical methods. Numerous papers attest to the spectacular impact of chemical-shift reagents on n.m.r. spectroscopy. Several major reviews have already appeared[72], and the broad applications of shift reagents in ^1H n.m.r. and (to a less extent) ^{13}C n.m.r. spectroscopy will not, therefore, be discussed here. Their specific use in the determination of enantiomeric purities is described in Section 3.2.2.2.

The value of conversion into O-trimethylsilyl ethers for the determination of hydroxy groups by ^1H n.m.r. techniques, by producing characteristic signals with nine-fold amplification, has been demonstrated with carbo-hydrates[73] and other hydroxy compounds[74]. The similar virtues of ^{19}F n.m.r. spectra of hexafluoroacetone adducts in quantitative determinations have also been confirmed[75], and the influence of structural and environmental factors (particularly hydrogen bonding) on these spectral characteristics have been studied further[76]. Cautionary notes on other n.m.r. methods for the classification of alcohols based on the multiplicity of the hydroxy proton resonance[77] and on acylation by trichloroacetyl isocyanate[78] have been published.

The advantages of chemical-ionisation mass spectrometry for alcohols include the convenient determination of hydroxy groups. During the ion–molecule interactions in which D_2O is used as the reagent gas, exchange of active hydrogens occurs[79], and these can be enumerated by comparing the spectra obtained with D_2O and with H_2O as reagent gas. The different classes of alcohols can be differentiated by using nitric oxide as the reagent gas[80].

3.2.2.2 *Determination of absolute configuration and enantiomeric purity*

Non-polarimetric methods based on n.m.r. spectroscopy and gas–liquid chromatography (g.l.c.) are widely used in the study of dissymmetric alcohols, and technical advances have been made in both areas.

Two terpenoid acids (chosen for the rigidity of their carbon skeletons) have given improved g.l.c. separations of the diastereoisomeric esters from several secondary alcohols[81]. Highly sensitive (down to 1 µmol) g.l.c. adaptations of Horeau's empirical method of determining the absolute configurations of chiral secondary alcohols have been validated[82,83]. Horeau's method involves the kinetic partial resolution of (±)-2-phenylbutanoic anhydride during esterification, in which alcohols of configuration (20) generally react prefer-

M

HO — C — L
H

M = medium-sized group
L = larger group

(20)

entially with the (R)-acyl group. In the g.l.c. adaptations, the experimental outcome is determined by treating the excess of the anhydride with (+)-(R)-1-phenylethylamine, to give the separable diastereoisomeric amides (without further kinetic resolution). The restriction of product analysis to the amides neatly avoids the problems associated with the variable esters. Schemes for the determination of optical purity, asymmetric yield and absolute configuration of alcohols, based on the analysis of esters by g.l.c. or n.m.r., have been outlined[84].

The established n.m.r. methods involving the formation of distinguishable diastereoisomers or the use of chiral solvents have been powerfully supplemented by the advent of chiral shift reagents. Following the initial studies with tris-(3-t-butylhydroxymethylene-(+)-camphorato)europium (21a), the related reagents (21b)[85], (21c)[86] and (21d)[87] have been shown to discriminate more effectively between enantiomeric alcohols. For example, the diastereoisomeric interactions of reagent (21b) with (±)-2-phenylbutan-2-ol produce a difference in chemical shift of 0.29 p.p.m. for the enantiotopic α-methyl ^1H singlets[85] at a molar ratio of reagent to substrate of 0.78 : 1. By using the spin-decoupling technique, resolution of the α-^1H signals adequate for the determination of the enantiomeric purities of alkylarylmethanols[88] can be

R
a; Bu^t
b; CF$_3$
c; C$_3$F$_7$

d;

(21)

achieved with lower proportions of reagent, thus minimising the problem of line broadening. Correlations between α-^{1}H n.m.r. shift data and absolute configurations have been noted for the interactions of the alkylarylmethanols with the reagent (21b)[88], and of 1-deuterio primary alcohols with the reagent (21c)[89]. Achiral shift reagents have proved to be satisfactory for checking the purity of 1-deuterio primary alcohols[89,90] after their conversion into diastereoisomeric esters. Although attention has mainly been focused on ^{1}H n.m.r. spectra, the possibility of determining enantiomeric purity by the application of chiral shift reagents in ^{13}C n.m.r. spectroscopy has been demonstrated[91].

Further progress has been made in establishing correlations between configuration and ^{1}H n.m.r. chemical shifts for several diastereoisomeric esters of secondary alcohols[92]. The configurational model proposed for the esters of 2-trifluoromethyl-2-methoxyphenylacetic acid (Mosher's reagent) is depicted in Figure 3.2.

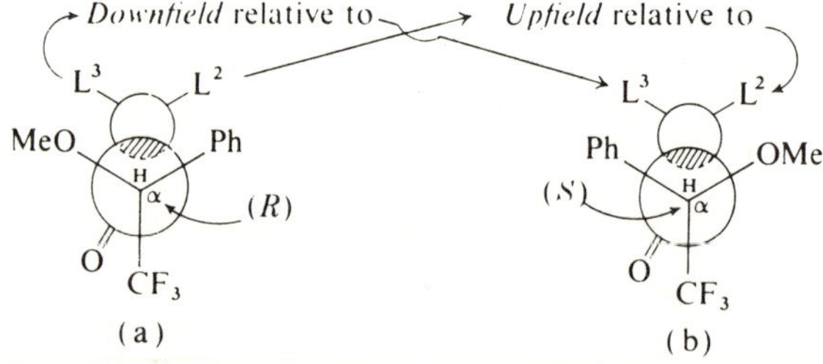

Figure 3.2 ^{1}H n.m.r. configurational correlation model for esters of chiral alcohols with (R)-2-trifluoromethyl-2-methoxyphenylacetic acid (a) and the (S)-isomer (b); L^2 and L^3 are the substituents for which relative chemical shifts were determined (From Dale and Mosher[92], by courtesy of the American Chemical Society)

3.2.3 Oxidation to carbonyl compounds

3.2.3.1 General methods

Two novel and efficient chromium(VI) oxidants have been introduced. Chromium trioxide suspended in dichloromethane readily forms a soluble complex, formulated as (22), with 3,5-dimethylpyrazole[93]. Oxidation may

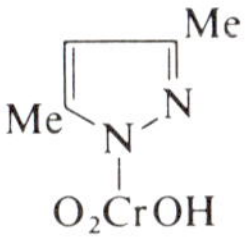

$$(22) \qquad (23)$$

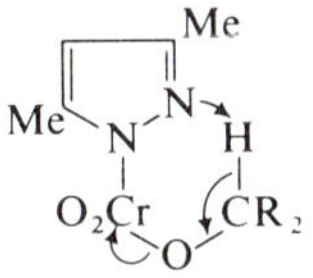

proceed by the intramolecular mechanism indicated (23). The intercalation of chromium trioxide in graphite makes the heterogeneous oxidation of alcohols technically very simple, although a prolonged reaction in boiling toluene is necessary[94]. According to the original report[94], the reagent selectively oxidises primary alcohols to aldehydes, but the commercial reagent (Seloxcette™) is claimed to be useful for the oxidation of secondary alcohols also.

Other new oxidative procedures have involved intermediate sulphonium complexes. Such complexes are readily formed by treating dimethyl sulphide with chlorine or (preferably) N-chlorosuccinimide (NCS) at 0 °C, followed by the addition of an alcohol [equation (3.18)]. Further treatment of the

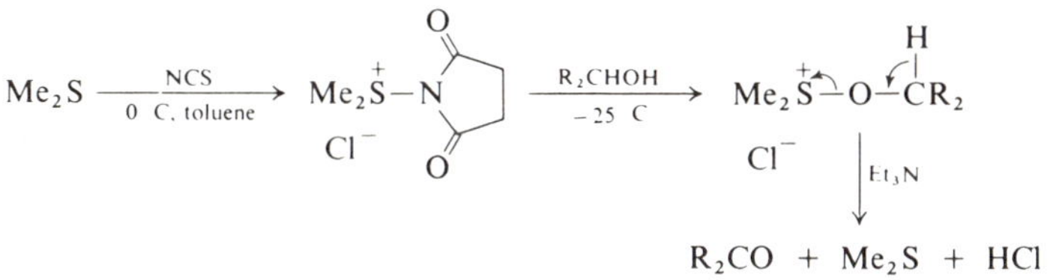

$$Me_2S \xrightarrow[\text{0 C, toluene}]{NCS} Me_2\overset{+}{S}-N \xrightarrow[\text{-25 C}]{R_2CHOH} Me_2\overset{+}{S}-O-CR_2 \tag{3.18}$$

$$R_2CO + Me_2S + HCl$$

complex with a suitable base gives the carbonyl compound (aldehyde or ketone) in high yield[95]. This mild and efficient process has already proved useful for the preparation of intermediates in prostaglandin synthesis[96]. One restriction is that alcohols which form stabilised carbocations are not oxidised, but are converted into halides (Section 3.2.6). This restriction does not apply to an analogous oxidant derived from chlorine and dimethyl sulphoxide[97], although this second oxidant does chlorinate alkenes. Very similar results are obtained with iodobenzene dichloride, which is only suitable[98] for the oxidation of saturated secondary alcohols [equation (3.19)].

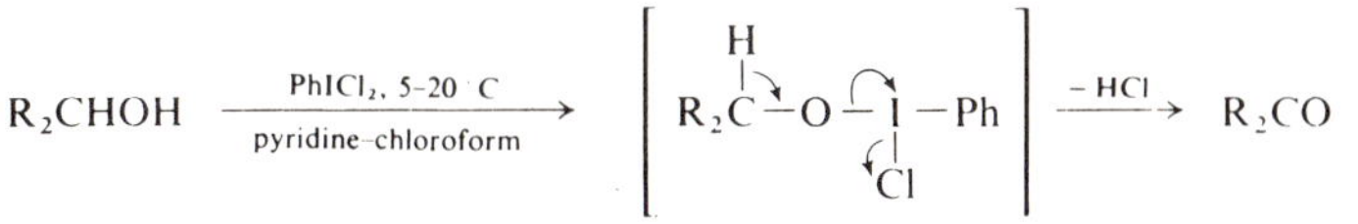

$$R_2CHOH \xrightarrow[\text{pyridine-chloroform}]{PhCl_2, \ 5-20 \ C} \left[R_2C-O-I-Ph \atop Cl \right] \xrightarrow{-HCl} R_2CO + PhI \tag{3.19}$$

Among other developments are the preparation and use of an insoluble polymer-based carbodi-imide for Moffatt oxidation with DMSO[99], and a simple laboratory adaptation of the procedure for vapour-phase oxidation of alcohols by copper(II) oxide[100]. The uses of *N*-bromo-*o*-sulphobenzoic imide (*N*-bromosaccharin)[101], its *N*-chloro analogue[102] and sodium ruthenate(VI)[103] as oxidants have been reported briefly.

3.2.3.2 *Mechanistic studies*

The recent literature again abounds with kinetic and mechanistic studies of chromium(VI) oxidations of hydroxy compounds. For oxidation of unhindered alcohols with chromic acid, the rate-limiting step is the decomposition of the intermediate chromate ester (24). This step is often represented as an electrocyclic process involving hydride transfer. However, the nature of the transition state of this two-electron process has remained controversial:

(24)   (25)

proton rather than hydride transfer and an acyclic mechanism involving the solvent have also been proposed. Support for a *symmetrical* rather than product-like planar transition state (25) for unhindered chromate esters can be obtained from measurements of the kinetic isotope effect (k_H/k_D) as a function of temperature[104]. By contrast, steric hindrance prevents the attainment of such a transition state in the oxidation of di-t-butylmethanol, and a non-planar, unsymmetrical transition state involving the solvent is proposed[104]. Both steric and electronic effects on the rates of oxidation of *ortho*-substituted benzhydrols have been studied[105]. It is concluded that the α-carbon remains largely sp^3 in character in the transition state, but that the degree of sp^2 character is increased by electron-donating substituents. Acceleratory steric effects are attributed to destabilisation of the reactant alcohol rather than to stabilisation (or modification) of the transition state.

Among other studies, the series of reports[106] by Roček's group on the co-oxidation of propan-2-ol and oxalic acid is of particular mechanistic interest. Together, both substrates are oxidised much more rapidly than individually. The kinetic results and product analyses can be rationalised on the basis of the formation and rate-limiting decomposition of a termolecular

$$\longrightarrow \quad Me_2CO \;+\; \overset{\cdot}{}CO_2 \;+\; \bar{O}-\overset{\cdot}{C}=O \;+\; Cr^{III} \qquad (3.20)$$

(26)

complex (26), or a similar acyclic structure, as shown in equation (3.20). In the presence of a free-radical scavenger (acrylonitrile), the stoichiometry of the oxidation accords with this equation. This novel one-step *three-electron* reduction of chromium(VI) was subsequently shown to characterise the co-oxidation of propan-2-ol with other bifunctional compounds, including hydroxy- and keto-acids[107].

Additional kinetic studies bearing on reaction mechanisms have included those of oxidations by bromine[108-110], *N*-bromosuccinimide[111], acid permanganate[112], potassium ferrate(VI)[113] and potassium hexacyanoferrate(III)[114].

3.2.3.3 *Thermal and photochemical oxidation*

Earlier studies by Ledwith and his colleagues on the involvement of short-lived alkoxyl radical intermediates in thermal and photochemical oxidations of alcohols have been consolidated[115]. The alternative processes [A and B in equation (3.21)] were differentiated by e.s.r. methods involving radical

$$\text{R}^1\text{R}^2\text{CHOH} \quad \begin{cases} \xrightarrow[\text{A}]{-\text{H}^\cdot} \text{R}^1\text{R}^2\dot{\text{C}}\text{OH} \\ \\ \xrightarrow[\text{B}]{-\text{e}} \text{R}^1\text{R}^2\text{CH}\overset{\cdot+}{\text{O}}\text{H} \xrightarrow{-\text{H}^\cdot} \text{R}^1\text{R}^2\text{CHO}\cdot \end{cases} \quad (3.21)$$

scavengers, principally the nitrone *N*-benzylidene-t-butylamine *N*-oxide (27) [equation (3.22)] and reagents known to produce hydroxyalkyl radicals.

$$\text{PhCH}=\overset{+}{\text{N}}\text{Bu}^t-\bar{\text{O}} \quad \begin{cases} \xrightarrow{\text{R}^1\text{R}^2\dot{\text{C}}\text{OH}} \text{R}^1\text{R}^2\text{C(OH)CHPhNBu}^t\dot{\text{O}} \\ \\ \xrightarrow{\text{R}^1\text{R}^2\text{CHO}\cdot} \text{R}^1\text{R}^2\text{CHOCHPhNBu}^t\dot{\text{O}} \end{cases} \quad (3.22)$$

$$(27)$$

Table 3.3 Oxidation of ethanol[115] in the presence of nitrone (27)[a]

Oxidant	Conc. /mol l^{-1}	Conditions[b]	Nitroxide radical splittings/mT		Radical trapped
			a_N	$a_{\beta\text{H}}$	
Paraquat dichloride	0.1	u.v.	1.449	0.268	EtO·
Pb(OAc)$_4$	0.01	u.v.	1.447	0.262	EtO·
UO$_2$(NO$_3$)$_2$	0.005	u.v.	1.550	0.360	MeĊHOH
K$_2$S$_2$O$_8$[c]	0.01	heat	1.450	0.272	EtO·
K$_2$S$_2$O$_8$[d]	0.01	heat	1.560	0.357	MeĊHOH
K$_2$S$_2$O$_8$	0.008 plus				
AgNO$_3$	0.004	r. temp	1.452	0.272	EtO·
EtONO	0.5	u.v.	1.447	0.259	EtO·
Bu^{t}OOBu^t	0.5	u.v.	1.536	0.362	MeĊHOH
Bu^{t}ON=NOBu^t	0.02	heat	1.537	0.361	MeĊHOH
Ph$_2$CO	0.03	u.v.	1.531	0.356	MeĊHOH

[a] Concentration of nitrone generally *ca.* 0.1 mol l^{-1}.
[b] Experiments in neat ethanol except those with paraquat dichloride, uranyl nitrate and potassium peroxydisulphate, in which aqueous ethanol solutions were used.
[c] [Nitrone] = 0.2–0.3 mol l^{-1}.
[d] [Nitrone] = 0.02 mol l^{-1}.

Depending to some extent on the alcohol and the reaction conditions, alkoxyl radicals can be detected during oxidations by paraquat dichloride, lead(IV) acetate, uranyl nitrate, potassium peroxydisulphate and the peroxydisulphate–silver(I) couple. Thus, each of these oxidants gives methoxyl radicals in neat or slightly aqueous methanol, but during photo-oxidation by uranyl acetate in 33% (v/v) aqueous methanol and in other alcohols (e.g. Table 3.3) only hydroxyalkyl radicals are detected. Alkoxyl radicals are not detected by matrix trapping techniques in the photo-oxidation of the higher primary alcohols by the uranyl ion[116]. The formation of alkoxyl radicals during the thermal, photochemical and silver(I)-catalysed oxidation of 2-phenylethanol by peroxydisulphate has been confirmed by product studies[115].

3.2.4　Oxidative cyclisation

During the treatment of higher alcohols with lead(IV) acetate under thermal or photolytic activation, direct oxidation to the corresponding carbonyl compounds competes with cyclisation and other reactions[117]. In a non-polar solvent such as benzene, cyclisation to tetrahydrofurans occurs preferentially with alcohols for which the intramolecular abstraction of a δ-hydrogen is possible, after the initial formation of an alkoxyl radical [equation (3.23)].

$$(3.23)$$

(28)

(29)

The predominant formation of the *trans* isomer of (29) in the oxidation of an unbranched secondary alcohol can be explained in terms of conformational preferences of the reaction intermediates[118]. Although the mechanism of the equivalent cyclisation induced by silver(I) oxide–bromine has been disputed, compelling evidence for a common mechanism for this and similar reactions utilising combinations of heavy-metal compounds with halogens ('hypohalite reactions') has been assembled[118]. According to this unified mechanism, each of these reactions proceeds via an intermediate δ-halogenohydrin [equation (3.24)]. The formation of the common intermediate (28) in oxida-

$$(28) \longrightarrow \xrightarrow{M^{n+}\ \text{or}\ HO^-} (29) \qquad (3.24)$$

$$X = Cl,\ Br\ \text{or}\ I$$
$$M = Pb^{IV},\ Hg^{II}\ \text{or}\ Ag^{I}$$

tions by lead(IV) acetate and by silver(I) oxide–bromine has been supported by studies of δ-deuterioalcohols[119,120].

The results of further studies of the effects of the solvent[121] and of the silver(I) salt[122] on the balance between cyclisation and carbonyl formation have been described. Cyclisation is almost completely suppressed by the use of silver(I) trifluoroacetate or by the addition of trifluoroacetic acid to the reaction mixture[122]. Improved procedures for the cyclisation of primary and secondary alcohols via hypochlorites and δ-chlorohydrins have been devised[118,123].

3.2.5 Dehydration to alkenes

Few new methods have been described during the period of this review; most studies under this section heading have concerned the mechanisms of established indirect methods of dehydration.

Anti-Saytzeff elimination occurs during the treatment of benzylic alkoxides with aluminium chloride, possibly by a concerted intramolecular *cis* mechanism[124]. The efficient dehydration of secondary alcohols can be achieved by reaction with methyltriphenoxyphosphonium iodide in HMTP[125]. Fuller accounts of the recently introduced dehydration methods utilising sulphuranes[126,127] or *N*-carbomethoxysulphamate esters[128] have been given.

Thermolysis of thiocarbonate *O*-esters such as (30b), prepared by the esterification of (30a) with *O*-4-methylphenyl chlorothiocarbonate, has proved useful for the dehydration of several sterically hindered alcohols

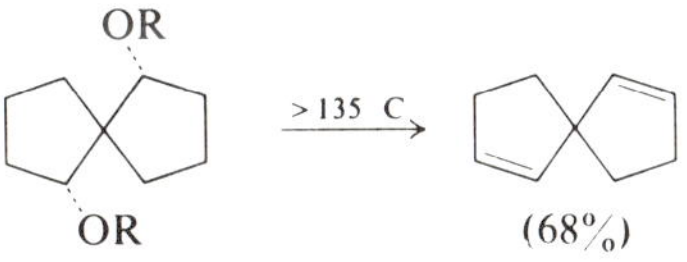

$$\text{OR} \quad \xrightarrow{> 135\ ^{\circ}C} \quad (68\%) \tag{3.25}$$

(30) a; R = H

b; R = CSO⟨C₆H₄⟩Me

[equation (3.25)][129,130]. Elimination during the thermolysis of thioacetate *O*-esters has been reported briefly[131], but not evaluated as a synthetically useful method. On the other hand, the smooth photolysis of thiobenzoate *O*-esters (31), in which R can conjugate with the alkene bond [equation (3.26)], should be a very useful, mild method for the dehydration of homoallylic alcohols[132]. A biradical mechanism is proposed for the reaction[133,134].

Further examples of rearrangements to dithiocarbonate di-*S*-esters during

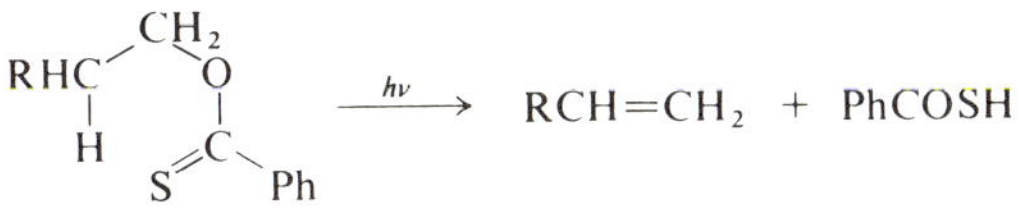

$$\text{RHC}\ \substack{\diagup CH_2 \\ \diagdown O} \quad \xrightarrow{h\nu} \quad RCH{=}CH_2 + PhCOSH \tag{3.26}$$

(31)

the thermolysis of xanthates (the Chugaev reaction) have been reported and possible mechanisms discussed[135,136]. Other papers on the thermolysis of esters contribute to the on-going debate about the nature of the transition state leading to *cis* elimination[137–140].

3.2.6   Replacement of the hydroxy group by halogen

The susceptibility of alkoxyphosphonium cations to nucleophilic attack at carbon constitutes the basis of useful methods for replacing hydroxy by other functional groups. A recent example of this approach to the conversion of alcohols into halides[141] utilises an *N*-halogeno-imide or -amide in the activation process [equation (3.27)]. The conversion of alcohols into car-

$$R^2OH + R^1{}_3P \xrightarrow{\ NXS\ } [R^1{}_3\overset{+}{P}OR\ \overset{-}{X}] \longrightarrow R^1{}_3P(O) + R^2X \qquad (3.27)$$

$$R^1 = Ph \text{ or } PhO$$
$$X = Cl, Br \text{ or } I$$

bazates ($ROCONHNH_2$), followed by oxidation with *N*-halogenosuccinimide in the presence of pyridine, is a new route to alkyl bromides and iodides. The reaction seems best suited to situations in which alkene formation (a side reaction) is difficult, and in which the conventional methods prove ineffective[142].

Secondary and tertiary alkyl fluorides can be prepared conveniently by treating alcohols with a pyridine–HF reagent (usually 30 : 70, w/w) in the presence of cyclohexane (to dissolve the product and minimise side reactions)[143]. Alternatively, fluorination may be carried out with a selenium tetrafluoride–pyridine complex [equation (3.28)][144]. In a third method, it was

$$PhCH_2OH \xrightarrow[CCl_2FCClF_2]{SeF_4\text{-pyridine}} PhCH_2OSeF_3 \xrightarrow{\ heat\ } \underset{100\%}{PhCH_2F + SeOF_2} \qquad (3.28)$$

initially reported[145] that *O*-trimethylsilyl ethers from all classes of alcohols give almost quantitative yields of alkyl fluorides on treatment with a fluoro-phosphorane (32) at or below room temperature [equation (3.29)]. However,

$$PhPF_4 + ROSiMe_3 \longrightarrow PhPOF_2 + RF + Me_3SiF \qquad (3.29)$$
$$(32)$$

in parallel studies[146], elimination was often found to compete seriously with fluorination, and this led to a more thorough investigation of the reaction[147]. Formation of the intermediate monoalkoxytrifluorophosphorane $PhPF_3(OR)$ (33) is almost always rapid and quantitative, and the spontaneous decomposition at low temperature (below $-50\,°C$) of the derivatives from tertiary alcohols leads to the alkyl fluorides in high yields. Heating is required for the more stable derivatives (33) from primary and secondary alcohols, and yields of the expected fluorides are usually about 40%. Various by-products were identified[147].

By treatment with sulphuryl chloride in pyridine–diethyl ether at $-60\,^{\circ}C$, benzylic alcohols are converted into the chlorides without the isolation of chlorosulphonates[148]. Inversion of configuration occurs during the chlorination of alcohols with thionyl chloride in HMPT, as a result of ionisation of the chlorosulphinate intermediates[149]. Some rearrangement occurs with allylic alcohols. The use of HMPT as solvent facilitates S_N2 displacements on neopentyl tosylate in the preparation of halides[150]. Two standard halogenation procedures have been recommended for the conversion of allylic alcohols into chlorides without rearrangement[151,152].

As mentioned in Section 3.2.3.1, sulphoxonium chlorides prepared from allylic or benzylic alcohols decompose to give the corresponding chlorides

$$Me_2S{-}OCHR^1R^2\ Cl \xrightarrow{-25\ \text{to}\ 0\ ^{\circ}C} Me_2SO + R^1R^2CHCl \qquad (3.30)$$

[equation (3.30)][153]. The selectivity of the chlorination is illustrated by the conversion of (34a) into (34b) in 87% yield. By using N-bromosuccinimide

$$\begin{array}{c}
Me \\ \ \ \ \ \ \ \ \ \diagdown \ \ \ \ \ \ \ \diagup H \\
\ \ \ \ \ \ \ \ \ \ \ \ C{=}C \\
HOCH_2CH_2 \ \ \ \ \ \diagup \ \ \ \ \diagdown CH_2X
\end{array}$$

(34) a; X = OH
b; X = Cl

for the preparation of the sulphoxonium salt, allylic and benzylic bromides can be prepared similarly. In related studies, good yields of bromides were obtained[154] from non-allylic alcohols by heating them with dimethylbromosulphonium bromide for 4–5 h at 80 °C. A useful technical improvement to the method of preparing alkyl bromides by the treatment of the methanesulphonates with anhydrous $MgBr_2$ in diethyl ether has been described[155]. The generality of the corresponding route to alkyl iodides by using MgI_2 and tosylate esters has been confirmed[156].

Fair yields of iodides can be obtained[157] by heating alcohols with cyanuric chloride in the presence of NaI. Good yields have been achieved by heating primary or secondary alcohols with N-methyl-N,N'-dicyclohexylcarbodiimidium iodide in THF or a hydrocarbon as solvent [equation (3.31)][158].

$$\begin{array}{c}
R^1 \\ \diagdown \overset{+}{N}{=}C{=}N\diagup R^1 \\ Me\diagup \ \ \ \ \ \ \ \ \ \ \ \ \ \ \ \\ I^-
\end{array}
\xrightarrow[35-50\ ^{\circ}C]{R^2R^3CHOH}
\left[\ Me\diagup N\diagdown C\diagup N\diagdown H \ \right]
\longrightarrow R^1NMeCONHR^1 + R^2R^3CHI \qquad (3.31)$$

3.2.7 Replacement of the hydroxy by the amino group

Tertiary alcohols can be converted into amines via alkylsulphamyl chlorides

$$\text{ROH} + \text{ClSO}_2\text{N}{=}\text{C}{=}\text{O} \xrightarrow{\text{hexane}} [\text{ROCONHSO}_2\text{Cl}] \qquad (3.32)$$

$$\downarrow \text{warm}$$

$$\text{RNHSO}_2\text{Cl}$$

$$(35)$$

(35), formed by the reaction with chlorosulphonyl isocyanate [equation (3.32)][159]. The intermediates (35) are degraded to amines via the formation and oxidation of the corresponding t-butyloxycarbonylhydrazides [equations

$$(35) + \text{NH}_2\text{NHCO}_2\text{Bu}^t \longrightarrow \text{RNHSO}_2\text{NHNHCO}_2\text{Bu}^t \qquad (3.33)$$

$$(36)$$

$$(36) \xrightarrow[\text{CH}_2\text{Cl}_2]{\text{Pb(OAc)}_4} \text{RNHSO}_2\text{N}{=}\text{NCO}_2\text{Bu}^t \xrightarrow{\text{H}^+} \text{RNH}_3{}^+ + \text{SO}_2 + \text{N}_2$$

$$+ \qquad (3.34)$$

$$\text{Me}_2\text{C}{=}\text{CH}_2$$

(3.33), (3.34)]. Another novel route to amines has emerged with the discovery[160] that alcohols can be converted into amides very rapidly and simply by treatment with chlorodiphenylmethylium hexachloroantimonate in nitriles as solvents, followed by quenching of the reaction with water. The probable mechanism of the reaction is shown in equation (3.35).

$$\text{R}^1{}_2\text{CHOH} \xrightarrow{\text{Ph}_2\overset{+}{\text{C}}\text{Cl}\ \overset{-}{\text{S}}\text{bCl}_6} \text{R}^1{}_2\text{CH}{-}\underset{\underset{\text{H}}{|}}{\overset{+}{\text{O}}}{-}\underset{\underset{\text{Cl}}{|}}{\text{CPh}_2} \xrightarrow{-\text{HCl}} \text{R}^1{}_2\text{CH}{-}\overset{+}{\text{O}}{=}\text{CPh}_2$$

$$\downarrow \text{R}^2\text{CN}$$

$$\text{R}^1{}_2\text{CH}{-}\overset{+}{\text{N}}{\equiv}\text{CR}^2 \qquad (3.35)$$

$$\downarrow \text{H}_2\text{O}$$

$$\text{R}^1{}_2\text{CHNHCOR}^2$$

Alkoxyphosphonium salts have been applied to the indirect conversion of alcohols into amines by the formation and reduction of azides. The decomposition of the salts [equation (3.27)] derived from primary alcohols and HMPT is retarded by the presence of dichloromethane or water[161], so that more powerful nucleophiles (such as azide ions) added to the reaction mixture can compete successfully with halide ions in displacement reactions. Alternatively, the alkoxyphosphonium chlorides can be converted into perchlorates, which can then be used in DMF for the monoalkylation of primary and secondary amines[162]. The scope of such reactions has been demonstrated further in selective conversions of the primary hydroxy groups of carbohydrates into amino and other functional groups[163,164].

3.2.8 Protection of hydroxy groups

An extensive review of protective groups has appeared[165]. The use of O-dimethyl-t-butylsilyl ethers is a promising new method[166] which has already made valuable contributions in prostaglandin and carbohydrate chemistry[166–168]. The derivatives are readily prepared by treating the hydroxy compounds with dimethyl-t-butylchlorosilane and imidazole in DMF. Selective removal of the protecting group can be achieved at room temperature by using aqueous acetic acid or tetra-n-butylammonium fluoride in THF. In these properties and in its stability to bases and to hydrogenolysis, the dimethyl-t-butylsilyl group usefully complements established protecting groups. Following the success of 4-methoxytetrahydropyran-4-yl (37a) as a protecting group, the properties of related sulphur-containing groups have been examined[169]. Compared with (37a), group (37b) is removed slightly

MeO   Me

X

(37) a; X = O
     b; X = S
     c; X = SO$_2$

more readily with acid, and group (37c) considerably less readily. The protection of alcohols as phosphoramide esters, $ROPO(NMe_2)_2$, has been suggested[170]; both the phosphorylation and regeneration steps require that any other functional groups be stable to butyl-lithium.

The selective degradation of alkyl allyl carbonates by nickel tetracarbonyl has been made the basis of a new method[171] of protection [equation (3.36)].

$$ROH + ClCO_2CH_2CH{=}CH_2 \xrightarrow[\text{pyridine}]{\text{THF}} ROCO_2CH_2CH{=}CH_2$$

$$\Big\downarrow \begin{array}{l} Ni(CO)_4 \text{ excess} \\ (Me_2NCH_2)_2,\ DMF, \\ 55\ ^\circ C,\ 4\ h \end{array} \qquad (3.36)$$

$$ROH$$
$$(87–95\%)$$

Allyloxy groups have found many applications in carbohydrate chemistry, and further studies in this area with the but-2-enyloxy and 2-methylallyloxy group have been reported[172]. The utility and complementarity of the various allylic groups are enhanced by the finding that the allylic ethers differ in their rates of isomerisation to acid-labile vinylic ethers in the presence of chlorotris(triphenylphosphine)rhodium(I) [equation (3.37)][173–175]. The efficient reductive cleavage of o-nitrobenzoates with Zn dust and NH_4Cl indicates that this acyl group may find protective applications[176]. Tritylpyridinium

$$ROCH_2CH{=}CH_2 \xrightarrow{(Ph_3P)_3RhCl} ROCH{=}CHMe \xrightarrow{pH\ 2} ROH + EtCHO \quad (3.37)$$

tetrafluoroborate appears to be a more reactive alternative to the familiar trityl chloride for the derivatisation of primary alcohols[177].

3.2.9 Miscellaneous reactions

A water-stable complex of $AlCl_3$ and cross-linked polystyrene has been introduced as a novel polymer-based reagent. The catalytic dehydrating action is manifested by the swelling of the polymer in a suitable solvent (e.g. benzene), and has been applied to the etherification[178] and esterification[179] of alcohols. The virtues of 4-N,N-dimethylaminopyridine as a catalyst for the acylation of sterically hindered alcohols have been extolled[180]. Dry HCl is an efficient catalyst for the conversion of alcohols into carbamates by reaction with cyanogen chloride; the corresponding mono-N-alkyl derivatives can be prepared by using a complex formed from cyanogen chloride, ferric chloride and the alkyl chloride[181]. The related reactions of diols with dichloromethyleneammonium salts are described in Section 3.3.6.

The remarkable replacement of hydroxy by methyl in alcohols which give relatively stable carbocations can be accomplished[182,183] by heating them with trimethylaluminium at 150–160 °C for about 20 h. The selective reduction of bromides, iodides and tosylates by sodium cyanotrihydridoborate ($NaBH_3CN$) in HMPT provides a rapid and convenient indirect method for the reductive deoxygenation of alcohols[184]. The reduction of phosphates by lithium–ethylamine [equation (3.38)] has been proposed as an alternative procedure[170].

$$ROH \xrightarrow[\text{(Me}_2\text{NCH}_2)_2]{\text{Bu}^n\text{Li, THF, X}_2\text{P(O)Cl}} P(O)X_2(OR) \xrightarrow{\text{Li, EtNH}_2\text{, Bu}^t\text{OH}} RH \qquad (3.38)$$

$$X = OEt \text{ or } NMe_2$$

3.3 DI- AND POLY-HYDRIC ALCOHOLS

3.3.1 Preparation of 1,2-diols

Useful improvements to three standard methods for the *cis* hydroxylation of alkenes have been reported. By using benzyltriethylammonium chloride as a 'phase-transfer catalyst', significantly better yields of the more lipophilic diols can be obtained[185] during heterogeneous hydroxylation with alkaline $KMnO_4$. The hazards associated with the use of osmium tetroxide can be minimised by the use of neutralised (acetic acid) solutions of potassium osmate and either H_2O_2 or sodium chlorate as oxidant[186]. The use of the expensive silver(I) acetate in Woodward's method, in which hydroxylation proceeds via an acetylated iodohydrin, is unnecessary[187]; various cheap and simple alternative procedures are equally effective.

The remarkably regio- and stereo-specific conversion of the allylic alcohol (38) into the *cis*-diol (39) via oxymercuriation–demercuriation [equation (3.39)] can apparently be attributed to the intramolecular addition of the hemiacetal hydroxy group generated by reaction in the presence of trichloro-acetaldehyde[188]. Unfortunately, the specificity of the reaction is not carried over to axial or acyclic alcohols.

The reductive coupling of carbonyl compounds with titanium(IV) chloride and zinc powder in THF is a good method for the preparation of pinacols[189]. Control of the temperature ($\leqslant 0\,^{\circ}\mathrm{C}$) is necessary to avoid an alternative reaction giving alkenes. Another reaction which can lead to pinacol formation is that between an aromatic carbonyl compound, magnesium and trimethylchlorosilane in HMTP[190]. Although yields of pinacols from benzaldehyde and acetophenone were *ca.* 90%, the scope of the reaction as a general method needs further study.

The stereoselectivity in the reduction of acyclic α-ketols with reagents of the aluminium hydride class has been investigated[191]. Although the predominant diols are the *erythro* diastereoisomers predicted by Cram's cyclic model for steric control of asymmetric induction, the degree of stereoselectivity with any particular reagent could not, in general, be correlated with the structure of the ketol. The simple method of enzymic reduction of α-ketols has been used for the preparation of several optically pure (R)-1,2-diols in satisfactory yields[192].

3.3.2 Structural assignments

N.m.r. methods are routinely used for the differentiation of diastereoisomeric diols. A general method for distinguishing between the *threo* and the *erythro* isomer with the partial structure $R^1MeC(OH)CHR^2(OH)$ has been described[193]. Irradiation of the methyl groups of the derived acetonides produces an intramolecular nuclear Overhauser effect on the adjacent methine proton in (40) but long-range W-type coupling in (41). Much enhanced separation of the ^{1}H n.m.r. methyl signals for isomeric 2,3-diarylbutane-2,3-diols, facilitating stereochemical assignments, is obtained by converting them into the phenylboronates[194]. An extensive study of the stereochemistry of various cyclic derivatives of 1,3-diols has been made[195].

(40)

(41)

3.3.3   Oxidation of 1,2-diols

An interesting method for the oxidation of secondary diols to α-ketols has been developed[196] from studies of de-acetalisation. The oxidation is effected by treating an acetonide (or other acetal) with trityl tetrafluoroborate in dichloromethane at room temperature [equation (3.40)]. The further oxidation of diols to α-diketones can be effected simply and conveniently by

$$ (3.40) $$

$$ R^2{}_2CO + R^1COCH(OH)R^1 $$

hydrogen transfer to monobenzylideneacetone with homogeneous catalysis[197] by dichlorotris(triphenylphosphine)ruthenium(II). Despite the rather drastic conditions (10 h at 195 °C), yields are generally satisfactory.

In addition to several kinetic and mechanistic studies of diol-cleaving reagents, new methods of preparative interest include oxidation with activated MnO_2 in dichloromethane at room temperature[198] and the use of thallium(III) salts[199]. The latter reagents are effective only with diols containing vicinal aryl substituents.

3.3.4   Conversion of 1,2-diols into alkenes

As part of a wider interest in stereospecific syntheses of alkenes, various new methods for the dehydroxylation (deoxygenation) of 1,2-diols have been explored. A convenient method[200] consists of heating the dialkoxide under reflux in THF with a tungsten(IV) salt. However, the conversions can be slow and the yields only moderate, whilst some isomerisation accompanies the predominantly *cis* elimination. Inferior stereospecificity was also found in a modification[201], in which iron pentacarbonyl was used for the desulphuri-sation–decarboxylation of cyclic thiocarbonates, of the Corey–Winter

$$PhCH{=}CHPh \qquad (3.41)$$

$$cis/trans \text{ ratio} = 0.31$$

method [equation (3.41)]. In a formally related method[202], alkenes were formed on vacuum thermolysis of cyclic carbonate tosylhydrazone salts. The salts were prepared by treatment of the diols with dichloromethylene-dimethylammonium chloride, followed by further reactions with tosylhydra-zine and NaH [equation (3.42)]. When the method is applied to pinacol

$$(3.42)$$

$$80\%$$

Ts = Tosyl

itself, the yield of 2,3-dimethylbut-2-ene is only 50% and the formation of the by-products (42) and (43) indicative of homolytic C—O cleavage in the intermediate carbene (44) suggests that the reaction may lack stereospecificity for acyclic diols.

The stereospecificity (*cis* elimination) associated with the Corey–Winter method (utilising trimethyl phosphite for desulphurisation of thiocarbonates), and with the methods of Eastwood and Whitham, has been shown to extend

$$CH_2{=}CMeCMe{=}CH_2 \qquad\qquad CH_2{=}CMeC(OH)Me_2$$

$$(42)\ 15\% \qquad\qquad\qquad (43)\ 5\%$$

$$(44)$$

to di-tertiary diols[203]. A fuller account of the scope and mechanism of alkene formation by fragmentation of 2-phenyl-1,3-dioxolans [Whitham's method, equation (3.43)] has been published[204]. Whereas the reaction is claimed to fail with pinacol and *meso*-hydrobenzoin[204], success was reported with *meso*-3,4-

$$\text{(3.43)}$$

dimethylhexane-3,4-diol and the isomeric 3-methylpentane-2,3-diols[203]. Additional applications of Eastwood's method [equation (3.44)] have been described[205]. Although unlikely to be a useful preparative procedure, the reaction of 1,2-diols with dichlorocarbene in the presence of a 'phase-transfer

$$MePhC(OH)C(OH)PhMe \xrightarrow{HC(OEt)_3} MePhC—CPhMe$$

meso

$$\downarrow PhCO_2H$$

$$\text{(3.44)}$$

$$MePhC=CPhMe$$

cis

catalyst' can lead to dehydroxylation in addition to other expected reactions[206]. Thus the probable alkoxychlorocarbene intermediate (45) in the reaction of pinacol may cyclise to the carbene intermediate (44) proposed for the Corey–Winter and related reactions.

(45)

A very simple but non-stereospecific conversion of 1,2-diols into alkenes is possible by treating the dimethanesulphonates with sodium anthracenide (or sodium naphthalenide) in THF or 1,2-dimethoxyethane[207]. This reaction should prove useful when other reducible functions and stereochemical complications are absent.

3.3.5 Cyclisation of 1,4-diols

Pyridinium chloride may be added[208] to the list of reagents[209] for effecting the cyclisation of 1,4-diols to tetrahydrofurans. The stereochemical and mechanistic features of these reactions have recently been scrutinised. In contrast to the similar reactions discussed in Section 3.2.4, the cyclisation of saturated di-secondary diols (catalysed by acid, DMSO or alumina) and of the derived monomethanesulphonates is stereospecific[209]. Thus *meso*-hexane-2,5-diol gives the *trans* product exclusively by means of an intramolecular

S_N2 displacement [equation (3.45)]. By first converting it into the disulphonate ester, the same diol can also be cyclised to the *cis* isomer, as this reaction gives rise to inversion at *both* chiral centres[209].

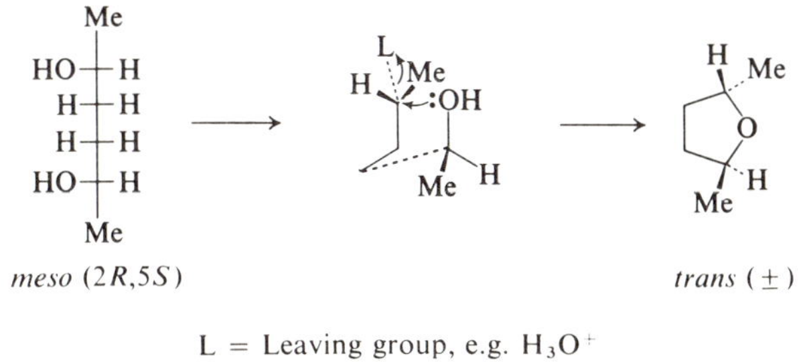

$$\text{meso } (2R,5S) \longrightarrow \longrightarrow \text{trans } (\pm) \tag{3.45}$$

L = Leaving group, e.g. H_3O^+

The stereospecificities reported above are not universal. Whereas configuration was retained during cyclisation of the mono- (presumed 1-) tosylate of $(-)$-(S)-4-methylhexane-1,4-diol[210], racemisation occurred during cyclisation of the diol with either acid or DMSO, indicating an S_N1 displacement at the tertiary C-4. The same mechanism apparently applies to the facile acid-catalysed cyclisation of 1-arylbutane-1,4-diols[211].

3.3.6 Miscellaneous reactions

The isomer distributions of the dienes formed during the dehydration of 3,4-dimethylhexane-3,4-diol with each of 11 different reagents have been recorded[212]. The acid-catalysed reaction of trimethylsilyl enol ethers with diols (1,2-, 1,3- or 1,4-) is recommended as a means of acetalisation[213].

The nucleophilic displacement reactions of alkoxyphosphonium salts have been used successfully for the replacement of only one of the two equivalent hydroxy groups of a neopentylic diol $[Me_2C(CH_2OH)_2]$[214]. Monoetherification of symmetrical, readily-available diols is practicable with the aid of a functionalised Merrifield resin to provide protecting acyl groups[215,216]. The formation of diesters during the protection stage is minimised by low-frequency functionalisation of the resin (about one acyl group to six phenyl residues) and by the use of an excess of diol.

Several novel procedures for the conversion of diols (1,2-, 1,3- and 1,4-) into the corresponding halogenohydrin esters have been announced in rapid succession. In an initial study by Newman's group[217], the diols are cyclised by reaction with an α-keto acid and the products treated with a phosphorus halide or thionyl chloride in dichloromethane at room temperature [equation

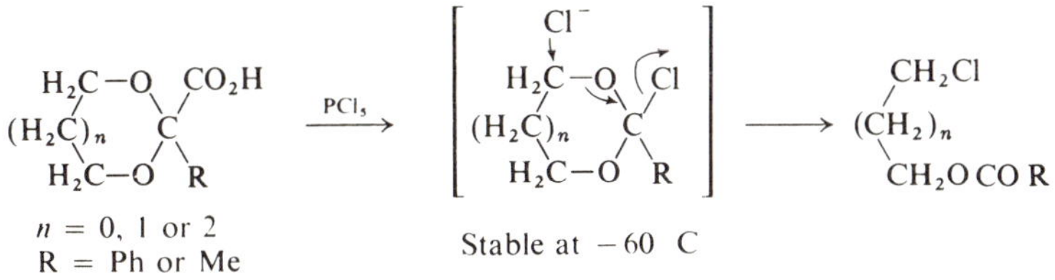

$$\tag{3.46}$$

$n = 0$, 1 or 2
R = Ph or Me

Stable at $-60\ °C$

(3.46)]. The regio- and stereo-specificities of such reactions involving unsymmetrical and optically active acetals are characteristic of S_N2 mechanisms. The weaknesses of the above route (modest yields of the initial acetals, and restrictions on the presence of other functional groups) are minimised and the advantages retained in similar reactions by utilising cyclic orthoacetates as the intermediates. Treatment of these derivatives with either trityl chloride[218] or trimethylsilyl chloride[219] under reflux in dichloromethane affords the chlorohydrin esters from 1,2-diols in high yields. The principal advantages of using trimethylsilyl chloride (economy, and greater ease of separating the products) are offset by inferior yields obtained with 1,3- and 1,4-diols. 1,3-Dioxolan-2-ylium cations are likely to be formed in the above reactions leading from 1,2-diols to the halogenohydrin esters, and evidence for such intermediate cations has been obtained with other reactions of this type. These reactions include the preparation of bromohydrin esters by the treatment of 2-phenyl-1,3-dioxolans with NBS[220], and the treatment of 1,2-diols themselves with HBr in acetic acid[221]. The latter process possesses the advantages of speed and simplicity, is apparently stereospecific with diols containing primary or secondary groups, and gives good regioselectivity with propane-1,2-diol (2-acetoxy-1-bromopropane, 94%; 1-acetoxy-2-bromopropane, 6%). However, some loss of stereospecificity was observed with 1-phenylethane-1,2-diol, while 1,2-diphenylethane-1,2-diol does not give the acetoxy-bromide[221]. Another general and direct route to chlorohydrin acetates (yields 60–90%) from diols (1,2-, 1,3- or 1,4-) which probably involves cyclic cationic intermediates consists of treating the diols with acetylsalicyloyl chloride in the absence of base[222].

A method for the direct conversion of the above classes of diols into the carbamates of chlorohydrins has emerged from further studies by Viehe and his colleagues of the interesting reactions of dichloromethyleneammonium salts[223] [the reaction with 1,2-diols has been exploited independently[202] in the dehydroxylation sequence given in equation (3.42)]. As in the formation of the carboxylates, unsymmetrical diols give mainly the product resulting from attack by the chloride ion at the least hindered carbon in the reaction intermediate (46).

$$\text{(46)}$$

3.3.7 Synthesis of simple glycerides

Methods in general use at about the time of the previous review have been summarised[224,225]. A method for the synthesis of 1-acylglycerols (particularly those derived from dicarboxylic acids) has been described[226]. Various approaches aimed at improved syntheses of *sn*-glycerol 2,3-cyclocarbonate, a valuable intermediate for the preparation of the corresponding lipids (D-1,2-diglycerides), have been explored[227]. A new route to 1,2-diacyl-*sn*-glycerols (L-1,2-diglycerides) from 2,5-*O*-methylene-D-mannitol, potentially applicable to lipids containing different and unsaturated fatty acids, has been pioneered[228].

The revived interest in the use of glyceryl halides in simple lipid syntheses has led to the $Ph_3P–CCl_4$ method being applied to the replacement of the 2-hydroxy group in 1,3-diacylglycerols[229]. As expected, the reaction proceeds with inversion at C-2, but *no* participation and migration of the neighbouring acyloxy groups (observed during the nucleophilic displacement of sulphonate by chloride) could be detected. Although acyl migration is not detected either in the nucleophilic displacement of chloride from 1,2-diacyl-3-chloro-3-deoxyglycerols by heating them with a sodium carboxylate in a dipolar aprotic solvent (thus providing a simple route to triglycerides)[230], the probability of its occurrence was highlighted by similar experiments with diacylglycerosulphonates[231].

3.4 ETHERS

3.4.1 Preparation

3.4.1.1 General methods

Several studies deal with variations on the Williamson synthesis of ethers from alcohols. The combination of NaH with MeI in THF has proved useful for the methylation of aromatic alcohols, but is of doubtful value with alkanols[232]. The use of a 'phase-transfer catalyst' ($Bu^n_4\overset{+}{N}\ \overset{-}{I}$) leads to generally high yields[233] of ethers in the methylation of alcohols with NaOH and Me_2SO_4. The preparation of several (unstable) trifluoromethanesulphonates and their use as potent alkylating agents have been described[234]. Other studies[235,236] re-emphasise the significant influence which the solvent may have in reactions of the Williamson type.

A new general synthesis of ethers involving the reductive coupling of carbonyl compounds and alcohols by using trialkylsilanes in the presence of acid has been developed [equation (3.47)][237,238].

$$R^1_2CO + R^2OH + R^3_3SiH \xrightarrow{H^+} R^1_2CHOR^2 + R^3_3SiOH \qquad (3.47)$$

Except with tertiary alcohols and secondary benzylic alcohols, good yields are obtainable. The reaction is applicable to both aldehydes and ketones, and many other functional groups are unaffected. The reduction of acetals to ethers via α-chloroethers by γ- or photoirradiation in the presence of trichlorosilane provides an alternative synthesis[239]. The utility of the route to ethers from alkenes via the formation and reduction of α-bromoethers may be improved by using tri-n-butylstannane as the reductant[240], although limitations are imposed by the modest regioselectivity of the alkoxybromination step[241].

3.4.1.2 Vinylic ethers

Two applications of metal–carbene complexes to the synthesis of vinylic ethers have been described. Thus, the ylide complex (47; R^1 = Ph) reacts rapidly with the Wittig ylide (48; R^2 = H) in diethyl ether at ambient temperature to give a high yield of methyl 1-phenylvinyl ether [equation (3.48)][242].

$$(CO)_5\bar{W}-\overset{+}{C}\overset{R^1}{\underset{OMe}{\diagup}} \quad + \quad R^2{}_2\bar{C}-\overset{+}{P}Ph_3 \quad \longrightarrow \quad R^2{}_2C=CR^1OMe \qquad (3.48)$$

$$\text{(47)} \qquad\qquad \text{(48)} \qquad\qquad\qquad + \ Ph_3PW(CO)_5.$$

Although the scope of the synthesis is not yet clear, the reaction fails for (48; R^2 = Me)[242] and for (47; R^1 = alkyl)[243]. This difficulty can be overcome[243] by using the less basic diazoalkanes in place of the phosphorane (48); methyl 1-methylvinyl ether is obtained in 93% yield by using diazomethane in this very facile reaction.

Certain functionalised vinylic ethers can be prepared by a conventional Wittig reaction in which a conjugated phosphorane is treated with a relatively reactive ester [equation (3.49)][244]. Ethers such as (49) can also be prepared

$$Ph_3\overset{+}{P}-\overset{-}{C}HPh + (CO_2Et)_2 \rightarrow PhCH=C(OEt)CO_2Et + Ph_3P(O) \qquad (3.49)$$

$$\text{(49)} \quad 85\%$$

by the reaction of phenylpropiolates with triphenylphosphine in the appropriate alcohol[245].

Recent developments in the preparation of β-hydroxy acids and the cyclisation of these to β-lactones form the basis of an excellent route from ketones to trialkylvinyl ethers, as indicated in the reaction sequence of equations (3.50) and (3.51)[246].

$$MeCH(OMe)CO_2H \xrightarrow[\text{THF}]{Bu^tNHLi} MeCLi(OMe)\dot{C}O_2Li$$

$$R_2CO \Big| THF \qquad\qquad (3.50)$$

$$R_2C(OH)CMe(OMe)CO_2H$$

$$\text{(50)}$$

$$\text{(50)} \xrightarrow[\text{pyridine, 0-5 C}]{PhSO_2Cl} R_2C\overset{\overset{\displaystyle O-CO}{|\quad\ |}}{-}CMe(OMe) \xrightarrow[\text{140-150 C}]{K_2CO_3} R_2C=CMe(OMe) \qquad (3.51)$$

3.4.2 Cleavage

A long-delayed fuller account[247] of the use of triphenylphosphine dibromide for the cleavage of ethers to produce alkyl bromides has appeared. The nature of the reagent imposes clear limitations on other functional groups which may be present, but good results are obtainable with simple primary and secondary aliphatic ethers. N,N-Dimethyltrichloromethylamine can be used for ether cleavage in similar situations[248]. Group VI metal carbonyls (particularly molybdenum hexacarbonyl) are effective catalysts for the thermal or photochemical cleavage of ethers by acyl halides[249].

Conditions for the electrophilic cleavage of ethers by reaction with benzynes have been reported, but this seems unlikely to be a process of preparative value[250,251]. On the other hand, nucleophilic cleavage by 1,1-diphenylhexyl-lithium [equation (3.52)] shows promise for ethers in which R^1 is an

$$R^1OR^2 + R^3Li \rightarrow R^1R^3 + R^2OLi \qquad (3.52)$$

allylic or vinylic group[252]. Studies of the palladium-catalysed exchange of allylic groups of ethers with active-hydrogen compounds and carboxylic anhydrides have been extended[253,254].

The disproportionation of alkyl trityl ethers during treatment with trityl salts can be used as an indirect method for the oxidation of alcohols to aldehydes or ketones[255]. The photolysis of several simple ethers by irradiation at 185 nm has been studied by von Sonntag's group[256,257]. Homolytic scission of the C—O bonds is the major primary process in each instance; the products of the reactions are identified and their quantum yields determined.

3.5 EPOXIDES

The literature on epoxides has been enriched by extensive reviews of stereochemical and mechanistic aspects of the synthesis[258] and ring-opening reactions[259] of these compounds. Other significant reviews deal with epoxidation reactions[260,261], rearrangements[262,263] and other ring-opening reactions[264,265].

3.5.1 Preparation

This account is restricted to methods of actual or potential importance in synthesis. For lack of space, preparations of novel and functionalised epoxides and reactions in which epoxides are incidental products have been excluded.

3.5.1.1 Reactions of peroxy compounds with alkenes

m-Chloroperoxybenzoic acid is widely favoured for the epoxidation of alkenes, and its suitability for reactions with passive and labile compounds has been enhanced. Unreactive alkenes can be epoxidised cleanly by treatment with the peroxy acid in 1,2-dichloroethane at 90 °C for 1–2 h in the presence of a radical inhibitor[266]. Yields of acid-labile epoxides can be improved by using the reagent in a two-phase system of dichloromethane and aqueous $NaHCO_3$ at ambient temperature[267]. Among new peroxy compounds described, p-methoxycarbonylperoxybenzoic acid has good stability and reactivity, and is readily prepared[268] by photo-oxidation of methyl *p*-formylbenzoate in CCl_4. Recent developments in epoxidations utilising alkyl hydroperoxides in the presence of metal ions are described in Section 3.6.2.

3.5.1.2 Cyclisation of halogenohydrins and their esters

The various new procedures for efficiently converting 1,2-diols into the corresponding halogenohydrin esters (Section 3.3.6) have increased the importance of these compounds as intermediates in stereospecific syntheses of epoxides. As inversion at one carbon atom occurs both during the formation of the halogeno-ester and during the subsequent ring closure with base, the epoxide and the parent diol have the same configuration. Thus, (R,R)-butane-2,3-diol is converted into (R,R)-2,3-epoxybutane by the cyclic ketal method [equation (3.53)][217], while optically pure $(-)$-(S)-1,2-epoxypropane

$$(3.53)$$

is prepared simply and efficiently (76% yield) from $(-)$-(S)-propane-1,2-diol by the HBr–acetic acid method[221].

A useful extension of the synthetic range is possible for reactions of 2-phenyl-1,3-dioxolans with NBS[220]. By carrying out this step with CCl_4 as solvent, *meso*-butane-2,3-diol is converted into *cis*-2,3-epoxybutane (80% overall yield) via the double inversion described above. However, when NBS is used in water, the diol monobenzoate is formed *without inversion*, and subsequent steps lead to racemic *trans*-2,3-epoxybutane (59% overall

$$(3.54)$$

Both enantiomers

yield) as shown in equation (3.54). An alternative to the conventional treatment with base for the cyclisation of halogenohydrins has been devised[269]. On treatment with tributylethoxytin, tributyl-2-halogenoalkoxytin compounds are formed which decompose on being heated (3 h at 120 °C) to give epoxides in excellent yields.

3.5.1.3 Alkylidene insertion reactions

The reactions of carbonyl compounds with ylides [e.g. (51) and (52), derived from sulphonium and oxosulphonium salts, respectively] are widely used for

$$Me_2\overset{+}{S}-\overset{-}{C}H_2 \qquad Me_2\overset{+}{S}(O)-\overset{-}{C}H_2$$
$$(51) \qquad\qquad (52)$$

the synthesis of epoxides. Although the ylide is normally formed and used under anhydrous conditions, two-phase systems incorporating aqueous NaOH are sometimes successful. Thus, the use of a surface-active sulphonium salt (53) gives good yields of epoxides with benzaldehyde [equation (3.55)], acetophenone and cyclohexanone[270]. Following a similar approach, a 'phase-

$$R\overset{+}{S}Me_2\overset{-}{Cl} + PhCHO \xrightarrow[\text{10 h, ambient temperature}]{\text{15 M NaOH, benzene}} PhHC\overset{O}{\overset{/\backslash}{-}}CH_2 + RSMe + NaCl \quad (3.55)$$
$$(53) \qquad\qquad\qquad\qquad\qquad (86\%)$$

$$R = \text{n-dodecyl}$$

transfer catalyst' $(Bu^n{}_4\overset{+}{N}\overset{-}{I})$ is included in a heterogeneous reaction mixture[271]. Despite the prolonged reaction time (48 h at 50 °C), styrene oxide is obtained in over 90% yield by methylene transfer from (51) to benzaldehyde, but low yields are obtained with the less reactive compound (52) and in reactions of ketones.

Following up earlier work[272] on the preparation of (54a) and validation of the compound as an active-methylene-transfer reagent, analogous reagents (55) derived from DMSO have now been prepared [equation (3.56)] and

$$Ar-\overset{\overset{\displaystyle O}{\|}}{\underset{\underset{\displaystyle NR^1{}_2}{|}}{\overset{+}{S}}}-\overset{-}{C}HR^2$$

$$(54) \quad a;\ Ar = Ph,\ R^1 = Me,\ R^2 = H$$
$$b;\ Ar = p\text{-tolyl},\ R^1 = Me,\ R^2 = H$$
$$c;\ Ar = p\text{-tolyl},\ R^1 = R^2 = Me$$

$$Me_2SO \xrightarrow[\text{CHCl}_3]{\text{NaN}_3,\ \text{H}_2\text{SO}_4} Me-\overset{\overset{\displaystyle O}{\|}}{\underset{\underset{\displaystyle NH}{\|}}{S}}-Me \xrightarrow{R_3\overset{+}{O}\ BF_4} Me-\overset{\overset{\displaystyle O}{\|}}{\underset{\underset{\displaystyle NR_2}{|}}{S}}-Me \xrightarrow[\text{DMSO}]{\text{NaH}} Me-\overset{\overset{\displaystyle O}{\|}}{\underset{\underset{\displaystyle NR_2}{|}}{\overset{+}{S}}}-\overset{-}{C}H_2 \quad (3.56)$$

$$(55)\ R = Me\ or\ Et$$

tested[273]. Both types of reagent react with carbonyl compounds on being heated at 50 °C for several hours. Optically active (dialkylamino)aryloxosulphonium alkylides (54) have also been employed in asymmetric syntheses of epoxides[274]. For example, the reaction of (R)-(54b) with benzaldehyde gives (R)-styrene oxide (yield 60%, optical purity 20%). Asymmetric induction occurs likewise in the reaction of a symmetrical ketone (4-t-butylcyclohexanone) with (R)-(54c) to give an epoxide, the chirality of which is derived

Table 3.4 Epoxides from ketones[281]

Ketone	Lithium reagent[a]	Hydroxy sulphide/ % isolated	Epoxide	Method[b]	Yield[c]/ % isolated
Hexadecan-2-one	A	56	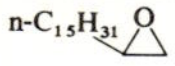	D	89
Cyclododecanone	B	65	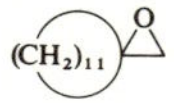	E	79
Di-t-butyl ketone	A	100		D	86
2,2,6,6-Tetramethylcyclohexanone	A B	91 92		E E	84 87
Benzyl t-butyl ketone	A	90	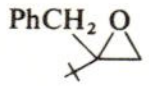	E	87
Deoxybenzoin	A	88	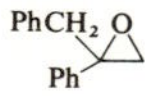	D	90
1,3-Diphenylpropan-2-one	A	41	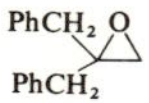	D	98
Cyclohexanone	C	81		D	92
Pentan-3-one	C	80	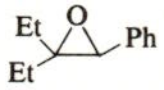	D	43[d]

[a] A, PhSCH$_2$Li(DABCO) in THF; B, CH$_3$SCH$_2$Li(TMEDA) in hexane; C, PhSCHLiPh in THF.
[b] D, β-hydroxy sulphide was alkylated with trimethyloxonium fluoroborate in methylene chloride followed by treatment of the solution with 0.5 N NaOH; E, trimethyloxonium fluoroborate in nitromethane, followed by treatment of the salt in methanol with 1.3 equiv of K$_2$CO$_3$ in methanol.
[c] For conversion of β-hydroxy sulphide into epoxide.
[d] The corresponding ketone, 4-phenylhexan 3-one (21 %), and alkene, 2-ethyl-1-phenylbut-1-ene (15 %), were also isolated.

from the nucleophilic carbon of the ylide rather than the carbonyl carbon. Other studies in this general area have included the use of ylides derived from benzyl(tetramethylene)sulphonium[275] and butadienyldimethylsulphonium[276] salts, and a stereospecific intramolecular methylene transfer [equation (3.57)][277].

$$\text{(3.57)}$$

Associated studies[278-282] on the conversion of β-hydroxy sulphides into epoxides [equation (3.58)] have been pursued. As the former compounds are readily accessible from ketones, this is an attractive alternative to the route utilising sulphur ylides (Table 3.4). In addition, mechanistic studies of the β-hydroxy sulphide reaction are also relevant to the ylide reaction, since both routes to epoxides converge on the betaine intermediate (56). The

$$\underset{\text{OH}}{R^1R^2C}-CHR^3-SR^4 \xrightarrow{R^5X} \underset{\text{OH}}{R^1R^2C}-CHR^3-\overset{+}{S}R^4R^5 \xrightarrow{\text{Base}} \underset{O^-}{R^1R^2C}-CHR^3-\overset{+}{S}R^4R^5$$

$$\text{(56)}$$

$$\text{(3.58)}$$

$$\underset{R^1R^2C-CHR^3}{\overset{O}{\diagup\diagdown}}$$

collapse of (56) apparently occurs with retention of configuration at the carbon bearing oxygen and inversion at the point of ring closure[280]. Thus, the synthesis of epoxides of high optical purity is possible by either route. On the other hand, oxosulphonium betaines are 'in equilibrium' with the corresponding ylides and carbonyl compounds[278,282]. Thus the optical purity of the styrene oxide obtained by the treatment of optically pure (57) with base is no better than that obtainable by the reaction of benzaldehyde with the appropriate optically active ylide.

A third procedure for the preparation of epoxides[283,284] by the transfer of an alkylidene group to a carbonyl compound consists of treating the latter with a geminal bromolithium reagent, prepared *in situ* from the corresponding dihalide [equation (3.59)]. Yields of epoxide are generally of the order of 60–70%.

$$\underset{H}{\overset{OH}{Ph-C}}-CH_2-\underset{NMe_2}{\overset{O}{\overset{\parallel}{S}}}{}^{+}-Ph\ \ BF_4^-$$

$$\text{(57)}$$

$$R^2CBr_2 \xrightarrow[\text{THF}]{\text{Li or Bu}^n\text{Li}} R^1R^2CBrLi \xrightarrow{R^3_2CO} \left[\begin{array}{c} OLi \\ | \\ R^1R^2C-CR^3_2 \\ | \\ Br \end{array} \right]$$

$$\downarrow \qquad\qquad (3.59)$$

$$R^1R^2C\overset{O}{\overbrace{\quad}}CR^3_2$$

3.5.2 Deoxygenation to alkenes

In addition to indirect methods proceeding via 1,2-diols (Section 3.3.4), more direct routes from epoxides to alkenes are of considerable interest both in synthesis and in the determination of structure. One new and efficient procedure[285] involves treatment of the epoxide (usually at ambient temperature or below) for several hours with a lower-valent tungsten halide. Even with acyclic epoxides, the retention of configuration is very high provided that LiI is included in the reaction mixture, when iodohydrins are likely intermediates. Other novel and mild procedures which are characterised by the retention of configuration are the treatment of epoxides with either triphenylphosphine sulphide[286] or selenide[287] in the presence of trifluoroacetic acid, probably following the route indicated in equation (3.60), and the treatment of epoxides with sodium (cyclopentadienyl)dicarbonylferrate and tetra-

$$(3.60)$$

$$X = S \text{ or } Se \qquad\qquad \text{(when } X = S\text{)}$$

fluoroboric acid, followed by decomposition of the resulting iron–alkene complex [equation (3.61)][288]. Although the steric demands of the organo-

$$(58) \xrightarrow[\text{THF}]{\text{Fp Na}} \quad \xrightarrow{\text{HBF}_4} \left[\begin{array}{c} \text{Fp–alkene} \\ \text{complex} \end{array} \right]^{+} \xrightarrow[\text{acetone}]{\text{NaI}} (59) \qquad (3.61)$$

$$Fp = [h^5\text{-}C_5H_5Fe(CO)_2]$$

metallic reagent in the latter method impede reaction with hindered epoxides, this can be utilised for selective monodeoxygenation of diepoxides (e.g. the conversion of 4-vinylcyclohexene diepoxide into 4-vinylepoxycyclohexane).

The complementary process of stereospecific inversion characterises the conversion of epoxides into alkenes via phosphorus betaines [equation (3.62)][289]. When allied to stereospecific methods of epoxidation, these reactions constitute a very useful means of inverting alkene isomers, although the use of strong base places restrictions on the presence of other functional groups.

$$(58) \xrightarrow[\text{THF}]{\text{Ph}_2\text{PLi}} \xrightarrow{\text{MeI}} \underset{R\ \ H}{\overset{\overset{-}{O}\ \ \overset{+}{\text{MePPh}}_2}{\text{H}\cdots\overset{|}{\diagup}\diagdown\cdots\text{R}}} \xrightarrow{25\ C} \underset{R\ \ H}{\overset{H\ \ R}{\diagup\diagdown}} + \text{Ph}_2\text{MePO} \qquad (3.62)$$

3.5.3 Reduction

An interesting study of the conjugate reduction of αβ-unsaturated epoxides to obtain either of the isomeric allylic alcohols has been reported briefly[290]. Thus, (60) gave the products shown [equation (3.63)]. Under the conditions

$$(3.63)$$

(60)

	Product ratio	Total yield
Method A, Bu$^i_2$AlH in boiling hexane	95 : 5	71%
Method B, Ca in liquid NH$_3$–THF	8 : 92	64%

described, only conjugate reduction was observed with (60), but competition from direct reduction occurred with compounds in which the double bond was relatively hindered.

The addition of triethylborane to a solution of LiAlH(OBu^t)$_3$ gives an exceptionally powerful reagent for the reductive ring-opening of cyclic ethers[291]. Although the effective reagent is a mixture of LiBHEt$_3$ and monomeric aluminium t-butoxide, the latter is unnecessary for the reduction of epoxides and its electrophilic intervention can lead to competitive rearrangements[292]. With both labile and hindered epoxides, reduction by pure LiBHEt$_3$ proceeds rapidly and cleanly, and with excellent regio- and stereo-specificity[292]. The stereochemical features of the reduction of the 2-phenyl-2,3-epoxybutanes by LiAlH$_4$, AlH$_3$ and AlHCl$_2$ and their deuterio-derivatives have been compared[293].

3.5.4 Rearrangement and isomerisation

3.5.4.1 Thermal and photochemical reactions

The characteristics of the thermal C—C bond fission of ethylene oxide have been discussed theoretically[294]; a preference for the biradical structure over the carbonyl ylide was inferred. With epoxycyclohexane the conrotatory motion which should follow such fission is prevented, and, in the temperature range 407–467 °C, compounds from C—O bond fission constitute ca. 97% of the primary products[295]. Supporting evidence for conrotatory motion, which is predictable from orbital symmetry considerations and has been demonstrated previously[296], has emerged from a study of the racemisation of optically active 2-phenyl-3-p-tolyloxiranes[297]. At 207 °C the *trans* isomer racemises sixty times faster than the *cis* isomer, by conrotatory formation of the more stable ylide intermediate [equation (3.64)]. In a further study[298],

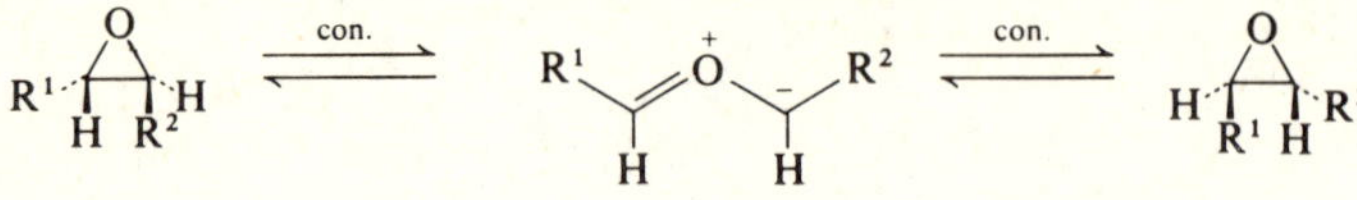

$$(3.64)$$

$$R^1 = p\text{-tolyl}; \quad R^2 = Ph$$

kinetic results for the facile racemisation of $(-)$-(R)-*trans*-2,3-divinyloxirane and the rearrangements to 2-vinyl-2,3-dihydrofuran and 4,5-dihydro-oxepin were obtained. The data support earlier conclusions[299,300] that carbonyl ylides are intermediates in the formation of these products. The formation of dihydro-oxepins from *cis*-epoxides is considered[300] to be a concerted [3,3]-sigmatropic rearrangement with a boat transition state, while the stereospecific formation of dihydrofurans requires disrotatory ring-closure of the intermediate carbonyl ylide. The thermal geometrical isomerisation of epoxides[298-300] can also be accommodated by this mechanistic scheme.

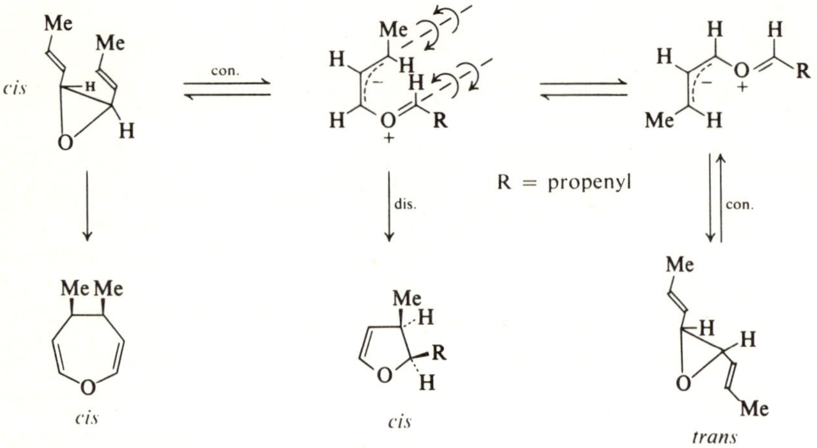

$$R = propenyl$$

Postulation of ylide intermediates is unlikely to explain the geometrical iso-merisation of the epoxide moiety in 3-isopropenyl-2-methyloxiranes[301] when irradiated at 300 nm in acetone. The rearrangement of 3,4-epoxy-alkenes to $\beta\gamma$-unsaturated ketones under these conditions has been des-cribed[301,302]. The ready formation of ylides by thermolysis of epoxides *gem*-disubstituted with electron-withdrawing groups has been applied in a new route to dioxoles [equation (3.65)][303].

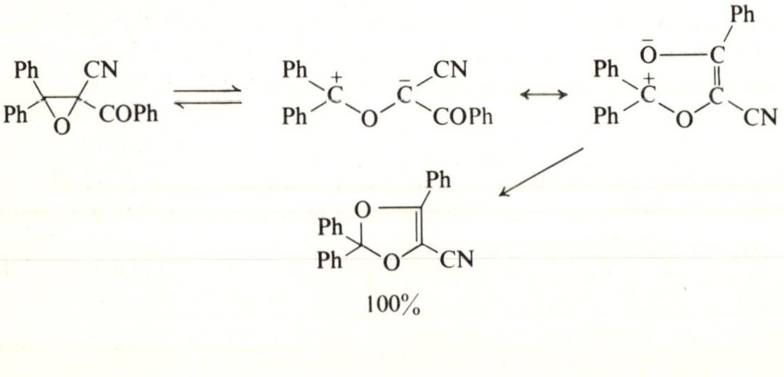

$$(3.65)$$

100%

3.5.4.2   *Acid-catalysed reactions*

Rearrangements with recyclisation with the participation of acetoxy or hydroxy substituents have been studied by Coxon and co-workers[304]. For acetoxy compounds the product composition (isomeric tetrahydrofurans) and the retention of configuration at the epoxy carbons can be explained by a mechanistic pathway involving orthoesters. Adventitious acid catalysis was invoked[305] to explain the facile thermal rearrangement (100 °C for 10–15 min) of aryl-substituted cyclopropyloxiranes to 3,6-dihydro-2*H*-pyrans [equation (3.66)]. The role of a rhodium(I) complex in catalysing the rear-

$$(3.66)$$

rangement of 2-vinyloxirane and related epoxides into αβ-alkenic aldehydes was also considered to be that of a Lewis acid[306]. The use of lithium salts for the catalysis of rearrangements (described in the previous review, p. 112) has been extended to the conversion of 2,3-epoxycyclohexanols into 1-formyl-cyclopentenes[307] and to studies of glycidic esters[308].

3.5.4.3   *Base-catalysed reactions*

Rickborn's group has continued its studies of rearrangements induced by lithium mono- and di-alkylamides[309,310]. By using epoxycyclohexane as a model, it was found that the yields of the allylic alcohol (cyclohex-2-en-1-ol) are maximal with $LiNR_2$ (R = primary alkyl), whereas bulky bases favoured the rearrangement to cyclohexanone[309]. Other studies by this group have involved epoxides with α- or β-phenyl substituents and diene monoepoxides[310]. The selective formation of allylic alcohols by treatment of epoxides with lithium phosphates has been confirmed[311,312].

3.5.5   **Nucleophilic substitution**

3.5.5.1   *Acid-catalysed reactions*

The glowing embers of controversy have been fanned by recent submissions[313-317] on the protic acid-catalysed hydrolysis of epoxides. The case for classical and distinct A1 and A2 mechanisms is expounded at length by Pritchard and Siddiqui[313]; the latter mechanism is held to apply to 1,2-epoxypropane (61) and 1-chloro-2,3-epoxypropane (epichlorohydrin), and the former mechanism to 1,2-epoxy-2-methylpropane (62). This last assignment is supported by a new value ($+8$ cal mol^{-1} K^{-1}) of $\Delta S^{\ddagger}$ for the hydrolysis[313]. On the other hand, Willi[316] has proposed an A2^{+} mechanism for both (61) and (62), in which the rate-limiting step is a bimolecular reaction between water and a carbonium ion formed relatively rapidly from the protonated epoxide. A suggestion that the hydrolysis of 2,3-dimethyl-2,3-epoxypropane proceeds via a kinetically free carbonium ion has been disputed[314]: contrary to an earlier report, the product was almost exclusively 2,3-dimethylbutane-2,3-diol (pinacol), with less than 1% of pinacolone formed. Attempts to explain the unexpected relative reactivities of hindered epoxides, from which 'abnormal' diols should arise by the A1 mechanism, were obscured by lack of evidence about the actual direction of ring-opening[315]. The *trans* addition characteristic of the A2 mechanism is confirmed for the alcoholysis of 2,3-epoxybutanes catalysed by BF_3 or a protic acid[318]. In an overview of acid-catalysis of ring-opening, in which the principal mechanistic criteria are reassessed, Wohl concludes that primary and secondary aliphatic epoxides react by the A2 mechanism but that further studies are necessary to resolve the issue with tertiary and monoaryl-substituted epoxides[317].

Macchia's group has continued its exploration of factors influencing the regio- and stereo-selectivity of ring-opening reactions of 1-aryl-1,2-epoxy-cyclohexanes with hydrogen halides and other acids[319]. This group has also noted that stable phosphonium salts are formed during ring-opening in HMPT[320]. The acid-catalysed concerted ring-opening of cyclopropyloxiranes [equation (3.67)] can be used to prepare halogen-substituted allylic alcohols with high yield and stereoselectivity[321]. An extended account of the prepara-

$$\text{(3.67)}$$

tion of hydroxyalkoxydimethylsulphonium salts by the acid-catalysed reaction of epoxides with DMSO has been given[322].

3.5.5.2 Reactions with organometallic compounds

Studies in this area are mainly in continuation of others reported during the period of the previous review. The principal reagents concerned are dialkyl-magnesiums[323,324], trialkylaluminiums[325,326] and lithium dialkylcuprates[327-329]. In general, the vagaries of the reactions (in terms of the nature, multiplicity, and stereochemistry of the products) mean that they are of greater mechanistic than preparative interest. The recent paper[327] by Johnson's group offers a useful summary of the synthetic scope of the reactions between epoxides and the organocuprates.

3.5.5.3 *Base-catalysed and uncatalysed reactions*

An MO theoretical study of the S_N2 ring-opening of ethylene oxide by a hydride ion has appeared[330]. The reactions of optically pure $(+)$-(R)-1,2-epoxybutane with several nucleophiles have been described: 'normal' ring-opening occurs in each instance[331].

Of particular synthetic interest is the use of an organoselenium reagent for nucleophilic substitution, leading to a stereospecific preparation of allylic alcohols [equation (3.68)][332]. The reagent is readily generated by the reduction

$$\text{(3.68)}$$

of diphenyl diselenide with $NaBH_4$. With acyclic epoxides, only the E-alcohol is apparently formed. Among the numerous other studies of nucleophilic substitutions may be noted the reactions of epoxides with carbanions from tertiary amides to give the hydroxyalkyl derivatives[333] and an improved experimental procedure for the preparation of amino-alcohols[334]. Additional studies of the relative reactivities towards nucleophiles of the epoxy and halogen moieties of epihalogenohydrins have been undertaken[335,336].

3.5.6 Miscellaneous reactions

The basic properties manifested by epoxides in the presence of halide ions have been surveyed[337] and utilised in dehydrohalogenations[338]. The extensive studies by Movsumzade and his colleagues of the concerted halogenation of epoxides and alkenes include bromochlorination[339] and iodobromination[340]. In contrast to previous experience with triphenylphosphine dichloride, the stereoselective conversion of epoxides into *cis*-1,2-dihalides can be achieved by using triphenylphosphine in carbon tetrachloride[341]. A novel route from epoxides to 1,4-diols is noted in Chapter 9 of this volume. The thermal reactions of gem-dicyanoepoxides with aldehydes[342] and Schiff bases[343] constitute further examples of reactions which can be attributed to the intermediate formation of carbonyl ylides. Conditions for the direct periodate oxidation of epoxides have been described[344].

3.6 HYDROPEROXIDES

The final part of the series on organic peroxides edited by Swern has appeared[345]. Association phenomena and their relevance to the decomposition of hydroperoxides have been reviewed[346]. These reactions have been discussed from an MO theoretical point of view[347,348].

3.6.1   Preparation

The full utilisation of alkyl groups in the autoxidation of organoboranes is possible with alkyldichloroboranes[349]. These compounds are oxidised rapidly in diethyl ether at $-18\,^\circ$C to give the hydroperoxides in high yields (*ca.* 90%).

3.6.2   Metal ion-catalysed epoxidation of alkenes and related reactions

Work in this area continues to dominate the literature on hydroperoxides. The industrial potential is reflected in the very many patents on epoxidation which have been disclosed, whilst many further studies of the kinetics and mechanisms of the reactions have been reported. Perhaps most notably, such reactions are playing a greater part in laboratory syntheses. The ability of catalysts based on molybdenum(VI) or especially vanadium(V) to complex with hydroxy compounds can lead to high regio- and stereo-selectivity, as well as acceleration, in the epoxidation of alkenols. Several high-yield syntheses have been based on these properties. Thus, the regioselective epoxidation of geraniol [equation (3.69)] and the stereoselective epoxida-

$$\text{(geraniol)} \xrightarrow[\text{benzene, boil, 4 h}]{\text{VO(acac)}_2,\ \text{Bu}^t\text{OOH}} \text{(6,7-epoxide)} \qquad (3.69)$$

$$\text{(cyclohex-3-en-1-ol)} \xrightarrow[\text{benzene, boil, 3 h}]{\text{Mo(CO)}_6,\ \text{Bu}^t\text{OOH}} \text{(3,4-epoxycyclohexan-1-ol)} \qquad (3.70)$$

tion of cyclohex-3-en-1-ol [equation (3.70)] have been described[350]. Although hydroxy groups are unaffected in these and other examples[351,352], the conversion of allyl alcohol into 2,3-epoxypropanal has been reported[353], and the oxidation of secondary alcohols to ketones by using a hydroperoxide in excess has been described[354]. For example, the oxidation of the allylic alcohol cholest-4-en-3β-ol to give cholest-4-en-3-one has been achieved[354]. Moreover, some alkenes can be oxidised through to α-ketols[355]. The scope of this reaction is not clear. When using MoCl$_5$ or Mo(CO)$_6$ as catalyst and t-pentyl hydroperoxide in benzene at 80 °C, Tolstikov's group failed to detect 2-hydroxycyclohexanone in the oxidation of cyclohexene, and state that only trisubstituted alkenes undergo this type of reaction. On the other hand, Sheldon[356] found that epoxycyclohexane gives the reaction on treatment with t-butyl hydroperoxide under very similar conditions, but only when MoCl$_5$ is used as the catalyst (possibly in a side reaction initiated by the production of HCl). During their extensive investigations of other metal-ion-catalysed oxidations, Tolstikov and his colleagues have also reported on the conversion of azomethines into oxaziridines[357,358], tertiary amines into N-oxides[357-359], nitrosoamines into nitroamines[357,358], sulphides into sulphoxides and sulphones[360], 1,4-dihydropyridines into pyridines[361] and the oxidation of condensed aromatic hydrocarbons, e.g. anthracene to give anthrone and anthraquinone[362]. The conversion of cyclohexylamine into cyclohexanone oxime[363] and that of azobenzenes into azoxybenzenes[364] are also possible.

The confines of this chapter prohibit an adequate account of the many studies devoted to the mechanism of the epoxidation reaction, but a set of leading references[356,365-374] is offered. Of special interest is the isolation of active catalysts from reaction mixtures incorporating molybdenum compounds. The catalysts were characterised[356] as Mo^{VI}–1,2-diol complexes related to the alkene being oxidised, e.g. (63) from propene, but not to the molybdenum compound used. A further point to be accommodated by the

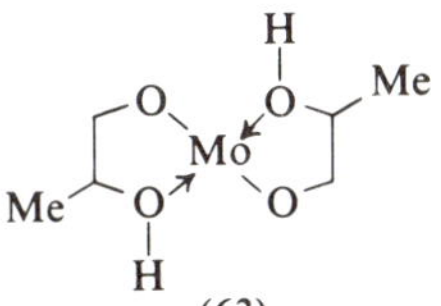

$$(63)$$

mechanistic schemes being disputed is the observation[371] that in the reactions of α-pinene and various steroids, epoxidation by the t-pentyl hydroperoxide–$MoCl_5$ combination proceeds with a stereospecificity the reverse of that for peroxy acids and covalent peroxides of molybdenum(VI).

3.6.3 Miscellaneous reactions

The base-catalysed oxidation of diarylphosphine oxides[375] and of sulphoxides[376] have been studied further, and the mechanisms of antioxidant action of cyclic phosphites have been investigated[377]. The decomposition of t-butyl hydroperoxide induced by 'positive halogen' compounds[378] is particularly rapid for ClO_2; t-butyl peroxyhypochlorite and di-t-butyl trioxide have been detected as reaction intermediates. The alkoxycarbonium ions proposed as intermediates in the acid-catalysed decomposition of hydroperoxides have been detected[379] by ^1H n.m.r. spectroscopy in reactions with FSO_3H–SbF_5 at $-40\,°C$. With t-butyl hydroperoxide, the ion $(Me_2\overset{+}{C}OMe)$ was formed immediately in $> 99\%$ yield.

3.7 DIALKYL PEROXIDES

The brevity of this section is less a measure of the interest in these compounds than of its specialised nature, the limited applications of these compounds in preparative organic chemistry, and the pressures generated by the wealth of material covered in the preceding sections of this chapter.

3.7.1 Preparation

Bloodworth's group has reported further on the peroxymercuriation of alkenes[380-382]. Epoxidation is a competing reaction[380] in the demercuriation of the products from terminal alkenes [equation (3.71)]. Whereas this side

$$R^1R^2C(OOBu^t)CH_2HgX \xrightarrow[\text{0-5 C}]{\text{NaBH}_4,\text{ NaOH}} R^1R^2C(OOBu^t)Me \ + \ R^1R^2C\underset{\displaystyle O}{\overset{\displaystyle \diagup\diagdown}{-}}CH_2 \quad (3.71)$$

reaction was not a serious factor in the preparation of secondary alkyl t-butyl peroxides ($R^1 = H$), it becomes the dominant reaction for di-tertiary alkyl peroxides. Starting from α-methylstyrene, the mixture of peroxide and epoxide contained 75 mol% of the latter. In the earlier studies by the same group of the peroxymercuriation of αβ-unsaturated esters and ketones, the demercuriation step was not attempted: this reaction has now been accomplished[381]. From isopropenyl methyl ketone and methyl methacrylate the novel α-carbonyl-substituted peroxides (64) were obtained [equation (3.72)],

$$CH_2{=}CMeCOY \ \longrightarrow \ XHgCH_2CMe(OOBu^t)COY \ \longrightarrow \ Me_2C(OOBu^t)COY \quad (3.72)$$

$$(Y = Me \text{ or } OMe) \qquad\qquad (64)$$

with relatively little interference from epoxidation (10 mol% for $Y = Me$, 30 mol% for $Y = OMe$), and surprisingly no reduction of the keto group by $NaBH_4$. Similarly, the β-methoxycarbonyl peroxides $R^1R^2C(OOBu^t)$-CH_2CO_2Me were obtained cleanly and in high yields (83–93%) from four αβ-unsaturated esters ($R^1R^2C{=}CHCO_2Me$), in which the mercuriation step has the reverse regiospecificity. Whereas this appears to be an excellent route to β-alkyloxycarbonyl peroxides, it seems to be inapplicable to β-keto-peroxides[381]. Peroxymercuriation has also been applied to αβ-unsaturated aldehydes, when peroxyacetal formation is a complicating factor[382].

3.7.2 Decompositions

The decomposition of di-t-butyl peroxide induced by aryl ether radicals (ArĊHOR) from ring-chlorinated ethers has been studied further by Goh[383]. The formation of the mixed acetals $ArCH(OR)(OBu^t)$, indicative of induced decomposition, is enhanced by the presence of chlorinated solvents (notably o-dichlorobenzene); decomposition products from the acetals have been identified[383]. Evidence for the induced decomposition of di-t-butyl peroxide by chlorotris(triphenylphosphine)rhodium(I) has been presented[384].

An ionic mechanism [equation (3.73)] is proposed[385] for the base-catalysed fragmentation of 2-t-butylperoxy-2-methylpropan-1-ol (65a) in aqueous methanol at 30 °C. A similar scheme had previously been proposed for the fragmentation of the corresponding acid (65b). Both compounds (65) undergo unusually fast thermal decomposition, and reaction pathways to explain the kinetic and product data have been elaborated[386]. For other di-t-alkyl peroxides, the reaction constant (ρ_I) for thermal decomposition in chlorobenzene at 150 °C was determined[386] as −0.291. Values for the polar sub-

$Me_2CXOOBu^t$

(65) a; $X = CH_2OH$
  b; $X = CO_2H$

$$Me_2\overset{|}{C} - O - OBu^t \xrightarrow{H_2O} Me_2CO + CH_2O + Bu^tOH \qquad (3.73)$$

[From (65a)]

stituent constant (σ^*, -0.520) and steric substituent constant (E_S, -1.96) of an alkylperoxy group (Bu^tOOCMe_2) have also been determined by Richardson and his colleagues[387].

References

1. Málek, J. and Černý, M. (1972). *Synthesis*, 217
2. Kříž, O. and Macháček, J. (1972). *Coll. Czech. Chem. Commun.*, **37**, 2175
3. Macháček, J., Čásenský, B. and Abrham, K. (1973). *Coll. Czech. Chem. Commun.*, **38**, 343
4. Kříž, O., Macháček, J. and Štrouf, O. (1973). *Coll. Czech. Chem. Commun.*, **38**, 2072
5. Hutchins, R. O. and Kandasamy, D. (1973). *J. Amer. Chem. Soc.*, **95**, 6131
6. Brown, H. C. and Krishnamurthy, S. (1972). *J. Amer. Chem. Soc.*, **94**, 7159
7. Brown, C. A. (1973). *J. Amer. Chem. Soc.*, **95**, 4100
8. Brown, C. A., Krishnamurthy, S. and Kim, S. C. (1973). *J. Chem. Soc. Chem. Commun.*, 391
9. Brown, H. C., Heim, P. and Yoon, N. M. (1972). *J. Org. Chem.*, **37**, 2942
10. Yoon, N. M., Pak, C. S., Brown, H. C., Krishnamurthy, S. and Stocky, T. P. (1973). *J. Org. Chem.*, **38**, 2786
11. Hall, P. L. and Perfetti, R. B. (1974). *J. Org. Chem.*, **39**, 111
12. Nakagawa, K. and Minami, K. (1972). *Tetrahedron Lett.*, 343
13. Nakagawa, K. (1973). *Jap. Pat.* 73 54 005
14. Herz, J. E. and de Márquez, L. A. (1973). *J. Chem. Soc. Perkin Trans. I*, 2633
15. Shanker, R. (1974). *Chem. Ind.*, 76
16. Lipowitz, J. and Bowman, S. A. (1973). *J. Org. Chem.*, **38**, 162; *Aldrichim. Acta*, **6**, 1
17. Larchevêque, M. and Cuvigny, T. (1973). *Bull. Soc. Chim. Fr.*, 1445
18. Larchevêque, M. and Cuvigny, T. (1973). *C.R. Acad. Sci., Ser. C*, **276**, 209
19. Ojima, I., Nihonyanagi, M. and Nagai, Y. (1972). *J. Chem. Soc. Chem. Commun.*, 938
20. Ojima, I., Kogure, T., Nihonyanagi, M. and Nagai, Y. (1972). *Bull. Chem. Soc. Jap.*, **45**, 3506
21. Corriu, R. J. P. and Moreau, J. J. E. (1973). *J. Chem. Soc. Chem. Commun.*, 38
22. Ojima, I., Nihonyanagi, M. and Nagai, Y. (1972). *Bull. Chem. Soc. Jap.*, **45**, 3722
23. Ojima, I., Kogure, T. and Nagai, Y. (1972). *Tetrahedron Lett.*, 5035
24. Pearce, P. J., Richards, D. H. and Scilly, N. F. (1972). *J. Chem. Soc. Perkin Trans. I*, 1655
25. Eaton, P. E., Cooper, G. F., Johnson, R. C. and Mueller, R. H. (1972). *J. Org. Chem.*, **37**, 1947
26. Buhler, J. D. (1973). *J. Org. Chem.*, **38**, 904
27. Kawabata, N., Nakamura, H. and Yamashita, S. (1973). *J. Org. Chem.*, **38**, 3403
28. Brown, H. C., Geoghegan, P. J., Lynch, G. J. and Kurek, J. T. (1972). *J. Org. Chem.*, **37**, 1941
29. Brown, H. C. and Geoghegan, P. J. (1972). *J. Org. Chem.*, **37**, 1937
30. Evans, D. A., Andrews, G. C. and Sims, C. L. (1971). *J. Amer. Chem. Soc.*, **93**, 4956

31. Evans, D. A., Andrews, G. C., Fujimoto, T. T. and Wells, D. (1973). *Tetrahedron Lett.*, 1385, 1389
32. Normant, J. F. (1972). *Synthesis*, 63
33. Mandeville, W. H. and Whitesides, G. M. (1974). *J. Org. Chem.*, **39**, 400 and references cited therein.
34. Corey, E. J. and Beames, D. J. (1972). *J. Amer. Chem. Soc.*, **94**, 7210
35. Normant, J. F., Cahiez, G., Chuit, C. and Villieras, J. (1973). *Tetrahedron Lett.*, 2407
36. Katzenellenbogen, J. A. and Lenox, R. S. (1973). *J. Org. Chem.*, **38**, 326
37. Claesson, A. and Bogentoft, C. (1973) *Synthesis*, 539
38. Cowie, J. S., Landor, P. D. and Landor, S. R. (1973). *J. Chem. Soc. Perkin Trans. I*, 720
39. Claesson, A., Olsson, L.-I. and Bogentoft, C. (1973). *Acta Chem. Scand.*, **27**, 2941
40. Ortiz de Montellano, P. R. (1973). *J. Chem. Soc. Chem. Commun.*, 709
41. Foster, A. M. and Agosta, W. C. (1972). *J. Org. Chem.*, **37**, 61
42. Monti, H., Leandri, G. and Bertrand, M. (1972). *C.R. Acad. Sci., Ser. C*, **274**, 734
43. Maurin, R. and Bertrand, M. (1972). *Bull. Soc. Chim. Fr.*, 2349
44. Ragonnet, B., Santelli, M. and Bertrand, M. (1973). *Bull. Soc. Chim. Fr.*, 3119
45. Santelli, M. and Bertrand, M. (1973). *Bull. Soc. Chim. Fr.*, 2326, 2331, 2335
46. Morrison, J. D. and Mosher, H. S. (1971). *Asymmetric Organic Reactions* (Englewood Cliffs, N.J.: Prentice-Hall)
47. Bonvicini, P., Levi, A., Modena, G. and Scorrano, G. (1972). *J. Chem. Soc. Chem. Commun.*, 1188
48. Tanaka, M., Watanabe, Y., Mitsudo, T., Iwane, H. and Takegami, Y. (1973). *Chem. Lett.*, 239
49. Solodar, A. J. (1973). *Ger. Offen.* 2 306 222
50. Ojima, I., Kogure, T. and Nagai, Y. (1973). *Chem. Lett.*, 541
51. Yamamoto, K., Hayashi, T. and Kumada, M. (1972). *J. Organometal. Chem.*, **46**, C65
52. Dumont, W., Poulin, J.-C., Dang, T.-P. and Kagan, H. B. (1973). *J. Amer. Chem. Soc.*, **95**, 8295
53. Tanaka, M., Watanabe, Y., Mitsudo, T., Yamamoto, K. and Takegami, Y. (1972). *Chem. Lett.*, 483
54. Yamaguchi, S., Mosher, H. S. and Pohland, A. (1972). *J. Amer. Chem. Soc.*, **94**, 9254
55. Yamaguchi, S. and Mosher, H. S. (1973). *J. Org. Chem.*, **38**, 1870
56. Andrisano, R., Angeloni, A. S. and Marzocchi, S. (1973). *Tetrahedron*, **29**, 913
57. Giongo, G. M., Di Gregorio, F., Palladino, N. and Marconi, W. (1973). *Tetrahedron Lett.*, 3195
58. Giacomelli, G. P., Menicagli, R. and Lardicci, L. (1971). *Tetrahedron Lett.*, 4135; (1973). *J. Org. Chem.*, **38**, 2370
59. Kretchmer, R. A. (1972). *J. Org. Chem.*, **37**, 801
60. Lardicci, L., Giacomelli, G. P. and Menicagli, R. (1972). *Tetrahedron Lett.*, 687
61. Borch, R. F. and Levitan, S. R. (1972). *J. Org. Chem.*, **37**, 2347
62. Tsuchihashi, G., Iriuchijima, S. and Ishibashi, M. (1972). *Tetrahedron Lett.*, 4605
63. Klyashchitskii, B. A. and Shvets, V. I. (1972). *Usp. Khim.*, **41**, 1315
64. Molotkovsky, J. G. and Bergelson, L. D. (1973). *Chem. Phys. Lipids*, **11**, 135
65. Vigneron, J. P., Dhaenens, M. and Horeau, A. (1973). *Tetrahedron*, **29**, 1055
66. Verbit, L. (1972). *Synthesis*, 254
67. Ahmadi, A., Herbert, M. and Pichat, L. (1972). *J. Label. Compounds*, **8**, 37
68. Bengsch, E., Corval, M. and Delaumeny, M. (1973). *Bull. Soc. Chim. Fr.*, 1788
69. Garnett, J. L., Halpern, B. and Kenyon, R. S. (1972). *J. Chem. Soc. Chem. Commun.*, 135
70. Günther, H., Biller, F., Kellner, M. and Simon, H. (1973). *Angew. Chem. Int. Ed. Engl.*, **12**, 146
71. Sawyer, C. B. (1972). *J. Org. Chem.*, **37**, 4225
72. Cockerill, A. F., Davies, G. L. O., Harden, R. C. and Rackham, D. M. (1973). *Chem. Rev.*, **73**, 553 and references cited therein
73. Bebault, G. M., Berry, J. M., Dutton, G. G. S. and Gibney, K. B. (1972). *Anal. Lett.*, **5**, 413
74. Hase, A. and Hase, T. (1972). *Analyst*, **97**, 998
75. Ho, F. F.-L. (1973). *Anal. Chem.*, **45**, 603

76. Leander, G. R. (1973). *Anal. Chem.*, **45**, 1700
77. Slocum, D. W. and Jennings, C. A. (1972). *Tetrahedron Lett.*, 3543
78. Samek, Z. and Novotný, L. (1972). *Tetrahedron Lett.*, 5167
79. Hunt, D. F., McEwen, C. N. and Upham, R. A. (1972). *Anal. Chem.*, **44**, 1292
80. Hunt, D. F. and Ryan, J. F. (1972). *J. Chem. Soc. Chem. Commun.*, 620
81. Brooks, C. J. W., Gilbert, M. T. and Gilbert, J. D. (1973). *Anal. Chem.*, **45**, 896
82. Brooks, C. J. W. and Gilbert, J. D. (1973). *J. Chem. Soc. Chem. Commun.*, 194
83. Gilbert, J. D. and Brooks, C. J. W. (1973). *Anal. Lett.*, **6**, 639
84. Weidmann, R. and Horeau, A. (1973). *Tetrahedron Lett.*, 2979
85. Goering, H. L., Eikenberry, J. N. and Koermer, G. S. (1971). *J. Amer. Chem. Soc.*, **93**, 5913
86. Fraser, R. R., Petit, M. A. and Saunders, J. K. (1971). *Chem. Commun.*, 1450
87. Whitesides, G. M. and Lewis, D. W. (1971). *J. Amer. Chem. Soc.*, **93**, 5914
88. Červinka, O., Maloň, P. and Trška, P. (1973). *Coll. Czech. Chem. Commun.*, **38**, 3299
89. Reich, C. J., Sullivan, G. R. and Mosher, H. S. (1973). *Tetrahedron Lett.*, 1505
90. Gerlach, H. and Zagalak, B. (1973). *J. Chem. Soc. Chem. Commun.*, 274
91. Fraser, R. R., Stothers, J. B. and Tan, C. T. (1973). *J. Magn. Resonance*, **10**, 95
92. Dale, J. A. and Mosher, H. S. (1973). *J. Amer. Chem. Soc.*, **95**, 512
93. Corey, E. J. and Fleet, G. W. J. (1973). *Tetrahedron Lett.*, 4499
94. Lalancette, J.-M., Rollin, G. and Dumas, P. (1972). *Can. J. Chem.*, **50**, 3058
95. Corey, E. J. and Kim, C. U. (1972). *J. Amer. Chem. Soc.*, **94**, 7586
96. Corey, E. J. and Kim, C. U. (1973). *J. Org. Chem.*, **38**, 1233
97. Corey, E. J. and Kim, C. U. (1973). *Tetrahedron Lett.*, 919
98. Wicha, J., Zarecki, A. and Kocór, M. (1973). *Tetrahedron Lett.*, 3635
99. Weinshenker, N. M. and Shen, C.-M. (1972). *Tetrahedron Lett.*, 3285
100. Sheikh, M. Y. and Eadon, G. (1972). *Tetrahedron Lett.*, 257
101. Bachhawat, J. M. and Mathur, N. K. (1971). *Indian J. Chem.*, **9**, 1335
102. Bachhawat, J. M., Koul, A. K., Prashad, B., Ramegowda, N. S., Narang, C. K. and Mathur, N. K. (1973). *Indian J. Chem.*, **11**, 609
103. Lee, D. G., Hall, D. T. and Cleland, J. H. (1972). *Can. J. Chem.*, **50**, 3741
104. Kwart, H. and Nickle, J. H. (1973). *J. Amer. Chem. Soc.*, **95**, 3394
105. Lee, D. G. and Raptis, M. (1973). *Tetrahedron*, **29**, 1481
106. Hasan, F. and Roček, J. (1972). *J. Amer. Chem. Soc.*, **94**, 3181, 8946, 9073
107. Hasan, F. and Roček, J. (1973). *J. Amer. Chem. Soc.*, **95**, 5421; *J. Org. Chem.*, **38**, 3813
108. Mason, J. G. and Baird, L. G. (1972). *J. Amer. Chem. Soc.*, **94**, 6116
109. Aukett, P. and Barker, I. R. L. (1972). *J. Chem. Soc. Perkin Trans. II*, 568
110. Banerji, K. K. (1973). *Indian J. Chem.*, **11**, 244
111. Srinivasan, N. S. and Venkatasubramanian, N. (1972). *Indian J. Chem.*, **10**, 1014
112. Banerji, K. K. (1973). *J. Chem. Soc. Perkin Trans. II*, 435; *Bull. Chem. Soc. Jap.*, **46**, 3623
113. Audette, R. J., Quail, J. W. and Smith, P. J. (1972). *J. Chem. Soc. Chem. Commun.*, 38
114. Radhakrishnamurti, P. S. and Mahanti, M. K. (1973). *Indian J. Chem.*, **11**, 762
115. Ledwith, A., Russell, P. J. and Sutcliffe, L. H. (1973). *Proc. Roy. Soc. (London)*, **A332**, 151; (1973). *J. Chem. Soc. Perkin Trans. II*, 630
116. Greatorex, D., Hill, R. J., Kemp, T. J. and Stone, T. J. (1972). *J. Chem. Soc. Faraday Trans. I*, **68**, 2059
117. Mihailović, M. Lj. and Partch, R. E. (1972). *Selective Organic Transformations*, Vol. 2, 97 (B. S. Thyagarajan, editor) (New York: Wiley-Interscience).
118. Mihailović, M. Lj., Gojković, S. and Konstantinović, S. (1973). *Tetrahedron*, **29**, 3675
119. Green, M. M., McGrew, J. G. and Moldowan, J. M. (1971). *J. Amer. Chem. Soc.*, **93**, 6700
120. Green, M. M., Moldowan, J. M. and McGrew, J. G. (1973). *J. Chem. Soc. Chem. Commun.*, 451
121. Deluzarche, A., Rimmelin, P. and Sommer, J.-M. (1973). *Bull. Soc. Chim. Fr.*, 1810
122. Roscher, N. M. and Jedziniak, E. J. (1973). *Tetrahedron Lett.*, 1049
123. Walling, C. and Bristol, D. (1972). *J. Org. Chem.*, **37**, 3514
124. Mead, T. J., Cum, G. and Uccella, N. (1972). *J. Chem. Soc. Chem. Commun.*, 679
125. Hutchins, R. O., Hutchins, M. G. and Milewski, C. A. (1972). *J. Org. Chem.*, **37**, 4190

126. Arhart, R. J. and Martin, J. C. (1972). *J. Amer. Chem. Soc.*, **94**, 5003
127. Kaplan, L. J. and Martin, J. C. (1973). *J. Amer. Chem. Soc.*, **95**, 793
128. Burgess, E. M., Penton, H. R. and Taylor, E. A. (1973). *J. Org. Chem.*, **38**, 26
129. Gerlach, H., Huong, T. T. and Müller, W. (1972). *J. Chem. Soc. Chem. Commun.*, 1215
130. Gerlach, H. and Müller, W. (1972). *Helv. Chim. Acta.*, **55**, 2277
131. Oele, P. C., Tinkelenberg, A. and Louw, R. (1972). *Tetrahedron Lett.*, 2375
132. Achmatowicz, S., Barton, D. H. R., Magnus, P. D., Poulton, G. A. and West, P. J. (1971). *Chem. Commun.*, 1014; (1973). *J. Chem. Soc. Perkin Trans. I*, 1567 and succeeding papers
133. Barton, D. H. R., Bolton, M., Magnus, P. D., West, P. J., Porter, G. and Wirz, J. (1972). *J. Chem. Soc. Chem. Commun.*, 632
134. Wirz, J. (1973). *J. Chem. Soc. Perkin Trans. II*, 1307
135. Rutherford, K. G., Tang, B. K., Lam, L. K. M. and Fung, D. P. C. (1972). *Can. J. Chem.*, **50**, 3288
136. Harano, K. and Taguchi, T. (1972). *Chem. Pharm. Bull.*, **20**, 2348, 2357
137. Tinkelenberg, A., Kooyman, E. C. and Louw, R. (1972). *Rec. Trav. Chim. Pays-Bas*, **91**, 3
138. Taylor, R. (1972). *J. Chem. Soc. Perkin Trans. II*, 165
139. Kwart, H. and Slutsky, J. (1972). *J. Chem. Soc. Chem. Commun.*, 1182
140. Chuchani, G., de Chang, S. P. and Lombana, L. (1973). *J. Chem. Soc. Perkin Trans. II*, 1961 and references cited therein
141. Bose, A. K. and Lal, B. (1973). *Tetrahedron Lett.*, 3937
142. Clive, D. L. J. and Denyer, C. V. (1971). *Chem. Commun.*, 1112
143. Olah, G. A., Nojima, M. and Kerekes, I. (1973). *Synthesis*, 786
144. Olah, G. A., Nojima, M. and Kerekes, I. (1974). *J. Amer. Chem. Soc.*, **96**, 925
145. Koop, H. and Schmutzler, R. (1971–72). *J. Fluorine Chem.*, **1**, 252
146. Robert, D. U. and Riess, J. G. (1972). *Tetrahedron Lett.*, 847
147. Robert, D. U., Flatau, G. N., Cambon, A. and Riess, J. G. (1973). *Tetrahedron*, **29**, 1877
148. Hedayatullah, M., Lévêque, J.-C. and Denivelle, L. (1972). *C.R. Acad. Sci., Ser. C*, **274**, 1937
149. Normant, J. F. and Deshayes, H. (1972). *Bull. Soc. Chim. Fr.*, 2854
150. Stephenson, B., Solladié, G. and Mosher, H. S. (1972). *J. Amer. Chem. Soc.*, **94**, 4184
151. Snyder, E. I. (1972). *J. Org. Chem.*, **37**, 1466
152. Bensimon, Y. and Ucciani, E. (1973). *C.R. Acad. Sci., Ser. C*, **276**, 683
153. Corey, E. J., Kim, C. U. and Takeda, M. (1972). *Tetrahedron Lett.*, 4339
154. Furukawa, N., Inoue, T., Aida, T. and Oae, S. (1973). *J. Chem. Soc. Chem. Commun.*, 212
155. Baumann, W. J., Gee, R. D., Madson, T. H. and Mangold, H. K. (1972). *Chem. Phys. Lipids*, **9**, 87
156. Gore, J., Place, P. and Roumestant, M. L. (1973). *J. Chem. Soc. Chem. Commun.*, 821
157. Sandler, S. R. (1971). *Chem. Ind.*, 1416
158. Scheffold, R. and Saladin, E. (1972). *Angew. Chem. Int. Ed. Engl.*, **11**, 229
159. Hendrickson, J. B. and Joffee, I. (1973). *J. Amer. Chem. Soc.*, **95**, 4083
160. Barton, D. H. R., Magnus, P. D. and Young, R. N. (1973). *J. Chem. Soc. Chem. Commun.*, 331
161. Castro, B. and Selve, C. (1971). *Bull. Soc. Chim. Fr.*, 2296
162. Castro, B. and Selve, C. (1971). *Bull. Soc. Chim. Fr.*, 4368
163. Castro, B., Chapleur, Y., Gross, B. and Selve, C. (1972). *Tetrahedron Lett.*, 5001
164. Castro, B., Chapleur, Y. and Gross, B. (1973). *Bull. Soc. Chim. Fr.*, 3034
165. Reese, C. B. (1973). *Protective Groups in Organic Chemistry*, 95 (J. F. W. McOmie, editor) (New York: Plenum)
166. Corey, E. J. and Venkateswarlu, A. (1972). *J. Amer. Chem. Soc.*, **94**, 6190
167. Ogilvie, K. K. and Iwacha, D. J. (1973). *Tetrahedron Lett.*, 317
168. Ogilvie, K. K. (1973). *Can. J. Chem.*, **51**, 3799
169. van Boom, J. H., van Deursen, P., Meeuwse, J. and Reese, C. B. (1972). *J. Chem. Soc. Chem. Commun.*, 766
170. Ireland, R. E., Muchmore, D. C. and Hengartner, U. (1972). *J. Amer. Chem. Soc.*, **94**, 5098

171. Corey, E. J. and Suggs, J. W. (1973). *J. Org. Chem.*, **38**, 3223
172. Gent, P. A., Gigg, R. and Conant, R. (1972). *J. Chem. Soc. Perkin Trans. I*, 1535; (1973). *ibid.*, 1858
173. Corey, E. J. and Suggs, J. W. (1973). *J. Org. Chem.*, **38**, 3224
174. Golborn, P. and Scheinmann, F. (1973). *J. Chem. Soc. Perkin Trans. I*, 2870
175. Gent, P. A. and Gigg, R. (1974). *J. Chem. Soc. Chem. Commun.*, 277
176. Barton, D. H. R., Coates, I. H. and Sammes, P. G. (1973). *J. Chem. Soc. Perkin Trans. I*, 599
177. Hanessian, S. and Staub, A. P. A. (1973). *Tetrahedron Lett.*, 3555
178. Neckers, D. C., Kooistra, D. A. and Green, G. W. (1972). *J. Amer. Chem. Soc.*, **94**, 9284
179. Blossey, E. C., Turner, L. M. and Neckers, D. C. (1973). *Tetrahedron Lett.*, 1823
180. Höfle, G. and Steglich, W. (1972). *Synthesis*, 619
181. Fuks, R. and Hartemink, M. A. (1973). *Bull. Soc. Chim. Belg.*, **82**, 23; (1973). *Tetrahedron*, **29**, 297
182. Meisters, A. and Mole, T. (1972). *J. Chem. Soc. Chem. Commun.*, 595
183. Salomon, R. G. and Kochi, J. K. (1973). *J. Org. Chem.*, **38**, 3715
184. Hutchins, R. O., Maryanoff, B. E. and Milewski, C. A. (1971). *Chem. Commun.*, 1097
185. Weber, W. P. and Shepherd, J. P. (1972). *Tetrahedron Lett.*, 4907
186. Lloyd, W. D., Navarette, B. J. and Shaw, M. F. (1972). *Synthesis*, 610
187. Mangoni, L., Adinolfi, M., Barone, G. and Parrilli, M. (1973). *Tetrahedron Lett.*, 4485
188. Overman. L. E. (1972). *J. Chem. Soc. Chem. Commun.*, 1196
189. Mukaiyama, T., Sato, T. and Hanna, J. (1973). *Chem. Lett.*, 1041
190. Chan, T. H. and Vinokur, E. (1972). *Tetrahedron Lett.*, 75
191. Katzenellenbogen, J. A. and Bowlus, S. B. (1973). *J. Org. Chem.*, **38**, 627
192. Guetté, J.-P., Spassky, N. and Boucherot, D. (1972). *Bull. Soc. Chim. Fr.*, 4217
193. Nakanishi, K., Schooley, D. A., Koreeda, M. and Miura, I. (1972). *J. Amer. Chem. Soc.*, **94**, 2865
194. Duncan, W. P., Eisenbraun, E. J. and Flanagan, P. W. (1972). *Chem. Ind.*, 78
195. Cazaux, L. and Maroni, P. (1972). *Bull. Soc. Chim. Fr.*, 773 and succeeding papers
196. Barton, D. H. R., Magnus, P. D., Smith, G., Streckert, G. and Zurr, D. (1972). *J. Chem. Soc. Perkin Trans. I*, 542
197. Regen, S. L. and Whitesides, G. M. (1972). *J. Org. Chem.*, **37**, 1832
198. Ohloff, G. and Giersch, W. (1973). *Angew. Chem. Int. Ed. Engl.*, **12**, 401
199. McKillop, A., Raphael, R. A. and Taylor, E. C. (1972). *J. Org. Chem.*, **37**, 4204
200. Sharpless, K. B. and Flood, T. C. (1972). *J. Chem. Soc. Chem. Commun.*, 370
201. Daub, J., Trautz, V. and Erhardt, U. (1972). *Tetrahedron Lett.*, 4435
202. Borden, W. T., Concannon, P. W. and Phillips, D. I. (1973). *Tetrahedron Lett.*, 3161
203. Guisnet, M., Plouzennec, I. and Maurel, R. (1972). *C.R. Acad. Sci., Ser. C*, **274**, 2102
204. Hines, J. N., Peagram, M. J., Thomas, E. J. and Whitham, G. H. (1973). *J. Chem. Soc. Perkin Trans. I*, 2332
205. Hiyama, T. and Nozaki, H. (1973). *Bull. Chem. Soc. Jap.*, **46**, 2248
206. Stromquist, P., Radcliffe, M. and Weber, W. P. (1973). *Tetrahedron Lett.*, 4523
207. Carnahan, J. C. and Closson, W. D. (1972). *Tetrahedron Lett.*, 3447
208. Egyed, J., Demerseman, P. and Royer, R. (1973). *Bull. Soc. Chim. Fr.*, 3014
209. Mihailović, M. Lj., Gojković, S. and Čeković, Ž. (1972). *J. Chem. Soc. Perkin Trans. I*, 2460
210. Jacobus, J. (1973). *J. Org. Chem.*, **38**, 402
211. Dana, G. and Girault, J.-P. (1972). *Bull. Soc. Chim. Fr.*, 1650
212. Reeve, W. and Reichel, D. M. (1972). *J. Org. Chem.*, **37**, 68
213. Larson, G. L. and Hernandez, A. (1973). *J. Org. Chem.*, **38**, 3935
214. Castro, B., Ly, M. and Selve, C. (1973). *Tetrahedron Lett.*, 4455
215. Leznoff, C. C. and Wong, J. Y. (1972). *Can. J. Chem.*, **50**, 2892
216. Wong, J. Y. and Leznoff, C. C. (1973). *Can. J. Chem.*, **51**, 2452
217. Newman, M. S. and Chen, C. H. (1972). *J. Amer. Chem. Soc.*, **94**, 2149; (1973). *J. Org. Chem.*, **38**, 1173
218. Newman, M. S. and Chen, C. H. (1973). *J. Amer. Chem. Soc.*, **95**, 278
219. Newman, M. S. and Olson, D. R. (1973). *J. Org. Chem.*, **38**, 4203
220. Seeley, D. A. and McElwee, J. (1973). *J. Org. Chem.*, **38**, 1691

221. Golding, B. T., Hall, D. R. and Sakrikar, S. (1973). *J. Chem. Soc. Perkin Trans. I*, 1214
222. Akhrem, A. A., Zharkov, V. V., Zaitseva, G. V. and Mikhailopulo, I. A. (1973). *Tetrahedron Lett.*, 1475
223. Le Clef, B., Mommaerts, J., Stelander, B. and Viehe, H. G. (1973). *Angew. Chem. Int. Ed. Engl.*, **12**, 404
224. Jensen, R. G. (1972). *Topics in Lipid Chemistry*, Vol. 3, 1 (F. D. Gunstone, editor) (London: Elek Science)
225. Shvets, V. I. (1971). *Usp. Khim.*, **40**, 625
226. Sagredos, A. N. (1973). *Ann. Chem.*, 87
227. Serebrennikova, G. A., Vtorov, I. B., Fedorova, G. N., Akivenson, O. S. and Evstigneeva, R. P. (1973). *Zh. Org. Khim.*, **9**, 488
228. Kabanova, M. A. and Shvets, V. I. (1972). *Zh. Org. Khim.*, **8**, 716
229. Aneja, R., Davies, A. P. and Knaggs, J. A. (1973). *J. Chem. Soc. Chem. Commun.*, 110
230. Mitchell, L. C. (1972). *J. Amer. Oil Chem. Soc.*, **49**, 281
231. Aneja, R. and Davies, A. P. (1972). *Tetrahedron Lett.*, 4497
232. Stoochnoff, B. A. and Benoiton, N. L. (1973). *Tetrahedron Lett.*, 21
233. Merz, A. (1973). *Angew. Chem. Int. Ed. Engl.*, **12**, 846
234. Beard, C. D., Baum, K. and Grakauskas, V. (1973). *J. Org. Chem.*, **38**, 3673
235. Leroux, Y., Larchevêque, M. and Combret, J.-C. (1971). *Bull. Soc. Chim. Fr.*, 3258
236. Chastrette, M. and Gauthier-Countani, H. (1973). *Bull. Soc. Chim. Fr.*, 363
237. Loim, L. M., Parnes, Z. N., Vasil'eva, S. P. and Kursanov, D. N. (1972). *Zh. Org. Khim.*, **8**, 896
238. Doyle, M. P., DeBruyn, D. J. and Kooistra, D. A. (1972). *J. Amer. Chem. Soc.*, **94**, 3659
239. Nakao, R., Fukumoto, T. and Tsurugi, J. (1972). *J. Org. Chem.*, **37**, 4349
240. Grady, G. L. and Chokshi, S. K. (1972). *Synthesis*, 483
241. Siouffi, A.-M. and Naudet, M. (1972). *Bull. Soc. Chim. Fr.*, 3976
242. Casey, C. P. and Burkhardt, T. J. (1972). *J. Amer. Chem. Soc.*, **94**, 6543
243. Casey, C. P., Bertz, S. H. and Burkhardt, T. J. (1973). *Tetrahedron Lett.*, 1421
244. Le Corre, M. (1973). *C.R. Acad. Sci., Ser. C*, **276**, 963
245. Wilson, I. F. and Tebby, J. C. (1972). *J. Chem. Soc. Perkin Trans. I*, 2830
246. Caron, G. and Lessard, J. (1973). *Can. J. Chem.*, **51**, 981
247. Anderson, A. G. and Freenor, F. J. (1972). *J. Org. Chem.*, **37**, 626
248 Kukhar', V. P., Lazukina, L. A. and Kirsanov, A. V. (1973). *Zh. Org. Khim.*, **9,** 304
249. Alper, H. and Huang, C.-C. (1973). *J. Org. Chem.*, **38**, 64
250. Hayashi, S. and Ishikawa, N. (1972). *Bull. Chem. Soc. Jap.*, **45**, 642
251. Richmond, G. D. and Spendel, W. (1973). *Tetrahedron Lett.*, 4557
252. Köbrich, G. and Baumann, A. (1973). *Angew. Chem. Int. Ed. Engl.*, **12**, 856
253. Takahashi, K., Miyake, A. and Hata, G. (1972). *Bull. Chem. Soc. Jap.*, **45**, 230
254. Takahashi, K., Hata, G. and Miyake, A. (1973). *Bull. Chem. Soc. Jap.*, **46**, 1012
255. Doyle, M. P., DeBruyn, D. J. and Scholten, D. J. (1973). *J. Org. Chem.*, **38**, 625
256. von Sonntag, C., Schuchmann, H.-P. and Schomburg, G. (1972). *Tetrahedron*, **28**, 4333
257. Schuchmann, H.-P. and von Sonntag, C. (1973). *Tetrahedron*, **29**, 1811, 3351
258. Berti, G. (1973). *Topics in Stereochemistry*, Vol. 7, 93 (N. L. Allinger and E. L. Eliel, editors) (New York: Wiley-Interscience)
259. Buchanan, J. G. and Sable, H. Z. (1972). *Selective Organic Transformations*, Vol. 2, 1 (B. S. Thyagarajan, editor) (New York: Wiley-Interscience)
260. Metelitsa, D. I (1972). *Usp. Khim.*, **41**, 1737
261. Rouchaud, J. (1972). *Ind. Chim. Belg.*, **37**, 741
262. Yandovskii, V. N. and Ershov, B. A. (1972). *Usp. Khim.*, **41**, 785
263. Dev, S. (1972). *J. Sci. Ind. Res.*, **31**, 60
264. Kirk, D. N. (1973). *Chem. Ind.*, 109
265. Matsumoto, K. (1973). *Kagaku No Ryoiki*, **27**, 148
266. Kishi, Y., Aratani, M., Tanino, H., Fukuyama, T., Goto, T., Inoue, S., Sugiura, S. and Kakoi, H. (1972). *J. Chem. Soc. Chem. Commun.*, 64
267. Anderson, W. K. and Veysoglu, T. (1973). *J. Org. Chem.*, **38**, 2267
268. Kawabe, N., Okada, K. and Ohno, M. (1972). *J. Org. Chem.*, **37, ** 4210

269. Delmond, B., Pommier, J.-C. and Valade, J. (1972). *J. Organometal. Chem.*, **35**, 91
270. Yano, Y., Okonogi, T., Sunaga, M. and Tagaki, W. (1973). *J. Chem. Soc. Chem. Commun.*, 527
271. Merz, A. and Märkl, G. (1973). *Angew. Chem. Int. Ed. Engl.*, **12**, 845
272. Johnson, C. R., Haake, M. and Shroeck, C. W. (1970). *J. Amer. Chem. Soc.*, **92**, 6594
273. Johnson, C. R. and Rogers, P. E. (1973). *J. Org. Chem.*, **38**, 1793
274. Johnson, C. R. and Shroeck, C. W. (1973). *J. Amer. Chem. Soc.*, **95**, 7418
275. Hetschko, M. and Gosselck, J. (1973). *Chem. Ber.*, **106**, 996
276. Braun, H., Huber, G. and Kresze, G. (1973). *Tetrahedron Lett.*, 4033
277. Matthews, R. S. and Meteyer, T. E. (1972). *Syn. Commun.*, **2**, 399
278. Johnson, C. R. and Shroeck, C. W. (1971). *J. Amer. Chem. Soc.*, **93**, 5303
279. Durst, T., Viau, R., van den Elzen, R. and Nguyen, C. H. (1971). *Chem. Commun.*, 1334
280. Townsend, J. M. and Sharpless, K. B. (1972). *Tetrahedron Lett.*, 3313
281. Shanklin, J. R., Johnson, C. R., Ollinger, J. and Coates, R. M. (1973). *J. Amer. Chem. Soc.*, **95**, 3429
282. Johnson, C. R., Shroeck, C. W. and Shanklin, J. R. (1973). *J. Amer. Chem. Soc.*, **95**, 7424
283. Cainelli, G., Ronchi, A. U., Bertini, F., Graselli, P. and Zubiani, G. (1971). *Tetrahedron*, **27**, 6109
284. Cainelli, G., Tangari, N. and Ronchi, A. U. (1972). *Tetrahedron*, **28**, 3009
285. Sharpless, K. B., Umbreit, M. A., Nieh, M. T. and Flood, T. C. (1972). *J. Amer. Chem. Soc.*, **94**, 6538
286. Chan, T. H. and Finkenbine, J. R. (1972). *J. Amer. Chem. Soc.*, **94**, 2880
287. Clive, D. L. J. and Denyer, C. V. (1973). *J. Chem. Soc. Chem. Commun.*, 253
288. Giering, W. P., Rosenblum, M. and Tancrede, J. (1972). *J. Amer. Chem. Soc.*, **94**, 7170
289. Vedejs, E. and Fuchs, P. L. (1971). *J. Amer. Chem. Soc.*, **93**, 4070; (1973). *ibid.*, **95**, 822
290. Lenox, R. S. and Katzenellenbogen, J. A. (1973). *J. Amer. Chem. Soc.*, **95**, 957
291. Brown, H. C., Krishnamurthy, S. and Coleman, R. A. (1972). *J. Amer. Chem. Soc.*, **94**, 1750
292. Krishnamurthy, S., Schubert, R. M. and Brown, H. C. (1973). *J. Amer. Chem. Soc.*, **95**, 8486
293. Guyon, R. and Villa, P. (1972). *Bull. Soc. Chim. Fr.*, 1375
294. Yamaguchi, K. and Fueno, T. (1973). *Chem. Phys. Lett.*, **22**, 471
295. Flowers, M. C., Penny, D. E. and Pommelet, J.-C. (1973). *Int. J. Chem. Kinet.*, **5**, 353
296. Dahmen, A., Hamberger, H., Huisgen, R. and Markowski, V. (1971). *Chem. Commun.*, 1192
297. MacDonald, H. H. J. and Crawford, R. J. (1972). *Can. J. Chem.*, **50**, 428
298. Crawford, R. J., Vukov, V. and Tokunaga, H. (1973). *Can. J. Chem.*, **51**, 3718
299. Paladini, J. C. and Chuche, J. (1971). *Tetrahedron Lett.*, 4383
300. Pommelet, J. C., Manisse, N. and Chuche, J. (1972). *Tetrahedron*, **28**, 3929
301. Paulson, D. R., Tang, F. Y. N. and Sloan, R. B. (1973). *J. Org. Chem.*, **38**, 3967
302. Paulson, D. R., Korngold, G. and Jones, G. (1972). *Tetrahedron Lett.*, 1723
303. Robert, A. and Moisan, B. (1972). *J. Chem. Soc. Chem. Commun.*, 337
304. Coxon, J. M., Hartshorn, M. P. and Swallow, W. H. (1973). *J. Chem. Soc. Chem. Commun.*, 261; (1973). *Aust. J. Chem.*, **26**, 2521
305. Donnelly, J. A., Hoey, J. G., O'Brien, S. and O'Grady, J. (1973). *J. Chem. Soc. Perkin Trans. I*, 2030
306. Adames, G., Bibby, C. and Grigg, R. (1972). *J. Chem. Soc. Chem. Commun.*, 491
307. Magnusson, G. and Thorén, S. (1973). *J. Org. Chem.*, **38**, 1380
308. Hartman, B. C. and Rickborn, B. (1972). *J. Org. Chem.*, **37**, 943
309. Kissel, C. L. and Rickborn, B. (1972). *J. Org. Chem.*, **37**, 2060
310. Thummel, R. P. and Rickborn, B. (1972). *J. Org. Chem.*, **37**, 3919, 4250
311. Sheng, M. N. (1972). *Synthesis*, 194
312. Imanaka, T., Okamoto, Y. and Teranishi, S. (1972). *Bull. Chem. Soc. Jap.*, **45**, 1353
313. Pritchard, J. G. and Siddiqui, I. A. (1973). *J. Chem. Soc. Perkin Trans. II*, 452
314. Carr, M. D. and Stevenson, C. D. (1973). *J. Chem. Soc. Perkin Trans. II*, 518
315. Durand, R., Geneste, P., Lamaty, G. and Roque, J.-P. (1973). *C.R. Acad. Sci., Ser. C*, **277**, 1395

316. Willi, A. V. (1973). *Helv. Chim. Acta*, **56**, 2094
317. Wohl, R. A. (1974). *Chimia*, **28**, 1
318. Shvets, V. F. and Tyukova, O. A. (1971). *Zh. Org. Khim.*, **7**, 1842
319. Balsamo, A., Crotti, P., Macchia, B. and Macchia, F. (1973). *Tetrahedron*, **29**, 2183 and preceding studies cited
320. Anselmi, C., Berti, G., Macchia, B., Macchia, F. and Monti, L. (1972). *Tetrahedron Lett.*, 1209
321. Nakamura, H., Yamamoto, H. and Nozaki, H. (1973). *Tetrahedron Lett.*, 111
322. Khuddus, M. A. and Swern, D. (1973). *J. Amer. Chem. Soc.*, **95**, 8393
323. Abenhaim, D., Boireau, G. and Namy, J.-L. (1972). *Bull. Soc. Chim. Fr.*, 985
324. Boireau, G., Namy, J.-L. and Abenhaim, D. (1972). *Bull. Soc. Chim. Fr.*, 1042
325. Abenhaim, D. and Namy, J.-L. (1972). *Tetrahedron Lett.*, 1001
326. Namy, J.-L. and Abenhaim, D. (1972). *C.R. Acad. Sci., Ser. C*, **274**, 803; (1972). *J. Organometal. Chem.*, **43**, 95
327. Johnson, C. R., Herr, R. W. and Wieland, D. M. (1973). *J. Org. Chem.*, **38**, 4263
328. Hartman, B. C., Livinghouse, T. and Rickborn, B. (1973). *J. Org. Chem.*, **38**, 4346
329. Normant, J.-M. (1973). *C.R. Acad. Sci., Ser. C*, **277**, 1045
330. Fujimoto, H., Katata, M., Yamabe, S. and Fukui, K. (1972). *Bull. Chem. Soc. Jap.*, **45**, 1320
331. Coke, J. L. and Shue, R. S. (1973). *J. Org. Chem.*, **38**, 2210
332. Sharpless, K. B. and Lauer, R. F. (1973). *J. Amer. Chem. Soc.*, **95**, 2697
333. Sucrow, W., Slopianka, M. and Winkler, D. (1972). *Chem. Ber.*, **105**, 1621
334. McManus, S. P., Larson, C. A. and Hearn, R. A. (1973). *Syn. Commun.*, **3**, 177
335. Asinger, F., Fell, B., Pfeifer, J. and Saus, A. (1972). *J. Prakt. Chem.*, **314**, 71
336. Bouchet, P. and Coquelet, C. (1973). *Bull. Soc. Chim. Fr.*, 3153, 3159
337. Buddrus, J. (1972). *Angew. Chem. Int. Ed. Engl.*, **11**, 1041
338. Buddrus, J. and Kimpenhaus, W. (1973). *Chem. Ber.*, **106**, 1648
339. Movsumzade, M. M., Shabanov, A. L., Babakhanov, R. A., Gurbanov, P. A. and Movsum-Zade, R. G. (1973). *Zh. Org. Khim.*, **9**, 1998
340. Movsumzade, M. M., Gurbanov, P. A., Khodzhaev, G. Kh., Shabanov, A. L. and Efendieva, D. R. (1972). *Azerb. Khim. Zh.*, 39
341. Isaacs, N. S. and Kirkpatrick, D. (1972). *Tetrahedron Lett.*, 3869
342. Robert, A., Pommeret, J. J. and Foucaud, A. (1972). *Tetrahedron*, **28**, 2085
343. Robert, A., Pommeret, J. J., Marchand, E. and Foucaud, A. (1973). *Tetrahedron*, **29**, 463
344. Nagarkatti, J. P. and Ashley, K. R. (1973). *Tetrahedron Lett.*, 4599
345. (1972). *Organic Peroxides*, Vol. 3 (D. Swern, editor) (New York: Wiley-Interscience)
346. Yablonskii, O. P., Belyaev, V. A. and Vinogradov, A. N. (1972). *Usp. Khim.*, **41**, 1260
347. Ohkubo, K. and Kanaeda, H. (1972). *Bull. Chem. Soc. Jap.*, **45**, 322, 2646
348. Ohkubo, K., Ninomiya, T. and Kanaeda, H. (1972). *Tetrahedron*, **28**, 2969
349. Midland, M. M. and Brown, H. C. (1973). *J. Amer. Chem. Soc.*, **95**, 4069
350. Sharpless, K. B. and Michaelson, R. C. (1973). *J. Amer. Chem. Soc.*, **95**, 6136
351. Sheng, M. N. and Zajacek, J. G. (1970). *J. Org. Chem.*, **35**, 1839
352. Farberov, M. I., Mel'nik, L. V., Bobylev, B. N., Podgornova, V. A. (1971). *Kinet. Katal.*, **12**, 1144
353. Farberov, M. I., Bobylev, B. N., Podgornova, V. A. and Mel'nik, L. V. (1971). *Nauch. Konf. Yaroslav. Tekhnol. Inst. 22nd*, 133
354. Tolstikov, G. A., Dzhemilev, U. M. and Yur'ev, V. P. (1972). *Zh. Obshch. Khim.*, **42**, 1611
355. Tolstikov, G. A., Dzhemilev, U. M. and Yur'ev, V. P. (1972). *Zh. Org. Khim.*, **8**, 1190
356. Sheldon, R. A. (1973). *Rec. Trav. Chim. Pays-Bas*, **92**, 367
357. Tolstikov, G. A., Jemilev, U. M., Jurjev, V. P., Gershanov, F. B. and Rafikov, S. R. (1971). *Tetrahedron Lett.*, 2807
358. Tolstikov, G. A., Dzhemilev, U. M. and Yur'ev, V. P. (1972). *Zh. Org. Khim.*, **8**, 1186
359. Tolstikov, G. A., Dzhemilev, U. M., Yur'ev, V. P., Pozdeeva, A. A. and Gerchikova, F. G. (1973). *Zh. Obshch. Khim.*, **43**, 1360
360. Tolstikov, G. A., Dzhemilev, U. M., Novitskaya, N. N. and Yur'ev, V. P. (1972). *Izv. Akad. Nauk SSSR*, 2744
361. Tolstikov, G. A., Dzhemilev, U. M. and Yur'ev, V. P. (1972). *Izv. Akad. Nauk SSSR*, 670

362. Yur'ev, V. P., Tolstikov, G. A. and Gailyunas, I. A. (1973). *Zh. Obshch. Khim.*, **43,** 215
363. Koshel, G. N., Farberov, M. I., Zalygin, L. L. and Krushinskaya, G. A. (1971). *Uch. Zap. Yaroslav. Tekhnol. Inst.*, **27,** 45
364. Johnson, N. A. and Gould, E. S. (1974). *J. Org. Chem.*, **39,** 407
365. Sheldon, R. A. (1973). *Rec. Trav. Chim. Pays-Bas*, **92,** 253
366. Sheldon, R. A. and van Doorn, J. A. (1973). *J. Catal.*, **31,** 427
367. Sheldon, R. A., van Doorn, J. A., Schram, C. W. A. and de Jong, A. J. (1973). *J. Catal.*, **31,** 438
368. Farberov, M. I., Stozhkova, G. A., Bondarenko, A. V. and Kirik, T. M. (1972). *Kinet. Katal.*, **13,** 291
369. Markevich, V. S. and Shtivel, N. Kh. (1973). *Neftekhimiya*, **13,** 240
370. Adzhamov, K. Yu, Alkhazov, T. G., Dolgov, V. Ya., Krylov, O. V. and Margolis, L. Ya. (1973). *Dokl. Akad. Nauk SSSR*, **209,** 1127
371. Tolstikov, G. A., Dzhemilev, U. M., Yur'ev, V. P. and Rafikov, S. R. (1973). *Dokl. Akad. Nauk SSSR*, **208,** 376
372. Su, C.-C., Reed, J. W. and Gould, E. S. (1973). *Inorg. Chem.*, **12,** 337
373. Baker, T. N., Mains, G. J., Sheng, M. N. and Zajacek, J. G. (1973). *J. Org. Chem.*, **38,** 1145
374. Descotes, G. and Legrand, P. (1972). *Bull. Soc. Chim. Fr.*, 2937, 2942
375. Curci, R. and Furia, F. Di (1972). *Tetrahedron*, **28,** 3905
376. Ogata, Y. and Suyama, S. (1971). *Chem. Ind.*, 707; (1973). *J. Chem. Soc. Perkin Trans. II*, 755
377. Humphris, K. J. and Scott, G. (1973). *J. Chem. Soc. Perkin Trans. II*, 826, 831
378. van Ham, J., Schors, A. and Kooyman, E. C. (1973). *Rec. Trav. Chim. Pays-Bas*, **92,** 393
379. Sheldon, R. A. and van Doorn, J. A. (1973). *Tetrahedron Lett.*, 1021
380. Bloodworth, A. J. and Bylina, G. S. (1972). *J. Chem. Soc. Perkin Trans. I*, 2433
381. Bloodworth, A. J. and Bunce, R. J. (1972). *J. Chem. Soc. Perkin Trans. I*, 2787
382. Bloodworth, A. J. and Bunce, R. J. (1973). *J. Organometal. Chem.*, **60,** 11
383. Goh, S. H. (1972). *J. Org. Chem.*, **37,** 3098
384. Kim, L. and Dewhirst, K. C. (1973). *J. Org. Chem.*, **38,** 2722
385. Richardson, W. H. and Heesen, T. C. (1972). *J. Org. Chem.*, **37,** 3416
386. Richardson, W. H., Yelvington, M. B., Andrist, A. H., Ertley, E. W., Smith, R. S. and Johnson, T. D. (1973). *J. Org. Chem.*, **38,** 4219
387. Richardson, W. H., Smith, R. S., Snyder, G., Anderson, B. and Kranz, G. L. (1972). *J. Org. Chem.*, **37,** 3915

4
Nitrogen Compounds

D. F. EWING
University of Hull

4.1   INTRODUCTION

This chapter differs in several aspects from the corresponding chapter in Series One[1a]. The selection of material has been more rigorously restricted to the realm of aliphatic chemistry. A different sub-division of material has also been adopted to reflect the changing emphasis evident in published work. Aliphatic nitrogen chemistry covers a wide field, including some eight major functional groups and about twice as many minor ones; pressure on space has inevitably meant a stringent selection of the topics covered. The patent literature has been excluded entirely since it is impracticable to evaluate the novelty of claims over such a wide field. Major emphasis is placed on the synthesis and reactions of the various classes of compounds; aspects of photochemistry, kinetics and spectroscopy are included only when particularly relevant to organic chemistry. N.m.r. studies of rotational isomerism about a partial double bond between carbon and nitrogen have not been covered since this subject is almost large enough to require a chapter in its own right and has recently been reviewed[1b].

4.2   NITRO COMPOUNDS

4.2.1   Synthesis

No really novel methods are noteworthy, but several improvements to established procedures have been reported. The well-known conversion of a nitroalkane into a higher homologue by successive reaction with an aldehyde, an acylating agent and sodium borohydride has been achieved in a 'one-pot' process[2]. Astute selection of reaction conditions gives yields of 75–80% for what was previously a laborious four-step synthesis with an overall yield of *ca.* 30%. To produce the highly hindered *gem*-dinitro compound (1), of theoretical interest, mononitroneopentane was treated with tetranitromethane in presence of base[3].

$$\mathrm{Bu^tCH_2NO_2 \xrightarrow[\mathrm{KOH}]{\mathrm{C(NO_2)_4}} Bu^tCH(NO_2)_2}$$
$$(1)$$

Compound (1) has unusually low acidity for a *gem*-dinitroalkane ($pK_a = 8.24$), probably because of steric inhibition of resonance stabilisation of the anion; the angle between the planes of the nitro-groups is probably *ca.* 80°. Oxidative dimerisation of nitroalkane salts in DMSO provides a valuable route to vicinal dinitro compounds[4]. The oxidant used is CCl_4, a compound much underrated in this role, and the mechanism involves one-electron transfer from the nitroalkane anion to CCl_4. A detailed study[5] of the conversion of tetranitroethane into hexanitroethane with HNO_3–H_2SO_4 has revealed a reaction first-order in substrate, nitronium ion, and base (H_2O or HSO_4^-). This confirms earlier ideas about the mechanism which involves simultaneous attack of NO_2^+ and base on the substrate.

Olah[6] has developed a new hydrofluorinating reagent (70% HF in pyridine) which reacts with alkenes in the presence of $NO_2^+BF_4^-$ to give a fluoronitro-alkane in 40–80% yield.

4.2.2 Reactions

4.2.2.1 *Alkylation and related reactions*

An interesting reaction[7] of the side chain of 9-trimethylammoniomethylanthracene chloride (2) involves alkylation at carbon in the anion of the lithium salt of 2-nitropropane (in contrast to the more usual *O*-alkylation of nitroalkane anions). The formation of (4) does not simply involve a nucleophilic displacement of trimethylamine. Initial attack occurs at C-10 in the anthracene ring to give a stable intermediate (3), which rearranges to (4) on being heated. An intermolecular mechanism for the rearrangement has been established by isolation of the intermediate and treating it with a different nitroalkane salt to give mixed alkylation products.

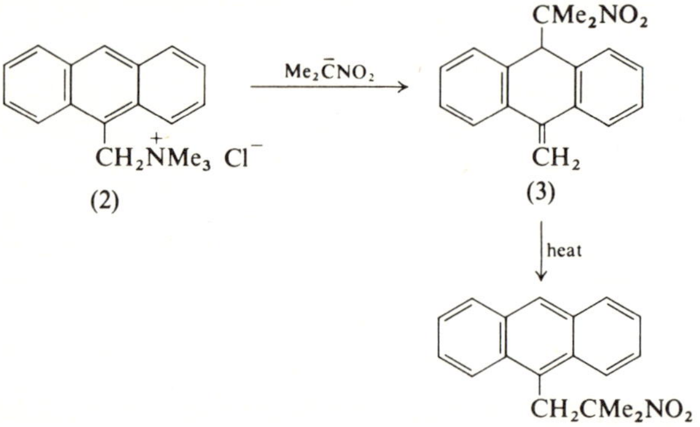

Trinitromethyl halides in DMSO solution exist entirely in the ionic form[8] $[(NO_2)_3C^-]$ $[X^+,S]$, the halogenonium ion (X = Cl, Br or I) being stabilised by solvation (S = solvent). Addition of an alkyl halide gives the corresponding *C*-alkylated species. This solvent-enhanced ionisation of $(NO_2)_3CX$ increases in the order $Cl < NO_2 < Br < I$; several polar solvents are effective. Another typical *C*-alkylation of nitroalkanes, viz. Michael addition to ethyl acrylate, is catalysed[9] by the very weak base tri-n-butylphosphine. Amines are totally ineffective catalysts, thus suggesting that the greater nucleophilicity of phosphorus is important. A reasonable mechanism involves initial formation of a phosphonium species which then generates the anion of the carbon acid by proton abstraction. Formation of the Michael adduct then follows in a chain reaction:

$$R_3P + CH_2{=}CHCO_2Et \rightleftharpoons R_3\overset{+}{P}CH_2\overset{-}{C}HCO_2Et \overset{YH}{\rightleftharpoons} R_3\overset{+}{P}CH_2CH_2CO_2Et + Y^-$$

$$Y^- + CH_2{=}CHCO_2Et \rightarrow YCH_2\overset{-}{C}HCO_2Et \overset{YH}{\longrightarrow} YCH_2CH_2CO_2Et + Y^-$$

Catalysis by tertiary phosphines appears[9] to be general for Michael additions of carbon acids.

Silylation of nitro compounds at oxygen can be achieved[10,11] directly by reaction with N',N''-diphenyl-N'-trimethylsilylurea or indirectly by treatment of the silver salt with trimethylsilyl chloride. With dimethyl nitromalonate the product (5) is stable, but reaction with water, alcohols or amines regenerates the nitro compound. The silylated species from the trinitromethane

$$(MeO_2C)_2C{=}\overset{+}{N}(\overset{-}{O})OSiMe_3 \qquad RO\overset{+}{N}(\overset{-}{O}){=}C(NO_2)_2 \qquad ClC(NO_2){=}NOAc$$
$$(5) \qquad\qquad\qquad (6) \qquad\qquad\qquad (7)$$

(6; $R = SiMe_3$) decomposes spontaneously. In contrast, acylation gives a stable product (6; $R = Ac$), which reacts with halogens to form the trinitromethyl halide[12]. O-Acyl derivatives of dinitromethane are unstable[13] and eliminate acetic acid; they react further with acetyl chloride to give O-acetylchloronitroformaldoxime (7). O-Alkylation of nitroalkanes with alkynes[14] requires the presence of a strong acid for an initial protonation. Ynamines are more reactive and react directly[15]. In both cases isomerisation to compounds of type (8) occurs, probably via a cyclic five-membered transition state.

$$R^1C{\equiv}CH + R^2CH_2NO_2 \rightarrow R^1COCH_2ON{=}CHR^2$$
$$(8)$$

Aryldiazonium and dinitroalkane salts react[16] to form azo compounds (9), which ionise to some extent in aqueous solution to form the salts (10).

$$R^1C_6H_4N{=}NCR^2(NO_2)_2 \rightleftharpoons [R^1C_6H_4\overset{+}{N}{\equiv}N]\,[\overset{-}{C}R^2(NO_2)_2]$$
$$(9) \qquad\qquad\qquad (10)$$

4.2.2.2 α,β-*Unsaturated nitro compounds*

Nucleophilic attack at the β-position of nitroalkenes has been of considerable interest recently and some novel nucleophiles have been studied. Diethyl phosphite adds[17] in the expected way to give a β-diethoxyphosphinylnitroalkane or an ester (11; $R^1 = H$ or CO_2Et). Initial attack by triethyl phosphite is analogous[17], but in the absence of an ionisable hydrogen atom a five-membered cyclic intermediate (12) is formed. The heterocyclic compound (12) may be isolated in certain cases, but it is usually transformed into a diethoxyphosphinylalkene (13) with loss of EtONO. t-Butyl and cyclohexyl isocyanides attack nitroalkenes at the β-position and, in an internal oxidation–reduction step, the oxide of an α-cyanoamide is formed. This is reduced

$$R^2CH(NO_2)CHR^1P(O)(OEt)_2$$

$(EtO)_2PH(O) \nearrow \qquad\qquad (11)$

$$R^1CH=C(NO_2)R^2$$

$(EtO)_3P \searrow$

$$\underset{(12)}{\begin{array}{c} R^1HC-CR^2 \\ (EtO)_3P \diagdown_O \diagup N^+ \diagdown O^- \end{array}} \longrightarrow R^2CH=CR^1P(O)(OEt)_2 + EtONO$$

(13)

by a second mole of isocyanide, which is oxidised to an isocyanate. Further reaction[18] of the α-cyanoamide with the nitroalkene may occur to afford the Michael adduct.

$$R^1CH=CHNO_2 + R^2NC \longrightarrow \left[\begin{array}{c} H \quad O^- \\ C=N^+ \\ R^1CH \quad O^- \\ C\equiv N^+R^2 \end{array}\right] \to\to \left[\begin{array}{c} C\equiv N^+-O^- \\ R^1CH \\ CONHR^2 \end{array}\right]$$

$R^2NC \diagup$

$$R^2N=C=O + R^1CH(CN)CONHR^2$$

Further studies[19] of the reactions of trinitromethyl halides with diazo-alkanes have shown that the 3-nitroisoxazoline *N*-oxides (16), which are formed under certain conditions[20], are secondary products. The principal primary products are a trinitroalkyl halide (14) and a *gem*-dinitroalkene (15), both probably being formed from the same intermediate, although an alternative pathway to the alkene may be possible[19,20]. With an excess of diazoalkane, further addition to the alkene (15) gives the isoxazoline deriva-

$$R^1R^2C=N_2 + (NO_2)_3CX \xrightarrow{-N_2} R^1R^2C \underset{C(NO_2)_3}{\overset{X}{\cdots}} \longrightarrow R^1R^2CXC(NO_2)_3$$

(14)

$$R^1R^2\bar{C}-C(NO_2)_3 \longrightarrow R^1R^2C=C(NO_2)_2$$

(15)

$R^3COCH=N_2 \diagup \atop (R^2 = H)$ $\qquad\qquad$ $\diagdown R^1R^2C=N_2$

$$\underset{(17)}{\begin{array}{c} R^1C-C-NO_2 \\ R^3COC \diagdown_{N} \diagup_{N} \\ H \end{array}} \qquad\qquad \underset{(16)}{\begin{array}{c} R^1R^2C-C-NO_2 \\ R^1R^2C \diagdown_O \diagup N^+ \diagdown O^- \end{array}}$$

tive (16) with elimination of N_2. Diazoketones or diazoesters lead to a pyrazole (17) with elimination of nitrous acid .The proportions of products of the types (14), (15) and (16) depends upon reaction conditions and the nature of the halogen X and the substituents R^1 and R^2 in the diazo compound.

4.2.2.3 β-*Nitroalkyl nitrates*

Base-catalysed elimination of the elements of nitric acid from β-nitroalkyl nitrates has been shown[21] to proceed via an $E1$ cb mechanism when the nitro and the nitrate groups are attached to a primary and to a secondary carbon atom, respectively. Eliminations of this type are sensitive to leaving-group abilities and to the acidity of the nitromethylene group. With sodium azide, β-nitroalkyl nitrates undergo[22] substitution to give β-nitroalkyl azides in high yield ($> 70\%$). Thermal decomposition of β-nitroalkyl peroxynitrates (18) is complex[23], but the initial step is probably homolysis of the O—O bond to yield an alkoxyl radical (19), the fate of which depends upon the

$$R^1{}_2C(NO_2)CMeR^2OONO_2 \rightarrow R^1{}_2C(NO_2)CMeR^2O \cdot \rightarrow \text{Products}$$
$$(18) \qquad\qquad\qquad (19)$$

nature of the substituents R^1 and R^2. With $R^2 = $ Me and $R^1 = $ H or Me, this radical dimerises, reacts with NO_2 to give a β-nitroalkyl nitrate or undergoes β-fission to form acetone. When $R^2 = Bu^t$, nitroacetone is formed by elimination of the bulky t-butyl group.

4.2.2.4 *Miscellaneous reactions*

Interest in the conversion of nitro compounds into ketones is reflected in the continued search for mild non-acidic conditions for this reaction. A full report of the use of titanium(III) chloride for this reduction has now appeared[24]. With this very versatile reagent the corresponding keto compounds can be obtained in high yield (usually *ca.* 90%) from, for example, nitroalkanes, nitroketones, nitroalkyl cyanides, nitroesters and nitroacetals. Reduction with $TiCl_3$ in aqueous NH_4OAc buffer is a sufficiently non-acidic method for most compounds, but for the most sensitive compounds the sodium salt of the nitro compound is formed first. A second mild non-acidic route[25] from nitro compounds to ketones involves their treatment with sodium nitrite and an alkyl nitrite. Yields are high (70–90%), but the full scope of this reaction remains to be established.

A convenient route to α-nitrosulphones[26] requires α-iodination of a nitroalkane with iodine followed by treatment of the product with a sodium sulphinate. Detailed study[27] has revealed three steps in the well-known conversion of a nitroalkane into a substituted hydroxylamine with a Grignard reagent, each step requiring one mole of RMgBr. A variety of nitro-substituted acid chlorides react[28] with diazomethane to give some new types of stable diazonitroketones.

4.2.3 Physical and theoretical studies

The stability of carbanions from nitroalkanes is usually rationalised in terms of delocalisation of the negative charge through the π molecular orbitals which were thought to have higher energies than the σ-orbitals. The corresponding valence-bond description also embodies this familiar concept of mobile π-electrons, but some recent SCF–MO calculations[29] in the CNDO/2 approximation clearly indicate that this simple picture is too naive. In the anion (20) the highest occupied σ molecular orbitals are of higher energy

$$(NO_2)_2\bar{C}R \qquad (R = H, Me, Cl \text{ or } CN)$$
$$(20)$$

than the lowest occupied π molecular orbitals, and this mixing of electronic levels reflects an approximately equal mobility for the σ- and the π-electrons, i.e. variation of the substituent R produces significant changes in both σ and π charge densities in the nitro group.

The acidic properties of simple nitroalkanes are of special interest because of the reversal of the usual trend in pK consequent upon replacing H by Me (i.e. a decrease in pK with each substitution; *cf.* formic and acetic acid). For the nitroalkanes, pK decreases with methyl substitution (nitromethane 10.2, nitroethane 8.5, 2-nitropropane 7.7), an effect which has been known for many years. Confirmation of the unusual nature of these compounds has come[30] from calorimetric measurements of $\Delta H_i{}^\circ$. Corresponding values of $\Delta G_i{}^\circ$ were obtained from equilibrium data and the inferred $\Delta S_i{}^\circ$ values are also unusual. These ΔS° values become more negative as ΔG° and ΔH° decrease, a trend opposite to that observed for carboxylic acids. The authors of this work were unable to offer any explanation for this peculiar property of the nitroalkanes beyond that advanced 30 years ago[31], in terms of differing effective dielectric constant for the medium between the nitro-group dipole and the ionising hydrogen atom.

4.3 NITRILES

Some aspects of nitrile chemistry have been briefly reviewed[32].

4.3.1 Synthesis

4.3.1.1 Dehydration and related reactions

As a route to nitriles, dehydration of amides and oximes has been an important reaction for a long time, but new reagents to affect this elimination continue to be discovered. Phosphonitrilic chloride, $(NPCl_2)_3$, previously used to obtain the nitrile from the sodium salt of a carboxylic acid, has now been applied to the reactions of amides[33], thioamides[33] and aldoximes[34]. The reaction proceeds rapidly in various solvents under mild conditions. Ready availability of the reagent and high yields of product (75–100%) mean an attractive alternative to traditional methods. Another novel dehydrating agent of wide applicability[35] is dichloromethylene, Cl_2C:. This carbene must be generated *in situ* and since the medium is essentially heterogeneous, a phase-transfer catalyst, benzyltriethylammonium chloride, is required. Carbene attack at the hydroxy group of the imino form of an amide, a urea or an aldoxime is thought to be the key step in the mechanism. Clearly amides containing sensitive functional groups (OH, NH_2, CN) are excluded; otherwise yields are 45–75%[35]. In a preliminary paper[36] the conversion of aldoximes into nitriles in high yield (> 95%) under neutral conditions at room temperature by 1,1'-dicarbonylbi-imidazole in dichloromethane is described. Cyanuric chloride in pyridine has also been suggested[37] as a reagent for this dehydration, perhaps being of particular value for acid-sensitive compounds. Dehydration of acrylamide with PCl_5 is accompanied by hydrochlorination, giving 3-chloropropionitrile in quantitative yield[38].

N-Trimethylsilylation of unsubstituted amides is readily achieved by treatment with chlorotrimethylsilane in the presence of triethylamine. Reaction[39] of these substituted amides, e.g. (21), with an acyl chloride regenerates the chlorosilane and produces a nitrile by loss of one mole of

$$Me_3SiNHCOMe + RCOCl \rightarrow MeCN + RCO_2H + Me_3SiCl$$
$$(21)$$

acid. Yields in excess of 90% have been obtained for aliphatic, alkenic and aromatic nitriles. The intermediacy of an isocyanide is postulated for the formation of nitriles in the high temperature (850 °C) thermolysis of chloro-acetamides, but this mechanism is difficult to confirm[40].

4.3.1.2 *From carbonyl compounds*

Reaction of a cyanomethylphosphonate with an aldehyde in the presence of base is a very convenient 'one-pot' method of preparing acrylonitriles[41]. The initial adduct readily eliminates diethyl hydrogen phosphate upon hydrolysis, giving a mixture of *cis*- and *trans*-nitriles in reasonable yield. The required phosphonates are obtained from triethyl phosphite and an α-chloronitrile.

$$ClCH_2CN + (EtO)_3P \rightarrow (EtO)_2P(O)CH_2CN \xrightarrow[\text{base}]{RCHO} RCH(OH)CH(CN)P(O)(OEt)_2$$
$$\xrightarrow[H_2O]{\text{heat}} RCH{=}CHCN + (EtO)_2P(O)OH$$

A quaternary iminium ion appears to be the key intermediate in the Strecker synthesis of α-aminonitriles[42]:

$$R^1{}_2CO + R^2{}_2NH \rightarrow R^1{}_2C(OH)NR^2{}_2 \xrightarrow{-H_2O} R^1{}_2C{=}\overset{+}{N}R^2{}_2 \xrightarrow{CN^-} R^1{}_2C(CN)NR^2{}_2$$

Pyrrolidine, morpholine and piperidine exhibit increasing activity, following their ability to show exocyclic double-bond character. Cyanosilylation of aldehydes and ketones with trimethylsilyl cyanide usually occurs in high yield (> 90%)[43]. The adduct, $R_2C(CN)OSiMe_3$, is stable in non-aqueous solvents, but hydrolysis with dilute acid or base regenerates the carbonyl compound.

Hydrocyanation of α,β-unsaturated ketones and similar compounds has long been a desirable but difficult procedure. By using either of two important new reagents, Et_3Al–HCN and Et_2AlCN, developed by Nagata and co-workers[44,45], this reaction can now be achieved under mild conditions to give a clean product in high yield. Both reagents are effective at or below room temperature in aprotic solvents; exhaustive studies with steroidal compounds have established that $EtAl_3$–HCN catalyses an irreversible ionic reaction, whereas Et_2AlCN is required in stoichiometric quantity for a reversible, solvent-dependent, non-ionic reaction. Details of the preparation of Et_2AlCN have been given[46].

A promising new method[47] for the conversion of a ketone (R_2CO) into a nitrile (R_2CHCN) involves its treatment with tosylmethyl isocyanide (p-$MeC_6H_4SO_2CH_2NC$). Usually yields are 75–85% but fall to 35% for sterically hindered ketones.

4.3.1.3 Miscellaneous methods

Alkylation of the cyanide ion can be achieved with benzylamines (in place of the more usual benzyl chlorides) but the presence of an *ortho*- or *para*-hydroxy group is essential, presumably to stabilise the transition state. Otherwise the amine must be quaternised before alkylation can occur[48]. A two-step reaction of cyanogen chloride with primary amines gives an alkyl dicyanamide[49], a type of compound difficult to prepare by other methods. Ketones are readily converted into β-cyanoketones by treatment with chlorosulphonyl isocyanate[50]. A general survey[51] of the synthesis and reactions of derivatives of carbonic acid includes discussion of cyanogen derivatives XCN where X = Cl, RO, RS, ClS or R_2N.

4.3.2 Reactions

4.3.2.1 Hydration

Interest in the conversion of nitriles into amides is reflected in reports of a number of new methods which all involve catalysis by metallic compounds of various types.

The action of the anionic planar complex Pt(cyclohexyne)-($Et_2PCH_2CH_2PEt_2$) converts acetonitrile homogeneously into acetamide in high yield[52]. The catalytic activity of complexes of this type depends upon their reaction with water to give a species with a hydroxy group directly attached to the metal. The suggested mechanism involves nitrile insertion into the M—OH bond to give an N-bonded imino alcohol complex (22), which rapidly rearranges to the amide complex (23). Certain tertiary phosphine complexes of iridium and rhodium are also effective hydration catalysts[52].

$$MOH + RCN \rightarrow M-N=C(OH)R \rightarrow M-NHCOR \xrightarrow{H_2O} MOH + RCONH_2$$

$$(22) \qquad\qquad\qquad (23)$$

Reaction of nitriles with aqueous acetic acid would normally cause substantial hydrolysis of the amide to the acid. The presence of titanium(IV) chloride suppresses the hydrolysis[53], and the reaction thus gives a reasonable yield of amide (65–90%). A very convenient inexpensive method[54] of some industrial potential involves the use of MnO_2 as catalyst in a simple fixed-bed continuous reactor. Yields are essentially quantitative.

Usual hydration methods (acid or base catalysis) are not effective for allenic nitriles, since attack occurs[55] preferentially at the β-carbon atom. However, two other established methods, alkaline hydrogen peroxide hydration and the Ritter reaction with t-butyl alcohol both give amides (60–70%), free from Michael addition and polymerisation products.

4.3.2.2 Decyanation

Two very similar papers by Cuvigny *et al.*[56,57] report on the reductive decyanation of nitriles by potassium in HMPT. This reagent is highly specific for formation of the hydrocarbon, very little amine is produced, and the 70–90% yields represent a substantial improvement over the method involving an alkali metal in liquid ammonia. Primary and secondary nitriles require t-butyl alcohol as co-solvent, and aromatic nitriles react in a complex way.

As a route to branched-chain alkenes, dehydrocyanation with soda-lime at 200 °C may offer a useful alternative to dehydrochlorination. 1-Cyanocyclobutene has been obtained[58] in this way from 1,2-dicyanocyclobutane, but the generality of this reaction has not been established.

4.3.2.3 Cycloaddition

Both ethylene and acrylonitrile are known to undergo [2 + 2] cyclodimerisation, but conditions for the formation of mixed cycloaddition products, e.g. cyanocyclobutane, have only recently been discovered[59]. A high pressure (1000 atm) of ethylene is required to reduce nitrile dimerisation and a high temperature (300–345 °C) is essential for a satisfactory reaction rate. Weak catalysis by Ni^0 complexes is observed. The stereochemical course of the Diels–Alder cycloaddition of α,β-unsaturated nitriles to cyclopentadiene has been examined[60]. The ratio of *endo* to *exo* products decreases along the series acrylonitrile, *trans*-crotononitrile, methacrylonitrile, thus indicating that steric effects are important, but that other factors are also involved.

Modifications to the mechanism of the condensation of α-aminonitriles with ketones to give imidazolidines have been proposed[61] on the basis of a ^{15}N-labelling experiments. A dihydropyrrole is the usual cyclisation product obtained from tetracyanoethylene and β-dicarbonyl compounds, but with acetylacetone an intermediate dihydrofuran can be isolated[62].

4.3.2.4 'Nitrile salts' and N-alkylation

Reactions between nitriles and hydrogen halides have been known for over

100 years, but the products (so-called 'nitrile salts') have never been satis-factorily characterised. Yanagida and co-workers have now reported[63-67] the results of a thorough reinvestigation of the structure and chemistry of 'nitrile salts' and have shown that two possible products may be formed. At 0 °C in ether an unstable species $2RCN \cdot 3HCl$ is formed which slowly loses HCl to give the same stable dimer $2RCN \cdot 2HCl$ as is isolated directly from the reaction at higher temperatures. The amidinium chloride structure (27) was unequivocably established by n.m.r. and mass spectrometry. Hydrolysis[63] of (27) gives a diacylamine and reaction[64] with phosgene affords a pyrimidine

$$R_2CHCCl{=}NH \xrightarrow[100\,°C,\ C_6H_6]{COCl_2} R_2CHCCl{=}NCOCl \longrightarrow R_2C{=}CClN{=}C{=}O$$

$$(24) \qquad\qquad (25) \qquad\qquad (26)$$

$$R_2CHCN \xrightarrow[60\,°C]{HCl} R_2CHC\underset{NHCCl=CR_2}{\overset{NH_2}{\diagup}}{+}\ Cl^{-} \xrightarrow{COCl_2} (28)$$

(27)

$$R_2CHC\underset{NHCCl_2CHR_2}{\overset{NH_2}{\diagup}}{+}\ Cl^{-}$$

(28) in one step. Direct treatment of a nitrile with phosgene and HCl in benzene produces some pyrimidine, but the principal product is an α,β-unsaturated isocyanate (26)[65]. The role of the imidoyl chloride (24) in this complex series of reactions is still in doubt, but it is probably a precursor of the carbamoyl chloride (25). Imidoyl chlorides are known to react with acyl chlorides and inter- and intra-molecular reactions of this type have been reviewed[66,67]. Aliphatic nitriles with no α-hydrogen atoms do not react with HCl, but aromatic nitriles dimerise[68] or, in the presence of HCl and PCl_5, they trimerise to give *sym*-triazines[69].

Lewis acids complex with nitriles[70], probably producing ylide-like species (29) which are converted into nitrilium salts (30) with alkyl halides. These tetrachloroferrates can be isolated, and direct aminolysis provides a general

$$R^1C{\equiv}\overset{+}{N}{-}\bar{F}eCl_3 \xrightarrow{R^2Cl} [R^1C{\equiv}\overset{+}{N}R^2]\,[FeCl_4^{-}]$$

$$(29) \qquad\qquad\qquad (30)$$

route to amidines. Yields (from the nitrile) of 40–80% make this mild reaction a useful alternative to available methods.

4.3.2.5 *Miscellaneous reactions*

Reduction of nitriles with sodium in liquid ammonia often leads to sub-stantial amounts of decyanation. Substitution of calcium for sodium virtually eliminates this side reaction[71]. Kirilov *et al.*[72] have isolated and character-ised an organocalcium intermediate (31) from phenylacetonitrile.

$$(Ph\bar{C}HC\equiv N)_2\ Ca^{2+} \rightleftharpoons (PhC\equiv C\bar{N}H)_2\ Ca^{2+}$$
(31)

Treatment of (31) with diethyl phosphorochloridate produced an α-phosphorylated nitrile. Unsaturated nitriles are effectively reduced over nickel boride to saturated amines[73].

Nucleophilic displacement of the cyano group in α-amino-α-methoxy- or bis-(α-amino)-nitriles by MeO^- provides a new route to amide acetals and so-called aminal esters, $R^1C(OR^2)(NR^3_2)_2$[74]. Tetra-alkylammonium cyanides offer a convenient source of CN^-, as they are soluble in organic solvents such as THF. This quaternary salt has been identified[75] as the active species in the catalysis of alkene dimerisation and polymerisation by $(Et_4\overset{+}{N})_3$ $[Co(CN)_5O_2]^{3-}$, and is a more powerful nucleophile than sodium cyanide.

4.3.3 Malononitrile

Reaction of active methylene compounds such as malononitrile with alkyl halides in the presence of a sodium alkoxide is a well known reaction, which has now been extended by the use of triethylamine as the base[76]. Compounds (32) and (33) are obtained in nearly quantitative yield at 25 °C from malononitrile and benzyl chloride or methylmalononitrile and p-xylylene dichloride, respectively. Attempts to extend this modified procedure to the formation

$$(PhCH_2)_2C(CN)_2 \qquad MeC(CN)_2CH_2C_6H_4CH_2C(CN)_2Me$$
$$(32) \qquad\qquad\qquad (33)$$

of a new type of polymer by polycondensation of malononitrile with appropriate chloro compounds in the presence of Et_3N met with varied success. With p-xylylene dichloride a polymer (34) of an average chain length of 22 units was obtained, but this was the maximum chain length achieved. Other systems gave shorter-chain polymers, e.g. (35), or no polymerisation at all.

$$Cl-[CH_2C_6H_4CH_2C(CN)_2]_{22}-H \qquad Cl-[CH_2C_6H_4OC_6H_4CH_2C(CN)_2]_{12}-H$$
$$(34) \qquad\qquad\qquad\qquad (35)$$

Other reactions of malononitrile or cyanoacetic ester include cycloaddition to acyl and sulphonyl azides to give triazines[77], dimerisation (catalysed by platinum–phosphine complexes) to an enamine[78] and reductive dimerisation with $SOCl_2$ to yield a tetra-substituted ethylene[79].

N-Substituted aminomalononitriles (36; $R^1 = H$, $R^2 = Bu^tCH_2CMe_2$; $R^1 = R^2 = Me$) undergo base-catalysed[80] or thermal[81] decomposition in a very complex fashion to yield many products. The stable radical (37), formed via an aminocyanocarbene, is an important intermediate in this decomposition, particularly in the presence of base. An extensive study[82] of the action

$$R^1R^2NCH(CN)_2 \qquad t\text{-}C_8H_{17}NH\overset{\bullet}{C}(CN)_2$$
$$(36) \qquad\qquad\qquad (37)$$

of nucleophiles on O-tosylisonitrosomalononitrile, $p\text{-}MeC_6H_4SO_3N{=}C(CN)_2$, indicates that reaction may occur at carbon, sulphur or nitrogen, involving displacement of CN, $ArSO_2$ or $ArSO_3$, respectively (see Section 4.7.6).

4.3.4 Diaminomaleonitrile and di-iminosuccinonitrile

HCN is known to undergo self-addition to form 'polymers' and this reaction may have played an important part in the chemistry of the primitive Earth since the polymer can be hydrolysed to a wide variety of amino acids[83]. There has been some doubt about the mechanism of HCN oligomerisation. This reaction is difficult to study directly, but Ferris[84] has now shown that *N*-alkyliminoacetonitriles (38; R = alkyl) do not exist in the carbene form and do not dimerise directly to an *N'*,*N''*-dialkylaminomaleonitrile (39) except at high temperatures[85]. Addition of HCN to (38; R = alkyl) occurs stepwise and it seems likely that HCN oligomerisation is similar. Furthermore HCN 'polymer' is probably a mixture of low molecular weight compounds. Diaminomaleonitrile (39; R = H) is a key compound in these reac-

$$2\,RCN \longrightarrow RN{=}CHCN \xrightarrow{\text{HCN}} RNHCH(CN)_2$$

$$(38)$$
$$(R = H \text{ or alkyl})$$

$$RNHC(CN){=}C(CN)NHR \longrightarrow HCN \text{ polymer}$$
$$(39)$$
$$HCN \Big\updownarrow [O]$$

$$HCN + (CN)_2 \xrightarrow[-40\,°C]{\text{Base}} RN{=}C(CN)C(CN){=}NR$$
$$(40)$$
$$(R = H \text{ or alkyl})$$

tions since it can be hydrolysed to glycine[84] and oxidised to di-iminosuccino-nitrile (40; R = H), which is hydrolysed to urea[86]. This last reaction is particularly significant for prebiotic chemistry because the oxidation step can occur in the absence of molecular oxygen. Webster *et al.*[87] have found that the reverse reaction, reduction of (40) to (39), occurs in the presence of an excess of HCN, confirming that these two compounds may constitute a redox system of fundamental importance for primitive biochemistry. The conversion of (40) into (39) may also be accomplished by a variety of standard chemical reducing agents. Di-iminosuccinonitrile has a rich chemistry and reactions with nucleophiles, acyl halides and carbonyl compounds have been examined[88].

4.3.5 Nitrile oxides

A recent book[89] covers the synthesis and reactions of this class of compound. Isomerisation of a nitrile oxide to give the corresponding isocyanate proceeds with complete retention of configuration[90]. Lack of crossover products in a suitable reaction with labelled compounds indicates an intramolecular process.

$$p\text{-}^2HC_6H_4C\equiv\overset{+}{N}\text{—}O + Ph^{13}C\equiv\overset{+}{N}\text{—}\overset{-}{O} \rightarrow p\text{-}^2HC_6H_4N{=}C{=}O + PhN{=}^{13}C{=}O$$

4.3.6 Theoretical studies

One of the simplest examples of a unimolecular isomerisation is the reaction MeNC $\rightarrow$ MeCN, and an accurate calculated potential energy surface for this process is of great theoretical interest. Earlier calculations suggested that the intermediate is ionic in nature (41), with a nearly planar Me$^+$ group, but this suggestion has now been refuted.

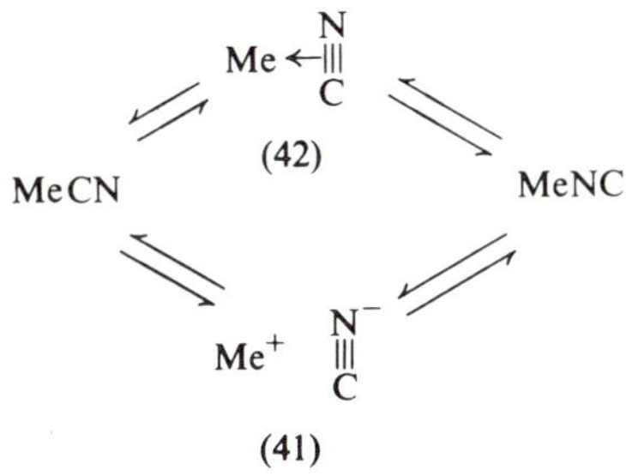

Both *ab initio* and semi-empirical SCF calculations[91–93] lead clearly to a transition state with a polarity not very different from that of the initial or the final state. The metastable intermediate (42) has a pyramidal methyl group and a structure similar to a π-complex[93], although the exact ionicity of (42) is still in dispute[91,92]. *Ab initio* SCF calculations[94] of the minimum-energy geometry of azodicarbonitrile, NC—N=N—CN, indicate a *trans* structure.

4.4 ISOCYANIDES

4.4.1 Synthesis

Preparative methods for isocyanides have been briefly reviewed[95]. To circumvent the use of phosgene, the Vilsmeier reagent chlorodimethylformiminium chloride (43) has been developed[96] as a dehydrating agent for the conversion of *N*-alkylformamides into isocyanides. The reagent (43), prepared *in situ* from DMF and SOCl$_2$, is treated with secondary amides at $-50\,^\circ$C to preserve the intermediate adduct (44), until the addition is complete. A wide variety of formamides have been dehydrated in this way to give isocyanides in 60–90% yield.

$$\begin{matrix} \text{HCONMe}_2 \\ + \\ \text{SOCl}_2 \end{matrix} \rightarrow \text{ClCH}{=}\overset{+}{N}\text{Me}_2\ \text{Cl}^- \xrightarrow{\text{RNHCHO}} \text{RN–CHOCHCINMe}_2 \rightarrow \begin{matrix} \text{RNC} \\ + \\ \text{HCONMe}_2 \\ + \\ \text{HCl} \end{matrix}$$
$$\qquad\qquad\qquad (43)\qquad\qquad\qquad (44)$$

4.4.2 Reactions

The full report has now appeared[97] of the formimidation of active methylene compounds with alkenyl isocyanides in the presence of catalytic amounts of copper(I) oxide. The initial imine (45 ; R^3 = alkenyl) tautomerises to an enamine (46) and this is the only product at low catalyst concentration. Increasing the concentration of Cu_2O gives rise to other products, including an alkene, a tetrahydropyrrole and a substituted cyclopropane. A copper–

$$R^1R^2CH_2 + R^3NC \xrightarrow{Cu_2O} R^1R^2CH\overset{\bullet}{C}H=NR^3 \to R^1R^2C\!=\!CHNHR^3$$

$$(45) \qquad\qquad (46)$$

isocyanide complex is almost certainly involved in this reaction, as for many other reactions in which the isocyanide moiety is not incorporated into the product but, as a metal complex, has a strictly catalytic function. Recent applications of this catalyst include the conversion of an alkene into a cyclopropane derivative[98], the esterification of carboxylic acids with alkyl halides[99,100] and the Michael reaction[101]. Sometimes the adduct between the substrate and the copper–isocyanide complex may be isolated[99].

Oxidative addition of amines or alcohols to isocyanides is catalysed by mercuric salts[102]. An initial oxymercuration followed by a redox step to give a urea or a urethane is postulated. The catalytic effectiveness of various mercury(II) salts follows the order of redox potentials, $Hg(OAc)_2 > Hg(NO_3)_2 > HgCl_2 > Hg(CN)_2$. Michael addition of ethyl isocyanoacetate, $EtO_2CCH_2\overset{+}{N}\!\equiv\!\overset{-}{C}$, to acrylonitrile gives mono- or bis-adducts, and sometimes cyclisation occurs to give a dihydropyrrole[103]. Oxidation of $Bu^tN\overset{+}{\equiv}\overset{-}{C}$ to $Bu^tN\!=\!C\!=\!O$ by nitroalkenes is discussed on pp. 171–172.

4.5 ISOCYANATES AND ISOTHIOCYANATES

4.5.1 Synthesis

In an excellent review[104] of recent developments in the synthesis of isocyanates, 25 different methods are discussed. A brief survey of the same area has appeared elsewhere[51]. The intense activity in this field reflects the commercial importance of this class of compounds.

4.5.1.1 The Curtius reaction

Thermal decomposition of an acid azide and rearrangement of the corresponding nitrene has long provided a facile synthesis of isocyanates, and trimethylsilyl azide appears to be a particularly valuable vehicle for the azide group[105–108]. Both acid chlorides[105,107,108] and acid anhydrides[106,107] may be used with this reagent, and yields of the isocyanate are high since the acid azide need not be isolated. Mixed anhydrides of a carboxylic and a sulphonic acid have also proved useful[105]. Another one-step conversion of an acid chloride into an isocyanate in quantitative yield involves[109] the use of hydrazoic acid in pyridine at 0–20 °C. Since this system is of the weak acid–weak base type, pyridinium azide is not precipitated and the mixture acts as a source of dissolved azide anion. The new polymerisable fluorovinyl isocyanates (47) and (48) have been obtained[110] from the corresponding acid fluoride by the Curtius reaction. It is essential to brominate the double bond first otherwise azide ion attack occurs preferentially at the β-carbon atom. Bromine is subsequently eliminated by treatment with zinc.

$$CF_2{=}CFN{=}C{=}O \qquad\qquad CF_2{=}C(CF_3)N{=}C{=}O$$
$$(47) \qquad\qquad\qquad\qquad (48)$$

4.5.1.2 Other methods

In the presence of a Lewis acid such as $FeCl_3$, cyanogen chloride can be alkylated by isopropyl chloride to give a highly reactive alkylimidoyl dichloride complex (49; $R = Pr^i$)[111]. I.r. spectroscopic evidence suggests that this complex does not exist as a nitrilium salt (50), in contrast to the analogous

$$(RN{=}CCl_2)\,(FeCl_3) \qquad\qquad [ClC{\equiv}\overset{+}{N}R]\,[FeCl_4]^-$$
$$(49) \qquad\qquad\qquad\qquad (50)$$

species produced by alkylation of a nitrile (Section 4.3.2.4). Conversion of (49) into an isocyanate is difficult with the usual reagents (e.g. Ag_2O, CaO), but when $R = Pr^i$ an 85% yield of isocyanate is obtained by reaction with ZnO. The generality of this reaction has not been established.

Thiophosgene adds to a ketimine to form the alkyl α-chloroisothiocyanate (51) by elimination of HCl[112]. For compounds with a β-hydrogen atom a

$$R^1CH_2CR^2{=}NH + CSCl_2 \xrightarrow{-HCl} R^1CH_2CR^2ClN{=}C{=}S \xrightarrow{-HCl} R^1CH{=}CR^2N{=}C{=}S$$
$$(51) \qquad\qquad\qquad\qquad (52)$$

second mole of HCl is eliminated, yielding an alkenyl isothiocyanate (52). The mechanism of this reaction may be compared with that proposed for the phosgenation of an imidoyl chloride (24), in which the initial species (25) does not correspond to addition of [COCl] across the C=N double bond but to substitution at nitrogen. In this latter reaction the presence of an α-chloro group in the substrate may be crucial and may indicate that elimination of HCl is concerted with attack of $COCl_2$. Further comparative study of these reactions is desirable.

Direct conversion of trichloroacetyl chloride into trichloroacetyl isothiocyanate can be very conveniently accomplished[113] with trimethylsilyl isothiocyanate, $Me_3SiN{=}C{=}S$.

4.5.2 Reactions

Some unconventional chemistry of isocyanates has been discussed in the first volume of a new series[114].

4.5.2.1 Nucleophilic addition to the isocyanate group

The structure of the isocyanate group favours nucleophilic attack at carbon. With reagents of the HX type simple addition across the C=N double bond is the usual reaction, but cyclisation may occur when either the substrate or the reagent contains other reactive groups. Often a range of products is obtained. The addition of phenylhydrazine to an acyl or an aroyl isocyanate is a good example[115]. The expected semicarbazides are often contaminated to a greater or lesser extent by other products, including diacyldiureas, substituted biurets, 3-hydroxytriazoles and triazolin-3-ones. With hydrazones and amidoximes, cyclisation is less prevalent and addition products only are obtained[116]. Aziridine gives[117] a normal urea and a ring-expanded product with acyl isocyanates. Carbamate thiol esters are formed with thiols[118] and sulphides[117], and methoxymethyl isocyanate is an excellent reagent for the reversible protection of the thiol group[118]. Isocyanates with a thiocarbonyl group in the α-position are particularly prone to cyclisation and various heterocyclic compounds have been obtained[119,120] from thiobenzoyl isocyanate by $[4 + 1]$ or $[4 + 2]$ cycloadditions with ketenes, ketenimines, nitrosobenzene, diazoalkanes and nitrones.

Nucleophilic carbon readily reacts with isocyanates; activated methylene compounds such as $(53; R^1 = CO_2Et^{121}, CN^{122}$ or $COCO_2Et^{123})$ have been

$$R^1CH_2CO_2Et + R^2N=C=O \rightarrow R^2NHCOCHR^1CO_2Et$$
$$(53)$$

studied. Cyclisation either directly or by further attack by isocyanate is observed[121,122]. With benzoyl isocyanate, diazoacetates give rise to α-carbamoyldiazoacetates which are surprisingly unreactive and do not cyclise[124]. Thermolysis of a similar adduct with a sulphonium ylide gives an oxazole[124]. Alkyl-lithium compounds and alkylmagnesium bromides are essentially sources of nucleophilic carbon and react with isocyanates to give the amide in high yield[125]. When combined with the Curtius reaction overall yields of amide are 57–97%, phenyl- and n-butyl-lithium giving slightly better results than methyl-lithium.

Unlike organotin hydrides, trialkylsilanes do not react directly with an isocyanate, but, in the presence of $PdCl_2$ or Pt–carbon, hydrosilylation can be achieved in high yield[126]. The trialkylsilyl group can attack either C or N, but methanolysis of both resultant silylamides gives a formamide. This is a novel route to such compounds.

4.5.2.2 With carbenes and nitrenes

Cycloaddition of an acylnitrene to an isocyanate can be expected to give an oxadiazolinone (54), but the formation of a urazole (55) is unexpected[127]. This product appears to result from azide addition followed by attack of a second molecule of isocyanate and is not formed when the nitrene is generated by α-elimination; a mechanism has been proposed[127]. A very similar reaction of dimethoxycarbene with two moles of an aryl isocyanate or iso-

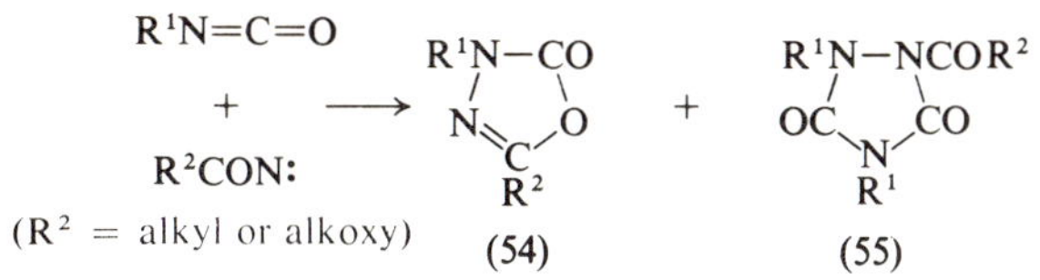

$$R^1N=C=O$$
$$+ \longrightarrow$$
$$R^2CON:$$
$$(R^2 = \text{alkyl or alkoxy})$$

(54)   (55)

thiocyanate has also been reported[128]. Ring closure without loss of N_2 from an azide group gives tetrazolin-5-ones[129].

4.5.2.3 Miscellaneous reactions

Both addition and [2 + 2] cyclisation are observed for the reaction of the N-trimethylsilyl ketimine (56) with an alkyl isocyanate. Addition predominates[130] and the product may have either the amide (57) or the imidate (58) structure; spectroscopic evidence suggests that both forms sometimes exist.

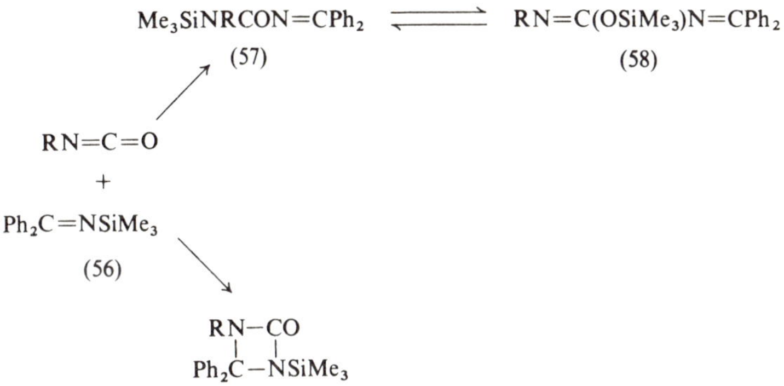

$$Me_3SiNRCON=CPh_2 \rightleftharpoons RN=C(OSiMe_3)N=CPh_2$$

(57)   (58)

$$RN=C=O$$
$$+$$
$$Ph_2C=NSiMe_3$$

(56)

$$RN-CO$$
$$Ph_2C-NSiMe_3$$

Formamidinyl isocyanate, $PhN=C(NR_2)N=C=S$, isomerises to a quinazoline and dimerises to a sym-triazine[131]. Reactions with other isocyanates have also been studied. The formation of an adduct between an isocyanate and sodium bisulphite in aqueous solution may have commercial significance as a method of making isocyanate prepolymers compatible with an aqueous solution[132]. Other studies include those of the reduction of $RN=C=O$ to $RNHCHO$ with lithium hydridotri-t-butoxyaluminate[96] and the decomposition of simple aliphatic isocyanates and isothiocyanates in the gas phase[133].

4.6 CYANATES AND THIOCYANATES

Although the $-OC\equiv N$ and $-SC\equiv N$ groups might be regarded as falling respectively into the areas of oxygen and sulphur chemistry, these functional groups are discussed in this chapter because of their isomeric relationship to other groups such as isocyanate and nitrile oxide. Cyanates are less stable than thiocyanates and there is little published work concerning them.

4.6.1 Synthesis

Thiocyanogen, $(SCN)_2$, must usually be synthesised under anhydrous conditions, but this restriction has been removed by the simple expedient of using a two-solvent system[134]. Passage of Cl_2 into an aqueous solution of NaSCN stirred with toluene gives an 80–90% yield of thiocyanogen in toluene, making this reagent much more economically attractive than before. Addition of $(SCN)_2$ to vinyl halides produces[134] the novel halogenoalkylene bis-thiocyanate $NCSCH_2CHXSCN$. Thiocyanogen cleavage of organotin compounds[135] is a convenient route to alkenyl thiocyanates free from contamination by the corresponding isothiocyanate. A cyclic intermediate may be involved and some rearrangement of the alkenyl group may occur.

At 25 °C in the absence of light and in the presence of a radical inhibitor, heterolytic addition of thiocyanogen chloride to alkenes in acetic acid has been observed[136]. The reaction is fast and stereospecific (*trans*), and probably proceeds through a cyanosulphonium ion intermediate (59) to give the expected products, β-chloro- and β-acetoxy-alkyl thiocyanates. β-Alkoxy-

$$
\begin{array}{ccc}
\text{(59)} & \text{RSCSCl} & \text{RSCSSCN} \\
 & \text{(60)} & \text{(61)}
\end{array}
$$

alkyl thiocyanates are obtained[137] from alkenes in 8–60% yield by treatment with thallium(III) acetate in ethanol followed by addition of KSCN. This reaction may have wide applicability. Nucleophilic displacement of Cl by SCN^- converts[138] the alkyl chlorodithiocarbonate (60) into the corresponding thiocyanate (61). Thermal isomerisation of (61) to the isothiocyanate proceeds in low yield. The previously unknown alkynyl thiocyanates[139] have been synthesised by the scheme shown:

$$ RC{\equiv}CLi \xrightarrow{S_8} RC{\equiv}CSLi \xrightarrow{ClCN} RC{\equiv}CSCN $$

4.6.2 Reactions

New reactions of cyanates have been surveyed[140]. Particular emphasis is given to reactions with acyl halides and to the formation of heterocyclic compounds. Methanolysis of ethyl and aryl thiocyanates to give thioimino-carbonates, $RSC(=NH)OMe$, is catalysed by CN^-, MeO^- and N_3^- but not by I^-, acetate or simple amines. A mechanism involving addition and substitution of CN^- is proposed[141]. Treatment of propynyl thiocyanate with *N,N*-dimethyl-lithamide (Me_2NLi) results[139] in displacement of the CN group to give a sulphenamide, $MeC{\equiv}CSNMe_2$. Other nucleophilic reagents were not studied, but this reaction may have wider scope. Formaldehyde or trichloroacetaldehyde reacts with two moles of an alkyl thiocyanate to give a bis-thiocarbamoyl derivative[142].

Photochemically induced isomerisation[143] of benzyl thiocyanates to benzyl isothiocyanates proceeds under milder conditions than the corresponding thermal process. Some by-products are formed by solvent capture of the SCN radical. Cyclohexyl thiocyanate does not isomerise and the products obtained suggest homolysis of the S—C bond.

4.7 IMINES* AND RELATED COMPOUNDS

This section covers the several types of compound containing the —C=N— group, including those with a heteroatom substituent at carbon (imidates, amidines) or at nitrogen (oximes). Two exceptions have been allowed; hydrazones, $R^1R^2C=NNR^3{}_2$, are included in the section on hydrazine derivatives, and diazo compounds, $R^1R^2C=\overset{+}{N}=\overset{-}{N}$, are treated separately in Section 4.8. The photochemistry of compounds containing the —C=N— group has been reviewed[144,145].

4.7.1 Synthesis

Lithiation of a primary amine has been investigated[146] as a novel route to ketimines. In the presence of an excess of an alkyl-lithium, a primary amine is dilithiated (62) and subsequent elimination of LiH or LiR alternates with addition of the alkyl-lithium. Non-hydrolytic work-up gives a 30–60% yield

$$R^1CH_2NH_2 \xrightarrow{2R^2Li} \underset{(62)}{R^1CH_2NLi_2} \longrightarrow R^1CH=NLi \xrightarrow{R^2Li} R^1R^2CHNLi_2$$

$$\underset{(63)}{R^1R^2C=NCH_2R^1} \xleftarrow{R^1CH_2NH_2} R^1R^2C=NLi \begin{array}{c} \xrightarrow{R^1R^2CO \; H_2O} \\ \xrightarrow{MeOH} \underset{(64)}{R^1R^2C=NH} \end{array}$$

of the imine (64), but if the alkyl-lithium is not in sufficient excess the imine is *N*-alkylated (63) by reaction with unchanged amine.

By using established functional group chemistry, three routes to α-amino-imines have been evaluated[147]. Condensation of a primary amine with an α-amino-aldehyde or -ketone is the most generally useful method, the amination of an α-bromoimine and the thermolysis of an α-amino-*gem*-diamine being of more limited scope. In the synthesis of some *N*-fluoro-vinylimines the C=C and the C=N double bonds were created simultaneously by elimination of HCl and HF, respectively[148]. Mercury photo-sensitisation allows photo-oxidation of amines to imines in high yield[149].

It is not possible to obtain simple unsubstituted aldimines by condensation of an aldehyde with ammonia. This has been appreciated for some time but

---

* The term imine has been used in preference to azomethine or Schiff's base.

the nature of the compounds which are formed has remained in doubt. From a detailed study of these compounds, Nielsen and co-workers[150] have shown that at low temperatures (-25 to $0\ ^\circ$C) an instantaneous precipitation of the hydrated 1-amino-alkan-1-ol (65) occurs. These very unstable amino-alcohols are dehydrated at 5–$10\ ^\circ$C to give triazines (66), again usually as hydrates. Most 'aldehyde-ammonias' of earlier workers are in fact triazine hydrates. With care, anhydrous triazines can be isolated and deaminated at

$$RCHO\ +\ NH_3\ \xrightarrow{\ -25\ ^\circ C\ }\ RCH(OH)NH_2 \cdot (H_2O)_x$$
$$(65)$$

$$\downarrow{\scriptstyle 5\ ^\circ C}$$

$$RCH(N{=}CHR)_2\ \xleftarrow{\ 40\ ^\circ C\ }\ (66)$$
$$(67)$$

40–80 $^\circ$C to yield di-imines (67). A series of diazapentadienes (67; R = substituted phenyl) were obtained[151] from the corresponding reaction of benzaldehydes in liquid ammonia. Cyclisation of these hydrobenzamides is discussed on p. 194.

4.7.2 Addition reactions

Iron dodecacarbonyl, $Fe_3(CO)_{12}$, catalyses the reduction of imines in refluxing benzene–methanol, presumably via an iron hydride complex. Yields are highest ($> 80\%$) for aromatic imines[152]. Reduction of imines by alkylsilanes (R_3SiH)[153] in the presence of $(Ph_3P)_3RhCl$ or $PdCl_2$ may also involve a transition-metal hydride. The N-silylamines can be isolated or the free amine can be obtained directly by methanolysis. Asymmetric induction has been noted[154] in the reduction of an imine with $LiAlH_4$.

Organo-lithium, -magnesium, -zinc and -aluminium reagents derived from halogenoalkynes or halogenoallenes add to imines[155,156] to form unsaturated amines, sometimes with marked stereoselectivity[156]. With α,β-unsaturated aldimines the ratio of 1,2- to 1,4-addition depends upon the organometallic reagent[157]. Addition of thiophosgene to a ketimine[112] and oxidation of an alkylimidoyl dichloride (49)[111] are discussed in Section 4.5.1.2. The imine (49) can be hydrolysed to the amine, or converted into the carbamate with ethanol. Complex kinetics are observed for the peroxyacid oxidation of imines, but a two-step mechanism appears to operate[158]. Some reactions of perfluorinated imines have been reported[159].

Metallation of ketimines such as $Ph_2C{=}NH$ with diethyl-lithamide permits N-alkylation with alkyl halides in an S_N2 reaction[160]. Sodamide is less effective and this reaction is mechanistically related to that on p. 187. In a similar reaction[161], N-methylketimines gave C-alkylated products $(Ph_2C{=}NCH_2R)$ with alkyl halides, and with aldehydes, $Ph_2C{=}NCH_2CH(OH)R$.

Product identification in the [2 + 2]-cycloaddition[162,163] of symmetrical dialkylcarbodi-imides (RC=N=CR) to other heterocumulenes such as isocyanates, isothiocyanates and N-sulphinylamides (RCON=S=O, RSO_2N=S=O) has always been a problem and methyl-t-butylcarbodi-imide has been developed as a diagnostic probe[163]. With the sulphinyl amides addition occurs[162] across the N=S double bond (together with some 1,4-addition) but otherwise the nature of the group R dictates the position of addition[163]. An elegant method of solving the problem of removing the reagents after completion of a reaction is to bind the reagent to an insoluble polymer. A polystyrene-bound carbodi-imide has been prepared[164] and successfully applied to the formation of acid anhydrides and the Moffat oxidation of alcohols. The polymer-bound reagent is easily regenerated with a slight reduction in activity. Synthetic methods for di-imides have been briefly surveyed[51].

4.7.3 Methyleneammonium salts

Recent years have seen an upsurge of interest in methyleneammonium salts, e.g. (68), especially the dichloro derivative phosgenimonium chloride (69). As a recent review indicates[165], these compounds have a rich chemistry;

$$CH_2=\overset{+}{N}Me_2\ Cl^- \qquad\qquad Cl_2C=\overset{+}{N}Me_2\ Cl^-$$
$$(68) \qquad\qquad\qquad\qquad (69)$$

some of it still relatively unexplored. Reactions of (69) with a range of compounds such as enamines[166], ynamines[167], secondary amides[168], dialkyl-cyanamides[167] and aliphatic diols[169] produce new products which themselves may often be regarded as versatile new reagents. Heterocyclic systems accessible from the reaction of (69) with an appropriate substrate[170] include indazole, triazole, oxadiazole, benzoxazole and benzothiazole. A triazine and an imidazoline are accessible from (68) and chloroethyl- and chloro-acetyl-hydrazine, respectively[171]. No stable products were formed with pyridine or triphenylphosphine[172].

4.7.4 Nitrones

Aliphatic nitrones (imine oxides) are not very stable and have not been extensively studied. Reaction of N-cyanoalkylnitrones with benzenethiol at 25 °C involves addition and cyclisation to an imidazole probably via a cyanoimine[173]. 1,3-Addition of an electron-deficient alkene such as EtO_2CCH=CHCO$_2$Et to the nitrone (70) to give an oxazoline is highly stereospecific[174]. The barrier to rotation about the C=N double bond in nitrones has been thought to be relatively low, but for compound (71) the

$$MeO_2CCH=\overset{+}{N}(OMe)\bar{O} \qquad\qquad ArCHPh=\overset{+}{N}(CH_2Ph)\bar{O}$$
$$(70) \qquad\qquad\qquad\qquad (71)$$

activation energy for isomerisation[175], measured by n.m.r. spectroscopy, is 33.6 kcal mol^{-1}.

4.7.5 Imidates and amidines

Alkylation of N-substituted phenylacetamides with $[Et_3O]^+ BF_4^-$ leads to an imidate, but cyanacetamides give[176] a tautomeric mixture of the imidate and the corresponding alkoxyenamine (Section 4.7.7). Nitrene insertion into a vinylic C—H bond seems a probable mechanism for the reported[177] interaction of an acyl or a sulphonyl azide and a vinyl ether to produce an imidate:

$$R^1R^2C{=}CHOR^3 + Cl_3CON_3 \rightarrow R^1R^2C{=}C(OR^3)NHCOCCl_3 \rightarrow$$

$$R^1R^2CHC(OR^3){=}NCOCCl_3$$

The first aliphatic hydrazonates (72) were obtained[178] by a similar insertion of ethoxycarbonylnitrene (73) into the N—H bond of the corresponding

$$MeC(OR){=}NNHCO_2Et \qquad EtOCON{:}$$
$$(72) \qquad\qquad (73)$$

imidate. Surprisingly, this bond is not very susceptible to insertion reactions, perhaps even less so than alkenic C—H.

Two novel procedures for formation of thioimidates have been reported. Thiol addition to N-arylketimines[179] and alkylation of thioamides with ethyl chlorothiocarbonate[180] provide useful alternatives to classical methods. N,N-dimethylamidines are produced by boiling secondary amides in HMPT[180a]. A general route to amidines is described in Section 4.3.2.4.

Peroxyacid oxidation of an imidate to an oxazirane has been studied[181] for strained and unstrained imidates. The oxazirane is hydrolysed in aqueous acid to an ester and an N-alkylhydroxylamine, thus providing a route for the transformation $R^1N{=}CHOR^2 \rightarrow R^1NHOH$ in high yield. Analysis of the kinetics and products of hydrolysis of ten imidates derived from amines of widely differing basicity reveals a surprising complexity in this reaction[182]. The pH dependence of the rates and products has been interpreted in terms of a mechanism involving cationic, neutral and anionic intermediates. Rearrangement of benzimidates to amides is catalysed by alkyl halides and other electrophilic species, thus suggesting that alkylation of neutral amides usually proceeds at the oxygen atom, with N-substitution arising from a

$$PhC(OR^3){=}NR^2 \xrightarrow[R^1X]{Heat} \left[\begin{array}{c} OR^3 \\ PhC{+} \qquad X^- \\ NR^1R^2 \end{array} \right] \longrightarrow \begin{array}{c} PhCONR^1R^2 \\ + \\ R^3X \end{array}$$

subsequent rearrangement[183]. Little loss in N-atom lone-pair conjugation upon protonation of a N,N',N''-trisubstituted amidine has been taken[184] to indicate that protonation must occur at the imino group, leaving the amino-nitrogen lone pair conjugated with the C=N double bond. This agrees with results from alkylation experiments.

4.7.6 Oximes and sulphenimines (thio-oximes)

Double decomposition between an oxime $R^1_2C=NOH$ and an alkoxyamine R^2ONH_2 presents a beautifully simple method of O-alkyl oxime synthesis[185]. Addition of acetic acid drives the reaction forward by precipitation of hydroxylammonium acetate. Facile displacement of hydroxylamine from simple oximes is confirmed[186] by instantaneous exchange with ^{15}N-labelled hydroxylamine, as shown by n.m.r. spectroscopy.

Lack of a convenient synthetic route to sulphenimines has left the chemistry of this class of compound relatively unexplored. S-Arylsulphenimines have now been prepared in one step from diaryl disulphides, $AgNO_3$, ammonia and an aldehyde or ketone[187,188]. Yields range from 30 to 98%, but aliphatic

$$(ArS)_2 + AgNO_3 + R_2C=O \xrightarrow{NH_3} ArSN=CR_2$$

disulphides do not react. A sulphenamide is a most likely intermediate since these compounds can be prepared independently. Sulphenimines are more resistant to hydrolysis than the corresponding imines, but much less stable than oximes. The barrier to rotation about the $C=N$ double bond is high enough to allow the existence of E and Z isomers[188].

A useful review[189] of less familiar reactions of oximes includes nitrosation at C, N and O, conversion into diazoalkanes, cyclisation and oxidation. The oxime derivative (74) undergoes a number of different substitution reactions with nucleophiles corresponding to attack at C, S or N[82]. Alcohols at $-70\,^\circ C$ give the mono- and di-substituted products (75) and (76), probably via addition and elimination. At $25\,^\circ C$ the arylsulphonyl group is displaced as the ester (77). A similar reaction is observed with nucleophiles such as Et_3N, F^-, Cl^- and Br^-. Malononitrile anion provides an example of attack

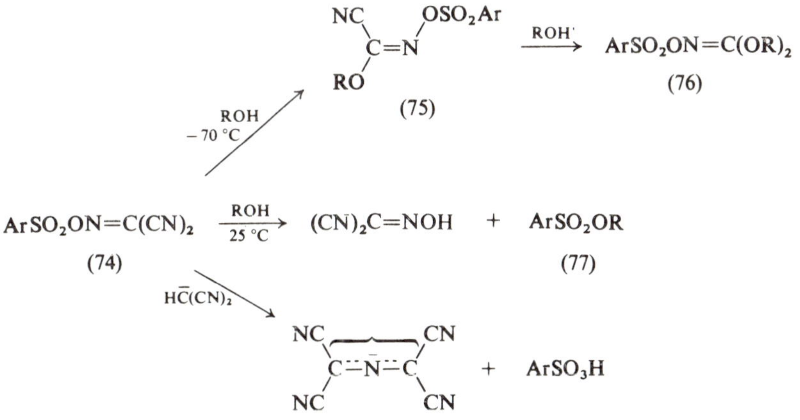

at nitrogen. Secondary reactions of some of these products were also investigated.

Condensation of an aldehyde with hydroxylamine in the presence of potassium cyanate has been reinvestigated[190,191] and the product is shown to be an O-carbamoyl oxime, $RCH=NOCONH_2$, and not a nitrone or oxazirane as previously suggested. The cycloaddition of α-amino-oximes to ethyl acetimidate or ethyl orthoacetate normally gives an imidazole, but sometimes a pyrazole may be formed[192]. The reaction with phosgene has also been studied[193]. A reappraisal[194] of the nitrous acid oxidation of oximes has led to a new mechanism to account for the formation of N_2, N_2O and NO. Further kinetic study of the competing cyclisation and elimination reactions of β-aminoketoxime anions has provided a rationalisation of the sensitivity of the product to reaction conditions[195]. Some asymmetric induction in the hydrogenolysis of a chiral oxime ester with diborane has been noted[196].

4.7.7 Imine–enamine tautomerism

Two brief surveys of this type of tautomerism have appeared[197,198]. Although undoubtedly a complex problem, the rationalisation of the effect of substituents on the tautomeric equilibrium between an imine (78) and an enamine (79) is an important and rewarding subject of study, since it is concerned with the fundamental way in which electronic and steric interactions affect the energy of relatively simple molecules. Analysis and interpretation of data is often complicated by the possibility of stereoisomerism in both the imine (*syn–anti*) and the enamine (*cis–trans*). A selection of some

$$R^1R^2CHCR^3{=}NR^4 \rightleftharpoons R^1R^2C{=}CR^3NHR^4$$

$$(78) \qquad\qquad (79)$$

recent results[199–201] which illustrate the complexity of this problem is given in Table 4.1. Generally substitution of Ph at any position stabilises the

Table 4.1 Effect of substitution on the imine–enamine equilibrium between (78) and (79)

R^1	R^2	R^3	R^4	*Enamine/%*	*Ref.*
Me	Me	H	Me	0	199
Me	Me	H	Ph	16	199
Me	H	Ph	Ph	4	200
Me	Me	Ph	Ph	30	200
Me	Me	Ph	Me	10	199
Me	Ph	H	Me	34	201
Me	Ph	H	$4\text{-MeOC}_6\text{H}_4$	73	201
Me	Me	H	$4\text{-MeOC}_6\text{H}_4$	0	199
Me	H	Ph	$4\text{-MeOC}_6\text{H}_4$	2	200
Me	Me	Ph	$4\text{-MeOC}_6\text{H}_4$	12	200

enamine form, but the effect of donor or acceptor substituents in the benzene ring is complex. With substituted aryl groups at the β-carbon atom, excellent correlations are found[201] between the equilibrium constants and Hammett σ-values, but only values for *meta*-substituents are correlated for aryl groups attached to nitrogen[200]. A substantial steric effect has been noted[202] for systems with a dialkylamino group at the β-carbon atom. For $R^1 = $ Ph, $R^3 = $ H and $R^4 = $ Me or Bu^n the amount of enamine changes from 5% for $R^2 = NMe_2$ to 100% for $R^2 = $ 2,6-dimethylpiperidino. Even more dramatic is the complete non-existence of the imine form for a range of compounds with $R^2 = CN$ (R^1, R^3 and R^4 from H, Me and Ph)[203]. Clearly, conjugation between the CN group and the double bond is important, and this probably accounts for the presence of 15% of the enamine form in the imidate (80; R = alkaloid residue), which is claimed to be the first report[204] of iminoether–enaminoether tautomerism. Two other novel variations on this

$$RCH(CN)C(OMe){=}NH \qquad\qquad R^1R^2CHC(SR^3){=}NR^4$$
$$(80) \qquad\qquad\qquad\qquad (81)$$

theme have also been reported, a thioimidate (81; R^1, R^2, R^4 = alkyl, R^4 = acyl)[205], and an iminomethyl disulphide (81; R^1, R^2 = alkyl, R^4 = o-$O_2NC_6H_4$, R^3 = thioalkyl)[206], both of which are in equilibrium with the corresponding enamine to an extent which depends upon the nature of the substituents.

Unusual stereochemistry has been noted[207] for the imine–imine (methylene-azomethine) tautomerism shown below. In complete contrast to all previous examples of this base-catalysed isomerisation, this particular system rearranges with a very low degree of stereospecificity. Detailed kinetic study has offered several explanations for this anomolous behaviour and a twisted carbanion may be involved.

$$PhCHMeN{=}CMeC_6H_4OMe\text{-}p \rightleftharpoons PhCMe{=}NCHMeC_6H_4OMe\text{-}p$$

4.7.8 Stereoisomerism and valence isomerism

The interconversion of *syn* and *anti* stereoisomers of imines and oximes has been studied intensively for many years in an attempt to distinguish between a planar inversion mechanism and a rotation mechanism. Most recent experimental evidence supports the inversion mode of isomerisation and further support now comes from several theoretical studies. The rotation barrier in methyleneimine has been calculated by *ab initio* methods[208,209] to be 55–57 kcal mol^{-1}, approximately twice the magnitude of the inversion barrier (26–28 kcal mol^{-1}). An even higher rotational barrier exists for the protonated imine[210] ($\Delta E > 72$ kcal mol^{-1}), and similar trends are apparent for the N-methyl and the N-phenyl derivative[211]. In formaldoxime both energy barriers are higher[212] but inversion (37.5 kcal mol^{-1}) is still favoured over rotation (75.5 kcal mol^{-1}).

Some n.m.r. studies[213] of N-alkylimines have shown that, for systems containing the HCC=N moiety, isomerisation via the enamine is a possible mechanism, which may predominate in certain cases. It is well established that substituents influence the *syn–anti* ratio, but it has been suggested[214] that the non-bonding electron pair on nitrogen may also have an effect. Isomer ratios for the imines from benzaldehyde and 1- and 2-naphthaldehyde can be rationalised in terms of a repulsive interaction between the lone-pair and the aromatic π-electrons, but further substantiation of this effect is required. Both direct and sensitised photoisomerisation of acetophenone oxime methyl ether proceed with a total quantum yield of less than unity, indicating that radiationless decay processes other than isomerisation must be involved[215]. The rearrangement of (82) to (83) has been found[216] to occur in benzene solution under reflux, surprisingly mild conditions for such a process. This is the first example of a [3,3] sigmatropic reaction of this type,

$$ArHC \overset{N}{\underset{N}{\bigcirc}} CHPy \quad \rightleftharpoons \quad ArHC \overset{N}{\underset{N}{\bigcirc}} CHPy$$

(82) Ar = Ph or o-HOC$_6$H$_4$ (83) Ar = Ph
                                  (84) Ar = o-HOC$_6$H$_4$

(Py = 2-pyridyl)

and has been termed a 'diazo-Cope' rearrangement by analogy with the well-known Cope rearrangement of biallyl systems. Replacement of phenyl by o-hydroxyphenyl displaces the equilibrium to the right such that the system exists entirely as (84). Hydrolysis of (84) gives a new diamine [2-PyCH(NH$_2$)]$_2$ in 68% yield, indicating a possible preparative route to such compounds. The stereochemistry of this 'diazo-Cope' rearrangement has not yet been fully elucidated, but the preliminary indications are that a boat form of the transition state may be favoured[216].

The first thorough study[217] of the prototropy and cyclisation of hydrobenzamide (85) reveals that both rearrangements probably involve the same intermediate (86). On the basis of substituent effects, hydrogen–deuterium exchange, isotope effects and solvent effects the intermediate carbanion (86)

$$PhCH=NCHPhN=CHPh \quad \rightleftharpoons \quad Carbanion \quad \rightleftharpoons \quad PhCH=NCPh=NCH_2Ph$$

(85)   (86)

$$PhHC \overset{N}{\underset{\underset{H}{N}}{\bigcirc}} CPh$$

probably has a W-shape, although other forms may be involved to a lesser extent. A possible complicating feature is the orbital symmetry requirement of a disrotatory motion involving rotation of at least one nitrogen atom through 180°.

In view of the rapid development in recent years of the orbital symmetry theory of electrocyclic reactions of dienes and trienes, the possibility of applying this theory to the analogous aza derivatives is of considerable interest. The perturbations consequent upon replacing C by N include the possibility of mixing σ- and π-orbitals in such a way that the highest occupied orbital and lowest unoccupied orbital do not possess π-symmetry, the existence of lone-pair interactions and n–π^* excitation and the influence of dipolar bonds on the reaction pathway. Molecular orbital calculations[218] in the Hückel and the CNDO/2 approximations reveal that none of these factors is important and that the aza compounds (87) and (88) are expected to demonstrate precisely the same symmetry control of cyclisation as do the corresponding hydrocarbons. The control exercised in thermal and photochemical process is summarised below:

(87)

(i) $h\nu$
(ii) Heat

(i) Disrotatory
(ii) Conrotatory

(88)

(i) $h\nu$
(ii) Heat

4.8 DIAZO COMPOUNDS

This section includes all compounds containing the $>C{=}\overset{+}{N}{=}\overset{-}{N}$ group, including diazoalkanes, diazoketones and diazoesters.

4.8.1 Synthesis

Two recently developed routes to diazo compounds, transformation of diazo compounds without loss of N_2, and diazo-group transfer from azides, have been thoroughly reviewed[219]. Other articles cover aspects of the preparation and reactions of diazoketones[220,221] and diazomalonic esters[222].

Base-catalysed decomposition of *N*-methyl-*N*-nitrosourea (89) to give diazomethane is not initiated by alkoxide attack at the nitroso group, as has been suggested. The first step is proton abstraction[223] followed by elimination of HNCO to give a methyldiazotate, which is converted into diazomethane by reaction with OCN^-. Deuteriated diazomethane can be conveniently obtained[224] by treatment of (89) with KOD in $D_2O{-}MeOCH_2CH_2OMe$. This procedure gives higher yields and higher isotopic purity than other methods. 1,2-Dimethoxyethane is also a suitable solvent for the generation of HNCO to give a methyldiazotate, which is converted into diazomethane for the decomposition of (89)[225]. The concentration of unreacted CH_2N_2 is always low and the only contaminant is KNCO.

The formation of bis(arylsulphonyl)diazomethane by diazo transfer from tosyl azide requires care in the addition of reagents in order to avoid secondary reactions[226]. Diazopropionates are not formed by diazo transfer to the ylide $Ph_3P{=}CMeCO_2Me$. Confusion has arisen over the product of this reaction, but it appears to a bis-azide, $MeC(N_3)_2CO_2Me$[227].

4.8.2 Reactions

Diazo compounds as a source of carbenes have been very extensively studied. No attempt will be made here to survey the general carbene chemistry of the common diazo compounds. A recent book[228] contains a chapter covering this area. The mechanism of the Wolff rearrangement of diazoketones has not yet been fully elucidated. Although evidence in favour of an oxirene

$$RCOC(N_2)R \longrightarrow RCO\ddot{C}R \longrightarrow R_2C{=}C{=}O$$

$$\begin{array}{c} O \\ / \ \backslash \\ RC{=}CR \end{array}$$

(90)

intermediate (90) for the photochemical reaction is now substantial[229-232], the position remains unresolved for thermal reactions, with evidence for[232] and against[231] the intervention of (90). Some simple MO calculations suggest that the oxirene is too high in energy to play a substantial role in the thermal rearrangement[233]. Direct photolysis of methyl diazothiolacetate, $MeSCOCH{=}N_2$, gives a very rapid Wolff rearrangement involving migration of the MeS group, but this reaction is entirely suppressed in the sensitised photolysis which gives products formed via radical reactions[234]. Decomposition of $p\text{-}O_2NC_6H_4COC(N_2)Me$ in the presence of silver salts results in the formation of $p\text{-}O_2NC_6H_4COCH{=}CH_2$ in high yield. Formation and decomposition of the diazonium ion is postulated[235]. A process of two-carbon homologation in yields $> 80\%$ involves[236] the interaction of diazoesters

$$N_2{=}CHCO_2Et + R_2BCl \rightarrow RCH_2CO_2Et$$

with dialkylchloroboranes at $-78\,^{\circ}C$, exceedingly mild conditions for formation of a C—C bond. The proposed mechanism requires initial formation of a quaternary boron species followed by elimination of nitrogen and alkyl group migration.

Several substituted diazo compounds can be obtained by reactions in which the diazo group is retained. β-Hydroxydiazoesters and α-carbamoyldiazoesters are accessible by reaction of a diazoester in the presence of OH^- with aldehydes[237] and isocyanates[124], respectively. Reaction with ketones is possible only in a non-hydroxylic solvent in the presence of a strong base such as di-isopropyl lithamide. Silver oxide catalysis gives poor yields of β-aminodiazo compounds from ketimines[237].

1,3-Cycloaddition of diazomethane to alkenes and alkynes exhibits substantial regioselectivity[238-240] and diazoethane is even more specific[240]. Houk[241] has applied perturbation theory to the evaluation of the coefficients of the frontier orbitals for several dipolar species (CH_2N_2, RN_3, N_2O) and dipolarophiles (alkenes substituted with CHO, CN, OH, NR_2 etc.) and rationalised some observed regioselectivities. More sophisticated calculations of frontier orbital perturbation and interaction should prove rewarding.

4.9 AMINES

4.9.1 Synthesis

4.9.1.1 Alkylation of primary and secondary amines

Two groups of workers[242,243] have independently discovered an essentially similar modification to the standard formaldehyde–formic acid (Clarke–Eschweiler) method for methylation of primary or secondary amines. Formic acid is replaced by $NaBH_4$ in methanol[242] or sodium cyanotrihydridoborate in acetonitrile[243], and the fully methylated amines are obtained in high yield (usually $> 80\%$) free from by-products. A wide range of amine basicities can be tolerated and sterically hindered amines react readily. Primary amines, RNH_2, when heated with a catalytic amount of strong base (NaH) are converted[244] into secondary amines, R_2NH, probably via an imine intermediate. Yields are not high (15–50%), but only one mixed secondary amine is formed when, for example, benzylamine and hexylamine are the reactants; this reaction may have some utility for special cases. The unusual photochemical product, $(Me_2N)_2CH_2$, has been isolated[245] in 70% yield from the irradiation of dimethylamine in CCl_4. Alkylation of secondary amines with buta-1,3-diene is catalysed[246] by rhodium(I) complexes, but some catalysts produce butadiene dimerisation before N-alkylation. π-Allylic complexes are thought to be involved. Amide acetals (91) and (92) may be obtained[247] from the

$$R^1R^2NCH(OEt)_2 \qquad \begin{array}{c} H_2C-O \\ Me_2C \diagup \quad \diagdown CHNR^1R^2 \\ H_2C-O \end{array}$$

(91)   (92)

corresponding quaternary triethylammonium tetrafluoroborates by exchange with secondary amines, R^1R^2NH (R^1 and R^2 are alkyl, aryl or alicyclic groups). Some amidinium salt is formed with aliphatic amines.

4.9.1.2 Other methods

The appropriate sections should be consulted for the formation of amines by reduction of the NO_2, the CN and the $C{=}NR$ groups. A full discussion has now appeared of the use of diborane for the reduction of amides[248]. This reagent is very selective and substituents such as NO_2, CO_2Et and Cl are unaffected, thus providing a synthetic procedure of remarkable utility in the approach to multifunctional amines. Tertiary amides are reduced rapidly and quantitatively whereas primary amides react slowly, but in neither case is there any evidence of C—N bond cleavage to give an alcohol, a typical by-product in a $LiAlH_4$ reduction. Reagents such as $LiAlH_4$, $LiAlH(OEt)_3$ or $AlHBu^i_2$ are usually considered to be strong reducing reagents but a controlled reduction at $-12\,°C$ in ether of α-aminoesters and α-aminoamides gives the corresponding α-aminoaldehydes[249]. A new general method[250] for the synthesis of secondary amines involves the reaction between an alkyl azide and an alkyldichloroborane ($RBCl_2$). Very high yields can be achieved (80–100%) for a wide variety of reactants. The mechanism probably involves formation of a tetrasubstituted borate species with subsequent alkyl group transfer to nitrogen, with complete retention of configuration:

$$R^1BCl_2 + R^2N_3 \longrightarrow Cl_2\bar{B}R^1NR^2\overset{+}{N}_2$$

$$R^1R^2NH \xleftarrow{H_2O-OH^-} Cl_2BNR^1R^2 + N_2$$

The conversion $OH \rightarrow NH_2$ for tertiary compounds, hitherto only available under the strongly acidic conditions of the Ritter reaction, can be achieved[251] by thermal decomposition of a chlorosulphonylurethane (93) via a carbonium ion intermediate, followed by reduction of the chlorosulphonamide (94). This final reduction is difficult with normal reducing agents and

$$RO_2CNHSO_2Cl \xrightarrow{heat} RNHSO_2Cl \xrightarrow{Reduction} RNH_2$$
$$\quad\quad (93) \quad\quad\quad\quad\quad (94)$$

it is necessary to use an indirect procedure [conversion into a t-butoxyhydrazide followed by oxidation with $Pb(OAc)_4$ and hydrolysis of the azo compound formed]. Further studies[252] of the mechanism of methanolysis of disubstituted cyanamides have revealed that the isourea is formed by base-catalysed addition to the CN group. Hydrolysis of the isourea produces a secondary amine. Other studies include the formation[253] of tris(chloromethyl)-amines by treatment of hexamethylenetetramine with PCl_5, and the Grignard alkylation of trichloromethylamines[254].

4.9.2 Reactions

4.9.2.1 *Alkylation, acylation and related reactions*

The synthesis and reactions of Mannich bases have been extensively surveyed[255]. A comprehensive analysis[256] of inductive and steric effects in the alkylation of amines by $PhCOCH_2Br$ has found a good correlation of rate constants for 33 amines with σ^* values and a steric factor. The bidentate nitrogen mustard (95) reacts with CO_2 to form an oxazolidinone (96). This reaction has been studied in detail with a view to elucidating the *in vivo* mechanism[257]. The reaction is complex, but the rate-determining step is ring closure of the carbamate anion. A silver acetate carbonyl complex

$$(ClCH_2CH_2)_2NH + CO_2 \longrightarrow$$
$$\quad (95)$$

(96)

provides[258] facile and quantitative carbonylation of primary and secondary amines to give the corresponding N',N''-dialkylurea and N',N''-tetra-alkyloxamide, respectively. The most likely mechanism involves initial coordination of the amine to silver. Subsequent insertion of a carbonyl group into the Ag—N bond gives a carbamoyl silver derivative.

Rapid acylation of primary amines in 90% yield with peroxyacetyl nitrate ($AcOONO_2$) is accompanied by evolution of oxygen and formation of nitrous acid[259]. Tertiary amines react to produce an unknown species with an appreciable chemiluminescence. Contrary to earlier reports, aliphatic amines are slowly formylated by DMF in the absence of a catalyst[260].

Acid-labile amino-protecting groups are well known, but few such groups are available which are acid-stable but alkali-labile. A series of benzyloxy-carbonyl derivatives of amino acids have been prepared[261] and the p-iso-propyloxycarbonyloxy compound (97) can be hydrolysed by 5% K_2CO_3 (1 h) or 0.1 M NaOH (10 min), but is quite stable to CF_3CO_2H in dichloro-methane. Hydrolysis under such mild conditions involves a 1,6-elimination promoted by the stability of the quinone methide (98). These novel carbamyl derivatives (97) are prepared from the amine by reaction with the N-methyl-

$$Pr^iOCO_2\text{---}\langle\text{C}_6\text{H}_4\rangle\text{---}CH_2OCONHR$$

(97)

$$\downarrow OH^-$$

$$Pr^iOCO_2H \ + \ O{=}\langle\text{C}_6\text{H}_4\rangle{=}CH_2 \ + \ CO_2 \ + \ RNH_2$$

(98)

imidazolinium chloride corresponding to (97). The interrelated problems of the protection, deprotection and monoalkylation of amines have been explored[262] by using the trifluoromethanesulphonyl group (CF_3SO_2). Triflamides (99) are readily obtained from the amine and the anhydride ($CF_3SO_2)_2O$ or the crystalline triflating agent $PhN(SO_2CF_3)_2$[263]. The protecting group is removed by reduction with $LiAlH_4$ for secondary amides, and Red–Al [$NaAlH_2(OCH_2CH_2OMe)_2$] for primary amides. Alkylation of the primary amides proceeds smoothly in high yield ($> 90\%$) with alkyl halides in K_2CO_3–acetone at 25 °C. A modification involving alkylation of N-benzyltriflamide followed by base-catalysed (NaH) elimina-tion[264] and hydrolysis gives a primary amine in a Gabriel-type reaction

$$R^1NH_2 \rightarrow R^1NHSO_2CF_3 \xrightarrow{R^2X} R^1R^2NSO_2CF_3 \rightarrow R^1R^2NH$$

(99)

without the use of a hydride reducing agent. The application of new amine protecting groups in organic synthesis has been reviewed[265].

4.9.2.2 Deamination

Studies of the deamination of 2-phenylethylamine and other amines by

NOCl in a range of solvents (CCl_4, C_6H_{12}, $CHCl_3$, Et_2O) have been described[266]. Based on the product ratios, the reactivity of carbonium ion intermediates is rationalised in terms of electronic and steric factors and solvation. Nitrous acid deamination of the amine (100) gives[267] the expected alcohol together with elimination and rearrangement products. 2-Methylcyclohexanone is shown to arise by two 1,2-hydride shifts rather than a 1,3-shift as has been suggested previously. Activation of an amine function by aβ-$PhSO_2$ group results[268] in quantitative elimination of a secondary amine in aqueous solution at 88 °C. Whereas the *threo* isomer undergoes a stereospecific *syn* elimination, the *erythro* compound gives a mixture of products, probably as a result of partial inversion of the carbanionic intermediate.

$$\text{(100)} \quad \xrightarrow{HNO_2} \quad + \quad \text{+ other products}$$

4.9.2.3 Miscellaneous reactions

The chemistry of α-peroxyamines has been reviewed[269], and a range of reactions of β-aminoalkyl tin compounds has been discussed in detail[270]. A very extensive survey[271] of the reactions of amines with photoexcited compounds containing unsaturated groups (C=O, C=C) highlights the wide range of compounds which can be photoreduced by amines.

A new class of superbases, potassium mono- and di-alkylamides, can be easily prepared in high yield from the amine and potassium hydride[272]. Potassium metal and sodium hydride react much more slowly with pyrrolidine, for instance. These new compounds are bases of exceptional reactivity for prototropic reactions, e.g. *N*-potassioethylenediamine aromatises limonene in 5 min at 25 °C, conditions under which the corresponding lithium compound is ineffective. There is likely to be extensive exploitation of these potassium amides and of the potassium salts of DMSO and hindered alcohols, all of which are accessible from reaction with potassium hydride.

An unusual photocyclisation[273] of phenylalkylamines (101) involves attack at the *meta* position in the ring by a methyl group. The 3-carbon side-chain

$$\text{(101)}$$

(101; $n = 3$) cyclises more readily than the four-carbon side-chain, but both reactions may be of value in the synthesis of alkaloids related to morphine. The intramolecular cyclisation of unsaturated amines in the presence of $HgCl_2$ has been studied in detail[274], in particular the effect of substituents on nitrogen and at the double bond, the effect of chain length and the role of the solvent. This aminomercuration is stereospecific and proceeds through a bridged transition state[275]. Treatment of a N,N-dimethylphenethylamine N-oxide with the iron(II) ion produces[276] a complex reaction, with formation of the corresponding tertiary amine, a demethylated amine, benzyl alcohol and a tetrahydroquinoline. The analogous reaction with the Ti^{III} ion was also investigated and overall results are consistent with an aminium radical cation as a common intermediate for all products.

4.9.3 Properties

Measured basicities of amines have always contained an unknown contribution from solvent interactions which makes simple rationalisations of apparent trends fraught with danger. The intrinsic basicities of some simple amines have now been estimated[277] from mass spectrometric measurements of the ion concentrations in a mixture of two bases. This method gives the ΔH° value for the proton-transfer reaction below and variable-temperature studies indicate that $\Delta G^\circ = \Delta H^\circ$, i.e. entropy changes are very small. The

$$MeNH_2 + NH_4^+ \rightleftharpoons MeNH_3^+ + NH_3$$

ΔG° differences relative to ammonia are a good measure of the differences in proton affinities and the results show that gas-phase basicity *increases* with successive alkyl substitution, as would be expected on the basis of inductive effects, in contrast to the order established for aqueous solutions. Aniline is a weaker intrinsic base than cyclohexylamine but a *stronger* base than ammonia.

An excellent review has appeared[278] of acidic properties of the amino group in amines, amides and amidines, a subject which has not been conveniently reviewed before. The methods of generating N-anions are surveyed and the physical and chemical properties of these ions are discussed.

A more detailed understanding of the nature of the interactions responsible for the separation of the optical isomers of chiral amines by g.l.c. on a chiral stationary phase has derived from a systematic study of the effect of substituent size and polarity in the chiral solute[279]. These results show that with an L-ureide as stationary phase, N-trifluoracetylamines[280] and aminoesters[279] consistently exhibit a longer retention time for that enantiomer which has a configuration (102) corresponding to a clockwise arrangement, in order of decreasing size, of the groups attached to the asymmetric carbon atom, viewed from carbon to nitrogen. In (102), P is a hydrogen-bonded molecule of the chiral stationary phase (solvent) and L, M and S are large, medium and small substituent groups, respectively. The difference in retention time between the solute molecule in (102) and the form with the opposite configuration of the L, M and S groups probably reflects a difference in stability of the respective diastereoisomeric association complexes. The magnitude of the resolution factor increases with difference in size between the groups L and M (assuming S is hydrogen). This rule for correlating the order of peak emergence on g.l.c. with configuration has been applied[280] to the determination of the configuration of two amines (103; R = H or Me), about which there was some confusion in the literature. Sulphenamides of the general formula $2,4\text{-}(NO_2)_2C_6H_3SNR^1R^2$ have a considerable barrier to rotation

R(CH$_2$)$_2$CHMeNH$_2$

(103)

(102)

about the S—N bond, sufficient in fact to make this bond a chiral axis at 25 °C. If R^1 and R^2 are identical chiral groups [($\pm$)-compound], the sulphenamide exhibits pseudoasymmetry, whereas if R^1 and R^2 are enantiomeric (*meso*-amine) the sulphenamide bond is neither asymmetric nor pseudoasymmetric[281]. The two different cases can be distinguished by n.m.r. spectroscopy.

4.9.4 Enamines and ynamines

4.9.4.1 *Synthesis*

A new method of wide applicability for the formation of enamines involves[282] *C*-alkylation of a tertiary amide with a Grignard reagent followed by elimination of MgBrOH (essentially a dehydration). Yields are reasonable (30–80%), but bulky groups in the Grignard reagent reduce the yield and increase the quantity of by-products; alkyl-lithium reagents were less effective. Enamines of type (104) are the product[283,284] of the reaction between a tertiary amine (Et$_3$N) and a sulphenyl chloride (RSCl). The yield of 50% is

$$Et_2NCH=C(SR)_2 \qquad NC—C{\equiv}C—NEt_2$$

(104)   (105)

quantitative on the basis of the proposed mechanism. Bis(dialkylamino)-ethylene is formed[285] from bromomalonodialdehyde by treatment with an excess of secondary amine, and allenic amines have been obtained by a Gabriel synthesis (60%) and a LiAlH$_4$–AlCl$_3$ reduction of an oxime (100%)[286]. Compound (105) is the product from the reaction of Cl$_2$C=CClNEt$_2$ with Bu^{n}Li and CNCl[287].

4.9.4.2 *Reactions*

Enamines are a very reactive group of amines with an extensive but specialised chemistry. Displacement of an α-H or an α-Cl in an enaminonitrile such as (106; R^3 = H or Cl) is accompanied[288] by reaction at nitrogen when the enamine is treated with phosgene to give the corresponding acid chloride. Amine displacement by OH with aqueous HCl is also reported[289] for the

$$R^1NHCR^2=CR^3CN \qquad (RO_2C)_2C=C(NH_2)CX_3$$

(106)   (107)

enamine (107; X = Cl or F), but the reaction with amines involves[290] displacement of the CX$_3$ group. Methanolysis of tetra-(*N,N*-dimethylamino)-

ethylene gives[291] the products corresponding to displacement of one or two amino groups by MeO, together with the addition compound $MeOCH(NMe_2)C(OMe)_2NMe_2$ and other minor products. In contrast, alcoholysis of 2-halogenoalkenamines (108) results[292] in addition with rearrangement to give the α-aminoacetal (109). A cyclic intermediate is probably involved, by analogy with the corresponding reaction with the iminium salt (110)[293]. Although the enamine can be formed from the iminium salt by elimination of HBr, it is not an intermediate in the alcoholysis of (110) to give (109)[292].

$$R^1R^2N\text{-}C(Br)=CH \xrightarrow{2MeO^-} R^1CH(NR^2{}_2)CH(OMe)_2$$

$$\text{(108)} \qquad \text{(109)}$$

$$R^1CHBrCH=\overset{+}{N}R^2{}_2\ Br^- \xrightarrow{MeO^-} R^1\overset{+}{C}H\text{-}\overset{\overset{\displaystyle NR^2{}_2}{|}}{}\text{-}CHOMe$$

$$\text{(110)}$$

Vilsmeier formylation of ethylenediamines affords[294] a route to the hitherto inaccessible (dialkylamino)malonodialdehydes which exist in the enol form (111). The hydrolysis of enamine (112) has been studied[295] as

$$R_2NC(CHO)\text{-}CH(OH) \qquad cis\text{-}NCCH_2NHCMe=CHCO_2Et$$

$$\text{(111)} \qquad\qquad \text{(112)}$$

part of an investigation of the decarboxylation of ethyl acetoacetate catalysed by aminoacetonitrile, a model for the corresponding enzyme-catalysed reaction. The kinetics of the hydrolysis of (112) to ethyl acetoacetate and aminoacetonitrile are complex, and show a substantial rate dependence on pH; a reasonable mechanism involves initial tautomerisation to the imine, followed by addition of water to the C=N double bond and elimination of the amine.

Salts of 1,3-diamino-1,3-dichloroallyl cations such as (113) have become readily available by reaction of N,N-disubstituted acetamides with dichloro-methyleneammonium chlorides (69). Treatment[296] of (113) with alkali gives an ynamine (114), and reaction with secondary amines gives the stable salt of the interesting 1,1,3,3-tetrakis(dialkylamino)allyl cation (115). Compound

$$R_2N\text{-}C\equiv C\text{-}CONR_2$$

$$\text{(114)}$$

$$\text{(113)} \xrightarrow{OH^-} \text{(114)}$$

$$\text{(113)} \xrightarrow{R_2NH} \text{(115)} \longrightarrow \text{(116)}$$

(113): $(R_2N)(Cl)C\cdots\overset{+}{C}H\cdots C(Cl)(NR_2)\ X^-$

(115): $(R_2N)_2C\cdots\overset{+}{C}H\cdots C(NR_2)_2\ X^-$

(116): $(R_2N)_2C=C=C(NR_2)_2$

(115) and the analogous 1,3-dialkylamino-1,3-dialkoxyallyl cation are readily deprotonated to the corresponding allenes (116) by the action of Bu^nLi or $NaNH_2$ in liquid ammonia. These allenes are electron-rich compounds, comparable in reactivity with ynamines and ethylenetetramines and some typical interactions are shown in the scheme for the tetra-aminoallene where $R = NEt_2$ [296].

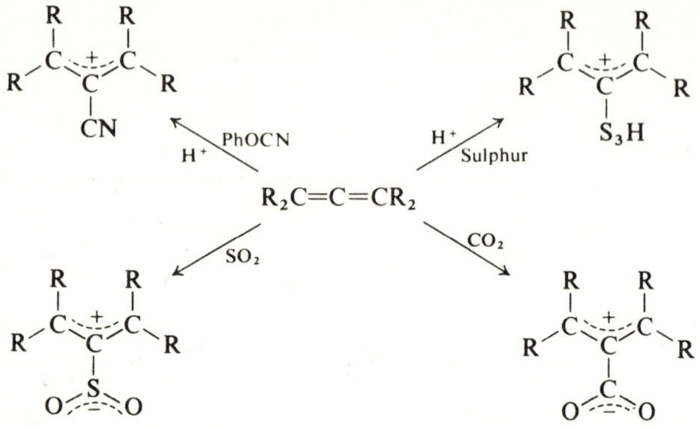

4.9.5 Quaternary ammonium compounds

A review has appeared[297] of the use of quaternary ammonium compounds as reagents and catalysts in organic synthesis, particularly in two-phase systems. Electroreduction of quaternary ammonium nitrates by stepwise two-electron transfer results in cleavage of a C—N bond to give a tertiary amine and an alkyl anion which is then protonated[298]. The alkyl groups can be benzyl, ethyl or methyl, or the nitrogen can be part of a saturated ring as in tetrahydroquinoline, but at least two methyl groups appear to be essential for reduction to occur.

Alkylammonium groups occur frequently in biologically important molecules and the size of the alkyl groups is often critical to the normal function of the compound. Some theoretical studies[299] of the interaction of water with $\overset{+}{N}Me_4$ and $\overset{+}{N}Et_4$ ions go some way to understanding the specificity associated with a methyl group. Interaction energy for close approach of a water molecule is lower for an α-methylene group than for a β-methylene group and the most favourable interaction involves approach to nitrogen along the N—C axis in such a way as to permit interaction with three α-methylene groups.

Current interest in stable nitrogen ylides (117) is reflected in the appearance of two reviews[300,301] of these compounds, usually called aminimides. The

$$R^1{}_3\overset{+}{N}—\overset{-}{N}R^2$$

(117)

group R^2 must be strongly electron withdrawing in order to stabilise the ylidic moiety. X-Ray studies[302] of $Me_3\overset{+}{N}—\overset{-}{N}—NO_2$ and $Me_3\overset{+}{N}—N—COPh$ confirm substantial negative charge delocalisation in these compounds.

N-Nitroaminimides (117; $R^2 = NO_2$) are obtained[303] by nitration of the corresponding hydrazinium nitrates with nitric acid or trifluoroacetyl nitrate. The effect of increasing the size of R in $RCO\bar{N}-\overset{+}{N}Me_2CH_2Ph$ increases the amount of Curtius rearrangement products (RNCO and Me_2NCH_2Ph) at the expense of the Stevens product, $RCON(CH_2Ph)NMe_2$, in the decomposition of these aminimides, but a simple steric effect alone cannot rationalise this trend[304].

4.9.6 Amines substituted by Cl, CN, NO, NO_2 or OH

The chemistry of N-chloroamines has been reviewed[305] (in Japanese) and a discussion of amination of aromatic compounds has appeared[306]. The first examples of a 1,2-shift of an alkyl group to positive nitrogen derived from an open-chain N-halogenoalkylamine have been observed[307] in an $AlCl_3$-catalysed reaction at $-30\,°C$. The proposed mechanism shown below

$$R_3CNCl_2 \xrightarrow{AlCl_3} R_3CN\underset{Cl\cdots AlCl_3}{\overset{Cl}{\diagup}} \longrightarrow R_2\overset{+}{C}NRCl \longrightarrow R_2CClNRCl \xrightarrow{H_2O,H^+} R_2CO + RNH_2$$

requires concerted migration of R and expulsion of $AlCl_4{}^-$; this is supported by a study[308] of migratory aptitudes, which show that the largest group migrates preferentially (Ph $\gg$ Bu^s > PhCH$_2$ > Bu^n > H > Me). A detailed comparison is made with other rearrangements such as the Schmidt and the Baeyer–Villiger reactions[309]. Synthesis of alkyldicyanamides (118) by a two-step reaction of cyanogen chloride with primary amines proceeds[49] in yields of 50–76%. Only one cyano group is activated in these compounds and reaction with H_2O, MeOH and Et_2N gives N-cyanoureas, isoureas and guanidines, respectively[309].

$$RN(CN)_2 \qquad R^1N(=CHR^2)NPh \dashrightarrow$$

(118) (119)

(120)

The photochemistry of N-nitrosamines has been reviewed[310]. Metallation of nitrosamines with Bu^{n}Li occurs at the α-carbon atom and subsequent reaction with an electrophile gives the C-alkylnitrosamine[311]. The use of lithium di-isopropylamide is more efficient since competitive nucleophilic attack at the nitroso group is completely suppressed. This latter reaction has been studied[312] in more detail for phenyl-lithium and it appears that the initial N-alkylated lithium salts undergo elimination to give an imine imide (119), which dimerises head-to-tail to give the *sym*-hexahydrotetrazine (120). Intermediate hydrazines may sometimes be isolated[312,313]. Two routes to *N*-fluoro-*N*-nitrobutylamine have been reported[314]. MO calculations of electron

$$MeN\overset{+}{=}\overset{-}{N}\diagup\overset{O^-}{\diagdown}_{O^-} \qquad M^+ \qquad\qquad Me\overset{-}{N}-\overset{+}{N}\diagup\overset{O^-}{\diagdown}_{O} \qquad M^+$$

(121)   (122)

distribution in nitramine salts confirm earlier ideas that these compounds have structure (121) and not (122)[315]. A convenient oxidation of an amine to the corresponding *N*-hydroxyamine can be accomplished with *m*-chloroperoxybenzoic acid[316].

4.10 HYDRAZINES AND HYDRAZONES

4.10.1 Synthesis

A convenient alkylation of hydrazines or hydrazones occurs[317] with an aldehyde and $NaBH_3CN$, a reaction which has also been applied to amines (see Section 4.9.1.1). Several alkyl groups may be introduced in the same reaction and the use of dialdehydes gives cyclic products, as shown below. Yields of pure products in small-scale preparations were 30–60%. Alkyl

$$MeNHNHMe \;+\; (CH_2)_3(CHO)_2 \xrightarrow{NaBH_3CN} (CH_2)_5\overset{NMe}{\underset{NMe}{|}}$$

exchange in *N,N*-dimethylhydrazides can be induced by thermolysis of the quaternary iodide, MeI being the most readily eliminated alkyl iodide[318]. Repeated quaternisation and thermolysis can introduce a second group to give, for instance, $RCONHNEtPr^n$. Reduction of CO_2Et to Me with $LiAlH_4$ has been used to obtain tetramethylhydrazine in 35% yield from the diester[319]. *N*-Amination of tertiary amines with *O*-mesitylenesulphonylhydroxylamine under mild conditions (5 min at 25 °C) is particularly useful for the formation of hydrazinium salts from alkaloidal amines, since an excess of amine is not required[320]:

$$R_3N + NH_2OSO_2C_6H_2Me_3 \rightarrow R_3\overset{+}{N}NH_2\;Me_3C_6H_2SO_3^-$$

(80–100%)

4.10.2 Reactions of hydrazines

N',N"-Di-isopropylhydrazine (123) is a new reagent for the protection of carboxylic acids. The derived hydrazide is stable to acid and base and sufficiently hindered to resist further acylation[321]. The acid can be quantitatively recovered by oxidation with $Pb(OAc)_4$ in pyridine at 25 °C, attractively mild conditions.

$$(Pr^iNH)_2 + RCOCl \rightarrow RCONPr^iNHPr^i$$
$$(123)$$

Cyclisation of unsaturated hydrazines with cyclohexanone involves a Fischer ring closure to yield an octahydrocarbazole[322]. Unsaturated hydrazides also form heterocyclic products by reaction with aldehydes[323]. Chloroformylation of diethyl N',N''-diphenylhydrazinecarboxylates gives[324] the compound (124), which cyclises in presence of $Cu(OAc)_2$ at 145 °C to give the anhydride (125). Photolysis of (125) produces azobenzene. A remarkably smooth rearrangement of 1,1-dimethyl-1-phenacylhydrazinium

$$PhN(COCl)NPhCO_2Et$$
$$(124)$$

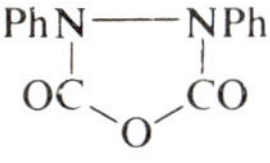

bromide to the heterocyclic compound (126) in the absence of base probably involves a Stevens type of N-to-N migration followed by decomposition to phenylglyoxaldimine which reacts further to give (126)[325]. Other reactions reported include condensation of hydrazines with β-ketoesters[326] and isocyanates[115].

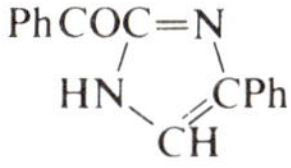

4.10.3 Reactions of hydrazones

The reactions of hydrazones with electrophilic and nucleophilic reagents have been reviewed[327]. Hexafluoracetone hydrazone is a novel compound of this class and a number of its reactions have been explored[328]. A new catalyst has been found for the oxidation of α-dihydrazones (127) to the corresponding alkynes. Copper(I) chloride in pyridine at 25 °C gives high yields (> 90%) of the hydrocarbon[329]. α-Bromohydrazones (128) react with the conjugate base of reactive methylene compounds to form substituted pyrazolines[330].

$$R^1C(=NNH_2)CR^2(=NNH_2) \rightarrow R^1C\equiv CR^2 \qquad RCBr=NNHPh$$
$$(127) \qquad\qquad\qquad (128)$$

Acid catalysed E–Z isomerisation of dinitrophenylhydrazones has been studied[331,332] for a wide range of compounds, including the derivatives of α-ketoacids[332]. The [3,3]-sigmatropic rearrangement of (129) to (130) has been reported[334] for a range of N-allylhydrazones. The di-imine (130) is unstable and eliminates N_2 to give (131); the highest yields of (131) were

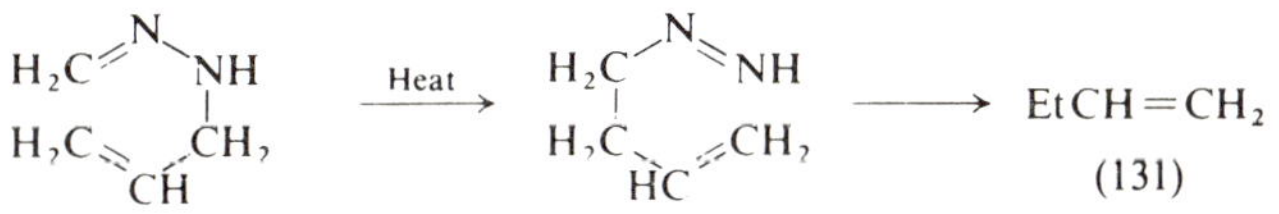

$$EtCH=CH_2$$
$$(131)$$

$$(129) \qquad\qquad (130)$$

found in the thermolysis of aromatic ketohydrazones. Radical fragmentation and recombination products are also formed.

Alkylation of amidrazones, $R^1C(=N^2R^2N^1R^3{}_2)N^3R^2{}_4$, usually occurs[335] at N^1 but N^3,N^3-dimethyl compounds are methylated at N^3. Reaction of such amidrazones with N',N'-diethylhydrazine gives[336] formazans (132). The first

$$RC(=NNMe_2)NHNEt_2 \qquad\qquad PhC(=NPh)\bar{N}\overset{+}{N}Me_3$$
$$(132) \qquad\qquad\qquad\qquad (133)$$

amidrazone ylide (133) has been prepared by the reaction of $PhCCl=NPh$ with the hydrazine ylide[337].

4.11 AMIDES

4.11.1 Synthesis

Alcohols are converted into amides in reasonable yield (50–87%) by treatment with chlorodiphenylmethylium hexachloroantimonate (134) in acetonitrile (or another nitrile solvent)[338]. Other methylium salts have also been

$$R^1R^2CHOH \;+\; [Ph_2\overset{+}{C}Cl][SbCl_6^-] \;\longrightarrow\; R^1R^2CH\overset{+}{O}HCPh_2Cl$$
$$(134)$$

$$R^1R^2CHNHCOR^3 \;\xleftarrow{\;H_2O\;}\; R^1R^2CH\overset{+}{N}{\equiv}CR^3 \;\xleftarrow{\;R^3CN\;}\; R^1R^2CH\overset{+}{O}=CPh_2$$

used. Diethyl phosphorocyanidate, $NCP(O)(OEt)_2$, has been developed[339] as a reagent for amide formation in peptide chemistry, since the reaction is rapid and clean and proceeds without racemisation. Reaction of alkyl halides with sodium tetracarbonylferrate produces an iron–alkyl complex which can rearrange to an acyl complex[340]. Treatment with iodine and an amine affords the corresponding amide in high yield, probably via an acyl iodide.

Lithiation of DMF with lithium di-isopropylamide provides a novel method of inserting the N,N-dimethylcarbamoyl (Me_2NCO) group into a molecule[341]. Reaction with ketones gives an α-hydroxyamide in variable yield (45–85%). Other work includes an improved synthesis of alkylidene-bis-amides[342] and the condensation of amides with trichloroacetaldehyde to give N-dichlorovinylamides[343].

The first general method[344] for the synthesis of N-hydroxyamides (hydroxamic acids) involves the oxidation of an N-trimethylsilylamide with the HMPT complex of MoO_5, followed by liberation of the hydroxy compound by treatment with EDTA. Anodic oxidation of N-methylformamide at a Pt electrode in an alcohol gives the corresponding N-alkoxyamide[345]. The synthesis and reactions of the amides (135) have been discussed[346,347]. The hydration of nitriles to give amides is discussed in Section 4.3.2.1.

$$RCONHCH=CHCHClNHCOR \;\rightleftharpoons\; RCON\overset{+}{H}CHCHCHNHCOR \quad Cl^-$$
$$(135)$$

4.11.2 Reactions

4.11.2.1 Hydrolysis and related reactions

The C—N bond in amides is strong and cleavage of this bond usually requires relatively severe conditions. Hence a new method[348] of breaking this bond in primary and secondary amides under anhydrous conditions at 25 °C is very welcome. The amide is first converted into its anion and this is treated with CS_2 to give a thiol acid and an isothiocyanate. The thiol acid can be recovered in yields of 65–90%, and the amino compound isolated as a

$$R^1CONHR^2 \xrightarrow{\text{NaH}} R^1CO\bar{N}R^2 \xrightarrow{CS_2} R^1CO\bar{S} + R^2N{=}C{=}S$$

thiourea or, after hydrolysis, as the amine salt in yields of 65–95%. Two examples of abnormal hydrolysis have appeared. Compounds of type (136) undergo[349] hydrolytic cleavage of the alkyl—N bond under mildly basic

$$R^1CONH{-}CHR^2C_6H_4N{=}NAr \longrightarrow R^1CONH_2 + R^2COC_6H_4NHNHAr$$

(136)

conditions. The presence of the azo group is essential and the postulated mechanism involves two subsequent tautomeric shifts via quinoid structures. Base-catalysed hydrolysis[350] of the anilides of α-nitroalkanoic acids results

$$Me_2C{-}CONHC_6H_4X \longrightarrow Me_2\bar{C}NO_2 + [XC_6H_4NCO] \xrightarrow{H_2O} XC_6H_4NH_2$$
$$O_2N$$

(137)

in C—C bond fission exclusively, e.g. (137), to give a carbanion and an isocyanate. Trifluoroacetylsulphonamides, $CF_3SO_2NR^1R^2$ (138), can be cleaved in two ways[264]. Nucleophilic displacement of $CF_3SO_2N^-$ can be achieved for acyl triflamides (138; R^1 = acyl, R^2 = H) by reaction with H_2O, EtOH or RNH_2. Elimination of the trifluoromethanesulphinate anion is effected by reduction with Red–Al for primary and with $LiAlH_4$ for secondary amides. Red–Al $[NaAlH_2(OCH_2CH_2OMe)_2]$ is generally useful for the regeneration of primary amines from sulphonamides[351], for which $LiAlH_4$ is ineffective.

4.11.2.2 Other reactions

The established procedure for converting disubstituted amides into ketones by treatment with an alkyl- or aryl-lithium derivative has been successfully applied[352] to the formation of 1,4- and 1,6-diketones, although yields are low for aliphatic compounds. A procedure for *in situ* formation of the organolithium compound has been described[353]. Electrolytic reduction of acetamide in lithium chloride–HMPT produces the *gem*-aminoalcohol initially, but subsequent reduction is acid sensitive[354]. At low acidity the

NH$_2$ group is eliminated to give ethanol whereas at high acidity ethylamine is formed.

Routes to amide thioacetals and aminal thioesters have been developed[355]. These compounds have very similar reactivity to their oxygen analogues. N-Formylsulphenamides, HCONHSR, are obtained by direct interaction of formamide and a sulphenyl chloride[356]. These sulphenamides can be dehydrated to the corresponding thiocyanate (RSCN). Some factors which limit the addition of amides to formaldehyde and glyoxal have been discussed[357].

4.11.3 Protonation and hydration

The site of protonation in simple amides has been the subject of some dispute, particularly with regard to the situation obtaining in dilute acidic media. Recent spectroscopic studies of N-methylacetamide[358] (n.m.r.) and N,N-dimethylacetamide[359] (i.r. and Raman) suggest protonation on oxygen, and this finds support in *ab initio* SCF calculations[360] of the total energies of the two conjugate acids of formamide. In the gas phase at least, the O-protonated form is more stable by 6.2 kcal mol^{-1}. Calculated potential energy curves for the interaction of HCONH$_2$ with H$^+$ also indicate preferential protonation at oxygen[361]. Amide hydrotetrafluoroborates have been isolated and spectroscopic examination is again consistent with O-protonation[362]. A problem related to the site of protonation is the site of preferential hydrogen bonding to water. This is of interest for studies of secondary structure in proteins, and because hydration may be the first stage in protonation. *Ab initio* MO calculations using a restricted basis set (STO–3G) indicate[363] that hydrogen bonding to the amino group is slightly more favourable than to the carbonyl group in formamide, but substitution of H by Me groups at N reverses this trend[359].

4.11.4 Ureas, isoureas, guanidines and carbamates

These four types of compound, incorporating respectively the moieties (139), (140), (141) and (142), are grouped together more on the basis of

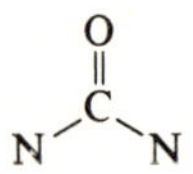

$$
\begin{array}{cccc}
\underset{\text{(139)}}{\text{N}-\overset{\displaystyle\overset{\text{O}}{\|}}{\text{C}}-\text{N}} &
\underset{\text{(140)}}{\text{N}-\overset{\displaystyle\overset{\text{N}}{\|}}{\text{C}}-\text{O}} &
\underset{\text{(141)}}{\text{N}-\overset{\displaystyle\overset{\text{N}}{\|}}{\text{C}}-\text{N}} &
\underset{\text{(142)}}{\text{N}-\overset{\displaystyle\overset{\text{O}}{\|}}{\text{C}}-\text{O}}
\end{array}
$$

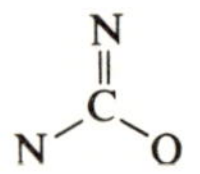

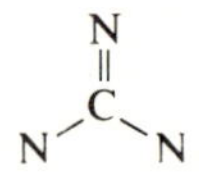

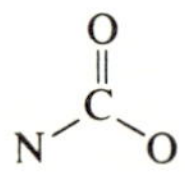

convenience than on any particularly strong chemical interrelationship, although some chemical similarities are evident.

4.11.4.1 Synthesis

Selenium has been found to catalyse the reaction of amines alone and with alcohols with CO. A selenocarbamate is probably an intermediate in the

$$R^1_2NH \;+\; CO \;+\; Se \xrightarrow[\text{R}^2\text{OH}]{\text{THF}} \begin{array}{l} R^1_2NCONR_2 \\[4pt] R^1_2NCO_2R^2 \end{array}$$

formation of both carbamates[364] and ureas[365]. Six general routes have been developed[366] to guanidines of types (143) and (144) bearing a wide variety of substituents. Four methods for the compounds of type (143) start from the dithiocarbonimidate (145) and the cyclic compounds result from the func-

$$R^1N=C \Big\langle \begin{array}{l} NR^2(CH_2)_x CO_2R^4 \\[4pt] NR^3(CH_2)_y CO_2R^4 \end{array}$$

(143)

$$R^1N=C \Big\langle \begin{array}{c} \overset{R^2}{N}-CH_2 \\ \qquad\quad CHCO_2R^4 \\ \underset{R^3}{N}-CH_2 \end{array}$$

(144)

$$TsN=C(SMe)_2$$

(145)

tionalisation of 2-tosylaminopyrimidine. The use of the tosyl protecting group reduces the effects of the strongly basic guanidine moiety. S-Alkyliso-thioureas (146) and their hydrobromides react with isopropyl nitrite to give the corresponding S-alkyl thiocarbamates[367]. The mild conditions required (2 h at 50 °C in benzene) make this an attractive method for the conversion

$$R^1_2NC(SR)=NH \xrightarrow{\text{Pr}^i\text{ONO}} R^1_2NCOSR^2$$

(146)

of C=N into C=O, although its scope is limited. The preparation of N-sub-stituted thioureas has been reviewed[368].

4.11.4.2 Reactions

Mannich-type condensation of aldehydes, ureas and sulphinic acids have been reported[369]. An unusual condensation of guanidine carbonate with benzaldehyde to give compound (147) is base-catalysed, and the proposed mechanism is similar to an intramolecular Cannizzaro reaction[370]. Three

$$PhCONH(=\overset{+}{N}H_2)NHCH_2Ph\; Cl^- \qquad\qquad R^1_2NC(=\overset{+}{N}R^2_2)SCOPh\; Cl^-$$

(147)   (148)

types of reaction of S-benzoylisothiouronium salts (148) have been exten-sively explored[371] and a detailed evaluation provided of the effect of substi-tuents and pH on the mode of reaction: thiol benzoate displacement, hydro-lysis to a thiourea or S-to-N migration to an N-benzoylthiourea. In solution N-hydroxyureas rearrange[372] to O-carbamoylhydroxylamines, probably by dissociation into hydroxylamine and isocyanate followed by a normal addition to the C=N double bond. The thermal isomerisation of (149) to (150) proceeds via a radical-cage mechanism, as indicated by a CIDNP

$$RMeN\overset{O}{\underset{S}{\diagdown}}{=}CNMe_2 \longrightarrow RMeN\diagdown_S\overset{O}{\diagup}CNMe_2$$

(149)   (150)

$$NH_2C(=\overset{+}{N}H_2)SO_2^-$$

(151)

effect in the n.m.r. spectra[373]. Both an aroyl (R = PhCO) and an alkyl (R = Bu^{t}CH$_2$) derivative rearrange in 120 h at 25 °C. Inversion of the azido group in carbamoyl azides at 80–90 °C has been established with ^{15}N-labelled compounds[374]. A reversible thermal dissociation to isocyanate and hydrazoic acid is involved rather than an intramolecular ion-pair mechanism. Most procedures for hydrogenolysis of carbamoyl azides result in elimination of the two outer nitrogen atoms. Some transformations of epoxy-N-nitrosocarbamates have been studied in detail[375]. Aminoiminomethanesulphinic acid (151) has been examined as a reducing agent for ketones[376]. Yields of secondary alcohols were 75–100%.

References

1. (a) Wilder, P. and Shepherd, J. M. (1973). *MTP International Review of Science, Organic Chemistry Series One*, Vol. 2, *Aliphatic Compounds*, 131 (N. B. Chapman, editor) (London: Butterworths); (b) Harris R. K. (1973). *Specialist Periodical Reports, Nuclear Magnetic Resonance*, Vol. 2, 216 (London: Chemical Society)
2. Bachman, G. B. and Maleski, R. J. (1972). *J. Org. Chem.*, **37**, 2810
3. Krylov, V. K. and Tselinkii, I. V. (1972). *Zh. Org. Khim.*, **8**, 233
4. Limatibul, S. and Watson, J. W. (1972). *J. Org. Chem.*, **37**, 4491
5. Lukashevich, O. V., Novatskii, G. N., Golod, E. L. and Bagal, L. I. (1972). *Zh. Org. Khim.*, **8**, 908
6. Olah, G. A. and Nojima, M. (1973). *Synthesis*, 785
7. Jaeger, C. W. and Kornblum, N. (1972). *J. Amer. Chem. Soc.*, **94**, 2545
8. Fridman, A. L., Surkov, V. D. and Mukhametshin, F. M. (1971). *Zh. Org. Khim.*, **7**, 1840
9. White, D. A. and Baizer, M. M. (1973). *Tetrahedron Lett.*, 3597
10. Ioffe, S. L., Kashutina, M. V., Shitkin, V. M., Yankelovitch, A. Z., Levin, A. A. and Tartakovskii, V. A. (1972). *Izv. Akad. Nauk SSSR, Ser. Khim.*, 1341
11. Ioffe, S. L., Kashutina, M. V., Shitkin, V. M. and Levin, A. A. (1973). *Zh. Org. Khim.*, **9**, 896
12. Erashko, V. I., Sankov, B. G., Shevelev, S. A. and Fainzil'berg, A. A. (1972). *Izv. Akad. Nauk SSSR, Ser. Khim.*, 1223
13. Sankov, B. G., Erashko, V. I., Shevelev, S. A. and Fainzil'berg, A. A. (1971). *Izv. Akad. Nauk SSSR, Ser. Khim.*, 2045
14. Roitburd, G. V., Smit, V. A., Semenovskii, A. V., Kucherov, V. F. and Chizhov, O. S. (1972). *Izv. Akad. Nauk SSSR, Ser. Khim.*, 2225
15. Ficini, J., Bonenfant, A. and Barbara, C. (1972). *Tetrahedron Lett.*, 41
16. Pan'kov, A. K., Pevzner, M. S. and Bagal, L. I. (1972). *Zh. Org. Khim.*, **8**, 913
17. Shin, C.-G., Yonezawa, Y., Katayama, K. and Yoshimura, J. (1973). *Bull. Chem. Soc. Jap.*, **46**, 1727
18. Saegusa, T., Kobayashi, S., Ito, Y. and Morino, I. (1972). *Tetrahedron*, **28**, 3389
19. Fridman, A. L., Gabitov, F. A. and Surkov, V. D. (1972). *Zh. Org. Khim.*, **8**, 2457
20. Fridman, A. L., Gabitov, F. A. and Nikolaeva, A. D. (1971). *Zh. Org. Khim.*, **7**, 1154
21. Cummings, W. M. and Kreuz, K. L. (1972). *J. Org. Chem.*, **37**, 3929
22. Svetlakov, N. V., Mikheev, V. V. and Fedotov, Yu. A. (1971). *Zh. Org. Khim.*, **7**, 2220

23. Duynstee, E. F. J., Hennekens, J. L. J. P., Housmans, J. G. H. M., Van Raayen, W. and Voskuil, W. (1973). *Rec. Trav. Chim. Pays-Bas*, **92**, 1272
24. McMurry, J. E. and Metton, J. (1973). *J. Org. Chem.*, **38**, 4367
25. Kornblum, N. and Wade, P. A. (1973). *J. Org. Chem.*, **38**, 1418
26. Kornblum, N., Kestner, M. M., Boyd, S. D. and Cattran, L. C. (1973). *J. Amer. Chem. Soc.*, **95**, 3356
27. Wawzonek, S. and Kempf, J. V. (1973). *J. Org. Chem.*, **38**, 2763
28. Fridman, A. L. and Ismagilova, G. S. (1972). *Zh. Org. Khim.*, **8**, 1126
29. Zacheslavskii, S. A., Melnikov, V. V., Tselinskii, I. V. and Gidospov, B. V. (1972). *Zh. Org. Khim.*, **8**, 449
30. Matsui, T. and Hepler, L. G. (1973). *Can. J. Chem.*, **51**, 1941
31. Wheland, G. W. and Farr, J. (1943). *J. Amer. Chem. Soc.*, **65**, 1433
32. Colvin, E. W. (1973). *Specialist Periodical Reports, Aliphatic, Alicyclic and Saturated Heterocyclic Chemistry*, Vol. 1, Part 1, Chap. 2 (London: Chemical Society)
33. Graham, J. C. and Marr, D. H. (1972). *Can. J. Chem.*, **50**, 3857
34. Rosini, G., Baccolini, G. and Cacchi, S. (1973). *J. Org. Chem.*, **38**, 1060
35. Saraie, T., Ishiguro, T., Kaweshima, K. and Morita, K. (1973). *Tetrahedron Lett.*, 2121
36. Foley, H. G. and Dalton, D. R. (1973). *J. Chem. Soc. Chem. Commun.*, 628
37. Chakrabarti, J. K. and Hotten, T. M. (1972). *J. Chem. Soc. Chem. Commun.*, 1226
38. Klein, H. A. and Latscha, H. P. (1972). *Chem. Ztg.*, **96**, 171
39. Hallensleben, M. L. (1972). *Tetrahedron Lett.*, 2057
40. Brown, R. F. C., Butcher, M. and Fergie, R. A. (1973). *Aust. J. Chem.*, **26**, 1319
41. Deschamps, B., Lefebvre, G. and Seyden-Penne, J. (1972). *Tetrahedron*, **28**, 4209
42. Stanley, J. W., Beasley, J. G. and Mathison, I. W. (1972). *J. Org. Chem.*, **37**, 3746
43. Evans, D. A., Truesdale, L. K. and Carroll, G. L. (1973). *J. Chem. Soc. Chem. Commun.*, 55
44. Nagata, W., Yoshioka, M. and Hirai, S. (1972). *J. Amer. Chem. Soc.*, **94**, 4635
45. Nagata, W., Yoshioka, M. and Murakami, M. (1972). *J. Amer. Chem. Soc.*, **94**, 4644, 4654
46. House, H. O. (Ed.), (1972). *Org. Synth.*, **52**, 90
47. Oldenziel, O. H. and van Leusen, A. M. (1973). *Tetrahedron Lett.*, 1357
48. Short, J. H., Dunnigan, D. A. and Ours, C. W. (1973). *Tetrahedron*, **29**, 1931
49. Benders, P. H. and Hackmann, J. T. (1972). *Rec. Trav. Chim. Pays-Bas*, **91**, 343
50. Rasmussen, J. K. and Hassner, A. (1973). *Synthesis*, 682
51. Kühle, E. (1973). *Angew. Chem. Int. Ed. Engl.*, **12**, 630
52. Bennett, M. A. and Yoshida, T. (1973). *J. Amer. Chem. Soc.*, **95**, 3030
53. Mukaiyama, T., Kamio, K., Kobayshi, S. and Takei, H. (1973). *Chem. Lett.*, 357
54. Haefele, L. R. and Young, H. J. (1972). *Ind. Eng. Chem.*, *Prod. Res. Develop.*, **11**, 364
55. Greaves, P. M., Landor, P. D. and Odyek, O. (1973). *Tetrahedron Lett.*, 209
56. Cuvigny, T., Larcheceque, M. and Normant, H. (1972). *C. R. Acad. Sci., Ser. C*, **274**, 797
57. Cuvigny, T., Larcheceque, M. and Normant, H. (1973). *Bull. Soc. Chim. Fr.*, 1174
58. Gale, D. M. and Cherkofsky, S. C. (1973). *J. Org. Chem.*, **38**, 475
59. Hall, H. K., Smith, C. D. and Plorde, D. E. (1973). *J. Org. Chem.*, **38**, 2084
60. Konovalov, A. I. and Kamasheva, G. I. (1972). *Zh. Org. Khim.*, **8**, 1831
61. Toda, T., Morimura, S. and Murayama, K. (1972). *Bull. Chem. Soc. Jap.*, **45**, 557
62. Ducker, J. W. and Gunter, M. J. (1973). *Aust. J. Chem.*, **26**, 1551
63. Yanagida, S., Fujita, T., Ohaka, M., Katagiri, I. and Komori, S. (1973). *Bull. Chem. Soc. Jap.*, **46**, 292
64. Yanagida, S., Fujita, T., Ohaka, M., Kumagai, R. and Komori, S. (1973). *Bull. Chem. Soc. Jap.*, **46**, 299
65. Ohaka, M., Yanagida, S., Sugahara, K., Fujita, T., Okahara, M. and Komori, S. (1973). *Bull. Chem. Soc. Jap.*, **46**, 1275
66. Yanagida, S. and Komori, S. (1973). *Synthesis*, 189
67. Simchen, G. and Entenmann, G. (1973). *Angew. Chem. Int. Ed. Engl.*, **12**, 119
68. Yanagida, S., Fujita, T., Ohaka, M., Katagiri, I., Mujake, M. and Komori, S. (1973). *Bull. Chem. Soc. Jap.*, **46**, 303
69. Yanagida, S., Yokoe, M., Katagiri, I., Ohoka, M. and Komori, S. (1973). *Bull. Chem. Soc. Jap.*, **46**, 306

70. Fuks, R. (1973). *Tetrahedron*, **29**, 2147, 2153
71. Doumaux, A. R. (1972). *J. Org. Chem.*, **37**, 508
72. Kirilov, M., Ivanov, D., Pétrov, G. and Golémchinski, G. (1973). *Bull. Soc. Chim. Fr.*, 3053
73. Russell, T. W., Hoy, R. C. and Cornelius, J. E. (1972). *J. Org. Chem.*, **37**, 3552
74. Kantlehner, W. and Speh, P. (1972). *Chem. Ber.*, **105**, 1340
75. White, D. A. and Baizer, M. M. (1973). *J. Chem. Soc. Perkin Trans. I*, 2230
76. Kawabata, N., Matsubara, K. and Yamashita, S. (1973). *Bull. Chem. Soc. Jap.*, **46**, 3225
77. Mertz, R., Van Assche, D., Fleury, J.-P. and Regitz, M. (1973). *Bull. Soc. Chim. Fr.*, 3342
78. Takahashi, K., Miyake, A. and Hata, G. (1971). *Bull. Chem. Soc. Jap.*, **44**, 3484
79. Ireland, C. J. and Pizey, J. S. (1972). *J. Chem. Soc. Chem. Commun.*, 4
80. De Vries, L. (1973). *J. Org. Chem.*, **38**, 2604
81. De Vries, L. (1973). *J. Org. Chem.*, **38**, 4357
82. Perchais, J. and Fleury, J. P. (1972). *Tetrahedron*, **28**, 2267
83. Lemmon, R. M. (1970). *Chem. Rev.*, **70**, 95
84. Ferris, J. P., Donner, D. B. and Lotz, W. (1972). *J. Amer. Chem. Soc.*, **94**, 6968
85. Dabek, H., Selvarajan, R. and Boyer, J. H. (1972). *J. Chem. Soc. Chem. Commun.*, 244
86. Ferris, J. P. and Ryan, T. J. (1973). *J. Org. Chem.*, **38**, 3302
87. Webster, O. W., Hartter, D. R., Begland, R. W., Sheppard, W. A. and Cairncross, A. (1972). *J. Org. Chem.*, **37**, 4133
88. Begland, R. W. and Hartter, D. R. (1972). *J. Org. Chem.*, **37**, 4136
89. Grundmann, C. and Gruenanger, P. (1971). *The Nitrile Oxides* (New York: Springer)
90. Grundmann, C., Kochs, P. and Boal, J. R. (1972). *Ann. Chem.*, **761**, 162
91. Liskow, D. H., Bender, C. F. and Schaefer, H. F. (1972). *J. Amer. Chem. Soc.*, **94**, 5178
92. Moffat, J. B. and Tang, K. F. (1973). *Theoret. Chim. Acta*, **32**, 171
93. Dewar, M. J. S. and Kohn, M. C. (1972). *J. Amer. Chem. Soc.*, **94**, 2704
94. Bak, B. and Jansen, P. (1972). *J. Mol. Struct.*, **12**, 167
95. Millich, F. (1972). *Chem. Rev.*, **72**, 107
96. Walborsky, H. M. and Niznik, G. E. (1972). *J. Org. Chem.*, **37**, 187
97. Saegusa, T., Murase, I. and Ito, Y. (1972). *Bull. Chem. Soc. Jap.*, **45**, 1884
98. Saegusa, T., Yonezawa, K., Murase, I., Konoike, T., Tomita, S. and Ito, Y. (1973). *J. Org. Chem.*, **38**, 2319
99. Saegusa, T., Murase, I. and Ito, Y. (1973). *J. Org. Chem.*, **38**, 1753
100. Saegusa, T. and Murase, I. (1972). *Synth. Commun.*, **2**, 1
101. Saegusa, T., Ito, Y., Tomita, S. and Kinoshita, H. (1972). *Bull. Chem. Soc. Jap.*, **45**, 496
102. Sawai, H. and Takizawa, T. (1972). *Tetrahedron Lett.*, 4263
103. Schoellkopf, V. and Porsch, P. H. (1972). *Angew. Chem. Int. Ed. Engl.*, **11**, 429
104. Ozaki, S. (1972). *Chem. Rev.*, **72**, 457
105. Kricheldorf, H. R. (1972). *Synthesis*, 551
106. Kricheldorf, H. R. (1972). *Chem. Ber.*, **105**, 3958
107. Washbourne, S. S. and Peterson, W. R. (1972). *Synth. Commun.*, **2**, 227
108. Washbourne, S. S. and Peterson, W. R. (1972). *J. Amer. Oil Chem. Soc.*, **49**, 694
109. Van Reijendam, J. W. and Baardman, F. (1973). *Synthesis*, 413
110. Middleton, W. J. (1973). *Tetrahedron*, **29**, 297
111. Fuks, R. and Hartemink, M. (1973). *Tetrahedron*, **29**, 297
112. Gorbatenko, W. I., Bondar, W. A. and Samaraj, L. I. (1973). *Angew. Chem. Int. Ed. Engl.*, **12**, 842
113. Fokin, A. V., Kolomiets, A. F., Studev, Yu. N., Rapkin, A. I. and Yakutin, V. I. (1972). *Izv. Akad. Nauk SSSR, Ser. Khim.*, 1210
114. Ulrich, H. (1971). *Advan. Urethane Sci. Technol.*, **1**, 33
115. Nuridzhanyan, K. A. and Kuznetsova, G. V. (1973). *Zh. Org. Khim.*, **9**, 1200
116. Nesterova, L. M., Nuridzhanyan, K. A., Mironova, N. E. and Aryutkina, N. L. (1972). *Zh. Org. Khim.*, **8**, 2502
117. Arbuzov, B. A., Zobova, N. N., Balabanova, F. B. and Ofitserov, E. N. (1973). *Dokl. Akad. Nauk*, **209**, 601

118. Tschesche, H. and Jering, H. (1973). *Angew. Chem. Int. Ed. Engl.*, **12**, 756
119. Goerdeler, J. and Schimpf, R. (1973). *Chem. Ber.*, **106**, 1496
120. Goerdeler, J., Schimpf, R. and Tiedt, M.-L. (1972). *Chem. Ber.*, **105**, 3322
121. Capuano, L., Boschat, P., Heyer, H. W. and Wachter, G. (1973). *Chem. Ber.*, **106**, 314
122. Capuano, L. and Zander, R. (1973). *Chem. Ber.*, **106**, 3070
123. Capuano, L., Kirn, H. R. and Kalweit, M. (1973). *Chem. Ber.*, **106**, 3677
124. Tsuge, O., Sakai, K. and Tashiro, M. (1973). *Tetrahedron*, **29**, 1983
125. LeBel, N. A., Cherluck, R. M. and Curtius, E. A. (1973). *Synthesis*, 678
126. Ojima, I., Inaba, S. and Nagai, Y. (1973). *Tetrahedron Lett.*, 4363
127. Hai, S. M. A. and Lwowski, W. (1973). *J. Org. Chem.*, **38**, 2442
128. Hoffmann, R. W., Steinbach, K. and Dittrich, B. (1973). *Chem. Ber.*, **106**, 2174
129. Vandensavel, J. M., Smets, G. and L'Abbe, G. (1973). *J. Org. Chem.*, **38**, 675
130. Matsuda, I., Itoh, K. and Ishii, Y. (1972). *J. Chem. Soc. Perkin Trans. I*, 1678
131. Abraham, W. and Barnikow, G. (1973). *Tetrahedron*, **29**, 691
132. Guise, G. B., Jackson, M. B. and Maclaren, J. A. (1972). *Aust. J. Chem.*, **25**, 2583
133. Baroeta, N. and Miralles, A. (1972). *J. Org. Chem.*, **37**, 2255
134. Welcher, R. P. and Cutrufello, P. F. (1973). *J. Org. Chem.*, **37**, 4478
135. Bullpitt, M. L. and Kitching, W. (1972). *J. Organometal. Chem.*, **34**, 321
136. Guy, R. G. and Pearson, I. (1973). *J. Chem. Soc. Perkin Trans. I*, 281
137. Mitani, H., Ando, T. and Yukawa, Y. (1972). *Chem. Lett.*, 455
138. Goerdeler, J. and Hohage, H. (1973). *Chem. Ber.*, **106**, 1487
139. Meijer, J. and Brandsma, L. (1971). *Rec. Trav. Chim. Pays-Bas*, **90**, 1098
140. Grigat, E. (1972). *Angew. Chem. Int. Ed. Engl.*, **11**, 949
141. Tanaka, K. (1972). *Bull. Chem. Soc. Jap.*, **45**, 834
142. Grochowski, E. and Hintze, B. (1973). *Rocz. Chem.*, **47**, 59
143. Parks, T. E. and Spurlock, L. A. (1973). *J. Org. Chem.*, **38**, 3922
144. Padwa, A., Dharan, M., Smolonoff, J. and Wetmore, S. I. (1973). *Pure Appl. Chem.*, **33**, 329
145. Read, S. T. (1973). *Specialist Periodical Reports, Photochemistry*, Vol. 4, 765, 825 (London: Chemical Society)
146. Erickson, W. F. and Richey, H. G. (1972). *Tetrahedron Lett.*, 2811
147. Duhamel, P., Duhamel, L., Legal, J. C. and Valnot, J. Y. (1972). *Bull. Soc. Chim. Fr.*, 3222
148. Parker, C. O. and Stevens, T. E. (1972). *J. Org. Chem.*, **37**, 922
149. Baum, A. A., Karnischky, L. A., McLeod, D. and Kasai, P. H. (1973). *J. Amer. Chem. Soc.*, **95**, 617
150. Nielsen, A. T., Atkins, R. L., Moore, D. W., Scott, R., Mallory, D. and La Berge, J. M. (1973). *J. Org. Chem.*, **38**, 3288
151. Hunter, D. H. and Sim, S. K. (1972). *Can. J. Chem.*, **50**, 669
152. Alper, H. (1972). *J. Org. Chem.*, **37**, 3972
153. Ojima, I. and Kogure, T. (1973). *Tetrahedron Lett.*, 2475
154. De Savignac, A., Bon, M. and Lattes, A. (1972). *Bull. Soc. Chim. Fr.*, 3426
155. Moreau, J.-L. and Gaudemar, M. (1973). *Bull. Soc. Chim. Fr.*, 2549
156. Moreau, J.-L. and Gaudemar, M. (1972). *C. R. Acad. Sci., Ser. C*, **274**, 2015
157. Mauzé, B. and Miginiac, L. (1973). *Bull. Soc. Chim. Fr.*, 1078, 1082
158. Ogata, Y. and Sawaki, Y. (1973). *J. Amer. Chem. Soc.*, **95**, 4687
159. Wright, C. D. and Zollinger, J. L. (1973). *J. Org. Chem.*, **38**, 1065, 1075
160. Hullot, P. and Cuvigny, T. (1973). *Bull. Soc. Chim. Fr.*, 2985
161. Hullot, P. and Cuvigny, T. (1973). *Bull. Soc. Chim. Fr.*, 2989
162. Minami, T., Fukuda, M., Abe, M. and Agawa, T. (1973). *Bull. Chem. Soc. Jap.*, **46**, 2156
163. Ulrich, H., Tucker, B. and Sayigh, A. A. R. (1972). *J. Amer. Chem. Soc.*, **94**, 3484
164. Weinshenker, N. M. and Shen, C.-M. (1972). *Tetrahedron Lett.*, 3281, 3285
165. Viehe, H. G. and Janousek, Z. (1973). *Angew. Chem. Int. Ed. Engl.*, **12**, 806
166. Viehe, H. G., van Vyre, T. and Janousek, Z. (1972). *Angew. Chem. Int. Ed. Engl.*, **11**, 916
167. Janousek, Z. and Viehe, H. G. (1973). *Angew. Chem. Int. Ed. Engl.*, **12**, 74
168. Janousek, Z., Collard, J. and Viehe, H. G. (1972). *Angew. Chem. Int. Ed. Engl.*, **11**, 917

169. Le Clef, B., Mommaerts, J., Stelander, B. and Viehe, H. G. (1973). *Angew. Chem. Int. Ed. Engl.*, **12**, 404
170. Hervens, F. and Viehe, H. G. (1973). *Angew. Chem. Int. Ed. Engl.*, **12**, 405
171. Böhme, H. and Martin, F. (1973). *Chem. Ber.*, **106**, 3540
172. Böhme, H. and Haake, M. (1972). *Chem. Ber.*, **105**, 2233
173. Masui, M., Suda, K., Yamauchi, M. and Yijima, C. (1972). *J. Chem. Soc. Perkin Trans. I*, 1955
174. Grée, R. and Carrié, R. (1972). *Tetrahedron Lett.*, 2987
175. Dobashi, T. S., Goodrow, M. H. and Grubbs, E. J. (1973). *J. Org. Chem.*, **38**, 4440
176. Ahlbrecht, H. and Vonderheid, C. (1973). *Chem. Ber.*, **106**, 2009
177. Ogloblin, K. A., Semenov, V. P. and Vasil'eva, E. V. (1972). *Zh. Org. Khim.*, **8**, 1613
178. Betkouski, M. F. and Lwowski, W. (1973). *J. Chem. Soc. Chem. Commun.*, 932
179. Barker, M. W., Lauderdale, S. C. and West, J. R. (1973). *J. Org. Chem.*, **38**, 3951
180. Razniak, S. L., Flagg, E. M. and Siebenthall, F. (1973). *J Org. Chem.*, **38**, 2242
180a. Pedersen, E. B., Vesterager, N. O. and Lawesson, S.-O. (1972). *Synthesis*, 547
181. Thomas, D. and Aue, D. H. (1973). *Tetrahedron Lett.*, 1807
182. Okuyama, T., Pletcher, T. C., Sahn, D. J. and Schmir, G. L. (1973). *J. Amer. Chem. Soc.*, **95**, 1253
183. Challis, B. C. and Frenkel, A. D. (1972). *J. Chem. Soc. Chem. Commun.*, 303
184. Kiro, Z. B., Teterin, Yu. A., Nikolenko, L. N. and Stepanov, B. I. (1972). *Zh. Org. Khim.*, **8**, 2573
185. Ioffe, B. V. and Koroleva, E. V. (1973). *Zh. Org. Khim.*, **9**, 1305
186. Somin, I. N. and Gindin, V. A. (1972). *Zh. Org. Khim.*, **8**, 1986
187. Davis, F. A., Slegeir, W. A. R. and Kaminski, J. M. (1972). *J. Chem. Soc. Chem. Commun.*, 634
188. Davis, F. A., Slegeir, W. A. R., Evans, S., Schwartz, A., Goff, D. K. and Palmer, R. (1973). *J. Org. Chem.*, **38**, 2809
189. Freeman, J. P. (1973). *Chem. Rev.*, **73**, 283
190. Dalton, D. R. and Foley, H. G. (1973). *J. Org. Chem.*, **38**, 4200
191. Dalton, D. R., Foley, H. G., Trueblood, K. N. and Murphy, M. R. (1973). *Tetrahedron Lett.*, 779
192. Gnichtel, H., Griebenow, W. and Lowe, W. (1972). *Chem. Ber.*, **105**, 1865
193. Gnichtel, H., Walentowski, R. and Sohuster, K.-E. (1972). *Chem. Ber.*, **105**, 1701
194. Kliegman, J. H. and Barnes, R. K. (1972). *J. Org. Chem.*, **37**, 4223
195. Conaill, R. J., Scott, F. L. and Hegarty, A. F. (1972). *Tetrahedron Lett.*, 1217
196. Busser, U. and Haller, R. (1973). *Tetrahedron Lett.*, 231
197. Kol'tsov, A. I. and Kheifets, G. M. (1972). *Usp. Khim.*, **41**, 877
198. Rajappa, S. (1972). *J. Sci. Ind. Res.*, **31**, 366
199. De Savignac, A., Bon, M. and Lattes, A. (1972). *Bull. Soc. Chim. Fr.*, 3167
200. Ahlbrecht, H. and Fischer, S. (1973). *Tetrahedron*, **29**, 659
201. Ahlbrecht, H., Hannisch, H., Funk, W. and Kalas, R. D. (1972). *Tetrahedron*, **28**, 5481
202. Duhamel, L., Duhamel, P. and Legal, J. C. (1972). *Bull. Soc. Chim. Fr.*, 3230
203. Deswarte, S., Bellec, C., Courteix, C. and Paris, M. C. (1972). *C. R. Acad. Sci., Ser. C*, **275**, 411
204. Toke, L., Blasko, G., Szabo, L. and Szantay, C. S. (1972). *Tetrahedron Lett.*, 2459
205. Walter, W. and Kohn, J. (1973). *Ann. Chem.*, 443
206. Walter, W. and Meyer, H.-W. (1973). *Ann. Chem.*, 462
207. Guthrie, R. D. and Hedrick, J. L. (1973). *J. Amer. Chem. Soc.*, **95**, 2971
208. Macaulay, R., Burnelle, L. A. and Sandorfy, C. (1973). *Theoret. Chim. Acta*, **29**, 1
209. Lehn, J. M. (1970). *Theoret. Chim. Acta*, **16**, 351
210. Kollman, P. A., Trager, W. F., Rothenberg, S. and Williams, J. E. (1973). *J. Amer. Chem. Soc.*, **95**, 458
211. Shanshal, M. (1972). *Tetrahedron*, **28**, 61
212. Arnaud, R., Faramond-Baud, D. and Gelus, M. (1973). *Theoret. Chim. Acta*, **31**, 335
213. Jennings, W. B. and Boyd, D. R. (1972). *J. Amer. Chem. Soc.*, **94**, 7187
214. Bjoergo, J., Boyd, D. R., Watson, C. G. and Jennings, W. B. (1972). *Tetrahedron Lett.*, 1747
215. Padwa, A. and Albrecht, F. (1972). *J. Amer. Chem. Soc.*, **94**, 1000

216. Vögtle, F. and Goldschmitt, E. (1973). *Angew. Chem. Int. Ed. Engl.*, **12,** 767
217. Hunter, D. H. and Sim, S. K. (1972). *Can. J. Chem.*, **50,** 678
218. Neiman, Z. (1972). *J. Chem. Soc. Perkin Trans. II*, 1746
219. Regitz, M. (1972). *Synthesis*, 351
220. Mangold, H. K. (1973). *Chem. Phys. Lipids*, **11,** 244
221. Fridman, A. L., Ismagilova, G. S., Zalesov, V. S. and Novikov, S. S. (1972). *Usp. Khim.*, **41,** 722
222. Peace, B. W. and Wulfman, D. S. (1973). *Synthesis*, 137
223. Hecht, S. M. and Kozarich, J. W. (1972). *Tetrahedron Lett.*, 5147
224. Hecht, S. M. and Kozarich, J. W. (1972). *Tetrahedron Lett.*, 1501
225. Hecht, S. M. and Kozarich, J. W. (1973). *Tetrahedron Lett.*, 1397
226. Heyes, G. and Holt, G. (1973). *J. Chem. Soc. Perkin Trans. I*, 189
227. Sohn, M. B., Jones, M., Hendrick, M. E., Rando, R. R. and Doering, W. E. (1972). *Tetrahedron Lett.*, 53
228. Jones, M. (1973). *Carbenes*, Vol. 1 (New York: Wiley-Interscience)
229. Zeller, K. P., Meier, H. and Müller, E. (1972). *Tetrahedron*, **28,** 5831
230. Zeller, K. P., Meier, H., Kolshorn, H. and Müller, E. (1972). *Chem. Ber.*, **105,** 1875
231. Fenwick, J., Frater, G., Ogi, K. and Strausz, O. P. (1973). *J. Amer. Chem. Soc.*, **95,** 124
232. Matlin, S. A. and Sammes, P. G. (1972). *J. Chem. Soc. Perkin Trans. I*, 2623
233. Csizmodia, I. G., Gunning, H. E., Gosari, R. K. and Strausz, O. P. (1973). *J. Amer. Chem. Soc.*, **95,** 133
234. Hixson, S. S. and Hixson, S. H. (1972). *J. Org. Chem.*, **37,** 1279
235. Duggleby, P. M., Holt, G., Hope, M. A. and Lewis, A. (1972). *J. Chem. Soc. Perkin Trans. I*, 3020
236. Brown, H. C., Midland, M. M. and Levy, A. R. (1972). *J. Amer. Chem. Soc.*, **94,** 3662
237. Wenkert, E. and McPherson, C. A. (1973). *J. Amer. Chem. Soc.*, **94,** 8084
238. El-Ghandour, N., Henri-Rousseau, O. and Soulier, J. (1972). *Bull. Soc. Chim. Fr.*, 2817
239. Helder, R., Doornbos, T., Strating, J. and Zwanenburg, B. (1973). *Tetrahedron*, **29,** 1375
240. Sabaté-Alduy, C. and Bastide, J. (1972). *Bull. Soc. Chim. Fr.*, 2764
241. Houk, K. N. (1972). *J. Amer. Chem. Soc.*, **94,** 8953
242. Sondengam, B. L., Heutchoya Hémo, J. and Charles, G. (1973). *Tetrahedron Lett.*, 261
243. Borch, R. F. and Hassid, A. I. (1972). *J. Org. Chem.*, **37,** 1673
244. Richey, H. G. and Erickson, W. F. (1972). *Tetrahedron Lett.*, 2807
245. Hancock, K. G. and Dickinson, D. A. (1973). *J. Chem. Soc. Chem. Commun.*, 783
246. Barker, R. and Halliday, D. E. (1972). *Tetrahedron Lett.*, 2773
247. Tritschler, W. and Kabusz, S. (1972). *Synthesis*, 32
248. Brown, H. C. and Heim, P. (1973). *J. Org. Chem.*, **38,** 912
249. Duhamel, L., Duhamel, P. and Siret, P. (1973). *Bull. Soc. Chim. Fr.*, 2460
250. Brown, H. C., Midland, M. M. and Levy, A. B. (1973). *J. Amer. Chem. Soc.*, **95,** 2394
251. Hendrickson, J. B. and Joffee, I. (1973). *J. Amer. Chem. Soc.*, **95,** 4083
252. Donetti, A. and Bellora, E. (1972). *J. Org. Chem.*, **37,** 3352
253. Fluck, E. and Meiser, P. (1973). *Chem. Ber.*, **106,** 69
254. Kukhar, V. P. and Pasternak, V. I. (1972). *Synthesis*, 611
255. Tramontini, M. (1973). *Synthesis*, 703
256. Litvinenko, L. M., Popov, A. F. and Gel'bina, Zh. P. (1972). *Dokl. Chem.*, **203,** 229
257. Robinson, C. B. and Herbrandson, H. F. (1972). *J. Amer. Chem. Soc.*, **94,** 7883
258. Tsuda, T., Isegura, Y. and Saegusa, T. (1972). *J. Org. Chem.*, **37,** 2670
259. Wendschuh, P. H., Fuhr, H., Gaffney, J. S. and Pitts, J. N. (1973). *J. Chem. Soc. Chem. Commun.*, 74
260. Kraus, M. A. (1973). *Synthesis*, 361
261. Wakselman, M. and Guibe-Jampel, E. (1973). *J. Chem. Soc. Chem. Commun.*, 593
262. Hendrickson, J. B. and Bergeron, R. (1973). *Tetrahedron Lett.*, 3839
263. Hendrickson, J. B. and Bergeron, R. (1973). *Tetrahedron Lett.*, 4607
264. Hendrickson, J. B., Bergeron, R., Giga, A. and Sternback, D. (1973). *J. Amer. Chem. Soc.*, **95,** 3412

265. Carpino, L. A. (1973). *Accounts Chem. Res.*, **6**, 191
266. Heublein, G. and Rang, D. D. (1972). *Tetrahedron*, **28**, 1873
267. Farcasiu, D., Kascheres, C. and Schwartz, L. H. (1972). *J. Amer. Chem. Soc.*, **94**, 180
268. Andrisano, R., Angeloni, A. S. and Fini, A. (1972). *Tetrahedron*, **28**, 2681
269. Hawkins, E. G. E. (1973). *Angew. Chem. Int. Ed. Engl.*, **12**, 783
270. Sato, Y., Ban, Y. and Shirai, H. (1973). *J. Org. Chem.*, **38**, 4373
271. Cohen, S. G., Parola, A. and Parsons, G. H. (1973). *Chem. Rev.*, **73**, 141
272. Brown, C. A. (1973). *J. Amer. Chem. Soc.*, **95**, 982
273. Bryce-Smith, D., Gilbert, A. and Klunkin, G. (1973). *J. Chem. Soc. Chem. Commun.*, 330
274. Perie, J. J., Laval, J. P., Roussel, J. and Lattes, A. (1972). *Tetrahedron*, **28**, 675
275. Laval, J. P., Perie, J. J., Roussel, J. and Lattes, A. (1972). *Tetrahedron*, **28**, 701
276. Lindsay Smith, J. R., Norman, R. O. C. and Rowley, A. G. (1972). *J. Chem. Soc. Perkin Trans. I*, 228
277. Briggs, J. P., Yandagni, R. and Keborle, P. (1972). *J. Amer. Chem. Soc.*, **94**, 5128
278. Pozharskii, A. F. and Zvezdina, E. A. (1973). *Usp. Khim.*, **42**, 65
279. Feibush, B., Tamari, T. and Gil-Av, E. (1972). *J. Chem. Soc. Perkin Trans. II*, 1197
280. Rubinstein, H., Feibush, B. and Gil-Av, E. (1973). *J. Chem. Soc. Perkin Trans. II*, 2094
281. Kost, D. and Raban, M. (1972). *J. Amer. Chem. Soc.*, **94**, 2533
282. Wickberg, B. and Hansson, C. (1973). *J. Org. Chem.*, **38**, 3074
283. Traynelis, V. J. and Rieck, J. N. (1973). *J. Org. Chem.*, **38**, 4339
284. Senning, A. and Kelly, P. (1972). *Acta Chem. Scand.*, **26**, 2877
285. Duhamel, L., Duhamel, P. and Truxillo, V. (1972). *C. R. Acad. Sci., Ser. C*, **275**, 225
286. Dulcere, J. P., Ragonnet, B., Santelli, M. and Bertand, M. (1972). *C. R. Acad. Sci., Ser. C*, **274**, 975
287. Kuehne, M. E. and Linde, H. (1972). *J. Org. Chem.*, **37**, 1846
288. Ohaka, A., Asada, K., Yanagida, S., Okahara, M. and Komori, S. (1973). *J. Org. Chem.*, **38**, 2287
289. Bodnarchuk, N. D., Momot, V. V. and Gavrilenko, B. B. (1972). *Zh. Org. Khim.*, **8**, 2493
290. Bodnarchuk, N. D., Momot, V. V. and Gavrilenko, B. B. (1973). *Zh. Org. Khim.*, **9**, 36
291. Norris, W. P. (1972). *Tetrahedron*, **28**, 1965
292. Duhamel, L., Duhamel, P., Collet, C., Haider, A. and Poivier, J. M. (1972). *Tetrahedron Lett.*, 4743
293. Duhamel, L., Duhamel, P., Collet, C. and Haider, A. (1971). *C. R. Acad. Sci., Ser. C*, **273**, 1461
294. Reichardt, C. and Schagerer, K. (1973). *Angew. Chem. Int. Ed. Engl.*, **12**, 323
295. Guthrie, J. P. and Jordan, F. (1972). *J. Amer. Chem. Soc.*, **94**, 9132, 9136
296. Viehe, H. G., Janousek, Z., Gompper, R. and Lach, D. (1973). *Angew. Chem. Int. Ed. Engl.*, **12**, 566
297. Dockx, J. (1973). *Synthesis*, 441
298. Wrobel, T. J. and Krawczyk, R. A. (1971). *Rocz. Chem.*, **45**, 1465
299. Port, G. N. J. and Pullmann, A. (1973). *Theoret. Chim. Acta*, **31**, 231
300. McKillip, W. J., Sedor, E. A., Culbertson, B. M. and Wawzonek, S. (1973). *Chem. Rev.*, **73**, 255
301. Timpe, H. J. (1972). *Z. Chem.*, **12**, 250
302. Cameron, A. F., Hair, N. J. and Morris, D. G. (1972). *J. Chem. Soc. Perkin Trans. II*, 1071
303. Epsztajn, J., Katritsky, A. R., Lunt, E., Mitchell, J. N. and Roch, G. (1973). *J. Chem. Soc. Perkin Trans. I*, 2622
304. Benecke, H. P. and Wikel, J. H. (1972). *Tetrahedron Lett.*, 289
305. Asahara, T., Ootani, N. and Seno, M. (1972). *Seisan-Kenkyu*, **24**, 37
306. Minisci, F. (1973). *Synthesis*, 1
307. Kling, T. A., White, R. E. and Kovacic, P. (1972). *J. Amer. Chem. Soc.*, **94**, 7416
308. White, R. E., Nazareno, M. B., Gleissner, M. R. and Kovacic, P. (1973). *J. Org. Chem.*, **38**, 3902
309. Benders, P. H. (1973). *Tetrahedron Lett.*, 3653

310. Chow, Y. L. (1973). *Accounts Chem. Res.*, **6**, 354
311. Seebach, D. and Enders, D. (1972). *Angew. Chem. Int. Ed. Engl.*, **11**, 301
312. Farina, P. R. and Tieckelman, H. (1973). *J. Org. Chem.*, **38**, 4259
313. Michejda, C. J. and Schluenz, R. W. (1973). *J. Org. Chem.*, **38**, 2412
314. Grakauskas, V. and Baum, K. (1972). *J. Org. Chem.*, **37**, 334
315. Avakyan, V. G., Shlyapochnikov, V. A. and Chekrygin, V. A. (1972). *Izv. Akad. Nauk SSSR, Ser. Khim.*, 948
316. Beckett, A. H., Coutts, R. T. and Ogunbona, F. A. (1973). *Tetrahedron*, **29**, 4149
317. Nelsen, S. F. and Weisman, G. R. (1973). *Tetrahedron Lett.*, 23241
318. Sokolova, T. A. and Osipora, I. N. (1971). *Zh. Org. Khim.*, **7**, 1826
319. Linke, K.-H., Turley, R. and Flaskamp, E. (1973). *Chem. Ber.*, **106**, 1052
320. Tamura, Y., Minamikawa, J., Kita, Y., Kim, J. H. and Ikeda, M. (1973). *Tetrahedron*, **29**, 1063
321. Barton, D. H. R., Girijavallabhan, M. and Sammes, P. G. (1972). *J. Chem. Soc. Perkin Trans. I*, 929
322. Sucrow, W., Slopianka, M. and Mentzel, C. (1973). *Chem. Ber.*, **106**, 745
323. Oppolzer, W. (1972). *Tetrahedron Lett.*, 1707
324. Henderson, W. A. and Zweis, A. (1972). *J. Chem. Soc. Chem. Commun.*, 169
325. Koga, M. and Anselme, J.-P. (1973). *J. Chem. Soc. Chem. Commun.*, 53
326. Favorskaya, T. A., Yakimovich, S. I. and Khrustalev, V. A. (1972). *Zh. Org. Khim.*, **8**, 899
327. Kitaev, Yu. P. and Buzykin, B. I. (1972). *Usp. Khim.*, **41**, 995
328. Weigert, F. J. (1972). *J. Fluorine Chem.*, **1**, 445
329. Tsuji, J., Takahashi, H. and Kajimoto, T. (1973). *Tetrahedron Lett.*, 4573
330. Shawali, A. S. and Hassanen, H. M. (1973). *Tetrahedron*, **29**, 121
331. Idoux, J. P. and Sikorski, J. A. (1972). *J. Chem. Soc. Perkin Trans. II*, 921
332. Kotsuki, H., Tanegashima, C., Tokushige, M. and Tanaka, S. (1972). *Bull. Chem. Soc. Jap.*, **45**, 813
334. Stevens, R. V., McEntire, E. E., Barnett, W. E. and Wenkert, E. (1973). *J. Chem. Soc. Chem. Commun.*, 662
335. Smith, R. F., Johnson, D. S., Abgott, R. A. and Madden, M. J. (1973). *J. Org. Chem.*, **38**, 1344
336. Gol'din, G. S., Poddubnyi, V. G., Orlova, E. V. and Kisin, A. V. (1973). *Zh. Org. Khim.*, **9**, 517
337. Smith, P. F. and Craig, T. A. (1973). *Tetrahedron Lett.*, 3941
338. Barton, D. H. R., Magnus, P. D. and Young, R. N. (1973). *J. Chem. Soc. Chem. Commun.*, 331
339. Yamada, S., Kasai, Y. and Shioiri, T. (1973). *Tetrahedron Lett.*, 1595
340. Collman, J. P., Winter, S. R. and Komoto, R. G. (1973). *J. Amer. Chem. Soc.*, **95**, 249
341. Banhidai, B. and Schöllkopf, V. (1973). *Angew. Chem. Int. Ed. Engl.*, **12**, 836
342. Gilbert, E. (1972). *Synthesis*, 30
343. Atavin, A. S., Mirskova, A. N. and Zorina, E. F. (1972). *Zh. Org. Khim.*, **8**, 708
344. Matlin, S. A. and Sammes, P. G. (1972). *J. Chem. Soc. Chem. Commun.*, 1222
345. Finkelstein, M. and Ross, S. D. (1972). *Tetrahedron*, **28**, 4497
346. Khasapov, B. N., Novikova, T. S., Lebedev, O. V., Khmel'nitskii, L. I. and Novikov, S. S. (1973). *Zh. Org. Khim.*, **9**, 20
347. Stovrovskaya, A. V., Protopopova, T. V. and Skoldinov, A. P. (1973). *Zh. Org. Khim.*, **9**, 699
348. Shahak, I. and Sasson, Y. (1973). *J. Amer. Chem. Soc.*, **95**, 3440
349. Stodola, F. H. (1972). *J. Org. Chem.*, **37**, 178
350. Sayo, H., Ohomori, H., Umeda, T. and Masui, M. (1972). *Bull. Chem. Soc. Jap.*, **45**, 203
351. Gold, E. H. and Babad, E. (1972). *J. Org. Chem.*, **37**, 2208
352. Owsley, D. C., Nelke, J. M. and Bloomfield, J. J. (1973). *J. Org. Chem.*, **38**, 901
353. Scilly, N. F. (1973). *Synthesis*, 160
354. Avaca, L. A. and Bewick, A. (1972). *J. Chem. Soc. Perkin Trans. II*, 1713
355. Brederick, H., Simchen, G. and Hoffmann, H. (1973). *Chem. Ber.*, **106**, 3725
356. Christophersen, C. and Carlsen, P. (1973). *Tetrahedron Lett.*, 211
357. Vail, S. L. and Pierce, A. G. (1972). *J. Org. Chem.*, **37**, 391

358. Martin, R. B. (1972). *J. Chem. Soc. Chem. Commun.*, 793
359. De Loze, C., Combelas, P., Bacelon, P. and Garrigou-Lagrange, C. (1972). *J. Chim. Phys. Physicochim. Biol.*, **69**, 397
360. Hopkinson, A. C. and Csizmadia, I. G. (1973). *Theoret. Chim. Acta*, **31**, 83
361. Bonaccorsi, R., Pullman, A., Scrocco, E. and Tomasi, J. (1972). *Chem. Phys. Lett.*, **12**, 622
362. Hecht, S. S. and Rothman, E. S. (1973). *J. Org Chem.*, **38**, 395
363. Johansson, A. and Kollman, P. A. (1972). *J. Amer. Chem. Soc.*, **94**, 6196
364. Kondo, K., Sonoda, N. and Tsutsumi, S. (1972). *Chem. Lett.*, 373
365. Kondo, K., Sonoda, N., Yashida, K., Koishi, M. and Tsutsumi, S. (1972). *Chem. Lett.*, 401
366. Bosin, T. R., Hanson, R. N., Rodricks, J. W., Simpson, R. A. and Rapoport, H. (1973). *J. Org. Chem.*, **38**, 1591
367. Akiba, K.-Y. and Inamoto, N. (1973). *J. Chem. Soc. Chem. Commun.*, 13
368. Mozolis, V. V. and Iokubaitite, S. P. (1973). *Usp. Khim.*, **42**, 587
369. Meijer, H., Tel, R. M., Strating, J. and Engberts, J. B. F. N. (1973). *Rec. Trav. Chim. Pays-Bas*, **92**, 72
370. Gund, P., Berkelhammer, G. and Wayne, R. S. (1972). *Tetrahedron Lett.*, 3983
371. Pratt, R. F. and Bruice, T. C. (1972). *J. Amer. Chem. Soc.*, **94**, 2823
372. Aurich, H. G. and Scharpenberg, H.-G. (1973). *Chem. Ber.*, **106**, 1881
373. Ankers, W. B., Brown, C., Hudson, R. F. and Lawson, A. J. (1972). *J. Chem. Soc. Chem. Commun.*, 935
374. Galland, R. and Heesing, A. (1973). *Chem. Ber.*, **106**, 2580
375. Padwa, A., Cimiluca, P. and Eastman, D. (1972). *J. Org. Chem.*, **37**, 805
376. Nakagawa, V. and Minami, K. (1972). *Tetrahedron Lett.*, 343

5
Phosphorus Compounds

R. B. KING
University of Georgia

5.1 INTRODUCTION

The literature of organophosphorus chemistry continues to be voluminous. The present review considers only aliphatic organophosphorus compounds with at least one carbon–phosphorus bond and excludes transition-metal coordination compounds. Even so, limitations of space prevent this review from being exhaustive.

Subjects of interest to organophosphorus chemists reviewed during 1972 and 1973 include phosphines and phosphonium salts[1], halogenophosphines and related compounds[2], addition reactions of tertiary phosphorus compounds with electrophilic alkenes and alkynes[3], reactions of derivatives of phosphorus(III) acids with carbonyl compounds[4], polytertiary phosphines and their metal complexes[5], heterocyclic phosphorus compounds[6], cyclic phosphonium ylides[7], formylalkyl phosphonates[8] and optically active organophosphorus derivatives[9]. Several reviews have also appeared on various aspects of five-coordinate phosphorus compounds[10–12].

Theoretical chemists have considered five-coordinate phosphorus species. The electronic structures of both the hypothetical [13]PH_5 and the real [14]PF_5 have been calculated. The energy differences found between D_{3h}, C_{4v} and C_s structures of PF_5 support Berry's pseudorotation mechanism[14]. The modes of rearrangement of trigonal bipyramidal molecules such as phosphoranes have been determined[15].

Some novel luminescence properties of organophosphorus compounds have been observed. Lithium dialkylphosphides, $LiPR_2$, prepared from butyllithium and the corresponding secondary phosphine, R_2PH, are chemiluminescent when exposed to traces of oxygen, particularly when R is an alicyclic group[16]. The colours of the light emitted by the cyclopentyl, the cyclohexyl and the cycloheptyl derivative are yellow, yellow-green and blue-green, respectively[17]. Hexaphenylcarbodiphosphorane, $Ph_3P{=}C{=}PPh_3$, shows a bright yellow triboluminescence when 0.5 g samples are ground up.

5.2 REACTIONS FORMING THE CARBON–PHOSPHORUS BOND

Numerous methods for forming new carbon–phosphorus bonds were developed during the period under review. Some are discussed later when specific organophosphorus compounds are mentioned. In addition, some of the highlights of new reactions for forming carbon–phosphorus bonds from phosphorus–hydrogen bonded species, from tricoordinate phosphorus species or from derivatives of active metals (e.g. Na, Mg, Al) are given below.

5.2.1 Additions of phosphorus–hydrogen compounds

The kinetics of the addition of dimethylphosphine to tetrafluoroethylene in the gaseous phase to form $Me_2PCF_2CHF_2$ have been studied. The rate is given by equation (5.1).

$$\frac{d\,(\text{product})}{dt} = k[Me_2PH]^{0.5}[C_2F_4]^{1.6} \tag{5.1}$$

A homogenous free-radical chain mechanism initiated by bimolecular abstraction of hydrogen from dimethylphosphine was suggested[18]. A similar free-radical mechanism was also proposed for the reaction of tetramethyl-bisphosphine with tetrafluoroethylene to give $Me_2PCF_2CF_2PMe_2$[19].

New organophosphorus compounds have been prepared by the addition of phosphorus–hydrogen compounds to the carbonyl group in ketones. Reaction of diphenylphosphine with hexafluoroacetone gives the alcohol $Ph_2PC(CF_3)_2OH$, which can be oxidised to the corresponding phosphine oxide, and which rearranges in the presence of base to give $Ph_2P(O)OCH(CF_3)_2$; in the latter reaction two new phosphorus–oxygen bonds are formed at the expense of a single phosphorus–carbon bond[20]. Similar products are formed from $(PhO)_2PH(O)$ and hexafluoroacetone. Reactions of $PhPH(CH_2)_4NHR^3$, available from the $LiAlH_4$ reduction of $PhPH(CH_2)_3CN$, with aldehydes and ketones R^1COR^2 give the heterocyclic compounds (1; R^1 = H, Et or Ph; R^2 = H or Me; $R^1 + R^2 = (CH_2)_4$ or $(CH_2)_5)$[21]. Reactions of $(RO)_2PH$ with various ketones followed by treatment of the product with methyl iodide give alkyl α-hydroxyphosphinates[22].

New examples of additions of compounds containing phosphorus–hydrogen bonds to vinyl compounds have been reported. Reactions of dimethylphosphine with the vinylsilicon derivatives $(Me_2N)_3SiCH=CH_2$ and $(Me_2N)_2Si(CH=CH_2)_2$ in the presence of $LiPMe_2$ give $(Me_2N)_3Si$-$(CH_2)_2PMe_2$ and $(Me_2N)_2Si(CH_2CH_2PMe_2)_2$, respectively[23]. Addition of phenylphosphine to the double bonds in $CH_2=CHCH_2NHR$ (R = H, Et, Bu^n or Ph)[24] and $CH_2=CH(CH_2)_nCO_2R$[25] proceed normally to give intermediates which can then be cyclised to compounds with phosphorus in the ring. A new phosphorus–hydrogen compound used in addition reactions is $(Me_3SiO)_2PH$, prepared by reaction of ammonium hypophosphite with Me_3SiCl and diethylamine at $0-10\,°C$; addition to the vinylsilane $MeEt_2SiCH=CH_2$ gives $(Me_3SiO)_2P(CH_2)_2SiMeEt_2$[26].

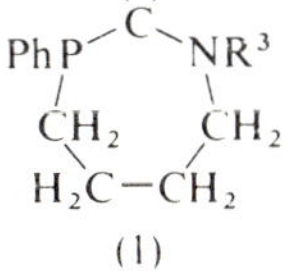

(1)

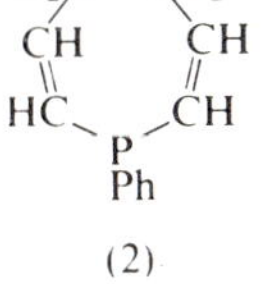

(2)

Additions of phosphorus–hydrogen compounds to alkynes have been studied. The nickel complex $(Ph_3P)_2Ni(CO)_2$ has been patented[27] as a catalyst for the addition of butylphosphine to pent-1-yne to give Bu^n-$PHCH=CHPr^n$. Free-radical addition of phenylphosphine to hexa-1,5-

diyne gives the phosphacycloheptadiene (2)[28]. Reaction of $(MeO)_2PH(O)$ with ethoxyacetylene in the presence of NaOMe gives the adduct $CH_2=C(OEt)P(O)(OMe)_2$ free from the isomer $EtOCH=CHP(O)(OMe)_2$; acid hydrolysis of the adduct regenerates $(MeO)_2PH(O)$ with liberation of ethyl acetate[29]. A similar addition of $(MeO)_2PH(O)$ to ethylthioacetylene proceeds in the opposite sense to give $EtSCH=CHP(O)(OMe)_2$[30].

Additions of phosphorus–hydrogen compounds to cyanides are also useful for forming carbon–phosphorus bonds. Two equivalents of $R_2PH(O)$ (R = Ph or Bu^n) react with $ClCH_2CN$ in the presence of a sodium alkoxide in a sealed tube at 100 °C to give the compound $R_2P(O)CH_2C(NH)P(O)R_2$[31]. Reaction of $Ph_2PH(O)$ with benzonitrile gives $Ph_2P(O)C(NH)Ph$. Addition of $R_2PH(O)$ (R = cyclohexyl[31], n-alkyl[32] or phenyl[32]) to malononitrile results in addition to one cyano group to give $R_2P(O)C(NH_2)=CHCN$[31]; a tautomeric form $R_2P(O)C(NH)CH_2CN$ also appears to exist.

Additions of phosphorus–hydrogen compounds to carbonyl groups are also useful. For example, addition of $(Bu^nO)_2PH$ to the aldehydes RCHO (R = Me, Et or Pr^n) followed by treatment with MeI gives, for example, $MeP(O)(OBu^n)CH(OH)Pr^n$[33]. Similarly, addition of $(RO)_2PH(O)$ to $AcCH_2CO_2Et$ in the presence of diethylamine gives $(RO)_2P(O)CMe(OH)CH_2CO_2Et$, which can be dehydrated to the vinylic derivative $(RO)_2P(O)CMe=CHCO_2Et$[34]. The reversibility of the addition of compounds having phosphorus–hydrogen bonds to carbonyl groups is shown by the thermolysis of $R^1CH(OH)P(O)(OR^2)_2$ in the presence of base to give $(R^2O)_2PH(O)$, with liberation of the aldehyde R^1CHO[35]. A greater tendency for phosphorus–hydrogen compounds to add to carbonyl then to cyano groups is indicated by the reaction of $(MeO)_2PH(O)$ with $PhCOCH_2CN$ to give $NCCH_2CPh(OH)P(O)(OMe)_2$ rather than $PhCOCH_2C(NH)P(O)(OMe)_2$[36].

Additions of phosphorus–hydrogen compounds to the carbon–carbon double bonds of nitroalkenes have been used. Reactions of ω-nitrostyrene with mono- and di-phenylphosphine give $PhP(CHPhCH_2NO_2)_2$ and $Ph_2PCHPhCH_2NO_2$, respectively[37]. (Additions of these phosphorus–hydrogen compounds to $PhCH=CHSO_2Ph$ proceed analogously). Similar reactions of $R^1CH=CHNO_2$ (R^1 = Me or Et) with $(R^2O)_2PH(O)$ in the presence of a sodium alkoxide give $(R^2O)_2P(O)CHR^1CH_2NO_2$, which can be used to synthesise aminoalkylphosphonic acids[38]. An adaptation of this reaction to nitroalkene–sugar compounds can be used to synthesise aminosugarphosphonates[39]. Addition of $(RO)_2PH(O)$ to a carbonyl group has also been used in carbohydrate chemistry[40].

5.2.2 Addition of tricoordinate phosphorus derivatives

Reactions of tricoordinate phosphorus derivatives with appropriate organic compounds can be used to form new carbon–phosphorus bonds, the coordination number of the phosphorus increasing to four or five during this process. Trialkyl phosphites are used most frequently in reactions of this type, although tris(dialkylamino)phosphines, $(R_2N)_3P$, and trialkylphosphines are also useful.

The formation of carbon–phosphorus bonds by reactions of trialkylphosphines with alkyl halides to give tetra-alkylphosphonium salts is, of course, well known. A similar reaction of tri-n-butylphosphine with the acyl halides RCOX (X = Cl or Br; R = Me or n-nonyl) at 0 °C in hexane gives the acylphosphonium salts $[Bu^n_3PCOR]^+ X^-$. The $v(CO)$ frequencies of these compounds are lower than those for aliphatic ketones[41]. Reactions of triphenylphosphine with the alkanesulphonyl chlorides, RSO_2Cl (R = Me or $PhCH_2$), in triethylamine give the corresponding phosphonium salts $[Ph_3PR^+]^-$ Cl; a sulphene intermediate is proposed for this reaction[42]. Conditions for the formation of 7-norbornenylphosphonium derivatives have been investigated[43]. Reactions of the salts $[R_2N=CH_2]^+ X^-$ with triphenylphosphine in dichloromethane give the phosphonium salts $[R_2NCH_2PPh_3]^+ X^-$ [44].

Tris(dimethylamino)phosphine and related compounds have also been used as carbon–phosphorus bond-forming reagents. Reaction of tris-(dimethylamino)phosphine with MeI in diethyl ether gives the phosphonium salt $[(Me_2N)_3PMe]^+ I^-$. However, reaction of $(MeO)_2PNMe_2$ with methyl iodide proceeds in an Arbusov-like manner to give $MeP(O)(OMe)(NMe_2)$[45]. Reaction of tris(dimethylamino)phosphine with benzaldehyde at 80 °C for 4 h in the presence of dimethylammonium chloride gives $(Me_2N)_2P(O)$-$CHPhNMe_2$, also obtained[46] from $(Me_2N)_2PH(O)$ and $PhCH(NMe_2)_2$ at 130–140 °C.

Trialkyl phosphites are familiar carbon–phosphorus bond forming reagents in their Michaelis–Arbusov reaction with alkyl halides. Various new reactions of trialkyl phosphites with numerous organic compounds forming new carbon–phosphorus bonds have been discovered during 1972 and 1973. Michaelis–Arbusov reactions of trialkyl phosphites with the cyclic fluoroalkenes (3; X = Cl) give the expected products (3; X = P(O)-$(OR)_2$)[47]. Ethylthioacetylene reacts with phosphites on being heated in the corresponding alcohol to give vinylic phosphorus compounds of the type $EtSCH=CHP(O)(OR)_2$[48]. Nitriles of the type $R^1_2CXC(CN)YZ$ (R^1 = Ph, X = CN or H, Y = Br, Z = CN or CO_2R) react with trialkyl phosphites to form products of the type $R^1_2CXCZ=CYN\overset{-}{\overset{+}{P}}(OR^2)_3$; a phosphorus–nitrogen rather than a phosphorus–carbon bond is formed in this reaction[49–52]. Dichloromethyl isocyanate reacts with triethyl phosphite to give a

$$(CF_2)_n \overset{X}{\underset{X}{\bigg\langle}}$$

(3) n = 2–4

mixture of $(EtO)_2P(O)CHClN=C=O$ (10–15% yield) and $ClCH=NCOP$-$(O)(OEt)_2$; the latter arises from the presence of the tautomer $Cl_2C=NCHO$[53]. Products of the type $[(EtO)_2P(O)]_nCCl_{3-n}(N=C=O)$ (n = 1, 2 or 3) arise from reactions of $Cl_3CN=C=O$ with triethyl phosphite in various molar ratios[54]. Trialkyl phosphites $(R^1O)_3P$ (R^1 = Me or Et) react with $(CF_3)_2C=NCH=CHR^2$ to give $(CF_3)_2CHNR^1CH=CR^2P(O)(OR^1)_2$[55]. Trimethyl phosphite reacts with $CCl_3CH=NCOPh$ to give $PhCONHC$-$(CCl_2)P(O)(OMe)_2$[55]. The phosphites $(RO)_3P$ (R = Me or Et) react with

$(CF_3)_2C=NN=C(CF_3)_2$ to give the novel heterocyclic compound $(4)^{56}$. Chloroacetonitrile reacts with triethyl phosphite in the normal Michaelis–Arbusov manner to form $(EtO)_2P(O)CH_2CN$; this is useful for cyanoalkene synthesis since it forms an alkene $RCH=CHCN$ ($R = Ph$, etc.) upon treatment with NaH followed by an aldehyde $RCHO^{57}$. Trialkyl phosphites $(R^1O)_3P$ react with orthoacetates $(R^2O)_3CMe$ in the presence of phosphorus trichloride and zinc chloride to give $(R^1O)_2P(O)CMe(OR^2)_2$, which lose an alcohol at 150–230 °C to form $(R^1O)_2P(O)C(OR^2)=CH_2{}^{58}$.

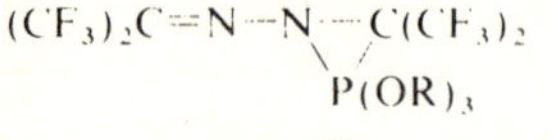

$$(CF_3)_2C=N-N-C(CF_3)_2 \qquad Me_2C-CO \qquad Me_2C-CO$$
$$\quad\quad\quad\backslash\quad/ \qquad\qquad\quad |\quad\quad| \qquad\qquad\quad |\quad\quad|$$
$$\quad\quad\quad P(OR)_3 \qquad\qquad\quad OC-CMe_2 \qquad\quad Me_2C=C-O$$
$$\qquad\quad (4) \qquad\qquad\qquad\qquad (5) \qquad\qquad\qquad\qquad (6)$$

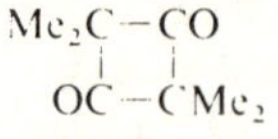

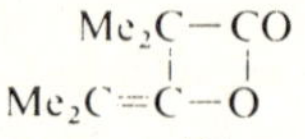

Reactions of the dimethylketene dimers (5) and (6) with trialkyl phosphites, $(RO)_3P$ ($R = Me$, Et, Pr^n, Bu^n, Pr^i or Bu^s) have received attention. The cyclobutanedione-type dimer (5) reacts to give a mixture of $RO_2CCMe_2CO-CMe_2P(OR)_2$ and $RO_2CCMe_2C(CMe_2)P(O)(OR)_2$; transient pentaco-ordinate phosphorus intermediates are proposed for this reaction^{59}. The latter product is also formed from the dimethylketene dimer (6) and the trialkyl phosphite. Treatment with tri-n-butylphosphine converts (5) into (6).

Reactions of alkyl halides or alkenes with phosphorus halides in the presence of aluminium halides can be used to form new carbon–phosphorus bonds. Reaction of Bu^tCl with phosphorus trichloride in the presence of aluminium chloride gives the salt $[Bu^tPCl_3]^+$ $[AlCl_4]^-$; the similar salt $[Bu^tPMeCl_2]^+[AlCl_4]^-$ is obtained analogously from $MePCl_2$, Bu^tCl and $AlCl_3{}^{60}$. Boron trichloride and tin(IV) chloride, but not the tetrachlorides of silicon or germanium, can be substituted for the aluminium chloride in this reaction. Reaction of the adduct $PhPCl_2 \cdot AlCl_3$ with Bu^tCl followed by treatment with an alcohol gives $Bu^tPhP(O)Cl^{61}$. Reaction of ethylene with phosphorus trichloride in the presence of aluminium chloride gives $ClCH_2CH_2PCl_2$; by-products form $(EtO)_2PH(O)$ with ethanol^{62}. The similar reaction of ethylene with phosphorus tribromide in the presence of aluminium bromide gives $BrCH_2CH_2PBr_2$, $(BrCH_2CH_2)_2PBr_3$, $AlBr_3$ and $P_2Br_4{}^{63}$. Reaction of $PSCl_3$ with ethylene in the presence of aluminium chloride followed by treatment with water or aqueous potassium chloride gives $ClCH_2CH_2P(S)Cl_2$; this can be converted into $CH_2=CHP(S)Cl_2$ and $CH_2=CHPCl_2{}^{64}$.

Carbon–phosphorus bonds can be formed by alkylation of elemental phosphorus or a species derived from it. Reaction of a mixture of phosphorus and iodine corresponding to P_4I_6 with benzyl chloride at 140–150 °C gives a mixture of PCl_3 and $(PhCH_2)_2PCl_2I_2$; the latter product, also obtained from iodination of $(PhCH_2)_3PCl_2$ in acetonitrile, is hydrolysed to give tribenzylphosphine oxide and presumably is a polyhalide with halogen–halogen bonds^{65}. Reaction of white phosphorus with formaldehyde in a mixture of water, MeOH and NaOH gives a complex mixture containing $HOCH_2P(O)(OH)_2$, $(HOCH_2)_2P(O)OH$, $HOCH_2PMe(O)OH$, $MeP(O)(OH)_2$ and $(HOCH_2)_3P(O)^{66}$. Products from the chlorination of this mixture with thionyl chloride include $ClCH_2P(O)Cl_2$, $(ClCH_2)_2P(O)Cl$ and

ClCH$_2$PMe(O)Cl; these can be hydrolysed to the corresponding P—OH compounds.

5.2.3 Reactions of active metal (Na, Mg, Al) derivatives of organophosphorus compounds

Several examples of the alkylation of organophosphorus compounds through alkali metal derivatives have been reported. Reaction of PhCH$_2$P(O)(OEt)$_2$ with NaH followed by addition of Me$_3$SiCH$_2$Cl gives the alkylated derivative (EtO)$_2$P(O)CHCH$_2$SiMe$_3$. However, a similar reaction of PhCH$_2$P(O)(OEt)$_2$ with potassium in 1,2-dimethoxyethane followed by addition of Me$_3$SiCH$_2$Cl results in elimination of *trans*-stilbene to give (EtO)$_2$P(OSiMe$_3$). Similarly, reaction of (RO)$_2$PH(O) or the sulphur analogue (R = Et or Pr^i but not Me because of too low solubility) with sodium in toluene followed by addition of benzyl chloride gives PhCH$_2$P(S)(OR)$_2$ or PhCH$_2$P(O)(OR)$_2$[67].

New reactions forming carbon–phosphorus bonds by using sodium bis-(2-methoxyethoxy)dihydridoaluminate, NaAlH$_2$(OCH$_2$CH$_2$OMe)$_2$ (Vitride), have been developed[68]. Reaction of PhMeP(O)OMe with Vitride in THF followed by addition of appropriate alkyl halides gives the phosphine oxides PhMeRP(O) (R = PhCH$_2$, Et or Pr^i). Similar reaction of PhCH$_2$P(O)(OEt)$_2$ with Vitride followed by addition of n-butyl halides results in dialkylation of the phosphorus to give the phosphine oxides PhCH$_2$P(O)Bu$^n_2$. Triethyl phosphate can be trialkylated with benzyl chloride by the Vitride method to give tribenzylphosphine oxide. Bromoalkyl derivatives of the type PhP(O)(OEt)(CH$_2$)$_n$Br (n = 4 or 5) can be cyclised with Vitride to give the cyclic phosphine oxides (7) (n = 4, 11%; n = 5, 28% yield). Cyclisation of o-(RO)$_2$P(O)CH$_2$C$_6$H$_4$CH$_2$Br with Vitride gives the tricyclic compound (8) (i.e. a dimerisation followed by phosphoryl elimination), but an analogous cyclisation of o-PhP(O)(OEt)CH$_2$C$_6$H$_4$CH$_2$Br gives the bicyclic compound (9)[69].

Recent examples of the use of magnesium in synthetic organophosphorus chemistry include the cyclisation of (EtO)$_2$P(O)(CH$_2$)$_2$CHBr(CH$_2$)$_2$Br with magnesium in THF to give the 1-phosphanorbornane derivative (10), and the syntheses of the trimethylsilylphosphines (Me$_3$Si)$_n$P(Bu^t)$_{3-n}$ (n = 1, 2 or 3) by reactions of the corresponding halides (Bu^t)$_{3-n}$PCl$_n$ with trimethylchlorosilane in the presence of magnesium with HMPT as solvent[71].

The cyclotetraphosphazane N$_4$P$_4$F$_8$ has been methylated step-wise with methyl-lithium[72]. A monomethyl derivative, all five possible dimethyl derivatives, only geminal trimethyl derivatives, a single tetramethyl derivative and the octamethyl derivative were isolated under various conditions.

5.3 C-FUNCTIONAL ORGANOPHOSPHORUS COMPOUNDS

5.3.1 Hydroxy, alkoxy and epoxy derivatives

The hydroxyalkyl derivative $Ph_2P(O)CMe_2OH$ has been prepared either by reaction of $Ph_2PH(O)$ with MeMgI in THF followed by treatment of the presumed $Ph_2POMgBr$ with acetone, or by reaction of Ph_2POEt with benzyl chlorocarbonate followed by treatment of the resulting $Ph_2P(O)CO_2CH_2Ph$ with NaI in acetone[73]. The conditions of the first preparative method are somewhat critical since if $Ph_2PH(O)$ is treated with *phenyl*magnesium bromide in THF followed by acetone, the phorone bis-adduct $[Ph_2P(O)CMe_2-CH_2]_2CO$ is formed.

Proton and ^{31}P n.m.r. spectroscopy have been used to study the following equilibrium:

$$[(HOCH_2)_4P]^+Cl + OH^- \rightleftharpoons (HOCH_2)_3P + CH_2O + H_2O + Cl^-$$

The n.m.r. signals of the phosphonium salt and those of the phosphine coalesce reversibly at higher temperatures[74]. Oxidation of $PhP(CH_2OH)_2$ with hydrogen peroxide in acetone at $-15\,^\circ C$ gives the corresponding oxide $PhP(O)(CH_2OH)_2$; it and the corresponding sulphide react with arene-sulphonyl chlorides in the normal manner to give the bis-arenesulphonates, $PhP(O)(CH_2OSO_2Ar)_2$ (Ar = p-tolyl, p-bromphenyl or p-nitrophenyl)[75]. Reaction of tris(hydroxymethyl)phosphine with cyanamide in the presence of formaldehyde and phosphoric acid at room temperature gives the novel

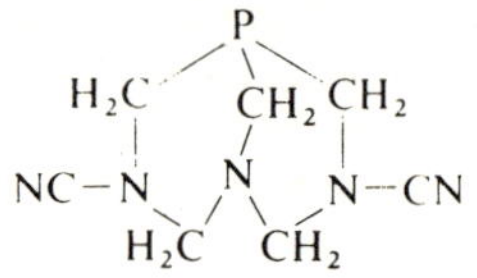

$$\begin{array}{c}
P \\
H_2C \quad CH_2 \; CH_2 \\
NC-N \qquad N \qquad N-CN \\
H_2C \quad CH_2
\end{array}$$

(11)

bicyclic derivative (11), which can be converted by routine methods into the corresponding oxide and methiodide[76].

Some reactions of $[(HOCH_2CH_2)_3\overset{+}{P}CH_2OH]\,Cl^-$, which contains both a hydroxymethyl and a 2-hydroxyethyl group, have been investigated[77]. Chlorination of this compound in aqueous solution at pH 5–7 results in elimination of the hydroxymethyl group to give $(HOCH_2CH_2)_3P(O)$, whereas aqueous chlorination at a lower pH (1–3) results in elimination of a 2-hydroxyethyl group to give $(HOCH_2CH_2)_2P(O)CH_2OH$; both hydroxy-alkylphosphine oxides can be converted into the corresponding chloroalkyl-phosphine oxides by standard methods.

The 2-hydroxyethyl derivative $(HOCH_2CH_2)_3P(O)$ is a useful starting material for the preparation of other 2-substituted ethylphosphorus derivatives[78]. Thus it reacts with phosphorus pentachloride, acetic anhydride or benzoyl chloride to give the corresponding 2-chloroethyl, 2-acetoxyethyl and 2-benzoyloxyethyl derivative, respectively. Reaction of $(ClCH_2CH_2)_3P(O)$ with a tertiary amine is useful for preparing trivinylphosphine oxide; nucleophilic substitution reactions of $(ClCH_2CH_2)_3P(O)$ with alcohols, thiols and secondary amines can also be carried out.

Several alkoxyalkylphosphorus derivatives have been prepared. Reactions of $(RO)_2P(O)CH_2OH$ (R = Et, Pr^n, Pr^i, Bu^n, Bu^t, n-amyl or isoamyl) with triethyl orthoformate at 130 °C give the compounds $(RO)_2P(O)CH_2OCH(OEt)_2$[79]. Additions of $(R^1O)_2P(O)CHR^2OH$ (R^1 = Me, Et or Pr^i; R^2 = H, Me or Et) to acrylonitrile or methyl acrylate derivatives in the presence of ethanolic NaOEt give $(R^1O)_2P(O)CHR^2OCH_2CHR^3R^4$ (R^3 = H or Me, R^4 = CN or CO_2Me)[80]. Reactions of $(R^1O)_2PCl$ with $MeCH(OR^2)Cl$ at 50–60 °C give $R^1OP(O)ClCHMeOR^2$ (R^1 = Et, R^2 = Et or Bu^n; R^1 = Bu^n or n-amyl, R^2 = Bu^n)[81]. Similarly, reactions of $ROPCl_2$ with EtOCHMeOAc give $ROP(O)ClCHMeOEt$ (R = Et or Pr^n)[82]. Nucleophilic substitution in $(ClCH_2)_3P$ with sodium alkoxides (alkyl, C_4 to C_{10}) give the bis(alkoxymethyl) derivatives $MeP(O)(CH_2OR)_2$[83]. Degradation of these compounds with KOH gives $MeP(O)(OH)CH_2OR$.

Reactions of the 3-hydroxypropyl derivative $PhPH(CH_2)_3OH$ with various aldehydes and ketones R^1COR^2 [R^1 = H, R^2 = Et, Pr^n or Ph; R^1 = Me, R^2 = Et or Bu^i; R^1 + R^2 = $(CH_2)_4$ or $(CH_2)_5$] give the heterocyclic compounds (12)[84].

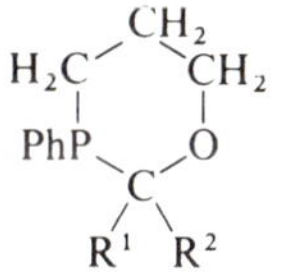

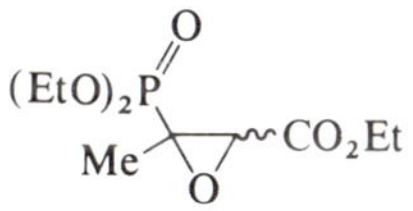

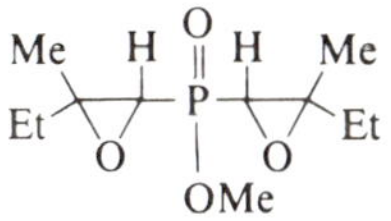

The epoxyalkyl derivatives (13) (both the *cis* and the *trans* isomer) have been prepared by reaction of $(EtO)_2PH(O)$ with $AcCHClCO_2Et$ in the presence of triethylamine[85], whereas the epoxyalkyl derivative (14) has been prepared from butanone and $(ClCH_2)_2P(O)(OMe)$ in the presence of $KOBu^t$ in Bu^tOH below 15 °C; the latter reaction also occurs similarly with cyclopropyl methyl ketone and cyclohexanone[86].

5.3.2 Ketones and aldehydes

Phosphorus-containing aldehydes remain rather rare; however, stable *trans*-enols (15) of $(EtO)_2P(O)CHXCHO$ have been reported[87].

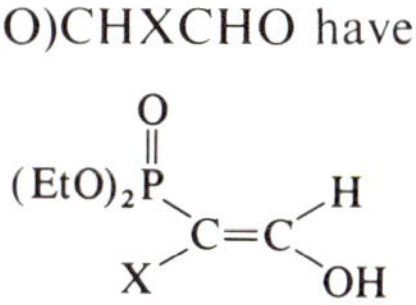

(15) X = Cl, Br

Considerable progress has been made with the preparation of phosphorus-containing ketones with the carbonyl group directly bonded to the phosphorus atom (i.e. acylphosphines or acylphosphonates). Both tricoordinate and tetracoordinate phosphorus derivatives of this type have been studied. Reaction of heptafluorobutanoyl chloride with di-n-butylphosphine in the presence of triethylamine in diethyl ether gives $C_3F_7COPBu^n_2$ (20%) and $C_3F_7C(OH)(PBu^n_2)_2$ (22%); the latter compound arises from addition of a second mole of di-n-butylphosphine to the former[88]. The dibutylphosphine can be replaced in this reaction by other secondary phosphines, R_2PH (R = Et, Pr^n or Pr^i). Reaction of di-n-butylphosphine with bis(trifluoromethyl)ketene, $(CF_3)_2C=C=O$, in dichloromethane below 30 °C gives $(CF_3)_2CHCOPBu^n_2$ (35%) and $(CF_3)_2C=C(PBu^n_2)_2$ (32%); the latter compound arises from the product of addition of a second mole of di-n-butylphosphine to the former. The phosphites $(RO)_2PH(O)$ react similarly to the above secondary phosphines with C_3F_7COCl and $(CF_3)_2C=C=O$[89,90].

Non-fluorinated acylphosphines and acylphosphonates can also be prepared. Reaction of $(RO)_2PCl$ with $(EtO)_2CHOAc$ at 50 °C gives the acetylphosphonate $(RO)_2P(O)Ac$, in addition to $(RO)_2P(O)CH(OEt)_2$[91]. Similarly, reactions of acetyl chloride with $(MeO)_2PH(O)$ and with $Me_2PPr^n(O)$ give $MeOPMeOAc$ and $Me_2P(O)Ac$, respectively[90]. Reactions of the silylphosphine Me_3SiPPh_2 with acid chlorides in diethyl ether at or below -78 °C apparently provide a reasonably general method for making acylphosphines of the type $RCOPPh_2$; acetyl chloride undergoes this reaction as well as the acid chlorides of the dibasic squaric, oxalic and phthalic acid[92]. An analogous reaction of Me_3SiPPh_2 with phosgene in diethyl ether at -100 °C gives yellow carbonyl bis(diphenylphosphide), $(Ph_2P)_2C=O$, which readily undergoes decarbonylation to tetraphenylbiphosphine[93].

Phosphorus-containing ketones with one methylene group separating the phosphorus atom from the carbonyl group have received some attention. An example of the preparation of such a compound is the metallation of methyldiphenylphosphine oxide followed by addition of benzonitrile to give, after hydrolysis, $PhCOCH_2P(O)Ph_2$[94]. New reactions of these compounds include the stepwise double metallation of $(MeO)_2P(O)CH_2Ac$ in THF, first with NaH and then with BuLi to give a 1,3 di-anion which after mono-alkylation and acidification gives the compounds $(MeO)_2P(O)CH_2COCH_2R$ (R = Me, Bu^n, Pr^i, allyl or $PhCH_2$); this sequence can be repeated to give dialkylated derivatives of the type $(MeO)_2P(O)CH_2COCHR^1R^2$[95]. Sodium iodide reacts with $AcCH_2P(O)(OMe)_2$ to give the anion $AcCH_2P(O)(OMe)O^-$; thermal decomposition of this anion gives acetone[96].

Compounds with carbonyl groups more remote from the phosphorus have also been studied. The compound $Ac(CH_2)_2P(O)(OMe)_2$ can be obtained from trimethyl phosphite and $Ac(CH_2)_2Cl$[97]. The compounds $AcCH_2CMe_2PR_2(O)$ (R = Ph or Et) are obtained from $AcCH_2CMe_2OH$ and the halides R_2PCl.

5.3.3 Carboxylic acids and their esters, amides and halides

The carboxylic acids $PhPHCHR^1CO_2H$ react with ketones R^2COR^3 to give

$$
\begin{array}{c}
R^1HC-CO \\
\diagup \quad \diagdown \\
PhP \qquad O \\
\diagdown \quad \diagup \\
C \\
\diagup \diagdown \\
R^2 \quad R^3
\end{array}
$$

(16)

the heterocyclic compounds (16)[98]. The acid $HO_2CCH_2P(O)(OH)_2$ can be esterified with ethanolic HCl; the ester is best isolated as the mono-potassium salt $K^+[EtO_2CCH_2PO_2H]^-$ [99]. This salt can be converted into $EtO_2CCH_2P(O)Cl_2$ and $EtO_2CCH_2P(O)(NR_2)_2$ (R = Me or Et) without destruction of the ester group.

Several other methods have been used to prepare organophosphorus esters with one methylene group separating the ester group from the phosphorus atom. Metallations of $R^1{}_2PH(O)$ with sodium in liquid ammonia or potassium in toluene followed by addition of the chloroacetates $ClCH_2CO_2R^2$ give 40–80% yields of $R^1{}_2P(O)CH_2CO_2R^2$ (R^1 = Me, Et, Pr^n or Bu^n; R^2 = H, Et or Bu^n); reactions of $R^1{}_2POBu^n$ with the bromoacetates $BrCH_2CO_2R^2$ can also be used[100]. These esters can be saponified with KOH in aqueous ethanol. The trivalent-phosphorus carboxylic ester $PhP(CH_2CO_2Et)_2$ can be prepared from $PhPCl_2$ and $Et_3SnCH_2CO_2Et$ in benzene; this ester can be converted into the corresponding oxide and sulphide by normal methods[101]. Reactions of Et_2PSEt with the chloroacetates $ClCH_2CO_2R$ give the phosphine sulphide carboxylic esters $Et_2P(S)CH_2$-CO_2R (R = Me, Et, Pr^n, Pr^i or n-amyl) as well as the bis(carboxyalkyl)-phosphonium salts $[Et_2P(CH_2CO_2R)_2]^+Cl^-$, EtCl and $Et_2P(S)SEt$. A similar reaction of the chloroacetates with $EtP(SEt)_2$ gives $EtP(S)(SEt)$-CH_2CO_2R [102].

The hydrogen atoms of a methylene group flanked by an alkoxycarbonyl group and a tetracoordinate phosphorus atom are relatively acidic and thus can be replaced by metal atoms to give useful synthetic reagents. For example, a second ethoxycarbonyl group can be introduced into $Ph_2P(O)CH_2CO_2Et$ by metallation with potassium followed by addition of ethyl chlorocarbonate, giving $Ph_2P(O)CH(CO_2Et)_2$ [103]. Similarly, metallation of $(EtO)_2P(O)CH_2$-CO_2Et with lithium in benzene, calcium in liquid ammonia or an isopropyl-magnesium halide in diethyl ether gives an intermediate which reacts with acyl chlorides to give $(EtO)_2P(O)CH(COR)CO_2Et$ (R = Me or Ph) containing both a carbonyl and an ethoxycarbonyl group, and with ketones R^1COR^2 to give $(EtO)_2PH(O)$ and an alkene, $R^1R^2C{=}CHCO_2Et$ [104]. A formyl group can be introduced into $Bu^n{}_2P(O)CH_2CO_2Et$ by metallation with sodium in diethyl ether followed by addition of ethyl formate to give $Bu^n{}_2P(O)CH(CHO)CO_2Et$ [105]. A benzyl group can be introduced into $(EtO)_2P(O)CH_2CO_2Et$ by condensation with benzaldehyde to give $PhCH{=}C(CO_2Et)P(O)(OEt)_2$; reduction with $NaBH_4$ gives $PhCH_2CH$-$(CO_2Et)P(O)(OEt)_2$ [106].

Carboxylic esters with two methylene groups between the ester group and the phosphorus atom have also been investigated. Esters of the type $(EtO)_{2-n}PEt_n(O)(CH_2)_2CO_2Et$ (n = 0, 1 or 2) can be prepared by various addition reactions to the carbon–carbon double bond of acrylic acid or its chloride, followed by esterification with ethanol (in some cases liberated from a reaction involving an ethoxyphosphorus compound[107,108]). In a more unusual reaction, apparently involving an oxyphosphorane intermediate, i.e. (17), treatment of $MeCHClP(O)(OMe)CH_2CO_2Me$ with methanolic sodium methoxide gives $(MeO)_2P(O)CHMeCH_2CO_2Me$ [109].

$$MeHC\!-\!CHCO_2Me$$
$$\diagdown \diagup$$
$$P$$
$$\diagup \diagdown$$
$$O \quad OMe$$

(17)

A reaction generally useful for the preparation of phosphorus-containing carboxylic acid derivatives with two methylene groups between the carbonyl group and the phosphorus atom is the addition of methyl hypophosphite, $MeOPH_2(O)$, to appropriate acrylic acid derivatives $CH_2{=}CHX$ (X = CO_2Et, $CONH_2$, $CONMe_2$ or CN) in the presence of NaOMe to give $MeOP(O)(CH_2CH_2X)_2$. The methyl hypophosphite necessary for this reaction can be generated from hypophosphorus acid and trimethyl orthoformate[110].

Several other reactions are useful for preparing phosphorus-containing carboxamides. Reactions of white phosphorus with the acrylamides $CH_2{=}CHCCNR^1R^2$ ($R^1 = R^2$ = Me or Et; R^1 = H, R^2 = Me or Et; $R^1 + R^2 = (CH_2)_n$, n = 4 or 5) in the presence of ethanolic KOH gives the derivatives $[R^1R^2NCO(CH_2)_2]_3P(O)$ [111]. The hygroscopic dimethylamides $[Me_2NCO(CH_2)_2]_3P(O)$ can also be prepared by reaction of the carboxylic acid $(HO_2CCH_2CH_2)_3P(O)$ with P_2O_5 and DMF[112]. Alkaline hydrolysis of the di-cation $PhP[CH_2C(\overset{+}{N}Me_2)NMe_2]_2$ gives the amide $PhP[CH_2CONMe_2]_2$, which can be oxidised to the corresponding oxide with hydrogen peroxide. Oxidation of $P[CH_2C(\overset{+}{N}Me_2)NMe_2]_3$ with hydrogen peroxide followed by alkaline hydrolysis gives $P(O)(CH_2CONMe_2)_3$. Acidic hydrolysis of these amides gives the corresponding carboxylic acids[113].

Hydroxamic acids of the type $R^1{}_2P(O)CH_2CONHOH$ (R^1 = Ph, p-ClC_6H_4, p-tolyl, Et, Pr^n or Bu^n) can react to give $R^1{}_2P(O)CH_2CONHOR^2$ (R^2 = $PhCH_2$, CO_2Me, CO_2Et or acyl) by metallation with potassium followed by addition of the corresponding halide[114].

Reactions of PCl_5 with the esters $MeCO_2R$ (R = Me or Et) give the acid chloride $Cl_2P(O)CCl_2COCl$; however, if a deficiency of PCl_5 is used in this reaction, two geometric isomers of $Cl_2P(O)CH{=}CCl(OEt)$ are obtained[115,116].

5.3.4   Organophosphorus derivatives with carbon–sulphur bonds

Compounds with $RS\!-\!CH_2P$ linkages are obtained by reactions of chloromethylphosphorus derivatives with sulphur-containing nucleophiles. Thus, reactions of $R^1R^2NP(O)(CH_2Cl)_2$ with sodium thiolates $NaSR^3$ (R^3 = Ph, p-tolyl or $PhCH_2$) at 60 °C give $R^1R^2NP(O)(CH_2SR^3)_2$ (R^1 = H or Et, R^2 = Et or Ph)[117]. Reactions of thiourea with appropriate chloromethylphosphonic acid derivatives give $H_2NC(NH)SCH_2PR(O)OH$, which upon treatment first with NaOH and then with HCl give $HSCH_2PR(O)OH$ (R = Ph, OH or Et)[118].

Some compounds containing both P—H and S—H linkages have been studied[119]. Reactions of PhPHNa (from phenylphosphine and sodium) with $Cl(CH_2)_3SH$ and with cyclohexene sulphide give $PhPH(CH_2)_3SH$ and the cyclohexane derivative (18), respectively. These compounds have been metallated with butyl-lithium or sodium; such metallations occur mainly at the S—H bond rather than the P—H bond. However, phenyl isocyanate and phenyl isothiocyanate both react at the P—H bond in preference to the S—H bond.

$$PhMe\overset{+}{S}CH_2P(O)(OEt)\bar{O}$$

(18) (19)

A phosphorus-containing sulphur ylide has been studied[120]. Reaction of $PhSCH_2P(O)(OEt)_2$ with MeI followed by addition of perchlorate gives the salt $[PhMe\overset{+}{S}CH_2P(O)(OEt)_2][ClO_4]^-$, which upon deprotonation with NaH gives the unstable oily ylide $PhMeS=CHP(O)(OEt)_2$. This ylide reacts with benzaldehyde to give alkenic sulphonium salts, $[PhCH=CHSMePh]^+ X^-$, with phenyl isocyanate to give $PhMeS=C[P(O)(OEt)_2]CONHPh$, and, after addition of H_2O, undergoes elimination of ethanol on silica to give the dipolar ion (19).

5.3.5 Aminoalkylphosphorus derivatives

Aminoalkylphosphorus derivatives with tricoordinate phosphorus are of potential interest as ligands in coordination chemistry. Such compounds with either one or two methylene groups separating the phosphorus and nitrogen atoms are known. Reaction of a mixture of a P—H compound, an N—H compound and formaldehyde can be used to synthesise a PCH_2N unit. For example, treatment of diphenylphosphine with $MeNH(CH_2)_2NMe_2$ and formaldehyde in benzene solution at 60 °C gives $Ph_2PCH_2NMe(CH_2)_2$-NMe_2; analogous reactions have been done with other ethylenediamine derivatives containing 2–4 N—H bonds[121]. Reactions of the acetates $R^1_nP(CH_2OAc)_{3-n}$ (n = 0, 1 or 2; R^1 = Et or Bu^n) with secondary amines R^2_2NH (R^2 = Me, Et, Pr^n or Bu^n) in methanolic KOH give good yields of the corresponding amines, $R^1_nP(CH_2NR^2_2)_{3-n}$[122]. Reaction of $[(HOCH_2)_4P]^+ Cl^-$ with aniline gives $[(PhNHCH_2)_4P]^+ Cl^-$, which reacts with ammonia to give the phosphine $(PhNHCH_2)_3P$, also obtained directly from $(HOCH_2)_3P$ and aniline[123]. Reactions of $[ClCH=NMe_2]^+ Cl^-$ with alkali metal dialkylphosphides MPR_2 (R = Me, Et, Bu^n or Ph) give $Me_2NCH(PR_2)_2$, also obtained by reaction of two moles of the secondary phosphine R_2PH with $HC(OMe)_2NMe_2$. Decomposition of Me_2NCH-$(OMe)PEt_2$ gives $Me_2NCH(PEt_2)_2$ with elimination of $HC(OMe)_2NMe_2$[124].

Aminoalkylphosphonic acids are phosphorus analogues of amino acids, which, of course, are of considerable biological importance. The synthesis of various aminoalkylphosphonic acids and their esters has been very actively investigated during 1972–73. Space does not allow presentation in full of all of the diverse methods which have been studied; however, the following methods are of particular interest.

(a) The synthesis of $H_2NCH_2P(O)(OH)_2$ through elimination of $CH_2=CMe_2$ from $Me_3CNHCH_2P(O)(OH)_2$ by hydrolysis[125] with hydrobromic acid at 175 °C.

(b) The synthesis (and resolution into optical isomers) of the hydrochloride of $PhCH(NH_2)P(O)(OEt)_2$ by reaction of $(EtO)_2PH(O)$ with $(PhCH=N)_2CHPh$ in the presence of triethylamine followed by acid hydrolysis[126].

(c) The synthesis of $R^1R^2C(NH_2)P(O)(OH)_2$ and $H_2N(CH_2)_2P(O)(OH)_2$ by successive reactions of the corresponding esters $R^1R^2C(CO_2Et)$-$P(O)(OEt)_2$ and $EtO_2C(CH_2)_2P(O)(OEt)_2$ with hydrazine and nitrous acid[127,128].

(d) The conversion of amides of the type $RCH(CONH_2)P(O)(OEt)_2$ into $RCH(NH_2)P(O)(OH)_2$ by reaction with alkaline sodium hypobromite followed by acid hydrolysis[129].

(e) The preparation of compounds of the type $PhCH_2NHCR^1R^2P(O)$-$(OH)_2$ by reaction of phosphorous acid with benzylamine and aldehydes or ketones of the type R^1COR^2 ($R^1 = R^2 = H$ or Me; $R^1 = Et$, $R^2 = H$ or Me) in the presence of HCl[130].

(f) Methods based on the addition of $(EtO)_2PH(O)$ to $Ar^1CH=NAr^2$ [131] or $Ar^1CH=NN=CHAr^2$ [132].

(g) Additional reaction sequences where the amino group of the aminoalkylphosphonic acid arises from catalytic hydrogenolysis or hydrogenation of a benzylamino, $PhCH_2NH$[133,134], a substituted benzylamino, $PhCHMeNH$[135], a cyano[128], an oximino[136] or a nitro[138] group.

These syntheses of aminoalkylphosphonic acids usually involve sequences of relatively well-known reactions in organic and organophosphorus chemistry.

Aminoalkylpolyphosphonic acids with two or three phosphonic acid residues have also been studied. An extensive series of geminal diphosphonic acids $R^1R^2NCR^3[P(O)(OH)_2]_2$ ($R^1 = H$ or C_{1-15} alkyl, $R^2 = H$ or C_{1-18} alkyl, $R^3 = H$, C_{1-18} alkyl, cyclohexyl or benzyl) have been prepared either by reactions of phosphorous acid with $[R^1R^2N=CR^3X]^+$ Cl^- or by reactions of PCl_3 with the amides $R^3CONR^1R^2$ in the presence of a limited amount of water[137]. Reaction of $Bu^tN=CH_2$ with $(EtO)_2PH(O)$ and formaldehyde gives $Bu^tN[CH_2P(O)(OEt)_2]_2$, which undergoes hydrolysis to give $HN[CH_2P(O)(OH)_2]_2$[138]. The triphosphonic acid with one methylene group between the nitrogen atom and each phosphorus atom, $N[CH_2P(O)(OH)_2]_3$, has been prepared by reaction of nitrilotriacetic acid, $N(CH_2CO_2H)_3$, with a mixture of phosphorous acid and phosphorus trichloride in tetrachloroethane[139]. This triphosphonic acid can be esterified quantitatively with triethyl orthoformate, but upon attempted conversion into the corresponding acid chloride with phosphorus pentachloride only mediocre yields of $N(CH_2Cl)_3$ and tars were obtained. The triphosphonic acid with two methylene groups between the nitrogen atom and each phosphorus atom, $N[CH_2CH_2P(O)(OH)_2]_3$, has been prepared by reaction of

$(ClCH_2CH_2)_3N$ with potassium phosphite in toluene followed by acidification[140].

5.3.6 Organophosphorus derivatives with other nitrogen-containing functional groups

Several organophosphorus derivatives containing C-cyano groups have been studied. The nitriles RCH_2CN react with PCl_3 in the presence of oxygen at -15 to $0\,°C$ to give $Cl_2P(O)CHRCN$ [141]. The related compound $(EtO)_2P(O)CH_2CN$ reacts with $Ph_2C{=}NH$ to give a mixture of $(EtO)_2P(O)CH(CN)CPh_2NH_2$ (25%) and $(EtO)_2P(O)C(CN){=}CPh_2$ (7%)[142], with aryl azides, in the presence of ethanolic NaOEt followed by hydrolysis, to give the triazoles (20) and with nitrile oxides, $RC{\equiv}\overset{+}{N}{-}O^-$, in the presence of NaOEt to give (21)[143].

$$H_2N-C{=}C-P(O)(OEt)_2$$

(20)

$$H_2N-C{=}C-P(O)(OEt)_2$$

(21)

2-Cyanoethylphosphorus compounds have also been studied. Reaction of $BrCH_2CHBrCN$ with $(EtO)_2PONa$ in diethyl ether gives a mixture of $(EtO)_2PH(O)$, $O[P(O)(OEt)_2]_2$ and $(EtO)_2P(O)(CH_2)_2CN$ [144]. 2-Cyanoethylphosphorus derivatives of the type $(EtS)_2P(Z)(CH_2)_2CN$ ($Z = O$ or S) can be obtained by reactions of a mixture of EtPClSEt and acrylonitrile with acetic acid or thiolacetic acid, respectively[145]. The 2-cyanoethylphosphonates $(RO)_2P(O)(CH_2)_2CN$ ($R = Pr^i$ or Bu^i) react with alcoholic HCl at $-10\,°C$ to give $(R^1O)_2P(O)(CH_2)_2C(OR^2){=}NH \cdot HCl$, which can be dehydrochlorinated[146] with triethylamine in dioxan at $-15\,°C$.

The isocyanomethylphosphorus derivative $(EtO)_2P(O)CH_2NC$ has been prepared by dehydration of the corresponding formamide with a mixture of $POCl_3$ and triethylamine in dichloromethane, a standard method of isocyanide formation. The required formamide, $(EtO)_2P(O)CH_2NHCHO$, is obtained by reaction of triethyl phosphite with $[Me_3\overset{+}{N}CH_2NHCHO]Br^-$ in nitromethane at $120\,°C$. The isocyanide $(EtO)_2P(O)CH_2NC$ can be metallated with butyl-lithium at $-78\,°C$ to give an anion which reacts with carbonyl compounds R^1COR^2 to give the vinyl isocyanide derivatives $R^1R^2C{=}CHNC$. Reactions of $(EtO)_2P(O)CH_2NC$ with aldehydes $RCHO$ in the presence of ethanolic NaCN give the heterocyclic compounds (22)[147]. Reaction of Et_3GePEt_2 with phenyl isocyanide gives the novel imine $PhN{=}C(PEt_2)(GeEt_3)$[148].

$$RHC-CHP(O)(OEt)_2$$

(22)

The nitrosation of phosphonium salts has been studied; the resulting nitrosoalkylphosphorus compounds are useful intermediates in nitrile syntheses. Successive reactions of $[Ph_3PCH_2CO_2Et]^+$ Cl^- with hydrogen chloride, isopropyl nitrite and sodium ethoxide give $Ph_3P=C(NO)(CO_2Et)$, which is degraded by silver nitrate to give silver cyanide and triphenylphosphine oxide. A similar nitrosation of $[Ph_3PCH_2Ac]^+$ Cl^- gives acetyl cyanide directly. Degradation of the similarly prepared C-nitroso derivative $[Ph_3PCH(NO)Ph]^+$ Cl^- with sodium carbonate in aqueous DMF gives benzonitrile and triphenylphosphine oxide[149].

The isocyanato derivative $Cl_2P(O)CCl_2N=C=O$ can be prepared from $CCl_2=NCOCl$ and $MeOPCl_2$ in the presence of iron(III) chloride; stepwise reaction of this compound with alcohols gives first $Cl_2P(O)CCl=NCO_2R$ and then $(RO)_2P(O)CCl=NCO_2R$ [150].

Diazoalkylphosphorus derivatives have been studied. The compounds $(EtO)_2P(O)C(N_2)CONHR$, containing both diazo and amide groups, have been prepared by reactions of azides with the esters $(EtO)_2P(O)CH_2CO_2Et$[143]. The compounds $(R^1O)_2P(O)CMe(OH)C(N_2)CO_2R^2$ (R^1 and R^2 = Me or Et) have been prepared by reactions of $(R^1O)_2P(O)Ac$ with diazoacetic esters, $N_2CHCO_2R^2$, at room temperature[151]. The diazomethylphosphorus derivative $(MeO)_2P(O)CH=N_2$ adds at its C—H bond to the carbon–carbon double bond of the enamine $R_2NCH=CMe_2$ at room temperature to give $(MeO)_2P(O)C(N_2)CH(NR_2)CHMe_2$, which forms a salt with perchloric acid[152].

The stable phosphorus free radicals (23) (R = Ph, PhCH=CH, vinyl or PhCCl=CH) have been prepared by conventional methods[153].

$$\left(\begin{array}{c} Me_2C-CH_2 \\ \dot{O}-N \qquad CHO \\ Me_2C-CH_2 \end{array} \right)_2 PR(O)$$

(23)

5.3.7 Halogenoalkylphosphorus derivatives

The chloromethyl derivative $ClCH_2PCl_2$ has been prepared by reaction of a mixture of phosphorus trichloride, aluminium chloride and dichloromethane at 80 °C for 4 days followed by reduction of the product with metallic antimony in the presence of diethyl phthalate. A similar reaction of $MePCl_2$ gives $ClCH_2PMeCl$. These compounds react with sulphur, selenium and o-chloranil in the normal manner for trivalent phosphorus compounds; however, catechol cleaves the chloromethyl group[154]. Substituted 1-chloroalkyl derivatives can be obtained by exposure of a mixture of phosphorus trichloride and an appropriate alkene to gamma rays[155].

The chloromethylphosphorus derivative $(ClCH_2)_2PCl$ reacts with acrylic acid to give, after treatment of the product with ethanol, the ester $(ClCH_2)_2P(O)(CCH_2)_2CO_2Et$[156]. Metallation of $R^1_2P(O)CH_2Cl$ with butyllithium in THF at −80 °C gives the lithium derivative $R^1_2P(O)CHLiCl$, which reacts with aldehydes R^2CHO to give $R^1_2P(O)CHClCH(OH)R^2$ after hydrolysis, or the epoxide (24) after being heated[157]. Dehydrobromination of the substituted 1-bromoalkyl derivative $(PhCHBr)_2P(O)Ph$ with 1,5-diazabicyclo[4,3,0]non-5-ene in benzene gives the oxyphosphirene (25), which gives diphenylacetylene on being heated to 120 °C at 10^{-5} mmHg and a 3 : 1 mixture of *cis*- and *trans*-stilbene upon treatment with sodium hydroxide[158].

$$(O)PR^1_2HC\!-\!CHR^2$$

(24)

(25)

Reactions of the dichloromethyl derivative $Cl_2CHP(O)Cl_2$ with phenols in the presence of pyridine give $Cl_2CHP(O)(OR)_2$ ($R = Ph$, p-MeC_6H_4, p-$MeOC_6H_4$, o- or p-ClC_6H_4 or p-$O_2NC_6H_4$) which show some evidence of aphicidal activity[159].

Reactions of the trichloromethyl compound $(CCl_3)_2PCl{=}NH$ have been studied. Chlorination of this compound in pyridine gives the N-chloro derivative $(CCl_3)_2PCl{=}NCl$ which is hydrolysed to $(CCl_3)_2P(O)NH_2$; the latter is converted into $(CCl_3)_2P(O)N{=}PCl_3$ with PCl_5. Direct treatment of $(CCl_3)_2PCl{=}NH$ with PCl_5 gives $(CCl_3)_2PCl{=}NPCl_4$, which reacts with SO_2 to give $(CCl_3)_2PCl{=}NPOCl_2$, also obtainable from $(CCl_3)_2PCl{=}NH$ and $POCl_3$ [160]. Reaction of $(CCl_3)_2PCl{=}NH$ with oxalyl chloride gives $(CCl_3)_2PCl{=}NCOCOCl$, which is decarbonylated at 120 °C to give $(CCl_3)_2PCl{=}NCOCl$, shown by i.r. spectroscopy to be in equilibrium with the isocyanate $(CCl_3)_2PCl_2N{=}C{=}O$ [161]. A mixture of $CCl_3P(O)(OEt)_2$ and triethylamine in DMF has been used for the phosphorylation of nucleosides[162].

Numerous syntheses of 2-chloroethylphosphorus derivatives have been described[163,164], as well as their dehydrochlorinations to the corresponding vinylphosphorus derivatives and their use for the synthesis of a non-inflammable polymer. Hydrolysis of $ClCH_2CH_2P(O)(OH)_2$ at 80 °C in the pH range 9–12 results mainly in the elimination of ethylene with the production of less than 5% of the 2-hydroxyethyl derivative $HO(CH_2)_2P(O)(OH)_2$ [165].

Reactions of the compound $(ClCH_2CH_2)_2P(O)CH_2Cl$, containing both chloromethyl and 2-chloroethyl groups, with nucleophiles under mild conditions results in the replacement of only the 2-chloroethyl chlorine atoms to give $(XCH_2CH_2)_2P(O)CH_2Cl$ (e.g. $X = OR$ or SR). However, under more vigorous conditions all the chlorine atoms are replaced to give products of the type $(XCH_2CH_2)_2P(O)CH_2X$. Reaction of $(ClCH_2CH_2)_2P(O)CH_2Cl$ with triethylamine results in dehydrochlorination of the 2-chloroethyl groups without destruction of the chloromethyl group to give the divinylphosphorus derivative $(CH_2{=}CH)_2P(O)CH_2Cl$ [166].

Reaction of $BrCH_2CH_2PBr_2$ with triethylamine in cold diethyl ether gives the relatively unstable $CH_2{=}CHPBr_2$. Reaction of $BrCH_2CH_2PBr_2$ with ethylene oxide gives $BrCH_2CH_2P(OCH_2CH_2Br)_2$, which on being heated rearranges to $(BrCH_2CH_2)_2P(O)OCH_2CH_2Br$. The conversion of

$BrCH_2CH_2PBr_2$ into $BrCH_2CH_2P(OH)_2$ and $BrCH_2CH_2P(OEt)_2$ is also described[167].

The 3-bromopropylphosphonium salt $[Br(CH_2)_3PPh_3]^+ Br^-$ cyclises with a stoichiometric amount of aqueous NaOH to give triphenylcyclopropylphosphonium bromide, which reacts with an excess of NaOH by elimination of a phenyl group to give cyclopropyldiphenylphosphine oxide[168].

Reaction of hexachlorobutadiene with trimethyl phosphite at 150° C for 24 h gives a mixture of the 1-pentachlorobutadienyl derivative $Cl_2C{=}CCl{-}CCl{=}CClP(O)(OMe)_2$ (31%) and the 2-pentachlorobutadienyl derivative $Cl_2C{=}CClC[P(O)(OMe)_2]{=}CCl_2$ (69%)[169].

5.3.8 Perfluoroalkylphosphorus derivatives

The chemistry of perfluoroalkylphosphorus derivatives, particularly trifluoromethylphosphorus derivatives, has been studied extensively during the past 15 years and continues to receive attention. The use of the relatively expensive CF_3I is by-passed by the generation of a $CF_3{-}P$ bond through the reactions of the phosphorus halides R_nPCl_{3-n} (R = Et or Ph, n = 1 or 2) with trifluoroacetic acid, followed by decomposition of the intermediate $R_nP(O_2CCF_3)_{3-n}$ on storage to give $[RP(O)(CF_3)]_2O$ or $R_2P(O)OPR_2$-$(CF_3)_2$[170]. The deuterides $(CF_3)_2PD$ and CF_3PD_2 have been prepared by treatment of $(CF_3)_2PI$ and CF_3PI_2, respectively, with DI in the presence of mercury[171]. Oxidative addition of $ClN(CF_3)_2$ to $(CF_3)_nPX_{3-n}$ (n = 1, X = F or Cl; n = 2, X = F, Cl or I) has been investigated[172]. Reaction of $(CF_3)_2PH$ with dimethylzinc gives the cyclic compound $(CF_3PCF_2)_2$ (26),

$$CF_3{-}P{-}CF_2$$
$$F_2C{-}P{-}CF_3$$

(26)

with elimination of methane and methylzinc fluoride. Bromination of (26) gives $CF_3P(CF_2Br)_2Br_2$, $CF_3P(CF_2Br)Br_3$ or CF_3PBr_4, depending on the molar ratio of Br_2 to (26)[173]. Fluorination of $(CF_3)_3P$ with sulphur tetrafluoride gives $(CF_3)_3PF_2$, which is useful as a source of difluorocarbene, itself used to prepare cyclopropane derivatives by reactions with fluorinated alkenes such as $CF_2{=}CFX$ (X = F or Cl) or, in lower yield, $CH_2{=}CF_2$[174]. The pentavalent P—H compounds $(CF_3)_3PH_2$, CF_3PHF_3 and $(CF_3)_2PHF_2$ have been prepared by reactions of the corresponding fluorides $(CF_3)_nPF_{5-n}$ (n = 3, 1 or 2, respectively) with trimethylsilane. The compounds $(CF_3)_3PH_2$ and CF_3PHF_3 were shown to be stereochemically non-rigid by n.m.r. spectroscopy and attain the limiting structures at -75 and -140 °C, respectively[175]. Reaction of $(CF_3)_2POP(CF_3)_2$ with F_2POPF_2 for 3 days at room temperature gives the 'mixed' derivative $(CF_3)_2POPF_2$; however, at higher temperatures $(CF_3)_2PF$ is formed. The very unstable sulphur analogue $(CF_3)_2PSPF_2$ could be prepared but not the methylimino analogue[176].

Some perfluoroalkyl phosphonitrilic derivatives have been prepared[177,178]. Chlorination of $(CF_3)_2PNH_2$ in pentane followed by addition of pyridine gives $[(CF_3)_2PN]_n$ ($n = 3$ or 4); the heptafluoropropyl derivative $[(C_3F_7)_2PN]_3$ can be similarly prepared. These compounds can also be prepared by reaction of sodium azide in acetone with the compound $(R_F)_2PCl$, or by reaction of ammonia with the $(R_F)_2PCl_3$ at $-40\,°C$ followed by heating the product to $200–250\,°C$.

Compounds with trifluoromethylphosophorus groups bonded to a variety of heteroatoms have been prepared. Reaction of $(CF_3S)_2PCF_3$ with LiB_5H_8 at $-45\,°C$ gives $2\text{-}(CF_3SPCF_3)B_5H_8$; this isomerises[179] to the 1-isomer after 1 day at $25\,°C$. Treatment of $(CF_3)_2PCl$ with $Me_3SiSSiMe_3$ gives $(CF_3)_2PSSiMe_3$, which is cleaved by HBr to give $(CF_3)_2PSH$ and Me_3SiBr, and which reacts with more $(CF_3)_2PCl$ to give $(CF_3)_2PSP(CF_3)_2$[180]. The novel tertiary–primary biphosphine $(CF_3)_2PPH_2$ has been prepared by reactions of R_3EPH_2 ($E = Si$, Ge or Sn) or $LiAl(PH_2)_4$ with $(CF_3)_2PX$ ($X = Cl$, Br or I). Reagents which cleave the P—P bond in $(CF_3)_2PPH_2$ include HBr, Br_2, Me_3SnH and $Mn_2(CO)_{10}$. Related bis(trifluoromethyl)-phosphorus derivatives include $(CF_3)_2PPF_2$ and $(CF_3)_2PAs(CF_3)_2$[182]. Reaction of Me_3SiPH_2 with $(CF_3)_2PH$ gives $Me_3SiP(CF_3)_2$; the germanium and tin analogues can be similarly prepared[183]. Reaction of divinyl-mercury (or, less desirably, dimethylmercury) with $(CF_3)_2PH$ gives pyrophoric $Hg[P(CF_3)_2]_2$ (60%). This mercurial gives $(CF_3)_2PP(CF_3)_2$ and mercury on being heated, the related mercurial $Hg[As(CF_3)_2]_2$ upon treatment with $(CF_3)_2AsH$, and $(CF_3)_2PH$ and HgI_2 upon reaction with hydrogen iodide[184].

Cyclopolyphosphines with C_2F_5 and C_3F_7 groups have been prepared by conventional methods. The pentafluoroethyl compound $(C_2F_5P)_5$ gives $(C_2F_5P)_4$ irreversibly at room temperature, but the heptafluoropropyl compound $(n\text{-}C_3F_7P)_5$ is stable under these conditions. The relative stabilities of cyclopolyphosphine rings thus depends significantly upon the nature of the substituents[185].

Perfluoro-2-methylpropene, $(CF_3)_2C{=}CF_2$, reacts with trivalent phosphorus derivatives of the type X^1X^2PCl (X^1 and X^2 are various combinations of Et, EtO, Me_2N and Et_2N) to give the derivatives $(CF_3)_2C{=}CFPFClX^1X^2$, which can be converted into other $(CF_3)_2C{=}CFP$ derivatives by conventional reactions[186].

5.3.9 Phosphorus compounds with tertiary alkyl substituents

Bulky substituents, such as t-butyl, can give organophosphorus derivatives special properties, although these effects seem to have received somewhat less attention during 1972–73 than in 1970–71. Details have been published for preparing the key compounds Bu^tPCl_2 and Bu^t_2PCl by reactions of t-butyl-magnesium chloride with PCl_3 in various molar ratios[187]. Reaction of Bu^t_2PH with Bu^t_2Hg at $80\,°C$ gives the yellow-green air-stable $Hg(PBu^t_2)_2$, which loses mercury upon u.v. irradiation to give $(Bu^t_2P)_2$[188].

The temperature dependence of the proton n.m.r. spectra of various t-butylphosphorus compounds has been investigated. At very low temperatures ($-120\,°C$) the normally single t-butyl resonance separates into two resonances of $2:1$ relative intensity, indicating restricted rotation about the t-butyl carbon–phosphorus bond at such temperatures. From the temperature dependence of the n.m.r. spectra the parameters $\Delta H^{\ddagger}$, $\Delta S^{\ddagger}$ and $\Delta G^{\ddagger}$ were determined for $Bu^t PCl_2$, $Bu^t_2 PCl$, $Bu^t_3 P$ and $Bu^t_3 P(:S)$[189].

5.3.10 Organophosphorus derivatives containing carbon bonded to elements such as silicon, germanium, tin and boron

Most of the work in this area during 1972–73 has been concerned with compounds containing both carbon–phosphorus and carbon–silicon bonds. Preparations of the compounds $R_2 P(E)SiMe_3$ (R = F or CF_3, E = O or S) have been described. Compounds of this type are unstable with respect to $R_2 PESiMe_3$ formed by an anti-Arbuzov rearrangement[190]. Reaction of the silacyclobutane $Me_2 Si(CH_2)_3$ with a mixture of PCl_3 and $AlCl_3$ followed by addition of $POCl_3$ gives $Me_2 SiCl(CH_2)_3 PCl_2$. Analogous methods can be used to prepare $MeSiCl_2(CH_2)_3 PCl_2$, $Cl_3 Si(CH_2)_3 PCl_2$ and $Me_2 SiCl(CH_2)_4 PCl_2$ [191]. Reaction of $(EtO)_2 PONa$ with $R_3 SiCH_2 Cl$ (R = Et, Pr^n or Bu^n) in boiling benzene gives $R_3 SiCH_2 P(O)(OEt)_2$, which is useful[192] for protecting steel from corrosion in 4 N HCl at 25, 60 and 100 °C. Reactions of $Me_3 SiCH_2 P(O)Cl_2$ with ethanol, ethylene oxide, sodium cyanate, potassium thiocyanate and ethylenimine have been described; these are all normal[193].

Reactions of phosphines of the type $R_3 EPEt_2$ (E = Si, Ge or Sn) have been described. Treatment of $R_3 EPEt_2$ (E = Si or Ge) with glyoxal, $(CHO)_2$, gives $R_3 EOCH(CHO)PEt_2$, which readily undergoes decarbonylation to $R_3 EOCH_2 PEt_2$; the latter can also be prepared from $R_3 EPEt_2$ and formaldehyde[194]. Reactions of $R^1_3 EPEt_2$ with the ketenes $R^2_2 C=C=O$ give $R^1_3 EOC(CR^2_2)PEt_2$, which undergo hydrolysis to $R^2_2 CHCOPEt_2$[195]. Reactions of $R_3 EPEt_2$ with β-propiolactone result in ring opening of the lactone to give $R_3 EOCOCH_2 CH_2 PEt_2$, as well as a polymer of propiolactone. Reaction of $R_2 SiHPEt_2$ with γ-butyrolactone gives (27)[196].

$$
\begin{array}{cc}
\text{(27)} & \text{(28)}
\end{array}
$$

Reactions of $Li[Me_2 P(BH_3)_2]$ with the methylammonium chlorides $[Me_n NH_{4-n}]^+ Cl^-$ give the adducts $(Me_n NH_{3-n})BH_2 PMe_2 BH_3$, which form (28) upon hydrolysis[197].

5.4 UNSATURATED ORGANOPHOSPHORUS COMPOUNDS

5.4.1 Vinylic derivatives

Several new vinylphosphorus compounds have been synthesised. Dehydrobromination of $(BrCH_2 CH_2)_2 P(O)CH_2 CH_2 Br$ with ethanolic KOH at

60 °C gives $(CH_2=CH)_2P(O)OCH_2CH_2Br$; the 2-bromoethoxy group can be removed by reaction with PCl_5 in benzene to give $(CH_2=CH)_2POCl$, also obtainable from $(BrCH_2CH_2)_2POCl$ and triethylamine[198]. Acetone reacts with R_2PCl (R = Et or Ph) to give $R_2P(O)CMe_2Cl$, which undergoes dehydrochlorination with aqueous NaOH to give $R_2P(O)CMe=CH_2$; similar products are also obtainable from the related ketones RCOMe (R = Et, Bu^n or Ph)[199]. Reaction of $(PhO)_2PCl$ with the ketones $R^1COCH_2R^2$ (R^1 = Me or Ph; R^2 = H, Me, Et or Pr^n) at 100–160 °C gives $(PhO)_2P(O)CR^1=CHR^2$ [200]. Reaction of three equivalents of phosphorus pentachloride with t-butyl alcohol in benzene followed by addition of SO_2 gives $Cl_2P(O)CH_2CMe_2Cl$, which undergoes dehydrochlorination with triethylamine to give $Cl_2P(O)CH=CMe_2$; the carbinol Me_2EtCOH reacts similarly[201]. Vinylphosphonates of the type $R^1R^2C=CHP(O)(OEt)_2$ can be prepared by reaction of Me_3SiCH_2Cl with triethyl phosphite to give $Me_3SiCH_2P(O)(OEt)_2$ followed by metallation with n-butyl-lithium and addition of the compounds R^1COR^2 (R^1 = Ph, R^2 = Ph or H; R^1 = R^2 = Me, etc.)[202].

Some additions of diazo compounds to vinylphosphorus derivatives have been investigated. Reactions of the phosphonium salt $[Ph_3PCH=CH_2]^+$ Br$^-$ with the diazomethyl ketones $RCOCHN_2$ (R = Ph, p-MeOC$_6$H$_4$ or p-tolyl) give the pyrazole derivatives (29), with elimination of the bisphosphonium salt $[Ph_3PCH_2]_2^+$ 2Br$^-$ [203]. The substituted vinylphosphonates $(RO)_2P(O)CMe=CH_2$ (R = Me or Et) react with diazomethane in diethyl ether to give the dihydropyrazole derivatives (30), which undergo rearrangement with sodium alkoxides to give (31)[204]. Reaction of $(MeO)_2P(O)CH=CH_2$ with 9-diazofluorene in benzene gives (32) through an intermediate Δ^1-pyrazoline[205].

Addition reactions of $[Ph_3PCH=CH_2]^+$ Br$^-$ are useful for heterocyclic syntheses, since reactions of $PhCOCR=NOH$ or $PhCOCPh=NNHR$ with

NaH in DMF followed by addition of triphenylvinylphosphonium bromide give the pyrroles (33; R = Ph, $EtCO_2$ or H) or the dihydropyrazines (34; R = Ph, $EtCO_2$ or Me), respectively.

Reactions of PCl_5 with appropriate organic compounds gives chlorovinyl-phosphorus derivatives. Thus treatment of t-butylacetylene with PCl_5 in benzene followed by addition of sulphur dioxide gives $Z\text{-}Me_3CCCl{=}CHPOCl_2$. Similarly, ethoxyacetylene and PCl_5 give $EtOCCl{=}CHPOCl_2$. However, the allylic phosphorus derivative $MeCH{=}CClCH_2POCl_2$ rather than a vinylic phosphorus derivative arises from a similar reaction sequence starting with but-1-yne[207]. Reactions of the methyl ketones $RCOMe$ (R = Me or Ph) with PCl_5 in $POCl_3$ followed by SO_2 give $RCCl{=}CHPOCl_2$; the reaction of acetone produces much $MeCCl{=}CHCl$ as a by-product[208].

Alcoholysis of dialkyl 2,3-epoxypropylphosphonates in the presence of a basic catalyst gives $(RO)_2P(O)CH{=}CHCH_2OH$ rather than an allylic phosphorus derivative[209]. Reaction of triethyl phosphite with the halides $XCH_2CH(OR)_2$ (R = Me, Et or nonyl; $2R = CH_2CH_2$; X = Cl or Br) give $(EtO)_2P(O)CH_2CH(OR)_2$, which undergo alcohol elimination on thermolysis to give $(EtO)_2P(O)CH{=}CHOR$ [210].

Aminovinylphosphorus compounds of the type $R^1R^2P(O)CH{=}CHNR^3_2$ can be obtained by reactions of the aldehyde $R^1R^2P(O)CH_2CHO$ with R^3_2NH ($R^1 = R^2 = $ Eto or Pr^iO; $R^3 = $ Et)[211].

Addition of Bu^t_2PH to the alkynes $HC{\equiv}CX$ (X = CN or CO_2Me) gives $Bu^t_2PCH{=}CHX$ (X = CN, *cis*-isomer; X = CO_2Me, *trans*-isomer)[212]. 1-Cyanovinylphosphonates of the type $R^1R^2C{=}C(CN)P(O)(OEt)_2$ add hydroxide to give (35), an intermediate in the Horner reaction since it is readily degraded to $R^1R^2C{=}CHCN$ [213]. Reactions of $(R^1O)_2P(O)CPh{=}C\text{-}(CN)CO_2R^2$ (R^2 = Me or Et) with the amines HNR^3R^4 ($R^3 = R^4 = $ H or Me; $R^3 = $ H, $R^4 = $ Me) give $R^3R^4NCPh{=}C(CN)CO_2R^2$ with elimination of the phosphoryl group. However, the phosphorus is retained in the reactions of $(R^1O)_2P(O)CPh{=}C(CN)CO_2R^2$ with cyanide ion and with the bromomalononitrile anion to give $(R^1O)_2P(O)CPh(CN)CH(CN)CO_2R$ and the cyclopropane derivative (36; X = CN or CO_2Me), respectively[214].

$$R^1R^2C(OH)C(CN)P(O)(OEt)_2$$

(35)

$$(NC)_2C{-}CXCN$$
$$\diagdown C \diagup$$
$$\diagup \quad \diagdown$$
$$Ph \quad P(O)(OR^1)_2$$

(36)

5.4.2 Allylic derivatives

Reactions of the trivalent phosphorus derivatives R^1R^2PX (X = Br or Cl; $R^1 = $ Et, Ph, $ClCH_2$, Pr^n or EtS; $R^2 = $ Cl, Br or EtS) with allyl thiol, $CH_2{=}CHCH_2SH$, form an allyl–phosphorus bond, giving $R^1R^2P(S)CH_2\text{-}CH{=}CH_2$, with elimination of HX. Similar reactions with allyl alcohol have been investigated[215]. Reaction of $CH_2{=}CHCH_2P(OBu^n)_2$ with $MeCH(OEt)_2$ at 125–130 °C gives $CH_2{=}CHCH_2P(O)(OBu^n)CH_2CH(OEt)_2$, which can be hydrolysed to the corresponding aldehyde with dilute HCl[216]. Compounds of the type $(CH_2{=}CHCH_2)_2P(O)X$ ($X = OCH_2CHMeNH_2 \cdot HCl$, $O(CH_2)_{11}Me$ or $OCH_2C{\equiv}CH$), prepared by routine methods, have some biological activity as depressants[217]. Reactions of compounds of the type $R^1{}_2PCH_2CH{=}CHR^2$ with the lithium amides $LiNR^3{}_2$ ($LiNR^3{}_2 = LiNEt_2$ or $LiNH(CH_2)_2NH_2$) give in successive steps $R^1{}_2PCH{=}CHCH_2R^2$ and $R^1{}_2PCH_2CH(NR^3{}_2)CH_2R^2$[218].

5.4.3 Alkynyl derivatives

Reaction of $MeC{\equiv}CLi$ with $BrPF_2$ gives the unstable $MeC{\equiv}CPF_2$, which forms a BH_3 adduct and which is fluorinated to give $MeC{\equiv}CPF_4$ upon reaction with BF_3 or $SbF_3{-}SbCl_5$[219].

Novel phosphorus-containing heterocyclic compounds have been prepared from alkynylphosphines. Reaction of $Ph_2PC{\equiv}CBu^t$ (from Ph_2PCl and $Bu^tC{\equiv}CLi$ by routine methods) with HCl in acetic acid gives the 1,4-diphosphoniacyclohexadiene (37; $R^1 = Ph$, $R^2 = Ph$, $R^3 = Bu^t$); related compounds can be similarly prepared. The 1,4-diphosphoniacyclohexadiene

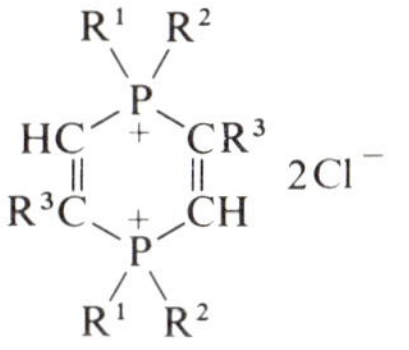

(37) 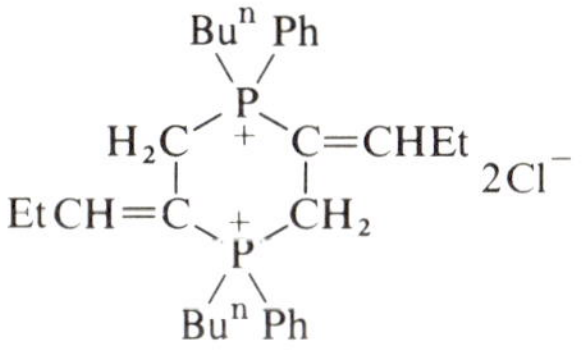 (38) 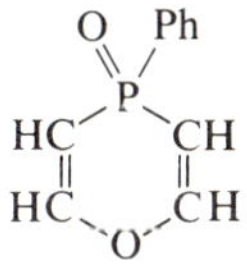 (39) 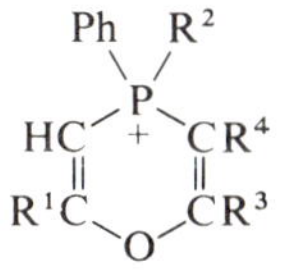 (40)

(37; $R^1 = Ph$, $R^2 = Bu^n$, $R^3 = Pr^n$) undergoes a rearrangement (endocyclic to exocyclic double bonds) upon reaction with acetic acid to form (38)[220]. Reactions of $(HC{\equiv}C)_2PPh(O)$ with primary amines RNH_2 ($R = Me$, Bu^t or Ph) give $RNHCH{=}CHPPh(O)C{\equiv}CH$; the derivative from methylamine ($R = Me$) is cyclised on alumina to give (39)[221]. The cyclic phosphonium salts (40) are obtained by reactions of $PhPR^2C{\equiv}CR^1$ with $R^3COCHBrR^4$ in triethylamine or HMPT[222].

Additions of 1,3-dipolar reagents to some alkynylphosphorus derivatives have been investigated. Phenyl azide reacts normally with the tetracoordinate phosphorus derivatives $R_2P(O)C{\equiv}CMe$ ($R = EtO$, Pr^nO, Et, Ph or Cl) to give the 1,2,3-triazoles (41)[223]. However, C,N-diarylnitrilimines, $R^1C{\equiv}NNR^2$, react abnormally with the tricoordinate phosphorus derivatives $R^3{}_2PC{\equiv}CR^4$ ($R^3{=}Ph$, $R^4{=}H$, Me, Ph or Bu^t; $R^1{=}Bu^nO$, $R^2{=}H$) with involvement of the phosphorus atom to give (42)[224,225].

$$\underset{\underset{\displaystyle (41)}{}}{\underset{\displaystyle PhN\diagdown_{N}\diagup N}{MeC=CP(O)R_2}}$$

$$(42)$$

Other reactions of alkynylphosphorus derivatives give vinylic phosphorus compounds. Reactions of $(R^1O)_2P(O)C\equiv CR^2$ with amines HNR^3R^4 give the enamine phosphonates $(R^1O)_2P(O)CH=CR^2NR^3R^4$ (R^1, R^2, R^3, R^4 are various alkyl or aryl groups); acid hydrolysis of $(EtO)_2P(O)CH=CR^1$-NHR^2 gives the ketones $(EtO)_2P(O)CH_2COR^1$ [226,227]. Additions of nucleophiles (e.g. Me_2NH, MeS^-, MeO^-), electrophiles (HCl, H_3O^+) or free radicals (e.g. $EtS\cdot$) to $R_2P(O)C\equiv CSMe$ ($R = Me$ or Et) give the corresponding substituted vinylphosphorus derivatives $R_2P(O)CH=C(SMe)X$ ($X =$ a group derived from adding reagent)[228]. The organo-copper reagents R_2CuI ($R = Me$ or Ph) add to the carbon–carbon triple bond in $Ph_2P(E)$-$C\equiv CR^2$ ($E = O$ or S, $R^2 = Me$ or Ph) to give an intermediate formulated as $Ph_2P(ECu)C=CR^1R^2$; this intermediate reacts with benzaldehyde to give, after hydrolysis with aqueous ammonium chloride, $Ph_2P(E)C(=$-$CR^1R^2)CH(OH)Ph$, which appears to exist in a hydrogen-bonded form (43)[229]. Reactions of $Ph_2PC\equiv CCF_3$ with the tetrahalogenometallates(II)

$$(43)$$

of palladium or platinum in ethanolic solution give the corresponding metal(II) complexes of the phosphine $Ph_2PCH_2C(CF_3)=CHPPh_2$, characterised by x-ray crystallography[230].

5.4.4 Allenyl and dienyl derivatives

Reactions of Ph_2PCl with the alcohols $R^1R^2C(OH)C\equiv CR^3$ ($R^1 = R^2 =$ alkyl or aryl; R^3 can be bromine) in the presence of a base such as pyridine or N-methylmorpholine give the allenylphosphorus derivatives $Ph_2P(O)CR^3=C=CR^1R^2$. The eclipsing tendency of the allene substituents has been investigated by n.m.r. spectroscopy. Compounds of the type $Ph_2P(O)CBr=C=CR^1R^2$ readily undergo a partial rearrangement to $Ph_2P(O)C\equiv CCBrR^1R^2$ [231,232]. Compounds of the type $Ph_2P(O)CR^1=C$-$=CMe_2$ ($R^1 = H$ or Me) readily add R^2OH, Ph_2PO^-, R^2MgX or R^2Li to the carbon–carbon double bond next to the phosphinyl group. The cyclohexylidene derivative $Ph_2P(O)CR=C=C(CH_2)_5$ can readily be converted into the cyclohexylidenecyclobutene derivative (44)[233].

$$\underset{(44)}{\overset{\displaystyle O \atop \displaystyle \|}{Ph_2P}} \!\!-\!\! \begin{array}{c} C_6H_{10} \\ \\ C_6H_{11} \end{array}$$

Reaction of but-1-en-3-yne with phosphorus pentachloride followed by addition of sulphur dioxide gives $Cl_2P(O)CH_2CCl{=}CHCH_2Cl$, which undergoes dehydrochlorination with triethylamine to give the 2-chlorobuta-1,3-dienylphosphoryl chloride, $CH_2{=}CHCCl{=}CHPOCl_2$. Similar reaction sequences can be carried out with $HC{\equiv}CCH{=}CHMe$ and $HC{\equiv}CC{-}Me{=}CH_2$ [234]. Reaction of isoprene with PCl_5 followed by addition of H_2S gives $ClCH_2CH{=}CMeCH_2P(S)Cl_2$, which is dehydrochlorinated by triethylamine to give $CH_2{=}CHCMe{=}CHP(S)Cl_2$ [235].

Reaction of potassium cyclopentadienide with $PBrF_2$ gives cyclopentadienyldifluorophosphine, $C_5H_5PF_2$, which is shown by n.m.r. spectroscopy to be a fluxional molecule [236].

5.5 COMPOUNDS WITH TWO OR MORE PHOSPHORUS ATOMS

5.5.1 Tricoordinate phosphorus derivatives

Compounds containing two or more tricoordinate phosphorus atoms are of potential interest as chelating or bridging ligands in coordination chemistry. For this reason details of the preparation of $H_2PCH_2CH_2PH_2$ [by $LiAlH_4$ reduction of $(EtO)_2P(O)CH_2CH_2P(O)(OEt)_2)$] and of the methylphosphorus derivatives Me_2PPMe_2 and $Me_2P(CH_2)_3PMe_2$ are given in *Inorganic Syntheses* [237,238]. Reaction of $(R_2N)_2PCl$ with $BrMgC{\equiv}CMgBr$ gives $(R_2N)_2PC{\equiv}CP(NR_2)_2$, which can be converted into the corresponding sulphide and oxide and which undergoes cleavage of the phosphorus–nitrogen bonds with hydrogen chloride to give the very unstable $Cl_2PC{\equiv}CPCl_2$ [239].

Numerous polyphosphines containing various combinations of primary, secondary and tertiary phosphorus atoms have been prepared generally by using vinylphosphonates as preparative reagents. For example, base-catalysed addition of $PhPH_2$ to two moles of $CH_2{=}CHP(O)(OEt)_2$ followed by $LiAlH_4$ reduction gives the diprimary tertiary phosphine $PhP(CH_2CH_2PH_2)_2$. Analogously prepared are $Ph_2PCH_2CH_2PHPh$, $P(CH_2CH_2PH_2)_3$ and $H_2P(CH_2)_2PPh(CH_2)_2PPh(CH_2)_2PH_2$. Reaction of $BrCH_2CH_2P(O)(OPr^i)_2$ with $PhP(OPr^i)_2$ followed by $LiAlH_4$ reduction gives $PhPHCH_2CH_2PH_2$ [240].

Polytertiary phosphines with methyl groups on the phosphorus atoms can be prepared by base-catalysed addition of phosphorus–hydrogen compounds to methylvinylphosphine sulphides followed by desulphurisation of the intermediates with $LiAlH_4$ in boiling dioxan. This method avoids handling obnoxious compounds such as $Me_2PCH{=}CH_2$ and $MeP(CH{=}CH_2)_2$.

As an example of this method, $Ph_2PCH_2CH_2PMe_2$ can be prepared by addition of Ph_2PH to $Me_2P(S)CH=CH_2$ to give $Me_2P(S)(CH_2)_2PPh_2$, which is desulphurised with $LiAlH_4$. Other potential ligands prepared by this method are the tritertiary phosphines $RP(CH_2CH_2PMe_2)_2$ (R = Ph or Me) and the tripod tetratertiary phosphine $P(CH_2CH_2PMe_2)_3$ [241].

5.5.2 Tetracoordinate phosphorus derivatives

Compounds containing two or more tetracoordinate phosphorus atoms have received considerable attention during 1972–73, and diverse preparative methods for these compounds have been studied. Reaction of 2-pyridyl-methyl-lithium with Ph_2POCl gives the pyridine derivative (45)[242]. Reaction of $HOCH_2P(O)(OEt)_2$ with benzenesulphonyl chloride followed by treatment with KOH at -10 to $0\,^\circ C$ and addition of $NaOP(OEt)_2$ gives $CH_2[P(O)-(OEt)_2]_2$ [243]. Lithiation of $R^1R^2MeP(O)$ with butyl-lithium in hexane followed by coupling of the lithium derivative $R^1R^2P(O)CH_2Li$ with a copper salt (e.g. $CuCl_2$) gives $R^1R^2P(O)CH_2CH_2P(O)R^1R^2$ [244]. Reaction of $Ph_2P(O)CH_2CH_2Cl$ with $Bu^n_2PH(O)$ in the presence of butyl-lithium gives $Ph_2P(O)CH_2CH_2P(O)Bu^n_2$ [245]. Reaction of $(EtO)_2PH(O)$ with a mixture

$$\text{(45)} \qquad \text{(46)}$$

of nickel(II) bromide and tri-n-butylphosphine at $100\,^\circ C$ followed by introduction of acetylene and heating the mixture to $170–180\,^\circ C$ gives $(EtO)_2P(O)CH_2CH_2P(O)(OEt)_2$ [246]. Reaction of P_4O_6 with chloroacetic acid gives $ClCH_2C(OH)[P(O)(OH)_2]_2$, which is hydrolysed by aqueous NaOH to the tetra-anion $[O_3PCH_2COPO_3]^{4-}$ [247]. Reaction of $(EtO)_2POSiMe_3$ with $AcP(O)(OEt)_2$ is exothermic and gives $[(EtO)_2P(O)]_2$-$CMeOSiMe_3$ [248]. Reactions of phosphites $(RO)_3P$ (R = Me, Et or $ClCH_2CH_2$) with $[(Me_2N)_2CCl]^+$ Cl^- gives the double zwitterion (46), the crystal structure of which has been determined. Reaction of (46) with NaOH followed by acid gives $Me_2NCOP(O)(OH)_2$ [249]. Sequences of reactions involving acetylenedicarboxylates and $(R^1O)_2PH(O)$ have been used to prepare compounds of the type $(R^1O)_2P(O)CH(CO_2R^2)CH(CO_2R^2)P(O)-(OR^1)_2$ and $(R^1O)_2P(O)C(CO_2R^2)=C(CO_2R^2)P(O)(OR^1)_2$ [250].

Reaction of $(EtO)_2P(O)CH_2P(O)(OEt)_2$ or $(EtO)_2P(O)CH_2P(O)(OEt)-CH_2P(O)(OEt)_2$ with benzaldehyde in the presence of sodium hydride gives $PhCH=CHP(O)(OEt)_2$ with elimination of $(EtO)_2PO_2^-$ or $(EtO)_2P(O)-CH_2P(OEt)O_2^-$, respectively[251]. Halogenation of $CH_2[P(O)Ph_2]_2$ with NBS or sulphuryl chloride gives $CHX[P(O)Ph_2]_2$ (X = Cl or Br)[252].

Compounds containing three tetracoordinate phosphorus atoms have been prepared. Reaction of $CH_2[P(O)Ph_2]_2$ with $Ph_2P(O)CH=CH_2$ in the presence of sodium in xylene at $180\,^\circ C$ gives $[Ph_2P(O)]_2CHCH_2CH_2P(O)Ph_2$ and an analogous reaction occurs with $(EtO)_2P(O)CH=CH_2$ [252]. Reaction of $PhN{=}CCl_2$ with $R^1_2POR^2$ ($R^1 = Ph$ or OEt, $R^2 = Me$ or Et) gives $PhN{=}C[P(O)R^1_2]_2$, which reacts with $R^3_2PH(O)$ to give $[R^1_2P(O)]_2CH$-$NPhP(O)R^3_2$ ($R^3 = Ph$, OMe or OEt), containing two different $P(O)R_2$ groups[253,254]. Reaction of R_2NCCl_3 with triethyl phosphite at $-40\,°C$ gives $[(EtO)_2P(O)]_3CNR_2$ ($NR_2 = NMe_2$ or morpholino)[255].

Reaction of $Ph_2P(O)CH_2OSO_2Ar$ ($Ar = p$-tolyl) with triphenylphosphine at $140\,°C$ for 8 h followed by addition of a tetraphenylborate gives the novel phosphonium salt $[Ph_2P(O)CH_2PPh_3]^+[BPh_4]^-$ [256]. Reactions of the bromoalkynes $R^1C{\equiv}CBr$ ($R^1 = H$ or Ph but not alkyl) with the phosphines R^2_3P ($R^2 = Ph$ or Bu^n) give the electrophilic phosphonium salts $[R^2_3PC{\equiv}CR^1]^+$ Br^-, which react with tri-n-butylphosphine in acetonitrile to give the bisphosphonium salts, $[Bu^n_3PCH{=}CR^1PR^2_3]^{2+}$ $2Br^-$ [257].

Several phosphinimine derivatives with two phosphorus atoms have been studied. Reactions of $(Ph_2P)_2CH_2$ with aryl azides give $ArN{=}PPh_2CH_2$-$PPh_2{=}NAr$ ($Ar =$ phenyl, m- or p-tolyl or m-ClC_6H_4)[258]. The phenyl derivative reacts with MeI in benzene to give $[MePhN{=}PPh_2CH_2PPh_2{=}NPh]^+I^-$, which upon acid hydrolysis followed by treatment with base gives aniline and N-methylaniline. Treatment of $PhN{=}PPh_2CH_2PPh_2{=}NPh$ with sodium in THF followed by MeI gives $[MePhNPPh_2{=}CHPPh_2$-$NMePh]^+$ I^- [259].

Reactions of $[Ph_3PCH_2C{\equiv}CH]^+Br^-$ with the phosphinimines $Ph_3P{=}NR$ ($R = Ph$ or $NHCO_2Et$) give $[Ph_3PCH_2C(NR)CH{=}PPh_3]^+$ Br^-, which are useful for preparing 1,3,5-trisubstituted pyrazoles and quinolines[260].

5.5.3 Compounds containing both tricoordinate and tetracoordinate phosphorus atoms

Oxidation of $Pr^i_2PP(OBu^n)_2$ with mercuric oxide gives $Pr^i_2PP(O)(OBu^n)_2$, also obtained by reaction of Pr^i_2PCl with $HP(O)(OBu^n)_2$, or of $HP(O)Pr^i_2$ with $ClP(OBu^n)_2$ in the presence of triethylamine. The oxidation of the less basic phosphorus atom by the mercuric oxide is remarkable[261]. The starting materials required for such preparations of compounds of this type can be obtained by reactions between R^1_2PCl and $(R^2O)_2PH$ in the presence of triethylamine[262].

Compounds of the type $R^1R^2P(O)(CH_2)_2PPh(CH_2)_2PPh(CH_2)_2P(O)$-$R^1R^2$ ($R^1 = Ph$, $R^2 = Ph$ or EtO) can be obtained by reactions of the disecondary phosphine $(PhPHCH_2)_2$ with the vinylphosphorus compounds $R^1R^2P(O)CH{=}CH_2$ [263,264].

Intermediates in the preparation of methylated polytertiary phosphines, such as $Ph_2P(CH_2)_2PMe_2(S)$, from Ph_2PH and $CH_2{=}CHPMe_2(S)$, are also examples of compounds containing both tricoordinate and tetracoordinate phosphorus atoms[241].

5.6 METHYLENEPHOSPHORANE (YLIDE) DERIVATIVES

The Wittig reaction, which involves treatment of appropriate methylenephosphoranes with carbonyl compounds to give alkenes, with elimination of tertiary phosphine oxides, is an important procedure in synthetic organic chemistry. In an attempt to obtain evidence regarding the mechanism of this reaction the ^{31}P n.m.r. spectrum of a mixture of the methylenephosphorane $Ph_3P=CHMe$ and cyclohexanone was obtained at $-70\,°C$. A resonance at 66.5 p.p.m. is assigned to a species (47), a presumed intermediate in the Wittig reaction[265].

New variations of the Wittig reaction have been investigated with either new types of methylenephosphoranes, new reactions conditions or use of polymeric methylenephosphoranes. Aldehyde bisulphite derivatives, $R^1CH(OH)SO_3Na$, were used in reactions with mixtures of $[Ph_3PCH_2R^2]^+$ and potassium t-butoxide to favour the *trans* isomer in the formation of alkenes of the type $R^1CH=CHR^2$ [266]. Reactions of triphenylphosphine with the 1,3-dibromides $BrCH_2CHRCH_2Br$ in boiling xylene give the phosphonium salts $[Ph_3PCH_2CHRCH_2Br]^+$ Br^-, which are cyclised with NaOEt to give the cyclopropylidenephosphoranes (48). Reactions of (48) or of the original phosphonium salt with NaH followed by addition of the ketones R^1COR^2 give methylenecyclopropane derivatives of the type (49) in a normal Wittig reaction[267]. Wittig reactions with the ylides $Ph_3P=C-(OMe)R$ (R = Me or Ph) have been investigated[268].

$$\begin{array}{ccc}
\underset{\text{(47)}}{\begin{array}{c} Ph_3P\!-\!O \\ |\quad\; | \\ MeCH\!-\!C\;\;(CH_2)_5 \end{array}}
&
\underset{\text{(48)}}{Ph_3P=C\!\!\underset{CHR}{\overset{CH_2}{<|}}}
&
\underset{\text{(49)}}{R^1R^2C=C\!\!\underset{CHR^3}{\overset{CH_2}{<|}}}
\end{array}$$

An organophosphorus compound acting as a Wittig reagent in acid solution has been developed. Metallation of $PhCH_2P(O)(OR)_2$ with lithium amide in liquid ammonia followed by addition of $Ph_2C=NCOPh$ and hydrolysis with HCl gives the amide $(RO)_2P(O)CHPhCPh_2NHCOPh$, which is cleaved to $Ph_2C=CHPh$ by aqueous perchloric acid[269]. Polymeric Wittig reagents based on polystyrene derivatives have been developed. These have the advantage that the resin can be regenerated after use[270-272].

Reaction of triphenylphosphine with $MeCBr=CH_2$ at elevated temperatures in benzonitrile in the presence of nickel(II) bromide gives the vinylic phosphonium salt $[Ph_3PCMe=CH_2]^+$ Br^-. This has no hydrogen on the carbon bonded to the phosphorus atom and thus cannot be converted into a methylenephosphorane by deprotonation. Treatment of this phosphonium salt with NaOEt followed by addition of benzaldehyde in the normal Wittig manner gives the alkene $PhCH=CMeCH_2OEt$. The methylenephosphorane $Ph_3P=CMeCH_2OEt$, arising from a Michael addition of ethoxide to the vinylic phosphonium salt, appears to be an intermediate in this reaction. A similar reaction of $[MePh_2PCMe=CH_2]^+$ I^- with NaOEt and benzaldehyde also gives $PhCH=CMeCH_2OEt$ and, in addition, appreciable amounts of $Ph_2P(O)CH=CHPh$ [273].

Reactions of methylenephosphoranes with reagents other than aldehydes and ketones have been investigated. Compounds of the type $Ph_3P=CR^1R^2$ (R^1 and R^2 are aryl groups) react with sulphur to give the thiones $R^1R^2C=S$ and triphenylphosphine sulphide. The methylenephosphorane $Ph_3P=CHPh$ reacts with sulphur to give (50) in addition to triphenylphosphine sulphide[274].

Reactions of methylenephosphoranes of the type $Ar_3P=CHR$ ($R = H$, Me or Et) with appropriate nickel complexes results in aryl migration from phosphorus to carbon to give the tertiary phosphine $Ar_2PCHRAr$ [275].

$$PhHC \overset{S}{\underset{S}{\diagup}} \overset{S}{\underset{\diagdown}{\big|}} \quad PhHC \overset{}{\underset{S}{\diagdown}} \overset{S}{\underset{}{\diagup}}$$

(50)

Hydrogen atoms on a carbon atom bonded to phosphorus in a methylenephosphorane can be replaced by other groups in various reactions. For example, the methylenephosphorane $Ph_3P=CH_2$ (from $[Ph_3PMe]^+ I^-$ and PhLi) reacts with hexafluorobenzene to give $Ph_3P=CHC_6F_5$, as shown by reaction of this product with aldehydes RCHO to give the alkenes $RCH=CHC_6F_5$ [276]. Similar reactions of $Ph_3P=CH_2$ with alkane sulphonyl fluorides (but not the corresponding chlorides) give the methylenephosphoranes $RSO_2CH=PPh_3$, also obtainable from triphenylphosphine and RSO_2CH_2Br followed by reaction with an appropriate base. Reaction of the methylenephosphorane $Ph_3P=CHCO_2Et$ with toluene-p-sulphonyl chloride results in reduction of the sulphonyl group to give $p\text{-}MeC_6H_4SC(CO_2Et)=$-$PPh_3$ [277].

Reactions of $Ph_3PCHCOPh$ with the tin halides R_3SnCl ($R = Me$ or Ph) give the phosphonium salts $[Ph_3PCH(SnR_3)COPh]^+ Cl^-$, which exhibit extremely low $v(CO)$ frequencies, indicative of strong coordination of the carbonyl group to the tin. These phosphonium salts undergo deprotonation with butyl-lithium to give the corresponding methylenephosphoranes, containing both carbon–phosphorus and carbon–tin bonds[278].

Methylenephosphoranes can give carbenes for preparative purposes. U.v. irradiation of $Ph_3P=CHCOPh$ gives triphenylphosphine and the carbene PhCOCH:, shown to be present by addition to cyclohexene[279]. Deprotonation of $[Ph_3PCHMe_2]^+ I^-$ with butyl-lithium in THF gives the red compound $Ph_3P=CMe_2$, a source of dimethylcarbene, since it reacts with alkenes of the type $RCH=CHCO_2Me$ to give the corresponding dimethylcyclopropanes[280]. Reaction of triphenylphosphine with CF_2Br_2 in triglyme followed by addition of KF generates difluorocarbene, as demonstrated by its addition to alkenes to give the *gem*-difluorocyclopropanes[281].

$$Ph_3P=CR-CHO \quad \longleftrightarrow \quad Ph_3\overset{+}{P}-CR=CH\overset{-}{O}$$

(51a)   (51b)

Acylmethylenephosphoranes are particularly stable since they can exist as the resonance hybrid (51). If the dipolar form (51b) predominates, then the possibility of *cis* and *trans* isomers exists. This has been investigated by ^1H and ^{31}P n.m.r. spectroscopy. Isomer mixtures are sometimes found with the proportion of the *trans* isomer increasing with increasing solvating power of the solvent[282]. Isomer inversion was found to be slow on the n.m.r. timescale[283].

Some special reactions of acylmethylenephosphoranes have been investigated. Vinyl azides react with acylmethylenephosphoranes to give 1-vinyl-1,2,3-triazoles with elimination of triphenylphosphine oxide[284]. The methyl group in $Ph_3PC=CHCOMe$ can be lithiated with butyl-lithium in THF at $-78\,°C$. The resulting lithium derivative $Ph_3P=CHCOCH_2Li$ reacts with alkyl halides to give the compounds $Ph_3P=CHCOCH_2R$ ($R = PhCH_2$, $n\text{-}C_5H_{11}$ or $n\text{-}C_9H_{19}$), and with benzophenone to give $Ph_3P=CHCOCH_2\text{-}CPh_2OH$ [285,286]. The acid–base properties and the ionisation constants of $(RCO)_2C=PPh_3$ ($R = Me$, Ph or OEt) have been determined by potentiometry in aqueous ethanol and in nitromethane; these compounds are protonated to form phosphonium salts[287].

Several novel stable methylenephosphorane derivatives have been prepared. Reaction of triphenylphosphine with $PhCOCOCH_2Br$ in THF followed by deprotonation of the resulting quaternary salt with aqueous NaOH gives the phosphorane $Ph_3P=CHCOCOPh$ [288]. Reaction of triphenylphosphine with dimethyl acetylenedicarboxylate (DMAD) at $-50\,°C$ gives the cyclopropene derivative (52), which readily rearranges to (53). Reaction

$$Ph_3P=C(CO_2Me)-\underset{\underset{CO_2Me}{|}}{\overset{\overset{CO_2Me}{|}}{C}}\overset{}{\underset{C}{\diagup\!\!\diagdown\,/\!/}}C-CO_2Me$$

(52)

(53)

of (52) with more triphenylphosphine gives the bis-ylide $Ph_3P=C(CO_2Me)C(CO_2Me)=C(CO_2Me)C(CO_2Me)=PPh_3$ [289]. Reaction of triphenylphosphine with DMAD in the presence of an alcohol, ROH, gives the methylenephosphorane $Ph_3P=C(CO_2Me)CH(OR)CO_2Me$ ($R = Me$ or Et); degradation of this phosphorane with aqueous NaOH gives $CH(CO_2Me)=CHCH_2CO_2Me$ in a *cis : trans* ratio of $1 : 6$ [290]. Reaction of triphenylphosphine with diphenylcyclopropenone gives the orange ketene ylide $Ph_3P=CPhCPh=C=O$, which undergoes facile alcoholysis with elimination of Ph_3P to give $PhCH=CPhCO_2R$ [291]. Addition of $Ph_3P=CHCO_2Et$ to the alkenes $RCH=CHNO_2$ gives the methylenephosphoranes $Ph_3P=C(CO_2Et)CHRCH_2NO_2$, which do not undergo the Wittig reaction with aldehydes[292]. The nitroalkenes $RCH=CBrNO_2$ ($R = Ph$, $p\text{-}MeC_6H_4$, $p\text{-}MeOC_6H_4$, $m\text{-}ClC_6H_4$, $m\text{-}O_2NC_6H_4$ or Me) react with an excess of triphenylphosphine by deoxygenation of the nitro group and dehalogenation to give the phosphonium salts $[Ph_3PCHRCN]^+$ Br^-, which react with aqueous NaOH to give the methylenephosphoranes $Ph_3P=C(CN)R$; the latter are inert in the Wittig reaction[293]. Reaction of triphenylphosphine with tetracyanomethane at room temperature gives $Ph_3P=C(CN)_2$ [294,295].

Some halogenated methylenephosphoranes have also been studied. Reaction of triphenylphosphine with CCl_4 gives $Ph_3P=CCl_2$, which undergoes the Wittig reaction with aroyl cyanides to give the compounds $RC(CN)=CCl_2$ [296]. Reaction of triphenylphosphine with hexafluorocyclobutene in diethyl ether at 25 °C gives hexafluorocyclobutylidenetriphenyl-phosphorane (54), the crystal structure of which has been determined[297].

$$Ph_3P=C\begin{smallmatrix}CF_2\\|\quad\ |\\CF_2\end{smallmatrix}CF_2 \qquad Ph_3\overset{+}{P}-\langle\ominus\rangle \longleftrightarrow Ph_3P=\langle\rangle$$

(54) (55a) (55b)

The crystal structure of triphenylphosphonium cyclopentadienylide (cyclopentadienylidene triphenylphosphorane) (55) has been determined as well as its phosphorus 2p photoelectron spectrum. The carbon–phosphorus bond length suggests that this phosphorane is a resonance hybrid with 86% of the 'ylide' form (55a) and 14% of the 'ylide' form (55b). This extensive π-electron delocalisation parallels the lack of reactivity of (55) in the Wittig reaction and its ease of electrophilic substitution[298].

Reaction of 3,5-dibromocyclopent-1-ene with tris(dimethylamino)phosphine in chloroform gives the tripolar methylenephosphorane phosphonium salt (56)[299].

Extensive ^{13}C n.m.r. results on methylenephosphoranes have been reported. The ^{13}C chemical shift of the ylide carbon of 30 p.p.m. relative to Me_4Si indicates a large charge density on this carbon atom[300].

$$\left[(Me_2N)_3\overset{+}{P}-\langle\ominus\rangle\overset{+}{P}(NMe_2)_3\right] Br^- \qquad Me_3\overset{+}{P}-\overset{-}{C}H_2$$

(56) (57)

The chemistry of the simplest methylenephosphorane, $Me_3P=CH_2$, has received considerable attention during 1972–73. Detailed n.m.r. data (^1H, ^{13}C and ^{31}P) on $Me_3P=CH_2$ suggest a dipolar structure (57) with sp^2 carbanion character of the carbon atom and sp^3 onium character of the phosphorus atom. (For convenience, however, the ylide carbon–phosphorus bonds in this review have been drawn as carbon–phosphorus double bonds)[301].

Alcohols react reversibly with $Me_3P=CH_2$ to give the pentacoordinate derivatives Me_4POR as colourless water- and oxygen-sensitive liquids, which can be distilled *in vacuo* and which give $Me_3P=O$ upon thermolysis[302]. Reaction of $Me_3P=CH_2$ or $Me_3P=CHSiMe_3$ with HF gives ionic $[Me_4P]^+ F^-$ as a solid which can be sublimed *in vacuo* and which etches moist glass. However, the corresponding reaction of $Et_3P=CHMe$ with HF at -80 °C gives Et_4PF, but this is a non-ionic liquid, sensitive to hydrolysis and can be distilled. The properties of compounds of the type R_4PF thus depend very greatly upon the nature of the alkyl group R [303,304].

Some novel methylenephosphoranes containing silicon have been prepared including $Me_2P(CH_2SiMe_3)=CHSiMe_2SiMe_3$, $Me_2P(CH_2SiMe_2SiMe_3)=$-$CHSiMe_2SiMe_3$ and $R_3P=CHSiMe_2SiMe_2CH=PR_3$ (R = Me or Ph). Reaction of $Me_3P=CHSiMe_3$ with $ClSiMe_2SiMe_2Cl$ and Bu^nLi gives the silicon-phosphorus compound (58)[305]. Reaction of $[Ph_3PMe]^+ Br^-$ with the

$$\begin{array}{c} CH_2 \\ Me_2P \diagdown \diagup SiMe_2 \\ \diagdown \diagup \\ C-SiMe_2 \\ \diagup \\ Me_3Si \end{array}$$

(58)

methylenephosphorane $Me_3P{=}CHSiMe_3$ results in a transylidation to give $[Me_4P]^+ Br^-$ and $Ph_3P{=}CHSiMe_3$. Metallation of the latter product with Bu^nLi followed by silylation with Me_3SiBr gives $Ph_3P{=}C(SiMe_3)_2$ [306].

Dimethylamino-substituted methylenephosphoranes of the type $(Me_2N)_{3-n}PMe_n{=}CH_2$ ($n = 1$ or 2) can be prepared by successive treatment of the corresponding $(Me_2N)_{3-n}PMe_n$ compounds with the chloramine RNHCl, with sodium amide, with methyl iodide, and again with sodium amide in liquid ammonia [307].

The trimethylphosphinemethylide anion $[Me_2P(CH_2)_2]^-$ has been shown to be an unusual ligand in transition-metal chemistry in the yellow nickel(II) complex $\{Me_2P(CH_2)_2\}_4Ni_2$ (59)[308] and the red chromium(III) complex $\{Me_2P(CH_2)_2\}_3Cr$ (60)[309]. Both of these compounds are very air-sensitive, but can be sublimed *in vacuo* only slightly above room temperature.

(59)   (60)

5.7 SOME REACTIONS OF ORGANOPHOSPHORUS COMPOUNDS AT THE PHOSPHORUS ATOM

Reactions of organophosphorus compounds have been reported during 1972–73 in which the main point of interest involves a change of the groups bonded to the phosphorus atom, with the organic portion being unaffected. Such reactions, which involve mainly phosphorus–fluorine compounds, phosphorus chalcogenides or phosphinimines, are now discussed.

Reactions of tertiary phosphines R_3P ($R = Bu^n$ or Ph) with CF_3OOCF_3 (or, less desirably, CF_3SSCF_3) in dichloromethane at $-78\,°C$ provide new routes to the corresponding R_3PF_2 derivatives. The phosphacyclobutane derivative (61), prepared by an analogous method, was shown by n.m.r. spectroscopy to be a slowly equilibrating mixture at low temperatures [310].

$$\begin{array}{c} H_2C-CMe_2 \\ | \qquad | \\ Me_2C-PF_2Ph \end{array}$$

(61)

Fluorination of the tertiary phosphines $PhPR_2$ ($R = C_{2-4}$ alkyl) with MoF_6 also gives the corresponding PhR_2PF_2 derivatives[311]. U.v. irradiation of P_2F_4 and C_2H_4 in the gas phase gives $F_2PCH_2CH_2PF_2$, which forms a bis-borane adduct[312]. Reactions of Me_2PF with methanol or methanethiol results in addition of HF to give Me_2PHF_2 in addition to forming the corresponding compound $Me_2P(E)Me$ ($E = O$ or S)[313]. The colourless hydrolysis-sensitive malodourous liquids $RPH(O)F$ ($R = Me$, Et or Ph), which can be distilled at atmospheric pressure, have been prepared by reaction of the corresponding compound $RPCl_2$, with HF and H_2O in diethyl ether[314].

Various transformations of dimethylphosphorus derivatives have been investigated. Reactions of Me_2PCl with the nucleophiles $NaEMe$ ($E = O$ or S) give the corresponding Me_2PEMe compounds, which undergo hydrolysis to give $MePH(O)$. The methoxy derivative Me_2POMe easily rearranges to $Me_3P(O)$; this process is catalysed by MeCl. Reaction of Me_2PSMe with methyl iodide gives $[Me_3PSMe]^+\ I^-$, which loses MeI on thermolysis to give $Me_3P(S)$. Methanolysis of Me_2PCl gives the quaternary salt $[Me_2PHOMe]^+\ Cl^-$, which loses MeCl above $-10\ ^\circ C$ to give $Me_2PH(O)$; this latter product in the presence of unreacted Me_2PCl or $[Me_2PHOMe]^+\ Cl^-$ reacts further to give $Me_2P(O)OH$, $Me_2P(O)OMe$, Me_2PH and $[Me_2PH_2]^+\ Cl^-$ [315]. The phosphorus–nitrogen bond in Me_2PNMe_2 [prepared from Me_2NPX_2 (X is preferably Cl) and $MeMgX$ or MeLi] is cleaved by HCN to give Me_2PCN [316].

Interconversions of phosphine oxides and phosphine sulphides have been studied. Boron trisulphide (B_2S_3) is useful for conversion of tertiary phosphine oxides into the corresponding sulphides with high stereospecificity and net retention of configuration at the phosphorus atom. Examples of optically active phosphine oxides which have been converted into the corresponding phosphine sulphides with retention of configuration by this method include $(+)$-(R)-$MePhPr^nP(O)$, $(+)$-(R)-$(cyclo$-$C_6H_{11})MePhP(O)$ and $(+)$-(R)-$EtMePhP(O)$[317]. The reverse transformation of tertiary phosphine sulphides to the corresponding phosphine oxides can be effected by DMSO with elimination of dimethyl sulphide and elemental sulphur. Tertiary phosphine oxides prepared from the corresponding sulphides by this method include $Ph_3P(O)$, $MePr^nPhP(O)$ and $(PhCH_2)MePhP(O)$[318]. Compounds of the type $RP(O)Cl_2$, $RP(S)Cl_2$ and $RPCl_2$ react rapidly and exothermically with DMSO to give products of the type $RPX(O)OPRX(O)$ ($X = Cl$ or OH). Sulphur-containing by-products of this reaction are $MeSCH_2Cl$, $(MeS)_2CH_2$, Me_2S and Me_2S_2 [319,320].

S
(S)RP⟨ ⟩PR(S)
S
(62)

S
(S)RP⟨ ⟩PR(S)
N
|
$SiMe_3$

(63)

Reactions of the compounds $[SPR(S)]_2$ (62; $R = Me$ or Et) with azido-trimethylsilane result in replacement of one of the bridging sulphur atoms by an Me_3SiN group to give (63; $R = Me$ or Et)[321].

Various reactions of silylated phosphinimines have been investigated. The lithium derivative $Me_3P=NLi$ reacts with $Me_3SiSiMe_2Cl$ to give $Me_3P=NSiMe_2SiMe_3$ and with $(ClSiMe_2)_2$ to give $Me_3P=NSiMe_2SiMe_2N=PMe_3$ [322]. Reactions of tertiary phosphines and phosphites R_3P ($R = Bu^n$, Ph, EtO, Pr^nO or Bu^nO) with $(Me_3Si)_2NCl$ in diethyl ether at 0–$10\,^\circ C$ gives $R_3P=NSiMe_3$ with elimination of Me_3SiCl [323]. Reactions of $Me_3P=NSiMe_3$ with PCl_3, $MePCl_2$ or Me_2PCl form new nitrogen–phosphorus bonds giving $(Me_3P=N)_3P$, $(Me_3P=N)_2PMe$ and $[Me_3P=N(PMe_2)_2]^+$ Cl^-, respectively[324]. Similarly, reactions of $R^1_3P=NSiMe_3$ ($R^1 = Me$, Pr^i or Ph) with the pentavalent phosphorus derivatives $R^2_2PF_3$ and R^2PF_4 ($R^2 = Me$ or Ph) give the salts $[R^2_2P(N=PR^1_3)_2]^+[R^2_2PF_4]^-$ and $[R^2PF(N=PR^1_3)_2]^+$ $[R^2PF_5]^-$, respectively[325].

5.8 SOME MECHANISTIC AND STEREOCHEMICAL CONSIDERATIONS

The hydrolysis of various tetracoordinate phosphorus derivatives has received considerable attention during 1972–73. Alkaline hydrolysis of salts of the type $[MePhP(OR^1)(SR^2)]^+$ $[SbCl_6]^-$ with 0.01 M NaOH in 50% aqueous dioxan gives a mixture of the corresponding $MePhP(O)(OR^1)$ and $MePhP(O)(SR^2)$ compounds. The ratio of these products is insensitive to the nature of the alkylthio group but is affected by a change in the alkoxy group. This work suggests a mechanism involving axial attack of OH^- on the face of the tetrahedral phosphonium salt opposite the alkoxy group, followed by a direct competition between loss of axial R^1O^- and isomerisation with loss of axial R^2S^- [326]. The stereochemistry of alkaline hydrolysis of the salts R- and S-$[MePhP(OMe)(OMen)]^+$ $[SbCl_6]^-$ (Men here and elsewhere signifies the menthyl group) and $[MePhP(OMe)(OEt)]^+$ $[SbCl_6]^-$ depends only upon the nature of the alkoxy groups. The reaction may proceed with direct OR^- loss or through pseudorotation[327]. Reactions of the salts $[MePhP(OMe)(OMen)]^+$ $[BF_4]^-$ with various nucleophiles to give the corresponding $MePhP(O)(OMen)$ compounds have also been investigated. Such reactions with the nucleophiles I^-, Br^-, Cl^-, aniline and pyridine proceed with retention of configuration whereas similar reactions with the nucleophiles F^-, Pr^iNH_2 and Et_3N proceed with epimerisation[328]. The reaction of $PhP(O)(SMe)(OMen)$ with methylmagnesium bromide to give $MePhP(O)(OMen)$ proceeds with retention of configuration whereas an apparently analogous reaction of $PhP(O)(SMe)(OMen)$ proceeds with inversion[329]. Reaction of $MeOP(O)(SMe)NH_2$ with hydroxide may result in either phosphorus–oxygen cleavage to give $^-OP(O)(SMe)NH_2$ or in phosphorus–sulphur cleavage to give $^-OP(O)(OMe)NH_2$. The first product is favoured in water whereas the second is favoured in 50% aqueous acetone. The second-order rate constants for these reactions have been measured[330].

Acid-catalysed hydrolysis of tetracoordinate phosphorus esters and amides has received considerable attention. A pentacoordinate intermediate is postulated for acid-catalysed hydrolysis of esters of the type $R^1_2P(O)OR^2$ ($R^1 = Me$, Et or Pr^i; $R^2 = Me$ or Et)[331]. The rates of acid-catalysed hydrolysis of the esters p-$XC_6H_4C(CH_2)OP(O)(OEt)_2$ ($X = H$, Me, OMe,

Cl or NO_2) in 0.1 N HCl in 40% aqueous ethanol are increased by electropositive substituents and lowered by electron-withdrawing substituents. An $A_{SE}2$ mechanism with rate-determining protonation on carbon is proposed for these reactions[332,333]. The kinetics of the perchloric acid-catalysed hydrolysis of the methyl esters p-$RC_6H_4PMe(O)(OMe)$ (R = H, Me or Cl) have been investigated at several perchloric acid concentrations and three temperatures. The hydrogen and methyl derivatives exhibit rate maxima at 6 to 7 M $HClO_4$, but the chlorine derivative exhibits no such rate maximum[334]. The acid-catalysed hydrolysis (2 M HCl in aqueous methanol) of amides of the type RPPh(O)NHPh is greatly dependent upon the nature of R. The cyclopropyl and 1-methylcyclopropyl derivatives give increased specific rates relative to those for the isopropyl and the t-butyl derivatives, respectively[335].

The acid-catalysed hydrolysis of amides of the type $R^1PPh(O)NHR^2$ (R^1 and R^2 are aryl groups) have been found to follow a rate law, rate = $k[amide]h_0^{0.59}$, where h_0 is an acidity function[336]. Alkaline hydrolytic cleavage of the nitrogen–phosphorus bond in amides of the type $R^1_2P(O)NR^2_2$ (R^2 = H or Me) is first order in hydroxide ion and occurs at rates comparable with those of the hydrolysis of analogous carboxamides[337].

The kinetics of the hydrolysis of the phosphetane derivative (64) have been

$$
\begin{array}{ccc}
\text{MeHC} & \!\!-\!\! & \text{CH}_2 \\
| & & | \\
\text{Me}_2\text{C} & \!\!-\!\! & \text{P}\cdots\text{Ph} \\
& & \backslash \\
& & \text{OMe}
\end{array}
$$

(64)

investigated by stopped-flow and u.v. spectrophotometric methods. The rates are proportional to the hydroxide ion concentration at pH > 9 but pH independent in dilute acid. In moderate to high acid concentration (pH < 1) the rate decreases with increasing acidity as a result of a decrease in the activity of water. In the pH range 3–8, a sigmoidal titration curve is obtained with the rate increasing with increasing pH. These observations can be ascribed to a change in the rate-determining step for hydrolysis from rate-limiting hydration around neutral pH to rate-limiting breakdown of a pentacoordinate intermediate in acid[338]. An investigation of the pH dependence of the hydrolysis of CHMePhOPH(O)Ph (I) reveals the rate law: rate = $k_1[I] + k_2[OH][I]$ with $k_1 = 1.58 \times 10^{-4}$ and $k_2 = 2.63 \times 10^2$ l mol^{-1} [339]. The alkaline hydrolysis of esters of the type $Ph_2P(O)OR$ (R = Me, Et, Pr^n, Pr^i, Bu^s or Me_3CCH_2) has also been investigated[340].

Migration of a $Ph_2P(O)$ group has been observed in reactions involving the carbonium ion $Ph_2P(O)CMe_2CH_2^+$ as an intermediate. Such reactions include the reaction of the amine $Ph_2P(O)CMe_2CH_2NH_2$ with nitrous acid, and reaction of the toluene-p-sulphonate $Ph_2P(O)CMe_2CH_2OTs$ with formic acid[341,342].

Optically active compounds with asymmetric phosphorus atoms have received some attention. Complexes of the type $trans$-$(diphos)_2OsCl_2$ have been prepared with the $meso$, the (+) and the (−) isomers of $PhMePCH_2$-

CH_2PMePh [343]. The racemic $Pr^iP(S)(OMe)OH$ has been resolved through diastereoisomeric salts with optically active α-phenylethylamine and ephedrine[344].

The configurations of acids of the type $R^2P(S)(OR^1)(OH)$ ($R^2 = $ Me, Et or Pr^i; $R^1 = $ Me, Et, Pr^n or Bu^n) have been correlated with that of $(+)$-(R)-$MeP(S)$-$(OPr^i)(OH)$ based on their syntheses via $R^2P(S)(OR^1)Cl$; $(+)$-$R^2P(S)$-$(OR^1)(OH)$ has the R-configuration. Alkaline hydrolysis occurs with inversion of configuration at the prosphorus atom[345].

The rotational barriers about the carbon–phosphorus bonds in Bu^tPCl_2 and Bu^t_2PCl have been investigated. Hindered rotation at low temperatures is indicated by a 3 : 2 : 1 n.m.r. pattern for the t-butyl protons in Bu^tPCl_2[346,347].

The pyramidal inversion barrier at phosphorus was found[348] to be less for $PhPr^iPSi(OMe)_3$ than for the corresponding trimethylsilyl derivative $PhPr^iPSiMe_3$, apparently from negative hyperconjugation involving contributions of structures such as (65).

$$PhPr^i\overset{+}{P}= Si(OMe)_2 \ ^-OMe$$

(65)

5.9 N.M.R. STUDIES ON ORGANOPHOSPHORUS COMPOUNDS

Vicinal P—C—C—H couplings for phosphonates have been determined as a function of the dihedral angle. Maxima were found for dihedral angles of 0° (*ca.* 18 Hz) and 180° (*ca.* 41 Hz) and a minimum for a 90° dihedral angle[349].

The europium(III) dipivaloylmethide-induced proton and ^{31}P n.m.r. chemical shifts have been determined for the potential ligands R_2NCH_2-CH_2PPh_2 ($R = $ H or Me)[350].

Phosphorus-31 n.m.r. spectroscopy has been used to differentiate between diastereoisomeric α-phenylethylamine salts of chiral phosphorus acids such as $MeP(S)(OPr^i)OH$ [351]. The coupling constants between adjacent phosphorus atoms, $^1J(PP)$, in pentamethylcyclopentaphosphine Me_5P_5 (66) are not all equivalent. If the phosphorus atoms are numbered as in (66),

(66)

$J_{12}(PP)$ and $J_{15}(PP)$ are -248.6 Hz, $J_{23}(PP)$ and $J_{45}(PP)$ are -236.4 Hz and $J_{34}(PP)$ is -310.3 Hz [352].

Carbon-13 n.m.r. spectroscopy continues to be useful in organophosphorus chemistry. The ^{31}P and ^{13}C n.m.r. spectra of various tertiary phosphines have been investigated. The coupling constant $^1J(CP)$ is in the range 12–25 Hz and increases with increasing numbers of t-butyl groups, suggesting that the CPC bond angle is a major influence on the magnitude of this coupling constant. The other coupling constants $^2J(CP)$ and $^3J(CP)$ are in the ranges 12–27 Hz and 11–13 Hz, respectively[353]. The $J(CP)$ values in various ethylenic phosphines of the type $R^1{}_2PCH=CR^2R^3$ (R^1 = EtO or NMe_2, $R^2 = Bu^t$ or H, R^3 = H or Ph) have also been determined[354]. Study of the ^{13}C n.m.r. spectra of various phosphetane oxides, sulphides, phosphetanium salts, phospholane oxides and phospholene oxides indicates fixed puckered conformations for some phosphetanes but rapidly interconverting forms for single symmetrically substituted derivatives[355,356]. Heterocyclic compounds of the types (67) and (68) have also been studied by ^{13}C

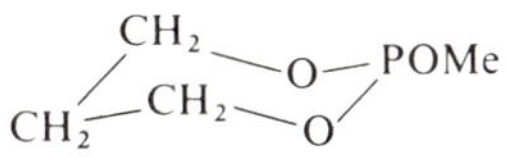

(67) 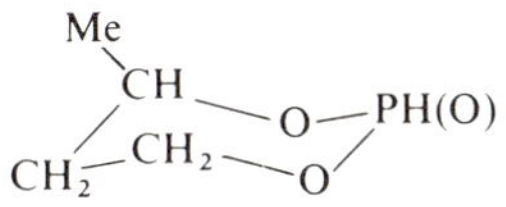   (68)

and ^{31}P n.m.r. spectroscopy. The chemical shifts of certain ring carbons in (67) vary by up to 3–4 p.p.m. depending upon whether the methoxy group is axial or equatorial[357]. The $^3J(CP)$ coupling constants for (68) and related compounds depend on the POCC dihedral angle and the orientation of the phosphorus–oxygen bond[358].

The ^{13}C n.m.r. spectra of compounds of the types $[Ph_3PCH_2X]^+\ Y^-$, $Ph_3P(O)$, $Ph_3P=CR^1CO_2R^2$ and of various triphenylphosphonium ylides have been investigated. In the phosphonium salts $[Ph_3PCH_2X]^+\ Y^-$ the methylene carbon chemical shifts parallel those in compounds of the type CH_3X, CH_3CH_2X and cyclo-$C_6H_{11}X$. The ylide carbon atoms have high shieldings typical of aliphatic carbons, suggesting high localisation of negative charge on basically sp^2 carbon atoms[359]. The low-temperature ^{13}C n.m.r. spectra of various dimethylaminophosphines of the types $PhP(NMe_2)Cl$ and $PhP(NMe_2)OMe$ have been investigated. The $^2J(PNC)$ coupling constant is large and positive when the carbon nitrogen bond is *cis* to the phosphorus lone pair, and small and negative when the carbon nitrogen bond is *trans* to the phosphorus lone pair[360].

Phosphorus–hydrogen decoupling in ^{31}P n.m.r. spectroscopy achieved by adding cobalt(II) chloride, iron(III) chloride or iron(III) acetylacetonate, is effective for the compounds $(RO)_3P(O)$ (R = Me, Et, Pr^i or Bu^n) and MeP-$(O)(OMe)_2$[361].

Selenium-77 chemical shifts have been determined by heteronuclear triple resonance for the phosphorus selenides R_3PSe (R = Me, Me_2N or MeO) and $Ph_2PH(Se)$. This involves irradiation of the ^{77}Se satellites in the ^{31}P n.m.r. spectrum at the appropriate proton frequency[362]. Proton-noise decoupling and digital sweep-time averaging have been used to measure the ^{31}P n.m.r. spectra of various triarylphosphines and their selenides. In these selenides the $^1J(PSe)$ coupling is 700–755 Hz[363].

Some novel concentration and solvent effects on the n.m.r. spectra of organophosphorus compounds have been observed. The ^{13}C n.m.r. spectra of the compound $(RO)_2P(O)CH_2P(O)(OR)_2$ and the coupling constants $J(PP)$ are concentration dependent[364]. Variations have also been noted in the n.m.r. spectra of phosphoryl and thiophosphoryl compounds in organic solvents of varying proton acidity[365]. Dramatic changes in the proton and ^{31}P n.m.r. spectra of the symmetrical biphosphines Me_2PPMe_2 and PhMePPMePh, but not in the n.m.r. spectrum of the unsymmetrical biphosphine Me_2PPPh_2, are observed in solvents of relatively high dielectric constant, and even in dichloromethane; intermolecular association is postulated to account for this observation. The ^{31}P n.m.r. spectrum has been used to follow the reaction between the symmetrical Me_2PPMe_2 and Ph_2PPPh_2 in dichloromethane at room temperature over a period of days to give the unsymmetrical Me_2PPPh_2[366].

References

1. Smith, D. J. H. (1973). *J. Organometal. Chem.*, **4**, 1
2. Miller, J. A. (1973). *Organophosphorus Chem.*, **4**, 51
3. Shaw, M. A. and Ward, R. S. (1972). *Topics Phosphorus Chem.*, **7**, 1
4. Konovalova, I. V. and Pudovik, A. N. (1972). *Usp. Khim.*, **41**, 799
5. King, R. B. (1972). *Accounts Chem. Res.*, **5**, 177
6. Vermeer, H. (1971). *Chem. Tech. (Amsterdam)*, **26**, 685
7. Davies, M. and Hughes, A. N. (1972). *J. Heterocycl. Chem.*, **9**, 1
8. Razumov, A. I., Liober, B. G., Moskva, V. V. and Sokolov, M. P. (1973). *Usp. Khim.*, **42**, 1199
9. Christol, H. and Criatan, H. J. (1971). *Ann. Chim. (Paris)*, **6**, 191
10. Trippett, S. (1973). *Organophosphorus Chem.*, **4**, 29
11. Ugi, I. and Ramirez, F. (1972). *Chem. Brit.*, **8**, 198
12. Gillespie, D., Ramirez, F., Ugi, I. and Marquarding, D. (1973). *Angew. Chem. Int. Ed. Engl.*, **12**, 91
13. Rauk, A., Allen, L. C. and Mislow, K. (1972). *J. Amer. Chem. Soc.*, **94**, 3035
14. Strich, A. and Veillard, A. (1973). *J. Amer. Chem. Soc.*, **95**, 5574
15. Musher, J. I. (1972). *J. Amer. Chem. Soc.*, **94**, 5662
16. Strecker, R. A., Snead, J. L. and Sollott, G. P. (1973). *J. Amer. Chem. Soc.*, **95**, 210
17. Zink, J. I. and Kaska, W. C. (1973). *J. Amer. Chem. Soc.*, **95**, 7510
18. Brandon, R., Haszeldine, R. N. and Robinson, P. J. (1973). *J. Chem. Soc. Perkin Trans. II*, 1295
19. Brandon, R., Haszeldine, R. N. and Robinson, P. J. (1973). *J. Chem. Soc. Perkin Trans. II*, 13
20. Janzen, A. F. and Vaidya, O. C. (1973). *Can. J. Chem.*, **51**, 1136
21. Issleib, K., Oehme, H. and Mohr, K. (1973). *Z. Chem.*, **13**, 139
22. Lutsenko, I. F., Proskurnina, M. V. and Karlstedt, N. B. (1973). *Phosphorus*, **3**, 55
23. Grobe, J. and Moeller, U. (1971). *J. Organometal. Chem.*, **33**, 13
24. Oehme, H. and Thamm, R. (1973). *J. Prakt. Chem.*, **315**, 526
25. van Reijendam, J. W. and Baardman, F. (1972). *Tetrahedron Lett.*, 5181
26. Voronkov, G. M., Marmur, L. Z., Dolgov, O. N., Pestunovich, V. A., Pokrovskii, E. I. and Popelis, J. (1971). *Zh. Obshch. Khim.*, **41**, 1987
27. Lin, K. C. (1973). *U.S. Pat.* 3 673 285
28. Märkl, G. and Dannhardt, G. (1973). *Tetrahedron Lett.*, 1455
29. Petrov, M. L. and Petrov, A. A. (1972). *Zh. Obshch. Khim.*, **42**, 1863
30. Petrov, M. L. and Petrov, A. A. (1972). *Zh. Obshch. Khim.*, **42**, 2345
31. Pudovik, A. N. and Sudakova, T. M. (1972). *Zh. Obshch. Khim.*, **42**, 1646
32. Pudovik, A. N., Sudakova, T. M., Raevskaya, O. E. and Fedechkina, V. A. (1972). *Zh. Obshch. Khim.*, **42**, 1727
33. Karlstedt, N. B., Proskurnina, M. V. and Lutsenko, I. F. (1972). *Zh. Obshch. Khim.*, **42**, 2418

34. Pudovik, A. N., Zimin, M. G., Sobanov, A. A., Vinogradov, L. I. and Samitov, Yu. Yu. (1972). *Zh. Obshch. Khim.*, **42**, 2167
35. Pudovik, A. N., Zimin, M. G., Sobanov, A. A. and Evstaf'ev, G. I. (1973). *Zh. Obshch. Khim.*, **43**, 1910
36. Ruveda, M. A. and de Licastro, S. A. (1972). *Tetrahedron*, **28**, 6013
37. Issleib, K. and von Malotki, P. (1973). *J. Prakt. Chem.*, **315**, 463
38. Lazaraeva, M. V., Perekalin, V. V. and Mastryukova, T. A. (1973). *Izv. Akad. Nauk SSSR, Ser. Khim.*, 1382
39. Paulsen, H. and Greve, W. (1973). *Chem. Ber.*, **106**, 2114
40. Paulsen, H. and Greve, W. (1973). *Chem. Ber.*, **106**, 2124
41. Yakshin, V. V. and Sokal'skaya, L. I. (1973). *Zh. Obshch. Khim.*, **43**, 440
42. King, J. F., Lewars, E. G. and Danks, L. J. (1972). *Can. J. Chem.*, **50**, 866
43. Schipper, P., Castenmiller, W. A. M., de Haan, J. W. and Buck, H. M. (1973). *J. Chem. Soc. Chem. Commun.*, 574
44. Böhme, H. and Haake, M. (1972). *Chem. Ber.*, **105**, 2233
45. Nifant'ev, E. E. and Ivanova, N. L. (1970). *Zh. Obshch. Khim.*, **40**, 1492; (1972). *Chem. Abstr.*, **76**, 3955
46. Nifant'ev, E. E. and Shilov, I. V. (1971). *Zh. Obshch. Khim.*, **41**, 2372
47. Park, J. D. and Furuta, O. K. (1973). *Daehan Hwahak Hwoejee*, **17**, 67; (1973). *Chem. Abstr.*, **79**, 5418
48. Petrov, M. L. and Petrov, A. A. (1973). *Zh. Obshch. Khim.*, **43**, 691
49. Leblanc, R., Corre, E. and Foucaud, A. (1972). *Tetrahedron*, **28**, 4039
50. Leblanc, R., Corre, E., Soenen-Svilarich, M., Chalse, M. F. and Foucaud, A. (1972). *Tetrahedron*, **28**, 4431
51. Corre, E., Chasle, M. F. and Foucaud, A. (1972). *Tetrahedron*, **28**, 5055
52. Svilarich-Soenen, M. and Foucaud, A. (1972). *Tetrahedron*, **28**, 5149
53. Shokol, V. A., Kozhushko, B. N., Doroshenko, V. V. and Kirsanov, A. V. (1973). *Zh. Obshch. Khim.*, **43**, 12
54. Shokol, V. A., Kozhushko, B. N. and Kirsanov, A. V. (1973). *Zh. Obshch. Khim.*, **43**, 54
55. Burger, K., Fehn, J., Albanbauer, J. and Friedl, J. (1972). *Angew. Chem. Int. Ed. Engl.*, **11**, 319
56. Burger, K., Fehn, J. and Thenn, W. (1973). *Angew. Chem. Int. Ed. Engl.*, **12**, 502
57. Deschamps, B., Lefebvre, G. and Seyden-Penne, J. (1972). *Tetrahedron*, **28**, 4209
58. Goldborn, P. (1973). *Synthesis*, 547
59. Bentrude, W. G., Johnson, W. D., Khan, W. A. and Witt, E. R. (1972). *J. Org. Chem.*, **37**, 631
60. Bullock, J. I., Taylor, N. J. and Parrett, F. W. (1972). *J. Chem. Soc. Dalton Trans.*, 1843
61. Legin, G. Ya. (1973). *Zh. Obshch. Khim.*, **43**, 210
62. Levin, Ya. A. and Pyrkin, R. I. (1973). *Zh. Obshch. Khim.*, **43**, 77
63. Pyrkin, R. I., Levin, Ya. A. and Gol'dfarb, E. I. (1973). *Zh. Obshch. Khim.*, **43**, 1705
64. Levin, Ya. A. and Pyrkin, R. I. (1973). *Zh. Obshch. Khim.*, **43**, 281
65. Zhuravleva, L. P. and Z'ola, M. I. (1972). *Zh. Obshch. Khim.*, **42**, 526
66. Maier, L. (1972). *Z. Anorg. Allgem. Chem.*, **394**, 117
67. (a) Prokof'eva, A. F., Mel'nikov, N. N., Vladimirova, I. L. and Einisman, L. I. (1971). *Zh. Obshch. Khim.*, **41**, 1702; (b) Novikova, Z. S., Zdorova, S. N. and Lutsenko, I. F. (1972). *ibid.*, **42**, 112
68. Wetzel, R. B. and Kenyon, G. L. (1972). *J. Amer. Chem. Soc.*, **94**, 1774
69. Chan, T. H. and Nwe, K. T. (1973). *Tetrahedron Lett.*, 3601
70. Wenzel, R. B. and Kenyon, G. L. (1972). *J. Amer. Chem. Soc.*, **94**, 9230
71. Schumann, H. and Roesch, L. (1973). *J. Organometal. Chem.*, **55**, 257
72. Ranganathan, T. N., Todd, S. M. and Paddock, N. L. (1973). *Inorg. Chem.*, **12**, 316
73. Cann, P. F., Warren, S. and Williams, M. R. (1972). *J. Chem. Soc. Perkin Trans. 1*, 2377
74. Ellzey, S. E., Jr., Connick, W. J., Jr., Boudreaux, G. J. and Klapper, H. (1972). *J. Org. Chem.*, **37**, 3453
75. Wegener, W. and Scholz, P. (1972). *Z. Chem.*, **12**, 178, 334
76. Daigle, D. J., Pepperman, A. B. Jr. and Normand, F. L. (1972). *J. Heterocycl. Chem.*, **9**, 715

77. Maier, L. (1972). *Phosphorus*, **1**, 237
78. Maier, L. (1972). *Phosphorus*, **1**, 245
79. Zarinov, R. K., Azerbaev, I. N. and Aimakov, V. A. (1973). *Zh. Obshch. Khim.*, **43**, 764
80. Pudovik, A. N., Zimin, M. G. and Kurguzova, A. M. (1971). *Zh. Obshch. Khim.*, **41**, 1964
81. Gazizov, M. B., Sultanova, D. B., Razumov, A. I., Ostanina, L. P. and Pusalkina, A. M. (1971). *Zh. Obshch. Khim.*, **41**, 2575
82. Gazizov, M. B., Sultanova, D. B., Razumov, A. I. and Ostanina, L. P. (1972). *Zh. Obshch. Khim.*, **42**, 1647
83. Mironova, Z. N., Tsvetkov, E. N., Nikolaev, A. V. and Kabachnik, M. I. (1973). *Zh. Obshch. Khim.*, **43**, 534
84. Issleib, K., Oehme, H. and Scheibe, M. (1972). *Syn. Inorg. Metal-org. Chem.*, **2**, 223
85. Pudovik, A. N. and Gareev, R. D. (1972). *Zh. Obshch. Khim.*, **42**, 1861
86. Dormidonov, I. A., Martynov, V. F. and Timofeev, V. E. (1972). *Zh. Obshch. Khim.*, **42**, 479
87. Yoffe, S. T., Petrovski, P. V., Goryunov, Ye. I., Yersheva, T. V. and Kabachnik, M. I. (1972). *Tetrahedron*, **28**, 2783
88. Knunyants, I. L., Bykhovskaya, E. G. and Sizov, Yu. A. (1972). *Zh. Vses. Khim. Obshchest.*, **17**, 346; (1972). *Chem. Abstr.*, **77**, 126 770
89. Knunyants, I. L., Bykhovskaya, E. G. and Sizov, Yu. A. (1972). *Zh. Vses. Khim. Obschest.*, **17**, 354; (1972). *Chem. Abstr.*, **77**, 114 504
90. Laskorin, B. N., Yakshin, V. V. and Sokal'skaya, L. I. (1972). *Zh. Obshch. Khim.*, **42**, 1261
91. Gazizov, M. B., Razumov, A. I. and Sekenin, E. A. (1973). *Zh. Obshch. Khim.*, **43**, 1407
92. Becher, H. J., Fenske, D. and Langer, E. (1973). *Chem. Ber.*, **106**, 177
93. Becher, H. J. and Langer, E. (1973). *Angew. Chem. Int. Ed. Engl.*, **12**, 842
94. Mathey, F. and Lampin, J. P. (1972). *Tetrahedron Lett.*, 1949
95. Grieco, P. A. and Pogonowski, C. S. (1973). *J. Amer. Chem. Soc.*, **95**, 3071
96. Kluger, R. (1973). *J. Org. Chem.*, **38**, 2721
97. Mukhametov, F. S., Rizpolozhenskii, N. I. and Goldfarb, E. I. (1971). *Izv. Akad. Nauk SSSR, Ser. Khim.*, 2221
98. Oehme, H., Issleib, K. and Leissring, E. (1972). *J. Prakt. Chem.*, **314**, 66
99. Malevannaya, R. A., Tsvetkov, E. N. and Kabachnik, M. I. (1972). *Zh. Obshch. Khim.*, **42**, 765
100. Malevannaya, R. A., Tsvetkov, E. N. and Kabachnik, M. I. (1971). *Zh. Obshch. Khim.*, **41**, 2359
101. Ismgailov, R. K., Razumov, A. I. and Yafarova, R. L. (1972). *Zh. Obshch. Khim.*, **42**, 1248
102. Razumov, A. I., Krasil'nikova, E. A., Noskova, N. A., Zykova, T. V. and Chemodanova, L. A. (1971). *Zh. Obshch. Khim.*, **41**, 2402
103. Ismagilov, R. K., Yafarova, R. L. and Razumov, A. I. (1972). *Zh. Obshch. Khim.*, **42**, 211
104. Kirilov, M. and Petrov, G. (1972). *Monatsh. Chem.*, **103**, 1651
105. Ioffe, S. T., Goryunov, E. I., Evshova, T. V., Petrovskii, P. V. and Kabachnik, M. I. (1971). *Zh. Obshch. Khim.*, **41**, 2664
106. Robinson, C. N., Li, P. K. and Addison, J. F. (1972). *J. Org. Chem.*, **37**, 2939
107. Gazizov, T. Kh., Pashinkin, A. P., Dmitrieva, G. V., Tuzova, L. L., Khairullin, V. K. and Pudovik, A. N. (1972). *Zh. Obshch. Khim.*, **42**, 1730
108. Gazizov, T. Kh., Unsyanina, M. A., Pashinkin, A. P., Anoshina, N. P., Gol'dfarb, Z. I. and Pudovik, A. N. (1971). *Zh. Obshch. Khim.*, **41**, 1957
109. Burns, P., Capozzi, G. and Haake, P. (1972). *Tetrahedron Lett.*, 925
110. Maier, L. (1973). *Helv. Chim. Acta*, **56**, 489
111. Maier, L. (1973). *Helv. Chim. Acta*, **56**, 1252
112. Daigle, D. J., Vigo, T. L. and Chance, L. H. (1973). *J. Chem. Eng. Data*, **18**, 108
113. Weingarten, H. and Wager, J. S. (1972). *Syn. Inorg. Metal-org. Chem.*, **2**, 7
114. Mukhacheva, O. A., Nikolaeva, V. G., Shchelkunova, M. A. and Razumov, A. I. (1973). *Zh. Obshch. Khim.*, **43**, 1240
115. Ismailov, V. M., Moskva, V. V., Novruzov, S. A., Razumov, A. I., Akhmedov, Sh. T., Zykova, T. V. and Salakhutdinov, R. A. (1973). *Zh. Obshch. Khim.*, **43**, 212

116. Ismailov, V. M., Zykova, T. V., Moskva, V. V., Novruzov, S. A., Razumov, A. I., Akhmedov, Sh. T. and Salakhutdinov, R. A. (1973). *Zh. Obshch. Khim.*, **43**, 1247. .
117. Wegener, W. and Scholz, P. (1972). *Z. Chem.*, **12**, 137
118. Ivasyuk, N. V. and Shermegorn, I. M. (1971). *Zh. Obshch. Khim.*, **41**, 2199
119. Issleib, K. and Franze, K. D. (1973). *J. Prakt. Chem.*, **315**, 471
120. Kondo, K. and Tunemoto, D. (1972). *J. Chem. Soc. Chem. Commun.*, 952
121. Grim, S. O. and Matienzo, L. J. (1973). *Tetrahedron Lett.*, 2951
122. Mironova, Z. N., Tsvetkov, E. N., Petrovskaya, L. I., Nearebetskii, V. V., Nikolaev, A. V. and Kabachnik, M. I. (1972). *Zh. Obshch. Khim.*, **42**, 2152
123. Frank, A. W. and Drake, G. L., Jr. (1972). *J. Org. Chem.*, **37**, 2752
124. Issleib, K. and Lischewski, M. (1972). *J. Organometal. Chem.*, **46**, 297
125. Moedritzer, K. (1972). *Syn. Inorg. Metal-org. Chem.*, **2**, 317
126. Rogozhin, R. V., Davankov, V. A. and Belov, Yu. P. (1973). *Izv. Akad. Nauk SSSR, Ser. Khim.*, 955
127. Berry, J. P., Isbell, A. F. and Hunt, G. E. (1972). *J. Org. Chem.*, **37**, 4396
128. Isbell, A. P., Berry, J. P. and Tansey, L. W. (1972). *J. Org. Chem.*, **37**, 4399
129. Soroka, M. and Mastalerz, P. (1973). *Tetrahedron Lett.*, 5201
130. Szczepaniak, W. and Siepak, J. (1973). *Rocz. Chem.*, **47**, 929
131. Giron-Forest, D. and Thomas, G. (1972). *Bull. Soc. Chim. Fr.*, **1**, 296
132. Rachon, J. and Wasielewski, C. (1973). *Z. Chem.*, **13**, 254
133. Tyka, R. (1972). *Pr. Nauk Inst. Chim. Org. Fiz. Politech, Wroclaw*, No. 4, 54; (1972). *Chem. Abstr.*, **77**, 114 505
134. Taube, D. O., Vovsi, B. A. and Ionin, B. I. (1972). *Zh. Obshch. Khim.*, **42**, 351
135. Gilmore, W. F. and McBride, H. A. (1972). *J. Amer. Chem. Soc.*, **94**, 4361
136. Asano, S., Kitahara, T., Ogawa, T. and Masano, M. (1973). *Agric. Biol. Chem.*, **37**, 1193; (1973). *Chem. Abstr.*, **79**, 66 468
137. Ploeger, W., Schindler, N., Wollmann, K. and Worms, K. H. (1972). *Z. Anorg. Allgem. Chem.*, **389**, 119
138. Moedritzer, K. (1973). *Syn. Inorg. Metal-org. Chem.*, **3**, 75
139. Krueger, F. and Bauer, L. (1972). *Chem. Ztg.*, **96**, 691; (1973). *Chem. Abstr.*, **78**, 84 492
140. (a) Maier, L. (1973). *Helv. Chim. Acta*, **56**, 1257; (b) Maier, L. (1971). *Phosphorus*, **1**, 67
141. Okamoto, Y., Takao, I. and Sakurai, H. (1973). *Yukogaku*, **22**, 81; (1973). *Chem. Abstr.*, **78**, 136 369
142. Petrova, I. P. and Kirilov, M. (1971). *Dokl. Bolg. Akad. Nauk*, **24**, 1183
143. Heep, U. (1973). *Annalen*, 578
144. Arbuzov, B. A., Novosel'skaya, A. D. and Vinogradova, V. S. (1972). *Izv. Akad. Nauk SSSR, Ser. Khim.*, 1153
145. Akamsin, V. D., Eliseenkova, R. M. and Rizpolozhenskii, N. I. (1973). *Izv. Akad. Nauk SSSR, Ser. Khim.*, 80
146. Shishkin, V. E., Elfimova, S. N. and No, B. I. (1972). *Zh. Obshch. Khim.*, **42**, 1165
147. Schöllkopf, U. and Schröder, R. (1973). *Tetrahedron Lett.*, 633
148. Satgé, J., Couret, C. and Escudié, J. (1973). *J. Organometal. Chem.*, **34**, 83
149. Shevchuk, M. I., Volyuskaya, E. M. and Dombrovskii, A. V. (1971). *Zh. Obshch. Khim.*, **41**, 1999
150. Doroshenko, V. V., Stukale, V. A., Shokol, V. A. and Kozhushko, B. N. (1971). *Zh. Obshch. Khim.*, **41**, 2155
151. Pudovik, A. N., Gareev, R. D., Remizov, A. B., Aganov, A. V., Evstaf'ev, G. I. and Shtil'man, S. E. (1973). *Zh. Obshch. Khim.*, **43**, 559
152. Welter, W. and Regitz, M. (1972). *Tetrahedron Lett.*, 3799
153. Rozantsev, E. G., Suskina, V. I., Ivanov, Yu. A. and Kaspruk, B. I. (1973). *Izv. Akad. Nauk SSSR, Ser. Khim.*, 1327
154. Wieber, M. and Eichhorn, B. (1973). *Chem. Ber.*, **106**, 2733
155. Zagorets, P. A., Shostenko, A. G. and Dodonov, A. M. (1971). *Zh. Obshch. Khim.*, **41**, 2171
156. Vasyanina, M. A., Khairullin, V. K., Dmitrieva, G. V. and Anoshina, N. P. (1972). *Izv. Akad. Nauk SSSR, Ser. Khim.*, 1834
157. Lavielle, G., Carpentier, M. and Savignac, P. (1973). *Tetrahedron Lett.*, 173
158. Koos, E. W., Vander Kooi, J. P., Green, E. E. and Stille, J. K. (1972). *J. Chem. Soc. Chem. Commun.*, 1085

159. Roy, N. K. and Mukerjee, S. K. (1972). *Indian J. Chem.*, **10**, 1159
160. Kozlov, E. S. and Gaidamaka, S. N. (1972). *Zh. Obshch. Khim.*, **42**, 106
161. Kolodyazhnyi, O. I., Samarai, L. I. and Gaidamaka, S. N. (1971). *Zh. Obshch. Khim.*, **41**, 1872
162. Holý, A. (1972). *Tetrahedron Lett.*, 157
163. Maier, L. (1971). *Phosphorus* **1**, 111
164. Maier, L. (1971). *Phosphorus*, **1**, 105
165. Audley, B. G. and Archer, B. L. (1973). *Chem. Ind. (London)*, 634
166. Maier, L. (1972). *Phosphorus*, **1**, 249
167. Levin, Ya. A. and Pyrkin, R. I. (1973). *Zh. Obshch. Khim.*, **43**, 283
168. Bestmann, H. J. and Eckart, K. (1972). *Chem. Ber.*, **105**, 2098
169. Gorzny, K. (1971). *Z. Naturforsch.*, **26b**, 1193
170. Sartori, P. and Thomzik, M. (1972). *Z. Anorg. Allgem. Chem.*, **394**, 157
171. Demuth, R. and Grobe, J. (1973). *J. Fluorine Chem.*, **2**, 263
172. Ang, H. G. (1973). *J. Fluorine Chem.*, **2**, 181
173. Kang, D. K. and Burg, A. B. (1972). *J. Chem. Soc. Chem. Commun.*, **7**, 763
174. Birchall, J. M., Haszeldine, R. N. and Roberts, D. W. (1973). *J. Chem. Soc. Perkin Trans. I*, 1071
175. Gilje, J. W., Braun, R. W. and Cowley, A. H. (1973). *J. Chem. Soc. Chem. Commun.*, 813
176. Cavell, R. G. and Sanger, A. R. (1973). *Inorg. Nucl. Chem. Lett.*, **9**, 461
177. Prons, V. N., Grinblat, M. P., Klebanskii, A. L. and Zaitsev, N. B. (1971). *Zh. Obshch. Khim.*, **41**, 2629
178. Prons, V. N., Grinblat, M. P. and Klebanskii, A. L. (1973). *Zh. Obshch. Khim.*, **43**, 692
179. Mishra, I. B. and Burg, A. B. (1972). *Inorg. Chem.*, **11**, 664
180. Gosling, K. and Miller, J. L. (1973). *Inorg. Nucl. Chem. Lett.*, **9**, 355
181. Demuth, R. and Grobe, J. (1973). *J. Fluorine Chem.*, **2**, 269
182. Demuth, R. and Grobe, J. (1973). *J. Fluorine Chem.*, **2**, 299
183. Ansari, S., Grobe, J. and Schmid, P. (1973). *J. Fluorine Chem.*, **2**, 281
184. Grobe, J. and Demuth, R. (1972). *Angew. Chem. Int. Ed. Engl.*, **11**, 1097
185. Ang, H. G., Redwood, M. E. and West, B. O. (1972). *Aust. J. Chem.*, **25**, 493
186. Knunyants, I. L., Bukhovskaya, E. G., Volkovitskii, V. N., Plotnikov, V. F., Galakhov, I. V. and Ragulin, L. I. (1972). *Zh. Vses. Khim. Obshchest.*, **17**, 598; (1973). *Chem. Abstr.*, **78**, 29 902
187. Fild, M., Stelzer, O. and Schmutzler, R. (1973). *Inorg. Syn.*, **14**, 4
188. Baudler, M. and Zarkadas, A. (1972). *Chem. Ber.*, **105**, 3844
189. Bushweller, C. H. and Brunelle, J. A. (1973). *J. Amer. Chem. Soc.*, **95**, 5949
190. Cavell, R. G., Leary, R. D., Sanger, A. R. and Tomlinson, A. J. (1973). *Inorg. Chem.*, **12**, 1374
191. Bugurenko, E. F., Petukhova, A. S. and Chernyshev, E. A. (1972). *Zh. Obshch. Khim.*, **42**, 168
192. Aliev, M. I., Israfilova, S. Z., Podovaev, N. I. and Israelyan, D. R. (1971). *Vop. Neftekhim.*, No. 3, 78; (1972). *Chem. Abstr.*, **76**, 113 404
193. Liptuga, N. I., Vasil'ev, V. V. and Derkach, G. I. (1972). *Zh. Obshch. Khim.*, **42**, 293
194. Couret, C., Satgé, J. and Couret, F. (1972). *Inorg. Chem.*, **11**, 2274
195. Couret, C., Satgé, J. and Couret, F. (1973). *J. Organometal. Chem.*, **47**, 67
196. Couret, C., Escudié, J. and Satgé, J. (1972). *Rec. Trav. Chim. Pays-Bas*, **91**, 429
197. Schwarz, L. P. and Keller, P. C. (1972). *Inorg. Chem.*, **11**, 1931
198. Levin, Ya. A. and Pyrkin, R. I. (1973). *Zh. Obshch. Khim.*, **43**, 578
199. Nurtdinov, S. Kh., Khairullin, R. S., Zykova, T. V., Tsivunin, V. S. and Kamai, G. Kh. (1971). *Zh. Obshch. Khim.*, **4**, 2158
200. Murtdinov, S. Kh., Khairullin, R. S., Tsivunin, V. S., Zykova, T. V. and Kamai, G. Kh. (1972). *Zh. Obshch. Khim.*, **42**, 123
201. Moskva, V. V., Bashirova, L. A. and Razumov, A. I. (1971). *Zh. Obshch. Khim.*, **41**, 2577
202. Carey, F. A. and Court, A. S. (1972). *J. Org. Chem.*, **37**, 939
203. Schweizer, E. E. and Labaw, C. S. (1973). *J. Org. Chem.*, **38**, 3069
204. Pudovik, A. N., Gareev, R. D., Strabrovskaya, L. A., Evstaf'ev, G. I. and Remizov, A. B. (1972). *Zh. Obshch. Khim.*, **42**, 80

205. Pudovik, A. N., Gareev, R. D., Stabrovskaya, L. A. and Aganov, A. V. (1973). *Zh. Obshch. Khim.*, **43**, 1236
206. Schweizer, E. E. and Kopay, C. M. (1972). *J. Org. Chem.*, **37**, 1561
207. Dogadina, A. V., Mingaleva, K. S., Ionin, B. I. and Petrov, A. A. (1972). *Zh. Obshch. Khim.*, **42**, 2186
208. Fokin, A. V., Kolomiets, A. F. and Shchennikov, V. S. (1972). *Zh. Obshch. Khim.*, **42**, 801
209. Rakov, A. P. and Alekseev, A. V. (1973). *Zh. Obshch. Khim.*, **43**, 276
210. Maier, L. (1972). *Z. Anorg. Allgem. Chem.*, **394**, 11
211. (a) Razumov, A. I., Liorber, B. G., Sokolov, M. P. and Zykova, T. V. (1971). *Zh. Obshch. Khim.*, **41**, 2106; (b) Moskva, V. V., Razumov, A. I., Sazonova, Z. Ya. and Zykova, T. V. (1971). *ibid.*, **41**, 1874; (c) Razumov, A. I., Liorber, B. G., Sokolov, M. P. and Zykova, T. V. (1971). *ibid.*, **41**, 2106
212. Kostyanovskii, R. G., El'natanov, Yu. I. and Plekhanov, V. G. (1971). *Izv. Akad. Nauk SSSR, Ser. Khim.*, 2355
213. Danion, D. and Carrie, R. (1972). *Tetrahedron*, **28**, 4223
214. Morel, G., Seux, R. and Foucand, A. (1972). *C.R. Acad. Sci., Ser. C*, **275**, 629
215. Rizpolozhenskii, N. I., Akamsin, V. D. and Eliseenkova, R. M. (1973). *Izv. Akad. Nauk SSSR, Ser. Khim.*, 85
216. Sokolov, M. P., Zykova, T. V., Salakhutdinov, R. A., Liorber, B. G. and Razumov, A. I. (1972). *Tr. Kazan. Khim.-Tekhnol. Inst.*, No. 50, 123; (1973). *Chem. Abstr.*, **79**, 18 807
217. Razumov, A. I., Liorber, B. G., Zaikonnikova, I. V., Urazaeva, L. G. and Tarzivolova, T. A. (1972). *Khim. Farm. Zh.*, **6**, 24; (1973). *Chem. Abstr.*, **78**, 72 296
218. Mathey, F. and Muller, G. (1972). *Tetrahedron*, **28**, 5645
219. Lines, E. L. and Centofanti, L. F. (1973). *Inorg. Chem.*, **12**, 598
220. Chattha, M. S. and Aguiar, A. M. (1973). *J. Org. Chem.*, **38**, 1611
221. Maumy, M. (1972). *Bull. Soc. Chim. Fr.*, 1600
222. Simalty, M. and Mebazaa, M. H. (1972). *Bull. Soc. Chim. Fr.*, 3532
223. Pudovik, A. N., Khusainova, H. G. and Nasybullina, Z. A. (1973). *Zh. Obshch. Khim.*, **43**, 1683
224. Tamm, L. A., Chistokletov, V. N. and Petrov, A. A. (1972). *Zh. Obshch. Khim.*, **42**, 1864
225. Tamm, L. A., Chistokletov, V. N. and Petrov, A. A. (1972). *Zh. Obshch. Khim.*, **42**, 192
226. Chattha, M. S. and Aguiar, A. M. (1973). *J. Org. Chem.*, **38**, 820
227. Chattha, M. S. and Aguiar, A. M. (1973). *J. Org. Chem.*, **38**, 2908
228. Hagens, W., Bos, N. J. T. and Arens, J. F. (1973). *Russ. Cast. Prod.*, **92**, 762; (1973). *Chem. Abstr.*, **79**, 92 328
229. Simalty, M. and Ramos, J. J. M. (1972). *C.R. Acad. Sci., Ser. C*, **274**, 2105
230. Simpson, R. T., Jacobson, S., Carty, A. J., Mathew, M. and Palenik, G. J. (1973). *J. Chem. Soc. Chem. Commun.*, 388
231. Guillerm, D. and Capman, M. L. (1972). *Tetrahedron*, **28**, 3559
232. Berlan, J., Capman, M. L. and Chodkiewicz, W. (1971). *C.R. Acad. Sci., Ser. C*, **273**, 1107
233. Horner, L. and Binder, V. (1971). *Phosphorus*, **1**, 17
234. Dogadina, A. V., Ionin, B. I. and Petrov, A. A. (1972). *Zh. Obshch. Khim.*, **42**, 1919
235. Mashlyakovskii, L. N., Zagudaeva, T. A., Ionin, B. I., Okhrimenko, I. S. and Petrov, A. A. (1972). *Zh. Obshch. Khim.*, **42**, 2648
236. Bentham, J. E., Ebsworth, E. A. V., Moretto, H. and Rankin, D. W. H. (1972). *Angew. Chem. Int. Ed. Engl.*, **11**, 640
237. Taylor, R. C. and Walters, D. B. (1973). *Inorg. Syn.*, **14**, 10
238. Kordosky, G., Cook, B. R., Cloyd, J. C., Jr. and Meek, D. W. (1973). *Inorg. Syn.*, **14**, 14
239. Kuchen, W. and Koch, K. (1972). *Z. Anorg. Allgem. Chem.*, **394**, 74
240. King, R. B. and Cloyd, J. C., Jr. (1972). *Z. Naturforsch.*, **27b**, 1432
241. King, R. B., Cloyd, J. C., Jr. and Hendrick, P. K. (1973). *J. Amer. Chem. Soc.*, **95**, 5083
242. Henning, H. G. and Forner, T. (1973). *Z. Chem.*, **13**, 55
243. Vizgert, R. V. and Volozhin, M. P. (1971). *Zh. Obshch. Khim.*, **41**, 1991

244. Maryanoff, C., Maryanoff, B. E., Tang, R. and Mislow, K. (1973). *J. Amer. Chem. Soc.*, **95**, 5839
245. Gloede, J. (1972). *J. Prakt. Chem.*, **314**, 281
246. Lin, K. (1900). *U.S. Pat.* 3 681 481
247. Quimby, D. T., Cilley, W. A., Prentice, J. B. and Nicholson, D. A. (1973). *J. Org. Chem.*, **38**, 1867
248. Pudovik, A. N., Batyaeva, E. S. and Zamaletdinova, G. U. (1973). *Zh. Obshch. Khim.*, **43**, 680
249. Birum, G. H. and Wilson, J. P. (1972). *J. Org. Chem.*, **37**, 2730
250. Nicholson, D. A. (1972). *Phosphorus*, **2**, 143
251. Gilmore, W. F. and Huber, J. W., III. (1973). *J. Org. Chem.*, **38**, 1423
252. Kulumbetova, K. Zh., Medved, T. Ya. and Kabachnik, M. I. (1971). *Izv. Akad. Nauk SSSR, Ser. Khim.*, 2747
253. Gross, H., Costisella, B., Brennecke, L. and Engelhardt, G. (1972). *J. Prakt. Chem.*, **314**, 969
254. Gross, H. and Costisella, B. (1972). *J. Prakt. Chem.*, **314**, 87
255. Kukhar, V. P., Pasternak, V. I. and Kirsanov, A. V. (1972). *Zh. Obshch. Khim.*, **42**, 1169
256. Wegener, W. and Scholz, P. (1972). *Z. Chem.*, **12**, 103
257. Dickstein, J. I. and Miller, S. I. (1972). *J. Org. Chem.*, **37**, 2168
258. Kovtun, V. Yu., Gilyarov, V. A. and Kabachnik, M. I. (1972). *Izv. Akad. Nauk SSSR, Ser. Khim.*, 2612
259. Kovtun, V. Yu., Gilyarov, V. A. and Kabachnik, M. I. (1971). *Izv. Akad. Nauk SSSR, Ser. Khim.*, 2217
260. Schweizer, E. E., Kin, C. S., Labaw, C. S. and Murray, W. P. (1973). *J. Chem. Soc. Chem. Commun.*, 7
261. Sveits, Yu. A., Borinsenko, A. A., Foss, V. L. and Lutsenko, I. F. (1973). *Zh. Obshch. Khim.*, **43**, 440
262. Foss, V. L., Veits, Yu. A., Kudinova, V. V., Borisenko, A. A. and Lutsenko, I. F. (1973). *Zh. Obshch. Khim.*, **43**, 1000
263. Kabachnik, M. I., Medved, T. Ya., Pisareva, S. A., Matrosov, E. I. and Petrovskii, P. V. (1973). *Izv. Akad. Nauk. SSSR, Ser. Khim.*, 119
264. Medved, T. Ya., Psiareva, S. A., Poliskarpov, Yu. M. and Kabachnik, M. I. (1971). *Izv. Akad. Nauk SSSR, Ser. Khim.*, 2839
265. Vedejs, E. and Snoble, K. A. J. (1973). *J. Amer. Chem. Soc.*, **95**, 5778
266. Koszmehl, G. and Bohn, B. (1973). *Angew. Chem. Int. Ed. Engl.*, **12**, 737
267. Utimoto, K., Tamura, M. and Sisido, K. (1973). *Tetrahedron*, **29**, 1169
268. Schubert, U. and Fischer, E. O. (1973). *Chem. Ber.*, **106**, 1062
269. Kirilov, M. and van Huyen, L. (1972). *Tetrahedron Lett.*, 4487
270. Heitz, W. and Michels, M. (1973). *Annalen*, 227
271. Heitz, W. and Michels, R. (1972). *Angew. Chem. Int. Ed. Engl.*, **11**, 298
272. McKinley, S. V. and Rakshys, J. W., Jr. (1972). *J. Chem. Soc. Chem. Commun.*, 134
273. Schweizer, E. E., Wehman, A. T. and Nycz, D. M. (1973). *J. Org. Chem.*, **38**, 1583
274. Tokunaga, H., Akiba, K. and Inamoto, N. (1972). *Bull. Chem. Soc. Jap.*, **45**, 506
275. Heydenreich, F., Mollbach, A., Wilke, G., Dreeskamp, H., Hoffman, E. G., Schroth, G., Seevogel, K. and Stempfle, W. (1972). *Isr. J. Chem.*, **10**, 293
276. Nesmeyanov, N. A., Berman, S. T. and Reutov, O. A. (1972). *Izv. Akad. Nauk SSSR, Ser. Khim.*, 605
277. Van Leusen, A. M., Reith, B. A., Iedema, A. J. W. and Strating, J. (1972). *Rec. Trav. Chim. Pays-Bas*, **91**, 37
278. Kato, S., Kato, T., Mizuta, M., Itoh, K. and Ishii, Y. (1973). *J. Organometal. Chem.*, **51**, 167
279. da Silva, R. R., Toscano, V. G. and Weiss, R. G. (1973). *J. Chem. Soc. Chem. Commun.*, 567
280. Grieco, P. A. and Finkelhor, R. S. (1972). *Tetrahedron Lett.*, 3781
281. Burton, D. J. and Naae, D. G. (1973). *J. Amer. Chem. Soc.*, **95**, 8467
282. Devlin, C. J. and Walker, B. J. (1972). *Tetrahedron*, **28**, 3501
283. Wilson, I. F. and Tebby, J. C. (1972). *J. Chem. Soc. Perkin Trans. I*, 31
284. Ykman, D., Mathys, G., L'abbe, G. and Smets, G. (1972). *J. Org. Chem.*, **37**, 3213
285. Cooke, M. P., Jr. (1973). *J. Org. Chem.*, **38**, 4082

286. Taylor, J. D. and Wolf, J. F. (1972). *J. Chem. Soc. Chem. Commun.*, 876
287. Mastryukova, T. A., Aladzheva, I. M., Matrosov, E. I. and Kabachnik, M. I. (1972). *Zh. Obshch. Khim.*, **42**, 1470
288. Shevchuk, M. I., Khalturnik, M. V. and Dombrovskii, A. V. (1971). *Zh. Obshch. Khim.*, **41**, 2146
289. Waite, N. E., Allen, D. W. and Tebby, J. C. (1971). *Phosphorus*, **1**, 139
290. Wilson, J. F. and Tebby, J. C. (1972). *J. Chem. Soc. Perkin Trans. I*, 2830
291. Hamada, A. and Takizawa, T. (1972). *Tetrahedron Lett.*, 1849.
292. Connor, D. T. and von Strandtmann, M. (1973). *J. Org. Chem.*, **38**, 1047
293. Devlin, C. J. and Walker, B. J. (1973). *J. Chem. Soc. Perkin Trans. I*, 1428
294. Douglas, J. E. (1972). *Inorg. Chem.*, **11**, 654
295. Zefirov, N. S. and Makhon'kov, D. I. (1973). *Zh. Org. Khim.*, **9**, 851; (1973). *Chem. Abstr.*, **79**, 5408
296. Soulen, R. L., Carlson, S. D. and Lang, F. (1973). *J. Org. Chem.*, **38**, 479
297. Howells, M. A., Howells, R. D., Baenziger, N. C. and Burton, D. J. (1973). *J. Amer. Chem. Soc.*, **95**, 5366
298. Ammon, H. L., Wheeler, G. L. and Watts, P. H., Jr. (1973). *J. Amer. Chem. Soc.*, **95**, 6158
299. Yoshida, Z. I., Yoneda, S., Yato, T. and Hazama, M. (1973). *Tetrahedron Lett.*, 873
300. Gray, G. A. (1973). *J. Amer. Chem. Soc.*, **95**, 5092
301. Schmidbaur, H., Buchner, W. and Scheutzov, D. (1973). *Chem. Ber.*, **106**, 1251
302. (a) Schmidbaur, H. and Stühler, H. (1972). *Angew. Chem. Int. Ed. Engl.*, **11**, 145; (b) Schmidbaur, H., Stühler, H. and Buchner, W. (1973). *Chem. Ber.*, **106**, 1238
303. Schmidbaur, H., Mitschke, K.-H. and Weidlein, J. (1972). *Angew. Chem. Int. Ed. Engl.*, **11**, 144
304. Schmidbaur, H., Mitschke, K.-H., Buchner, W., Stühler, H. and Weidlein, J. (1973). *Chem. Ber.*, **106**, 1226
305. Schmidbaur, H. and Vornberger, W. (1972). *Chem. Ber.*, **105**, 3173
306. Schmidbaur, H., Stühler, H. and Vornberger, W. (1972). *Chem. Ber.*, **105**, 1034
307. Issleib, K. and Lischewski, M. (1973). *Syn. Inorg. Metal-org. Chem.*, **3**, 255
308. Karsch, H. H. and Schmidbaur, H. (1973). *Angew. Chem. Int. Ed. Engl.*, **12**, 853
309. Kurras, E., Rosenthal, U., Mennenga, H. and Oehme, G. (1973). *Angew. Chem. Int. Ed. Engl.*, **12**, 854
310. De'Ath, N. J., Denney, D. Z. and Denney, D. B. (1972). *J. Chem. Soc. Chem. Commun.*, 272
311. Mathey, F. and Bensoum, J. (1972). *C.R. Acad. Sci., Ser. C*, **274**, 1095
312. Morse, K. W. and Morse, J. G. (1973). *J. Amer. Chem. Soc.*, **95**, 8469
313. Seel, F., Gombler, W. and Velleman, K. D. (1972). *Annalen*, **756**, 181
314. Ahrens, U. and Falius, H. (1972). *Chem. Ber.*, **105**, 3317
315. Seel, F. and Velleman, K. D. (1972). *Chem. Ber.*, **105**, 406
316. Dietz, E. A., Jr. and Martin, D. R. (1972). *Inorg. Chem.*, **12**, 241
317. Marynoff, B. E., Tang, R. and Mislow, K. (1973). *J. Chem. Soc. Chem. Commun.*, 273
318. Luckenbach, R. (1973). *Synthesis*, 307
319. Ruveda, M. A., Zerba, E. N. and de Moutier Aldao, E. M. (1972). *Tetrahedron*, **28**, 501
320. Ruveda, M. A., Zerba, E. N. and de Moutier Aldao, E. M. (1973). *Tetrahedron*, **29**, 585
321. Roesky, H. W. and Dietl, M. (1973). *Angew. Chem. Int. Ed. Engl.*, **12**, 425
322. Schmidbaur, H. and Vornberger, W. (1972). *Chem. Ber.*, **105**, 3187
323. Pinchuk, A. M., Suleimanova, M. G. and Filonenko, L. P. (1972). *Zh. Obshch. Khim.*, **42**, 2115
324. Wolfsberger, W., Pickel, H. H. and Schmidbaur, H. (1971). *Z. Naturforsch.*, **26b**, 979
325. Stadelmann, W., Stelzer, O. and Schmutzler, R. (1971). *Z. Anorg. Allgem. Chem.*, **385**, 142
326. DeBruin, K. E. and Johnson, D. M. (1973). *J. Amer. Chem. Soc.*, **95**, 4675
327. DeBruin, K. E. and Petersen, J. R. (1972). *J. Org. Chem.*, **37**, 2272
328. DeBruin, K. E. and Chandrasekaran, S. (1973). *J. Amer. Chem. Soc.*, **95**, 974
329. Farnham, W. B., Mislow, K., Mandel, N. and Donohue, J. (1972). *J. Chem. Soc. Chem. Commun.*, 120

330. Fahmy, M. A. H., Khasawinah, A. and Fukuto, T. R. (1972). *J. Org. Chem.*, **37**, 617
331. Cook, R. D. and Abbas, K. (1973). *Tetrahedron Lett.*, 519
332. Arcoria, A., Fisichella, S. and Sardo, F. (1972). *J. Org. Chem.*, **37**, 2612
333. Frampton, R. D., Tidwell, T. T. and Young, V. A. (1972). *J. Amer. Chem. Soc.*, **94**, 1271
334. Bunnett, J. F., Edwards, J. O., Wells, P. V., Brass, H. J. and Curci, R. (1973). *J. Org. Chem.*, **38**, 2703
335. Harger, M. J. P. (1973). *J. Chem. Soc. Chem. Commun.*, 774
336. Tyssee, D. A., Bausher, L. P. and Haake, P. (1973). *J. Amer. Chem. Soc.*, **95**, 8066
337. Koizumi, T. and Haake, P. (1973). *J. Amer. Chem. Soc.*, **95**, 8073
338. Gorenstein, D. G. (1973). *J. Amer. Chem. Soc.*, **95**, 8060
339. Noyce, D. S. and Virgilio, J. A. (1972). *J. Org. Chem.*, **37**, 1052
340. Cook, R. D., Turley, P. C., Diebert, C. E., Fiermon, A. H. and Haake, P. (1972). *J. Amer. Chem. Soc.*, **94**, 9260
341. Cann, P. F., Howells, D. and Warren, S. (1972). *J. Chem. Soc. Perkin Trans. II*, 304
342. Howelis, D. and Warren, S. (1973). *J. Chem. Soc. Perkin Trans. II*, 1472
343. Horner, L. and Mueller, E. (1972). *Phosphorus*, **2**, 77
344. Mikolajczyk, M. and Omelanczuk, J. (1973). *Phosphorus*, **3**, 47
345. Mikolajczyk, M. and Omelanczuk, J. and Para, M. (1972). *Tetrahedron*, **28**, 3855
346. Bushweller, C. H., Brunelle, J. A., Anderson, W. G. and Bilofsky, H. S. (1972). *Tetrahedron Lett.*, 3261
347. Robert, J. B. and Roberts, J. D. (1972). *J. Amer. Chem. Soc.*, **94**, 4902
348. Baechler, R. D. and Mislow, K. (1972). *J. Chem. Soc. Chem. Commun.*, 185
349. Benezra, C. (1973). *J. Amer. Chem. Soc.*, **95**, 6890
350. Taylor, R. C. and Walters, D. B. (1972). *Tetrahedron Lett.*, 63
351. Kotajczyk, M. M. and Omelanczuk, J. (1972). *Tetrahedron Lett.*, 1539
352. Albrand, J. P., Gagnaire, D. and Robert, J. B. (1973). *J. Amer. Chem. Soc.*, **95**, 6498
353. Mann, B. E. (1972). *J. Chem. Soc. Perkin Trans. II*, 30
354. Simonnin, M.-P., Lequan, R.-M. and Wehrli, F. W. (1972). *Tetrahedron Lett.*, 1559
355. Gray, G. A. and Cremer, S. E. (1972). *J. Org. Chem.*, **37**, 3458
356. Gray, G. A. and Cremer, S. E. (1972). *J. Org. Chem.*, **37**, 3470
357. Haemers, M., Ottinger, R., Zimmermann, D. and Reisse, J. (1973). *Tetrahedron*, **29**, 3539
358. Borisenko, A. A., Sergeyev, N. M., Nifant'ev, E. Ye. and Ustynyuk, Yu. A. (1972). *J. Chem. Soc. Chem. Commun.*, 406
359. Gray, G. A. (1973). *J. Amer. Chem. Soc.*, **95**, 7736
360. Simonnin, M. P., Lequan, R. M. and Wehrli, F. W. (1972). *J. Chem. Soc. Chem. Commun.*, 1204
361. Engel, R. and Gelbaum, L. (1972). *J. Chem. Soc. Perkin Trans. I*, 1233
362. McFarlane, W. and Rycroft, D. S. (1972). *J. Chem. Soc. Chem. Commun.*, 902
363. Pinnell, R. P., Megerle, C. A., Manatt, S. L. and Kroon, P. A. (1973). *J. Amer. Chem. Soc.*, **95**, 977
364. Fild, M. and Althoff, W. (1973). *J. Chem. Soc. Chem. Commun.*, 933
365. Steo, W. J., van Wazer, J. R. and Goodard, N. (1972). *J. Chem. Soc. Perkin Trans. II*, 463
366. McFarlane, H. C. E. and McFarlane, W. (1972). *J. Chem. Soc. Chem. Commun.*, 1189

6
Sulphur Compounds

D. R. HOGG
University of Aberdeen

6.1   COMPOUNDS OF DIVALENT SULPHUR

6.1.1   Thiols

6.1.1.1   *Preparation*

Methanolysis of 1-acetyl-2-alkyl-2-thiopseudourea hydroiodide, AcNHC-
$(:\overset{+}{N}H_2)SR\ I^-$, prepared from 1-acetyl-2-thiourea and the alkyl halide, gives[1]
the alkanethiol, methyl iodide and the substituted urea. The absence of a
hydrolytic step makes this a useful method for the synthesis of water-sensitive
or base-sensitive thiols. Convenient syntheses are reported[2] for dihydro-
asparagusic acid, $(HSCH_2)_2CHCO_2H$, which stimulates pyruvate oxidations
in asparagus and in *Streptococcus faecalis 10C1*, and shows inhibitory action
in higher plants.

6.1.1.2   *Nucleophilic substitution and addition reactions*

Thiolate ions react with bromonaphthalenes in boiling DMSO to give[3] the
alkyl naphthyl sulphides directly. In refluxing DMF, the ethylthiolate ion
rapidly and selectively demethylates[4] aryl methyl ethers. Aryl bromides are
not affected under these conditions and ethers of polyhydric phenols are
usually monodemethylated. α-Halogenoesters, $R^1R^2CXCO_2Et$ (X = Cl or
Br) are reduced with thiolate ions in a second-order process[5], which, it is
suggested, involves nucleophilic substitution at halogen, or successively at
carbon and at sulphur. Limitations of the rule of *trans* nucleophilic addition
of thiols to alkynes have been studied[6]. Activating substituents containing a
conjugated carbonyl group, e.g. CO_2Me, COMe, give rise to appreciable
amounts of the *cis*-adduct in methanolic solution, presumably by an initial
1,4-addition to give the enol. α-Mercaptoketones give[7] 2,5-dihydrothiophens

with triphenylvinylphosphonium bromide by nucleophilic addition and an intramolecular Wittig reaction of the intermediate ylide.

6.1.1.3 *Protection and blocking of thiol groups*

Selenol esters acylate[8] thiol groups in acidic media. Amino groups do not react and the method is suitable for molecules of biological interest. The selenols formed are removed by oxidation and the protecting group by aminolysis:

$$R^1COSeCH_2CH_2\overset{+}{N}Me_3 + R^2SH \rightarrow R^1COSR^2 + HSeCH_2CH_2\overset{+}{N}Me_3$$

The SH groups in cysteine are carbamoylated in 2 min at pH 4–5 in an aqueous medium at room temperature by methoxymethyl isocyanate[9]. The derivative decomposes at pH 9.6 within 30 min:

$$MeOCH_2N{=}C{=}O + RSH \rightarrow MeOCH_2NHCOSR$$

Cysteine and 4-picolyl chloride give[10] a derivative which is stable in acid solution and the 4-picolyl group can be removed by electrolytic reduction.

Mechanisms of radiation protection by aminothiols have been reviewed[11], as has hydrogen bonding involving the thiol group[12].

6.1.2 Sulphides

6.1.2.1 *Preparation*

Sulphides containing hydrolytically unstable functions can be prepared[13] in high yield by treating a thiol ester, having at least one α-hydrogen atom, with a strong non-nucleophilic base in ether or DMSO, and then with an alkylating agent. It is suggested that a ketene and a thiolate ion are formed initially:

$$R^1R^2CHCOSR^3 \xrightarrow[\text{DMSO}]{\text{Ph}_3\text{CLi}} R^1R^2C{=}C{=}O + R^3S^- \xrightarrow{\text{R}^4\text{X}} R^3SR^4$$

'Methanetrithiolactic acid', $CH(SCHMeCO_2H)_3$, has been synthesised[14] and three of the four optically active forms have been isolated.

6.1.2.2 *Nucleophilic reactions*

The decomposition of benzoyl peroxide is accelerated by aliphatic sulphides and, to a lesser extent, by disulphides, which are considered[15] to assist nucleophilically in the heterolysis of the oxygen–oxygen bond. This heterolytic reaction, which competes with the normal homolysis of the bond, gives the sulphoxide (1) or thiolsulphinate, respectively, and also the α-benzoyloxy derivative (2) if the sulphide has an α-hydrogen atom.

$$R^1CH_2SR^2 + (PhCO_2)_2 \longrightarrow R^1CH_2\overset{+}{S}R^2OCOPh\ \bar{O}_2CPh$$

via the ylide

$$R^2SCHR^1OCOPh + PhCO_2H \qquad R^1CH_2S(O)R^2 + (PhCO)_2O$$

$$(2) \qquad\qquad\qquad\qquad (1)$$

Anchimeric assistance by the β-sulphur atom in the solvolysis of *trans*-1,2-dimethyl-2-methylthiovinyl 2,4,6,-trinitrobenzenesulphonate and the intermediacy of thi-irenium ions has[16] been established. The energy difference between the linear and the cyclic structure of formula $C_2H_3S^+$ is calculated[17] to be *ca.* 4 kcal mol^{-1}. Three-membered cyclic sulphonium salts are formed more rapidly[18] than their five-membered analogues in the solvolysis of ω-halogenoalkyl *p*-tolyl sulphides, $p\text{-MeC}_6H_4S(CH_2)_nX$ (X = Cl or Br). The results are not in accord with the acceleration of three-membered ring formation by conjugation with the aryl substituent. The formation of bicarbamic esters (3) from sulphides and azodicarboxylic esters[19] is consistent with association of the sulphur atom with the azo linkage, intramolecular hydrogen transfer and migration of the nitrogen from sulphur to carbon.

$$R^1CH_2SR^2 + R^3O_2CN{=}NCO_2R^3 \longrightarrow R^1CH(SR^2)N(CO_2R^3)NHCO_2R^3$$

$$(3)$$

6.1.2.3 Carbanion reactions

Alkenes, e.g. (4), can[20] be obtained from sterically hindered or moderately acidic ketones by reaction with phenylthiomethyl-lithium, acylation and

$$R^1R^2CO \xrightarrow[\text{THF}]{\text{PhSCH}_2\text{Li}} R^1R^2C(\bar{O})CH_2SPh \xrightarrow{(PhCO)_2O} R^1R^2C(OCOPh)CH_2SPh$$

Li–NH$_3$

$$R^1R^2C{=}CH_2$$

$$(4)$$

reductive elimination. The method complements the Wittig reaction. The difficulties associated with alkylidene transfer from an ylide can be circumvented by using sulphur-stabilised anions[21]. Thus the intermediate from the addition of phenylthiomethyl-lithium to aldehydes or ketones, on alkylation gives the sulphonium salt which is converted into the epoxide by treatment with base. The first example of a Wittig-type rearrangement of sulphides and thiols has been reported[22]. Benzylthiotrimethylsilane gives the thiol

$PhCH(SH)SiMe_3$ with t-butyl-lithium; the starting material can be reformed by heating the product with a radical source.

6.1.2.4 Chlorosulphonium salts and α-chlorosulphides

Alcohols are oxidised[23] to aldehydes or ketones in the presence of triethylamine by the complexes (5) formed by addition of halogens to sulphides.

$$R^1SMe \xrightarrow[-25\,°C]{Cl_2-CCl_4} [R^1\overset{+}{S}MeCl]\;Cl^- \xrightarrow{R^2R^3CHOH} R^2R^3CHO\overset{+}{S}RMe$$
$$(5)$$

$$\downarrow{Et_3N}$$

$$R^2R^3CO$$

The reaction is unsuccessful with alcohols which can give stabilised carbocations, as alkyl chlorides are formed. This can be circumvented by using[24] a complex of chlorine and DMSO at $-45\,°C$ as the chlorinating agent. At $80\,°C$, bromodimethylsulphonium bromide brominates alcohols in good yield[25] with inversion of configuration, thus providing a useful synthetic method. α-Chlorosulphides react[26] with zinc and an α-bromoketone to give γ-ketosulphides. The formation of methylthiomethyl esters, RCO_2CH_2SMe, by using chloromethyl methyl sulphide is suggested[27] as a method for protecting carboxylic acids. The acid is recovered by treatment of the ester with trifluoroacetic acid at room temperature.

6.1.2.5 Unsaturated sulphides

The thioenol ether $Me(C≡C)_2CH=C(SMe)(CH=CH)_3H$, isolated[28] from *Buphthalmum salicifolium L.*, has been characterised and synthesised. Alk-1-ynyl allenyl sulphides and alk-1-ynyl alk-2-ynyl sulphides rearrange at room temperature to give[29] a thioketene (6), which can be trapped as the thioamide with a secondary amine. The thioamide cyclises to give a thiophen derivative (7) with base.

$$\begin{array}{c} R^1C≡C \\ \diagdown S \\ H_2C=C=CH \diagup \end{array} \xrightarrow[\text{rearrangement}]{\text{[3,3]-sigmatropic}} \left[\begin{array}{c} R^1C=C=S \\ | \\ H_2C-C≡CH \end{array}\right]$$
$$(6)$$

$$\downarrow{R^2{}_2NH}$$

$$\begin{array}{c} R^1C=CNR^2{}_2 \\ |\diagdown S \\ \diagup \\ HC=CMe \end{array} \xleftarrow[\text{HMPT}]{Bu^tO^-} \begin{array}{c} R^1CHCSNR^2{}_2 \\ | \\ CH_2-C≡CH \end{array}$$
$$(7)$$

The anions of alk-1-ynyl or allenyl alkyl sulphides react with benzyne to give[30] alk-3-ynyl aryl sulphides, which presumably are formed by reaction of the benzyne with the sulphur atom, intramolecular hydrogen transfer from the alkyl group and a [2,3]-sigmatropic rearrangement. 1,4-Dithi-ins are obtained[31] in good yield by treating dialk-1-ynyl sulphides with sodium sulphide in liquid ammonia–methanol.

Reaction of allyl vinyl sulphide with s-butyl-lithium in THF at $-78\,^{\circ}C$ gives the anion, which is successively alkylated and heated in aqueous dimethoxyethane containing calcium carbonate to give[32] the aldehyde (8) by a thio-Claisen rearrangement. 1-(Alkylthio)vinyl-lithium, $R^1CH=C$-

$$CH_2=CHCH_2SCH=CH_2 \xrightarrow[\text{(ii) RX}]{\text{(i) Bu}^s\text{Li}} CH_2=CHCHRSCH=CH_2$$

$$\xleftarrow[\text{CaCO}_3,\ \text{heat}]{\text{DME-H}_2\text{O}}$$

$$RCH=CH(CH_2)_2CHO$$

$$(8)$$

$(SR^2)Li$, reacts[32] smoothly with alkyl halides, aldehydes or epoxides to give, after hydrolysis with mercury(II) chloride in aqueous acetonitrile, or with titanium(IV) chloride in acetic acid[33] and then with water, ketones, acyloins or αβ-unsaturated ketones, respectively.

Dialkyl disulphides are inert to sulphurisation with sulphur in DMSO below 90 °C. The introduction of an allyl substituent gives rearranged allyl disulphides by an intramolecular [2,3]-sigmatropic shift. Some of the principal reactions proposed[34] for allyl methyl sulphide are shown below.

$$RS^{14}CH_2CH=CH_2 + S_8 \rightleftharpoons R\overset{+}{S}(^{14}CH_2CH=CH_2)\overset{-}{S}_8 \rightleftharpoons CH_2=CH^{14}CH_2SR=S$$

$$^{14}CH_2=CHCH_2\overset{+}{S}(SR)\overset{-}{S}_n \rightleftharpoons {}^{14}CH_2=CHCH_2SSR$$

6.1.3 Uses of 1,3-dithianes and related compounds in synthesis

6.1.3.1 1,3-Dithianes

The lithium derivatives of 1,3-dithianes (9) are readily alkylated and the products are subsequently hydrolysed to give ketones. This is illustrated by a synthesis[35] of 1,4-diketones and hence *cis*-jasmone. Reaction of the lithium derivatives with dimethyl disulphide gives[36] the corresponding orthothioformates, which give esters in good yield on ethanolysis in the presence of mercury(II) salts. Similarly, hydrolysis gives the acid.

(9)

2-Ethoxycarbonyl-1,3-dithiane, prepared from ethyl diethoxyacetate and propane-1,3-dithiol, is alkylated by successive treatment with NaH in DMF and an alkyl halide to give[37] the intermediate (10), which may be reduced to the ester or, more significantly, oxidatively hydrolysed to the α-keto-ester.

$$RCH_2CO_2Et \xleftarrow{Ni(R)} (10) \xrightarrow{NBS} RCOCO_2Et$$

1,3-Dithianes react with triphenylmethyl tetrafluoroborate to give 1,3-dithienium ions, which undergo Diels–Alder addition with dienes and thus provide a method[38] for their conversion into cyclopenten-4-ones (11).

Hg^{II}–$CaCO_3$ in aqueous $COMe_2$

(11)

6.1.3.2 Ketene thioacetals

Ketene thioacetals, e.g. (12), are obtained[39] by treating 1,3-dithianes or thioacetals successively with butyl-lithium, trimethylsilyl chloride, butyl-lithium and a wide variety of aldehydes or ketones. Hydrolysis gives acids but reduction and hydrolysis gives aldehydes. Vinylketene thioacetals,

(12)

H^+–H_2O

$R^1R^2CHCO_2H$

NBS–$MeCN$–H_2O

R^1R^2CHCHO

prepared as above by using αβ-unsaturated aldehydes and ketones, undergo cycloaddition with reactive dienophiles and hence, after hydrolysis, give[40] cyclohexenones. Michael-type addition occurs[41] with alkyl-lithium compounds and vinylketene thioacetals, giving lithio derivatives which may be further alkylated. α-Dimethylthiomethylene ketones (13) are obtained[42] by treating a ketone having an α-methylene group with lithium 4-methyl-2,6-di-t-butylphenolate and CS_2 followed by methylation. Synthetically the process could be used to protect the α-methylene group or to introduce an α-t-butyl, α-isopropylidene or an α-carboxy group. The use of NaH as an

$$(Ar = 2,6,4\text{-}Bu^t{}_2MeC_6H_2) \qquad (13) \qquad (96\%)$$

alternative to the phenolate ion has been suggested[43]. A wide variety of dethioacetalising agents have been reported[44].

The reaction of lactones with bis(dimethylaluminium)ethane-1,2-dithiolate gives ketene thioacetals, which rearrange to substituted 1,3-dithiolanes (14). These compounds are inert to a wide range of nucleophiles, but are converted into the lactone with neutral aqueous mercury(II) ions. The method is recommended[45] for the protection of lactones.

$$(14)$$

6.1.3.3 Thioacetal monoxides

Anions of thioacetal monoxides (15) may be used as carbonyl anion equivalents. Methyl methylthiomethyl sulphoxide gives α-hydroxyaldehydes[46] on successive treatment with butyl-lithium, an aldehyde or ketone, and acidic THF. Reaction with an aromatic aldehyde and Triton B followed by desulphuration in ethanolic acid gives[47] the arylacetic ester in good yield. The

$$R^1SCH_2SR^2=O \xrightarrow[\text{(ii) } R^3X]{\text{(i) } Bu^nLi} R^1SCHR^3SR^2=O \xrightarrow[\text{(ii) } CH_2=CHCO_2Me]{\text{(i) } Bu^nLi} R^1SCR^3(SR^2=O)(CH_2)_2CO\cdots$$

(15)

$$\downarrow \begin{array}{l}\text{(i) } Bu^nLi \\ \text{(ii) } R^4CHO\end{array}$$

$$R^1SCH(SR^2=O)CHR^4\bar{O} \xrightarrow[\text{(ii) } KOH \text{ benzene}]{\text{(i) } AcCl} R^1SC(SR^2=O)=CHR^4$$

monoalkylated anion undergoes[48] Michael addition reactions and is acylated[49] with acid chlorides but not with esters. Addition of the unsubstituted anions to aldehydes followed by acetylation and elimination gives ketene thioacetal monoxides, which act[50] as Michael acceptors. All products are hydrolysed smoothly to carbonyl compounds with mercury(II) chloride in aqueous THF.

Useful synthetic advances involving organosulphur compounds which have been developed since the publication of Houben-Weyl-Müller, Vol. IX, have been reviewed[51].

6.1.4 Disulphides

6.1.4.1 Preparation

Treatment of thiols with an excess of dimethyl disulphide in the presence of triethylamine gives[52] unsymmetrical methyl disulphides in good yields. Dithiols react with lead acetate, giving[53] the lead dithiolates which give disulphides with sulphur in benzene. This method is relevant to the synthesis of lipoic acid and related compounds. Sulphurated sodium borohydride, $NaBH_2S_3$, converts[54] aldehydes and ketones into disulphides, and epoxides into disulphide diols. The relatively inaccessible α-sulphinyl disulphides, $R^1SOCH_2SSR^2$, are readily obtained[55] by the thermolysis of thiolsulphinates in wet benzene. A sulphinylsulphonium ion, $R^1SO\overset{+}{S}(SR^2)Me$, is postulated as an intermediate. ($-$)-Bis-(3-acetoxyundec-5-enyl) disulphide has been isolated[56] from the Hawaiian seaweed of the genus *Dictyopteris*. It may be a precursor of the undeca-1,3,5-trienes found in the essential oil.

6.1.4.2 Disproportionation and nucleophilic substitution

For[57] the equilibrium:

$$R^1SSR^1 + R^2SSR^2 \rightleftharpoons 2R^1SSR^2$$

at 25 °C, $K \approx 4$ for R^1 = alkyl, except for R^1 = Bu^t. Sodium 4-(2-acetamidoethyldithio)butanesulphinate, $AcNH(CH_2)_2S_2(CH_2)_4SO_2Na$, disproportionates rapidly[58] compared with the sulphone or the sulphonate; equilibrium is attained in 30 min at 61 °C in water. This is attributed to neighbouring group participation by the sulphinate group, which gives an intermediate 1,2-dithiane 1,1-dioxide. Terminal carboxylate groups have a similar but weaker effect. The above intermediate reacts with Na_2S to give[59] the trisulphide, $[NaSO_2(CH_2)_4S]_2S$, which shows promise as an anti-radiation drug. 1,2-Dithiolane-4-carboxylic acid forms[60] a linear polymer, $-[SCH_2CH(CO_2H)CH_2S]_n-$, on being heated. Alkaline hydrolysis reforms the cyclic monomer and subsequently gives 2-(mercaptomethyl)acrylic acid, presumably by a β-elimination mechanism. Dithiobis(methylcyclopropane-1-carboxylic acid) is considered[60] to be hydrolysed by a mechanism involving direct substitution at sulphur. The reaction of disulphides with sodium methanesulphinate, which gives the thiolsulphonate, is catalysed[61] by silver nitrate, and thus provides another example of electrophilic–nucleophilic assistance in the fission of sulphur–sulphur bonds. Carbanions derived from esters may be sulphenylated with[62] dimethyl disulphide. Oxidation to the sulphoxide and thermolysis then gives the $\alpha\beta$-unsaturated ester (16).

$$R^1R^2CHCHR^3CO_2Et \xrightarrow[\text{(ii) MeSSMe}]{\text{(i) } LiNR^+_2,\ THF} R^1R^2CHCR^3(SMe)CO_2Et$$

$$\Big\downarrow \begin{array}{l}\text{(i) } NaIO_4 \\ \text{(ii) Heat}\end{array}$$

$$R^1R^2C{=}CR^3CO_2Et$$

$$(16)$$

6.1.4.3 *Unsaturated disulphides*

Dialk-1-enyl disulphides, some of which play an important role in the organoleptic quality of onion flavour, give[63] 3,4-dialkylthiophens at 150–200 °C in the presence of $KHSO_4$. It is suggested that the reaction involves a [3,3]-sigmatropic rearrangement to a dithial, which polymerises and eliminates H_2S.

6.1.5 Sulphenic acids and their derivatives

6.1.5.1 *Sulphenic acids*

Alkanesulphenic acids, obtained by the thermolysis of alkyl thiolsulphinates (17), can be trapped[64] by addition to alkynes or Michael acceptors. It is

$$MeS \xrightarrow[H_2C=CHCO_2Et]{96\,°C,\ 8\text{–}16\,h} MeS(O)(CH_2)_2CO_2Et$$
$$(90\%)$$

$$(17)$$

further suggested[65] that sulphenic acids are intermediates in the thermal disproportionation of unsymmetrical thiolsulphinates. Thiolsulphinates without α-hydrogen atoms undergo[64] a similar reaction to give a new class of compounds, thiosulphoxylic acids, e.g. Bu^tSSOH, which can be trapped similarly. Alkanesulphenic acids are also suggested[66] to be formed by the acid-catalysed ring-opening of episulphoxides, and are trapped as the disulphide with thiols. Further evidence has been adduced[67] that sulphenic acids are much stronger nucleophiles towards sulphenyl sulphur than water, and the reason that they are not normally observed as intermediates in the hydrolysis of sulphenyl derivatives is that they react with the sulphenyl derivative faster than it is hydrolysed.

6.1.5.2 *Sulphenyl halides*

Methanesulphenyl tetrafluoroborate, prepared from the sulphenyl chloride and silver tetrafluoroborate, reacts[68] with cyclohexene to give an episulphonium ion, which can react with a subsequently added nucleophile. A stable episulphonium salt is also obtained[69] by treating the adduct of methanesulphenyl chloride and α-methylacrylamide with silver tosylate in acetonitrile. In contrast, the reaction of 2,3-dimethylbut-1-ene or -2-ene with methanesulphenyl 2,4,6-trinitrobenzenesulphonate gave[70] 2,3,5,6,6,7-hexamethyl-5-thionia-oct-2-ene 2,4,6-trinitrobenzenesulphonate, $Me_2CHCMe_2\overset{+}{S}MeCH_2CMe{=}CMe_2$ TNBS$^-$, by a complex reaction. *trans*-Stilbene reacts[71] with an excess of methanesulphenyl chloride to give a mixture of *cis*- and *trans*-α-chloro-α′-methylthiostilbene, the formally dehydrogenated 1 : 1 adducts. A mechanism is proposed in which the initial step is the formation of a thiomethylsulphonium salt from the adduct and the sulphenyl chloride. The intermediate sulphenyl iodide obtained[72] from the reaction of pent-4-enyl thiosulphate, or the corresponding disulphide, with iodine in ethanol is trapped as 2-ethoxymethyltetrahydrothiophen in 80% yield by intramolecular addition to the double bond.

Sulphenyl chlorides have been suggested[73] to be intermediates in the conversion of 3-arylpropanoic acids into benzo[*b*]thiophens (18) with thionyl

$$ArCH_2CH_2CO_2H \xrightarrow[\text{pyridine}]{SOCl_2} ArCH_2CCl(SCl)COCl \longrightarrow ArCHClCCl(SCl)COCl$$

$$ArCH{=}CClCOCl \; + $$

(18)

chloride and pyridine. A similar reaction of 3-(2,6-dichlorophenyl)propanoic acid gave a mixture of the two sulphenyl halides, corresponding to those postulated above as intermediates, and the substituted acrylyl chloride. The β-chlorination is considered to involve elimination and addition rather than a radical process. Trichloromethanesulphenyl chloride reacts[74] with triethylamine to give 1,1-bis(trichloromethylthio)-2-(*N,N*-diethylamino)ethylene, $(Cl_3CS)_2C{=}CHNEt_2$. The initial reaction, it is suggested, is an α-chlorination, which is followed by elimination and sulphenyl halide addition. Methanesulphenyl chloride in the presence of tetraethylammonium chloride adds[75] to the cyano group of sulphonyl cyanides in THF to give *N*-methylthio-(alkylsulphonyl)formimidoyl chlorides, $RSO_2CCl{=}NSMe$. The sulphenylation of ketones by treating the lithium enolate with sulphenyl halides at $-100\,°C$, or with the disulphides at room temperature, has been explored[76].

(19)   (20)

(R¹ = H or Me)

Ethane- or propane-1,2-disulphenyl chloride react[77] with aldehydes in ethyl acetate to give 2-substituted 1,3-dithiolane-2-carboxaldehydes (19), which result from cyclic disulphenylation.

Chlorocarbonylsulphenyl chloride, ClCOSCl, reacts[78] with the amides of aliphatic acids to give 5-alkyl-1,3,4-oxathiazol-2-ones (20), which were previously prepared from trichloromethanesulphenyl chloride.

Thiols are converted[79] into alkyl chlorides in good yield by successive treatment with chlorocarbonylsulphenyl chloride and triphenylphosphine, thus making these functions synthetically equivalent:

$$\text{RSH} + \text{ClCOSCl} \rightarrow \text{RSSCOCl} \xrightarrow{\text{Ph}_3\text{P}} \text{RCl} + \text{Ph}_3\text{P(S)} + \text{COS}$$

Alternating copolydisulphides are obtained[80] by treating chlorocarbonylsulphenyl chloride (21) with dithiols and then with a dithiol in basic solution. An intermediate sulphenyl dithiocarbonate is postulated, which reacts with the thiol to liberate carbonyl sulphide.

$$\text{ClCOSCl} \xrightarrow{\text{Z(SH)}_2} \text{ClCOS}_2\text{ZS}_2\text{COCl} \xrightarrow[\text{Et}_3\text{N}]{\text{Y(SH)}_2} \left(\text{YS}-\overset{\text{O}}{\overset{\|}{\text{C}}}-\text{S}-\text{SZS}_2\overset{\text{O}}{\overset{\|}{\text{C}}}-\text{S}\right)_n$$

$$\text{SYSH}$$

$$(\text{YSSZSS})_n + \text{COS}$$

(Y and Z are alkylene chains)

6.1.5.3 Sulphenates

The ethanolysis of *p*-methoxybenzyl trichloromethanesulphenate, which gives[81] ethyl *p*-methoxybenzyl ether by an essentially unimolecular process, provides the first authentic example of carbon–oxygen bond fission in sulphenate solvolysis. The rearrangement of sulphoxides to sulphenates has been postulated as a key step in several mechanisms. Phenyl benzenethiolsulphinate and bis(methoxymethyl) ether, which are obtained[82] from the decomposition of methoxymethyl phenyl sulphoxide, are the expected decomposition products of the sulphenate. The formation of allyl dimethyl phosphate and methyl *p*-tolyl sulphide from the reaction[83] of allyl *p*-tolyl sulphoxide and trimethyl phosphite in THF is best explained by sulphenate formation followed by reaction of the phosphite at the oxygen atom. Similarly, the reaction[84] of secondary amines with bromomethyl *p*-tolyl sulphoxide to give the sulphenamides may involve rearrangement and substitution.

6.1.5.4 Other derivatives

Reaction[85] of benzylsulphinylmethylenemalonic acid with a mixture of acetic acid and hydrochloric acid gives the stable sulphenyl carboxylate (22). It

$$\text{(22)}$$

reacts with thiols to give unsymmetrical disulphides, which readily dispro-portionate on alkaline hydrolysis owing to anchimeric assistance. Sodium benzylthiosulphate reacts[86] with 1- or 2-naphthol, with or without zinc chloride, to give benzyl 1- or 2-naphthyl sulphide instead of the expected benzylthionaphthols. A mechanism similar to that of the Bucherer reaction is proposed. The 2-aminoethanethiosulphate ion reacts with the thiobenzoate ion to give[87] the N-benzoyl disulphide. An initial reaction at the sulphenyl sulphur atom by the thiobenzoate ion giving the benzoyl disulphide is proposed; this transfers the benzoyl group to the nitrogen atom by an intra-molecular process, leaving a hydrodisulphide ion which is reduced by the sulphite ion:

$$NH_2(CH_2)_2SSO_3{}^- \xrightarrow{\ PhCOS^-\ } [NH_2(CH_2)_2SSCOPh] \xrightarrow{\ SO_3{}^{2-}\ } [PhCONH(CH_2)_2S]_2$$

Unsymmetrical thiolsulphonates can be prepared[88] from the thiol and an excess of a trifluoromethyl alkanethiolsulphonate:

$$R^1SH + R^2SO_2SCF_3 \rightarrow R^2SO_2SR^1 + CF_3SH$$

Thiolsulphinates are obtained in good yield from[89] the photosensitised oxidation of dialkyl disulphides.

6.1.6 Thiono compounds

6.1.6.1 *Thioketones and related compounds*

The conversion of carbonyl compounds into thiono compounds by tetraphosphorus decasulphide is best[90] effected in acetonitrile containing sodium hydrogen carbonate. O,O-Diethyl hydrogen phosphorodithioate, $(EtO)_2P(S)SH$, in benzene at $80\,°C$ is suggested[91] as an alternative reagent. Thiophilic addition of the alkyl moiety in alkylmagnesium halides has been observed for thiopinacolone[92] and unsaturated thioketones[93]:

$$R^1R^2C{=}S + R^3MgX \rightarrow R^1R^2C(SR^3)MgX \rightarrow R^1R^2CHSR^3$$

The intermediate organomagnesium compound can[94] undergo homolytic fission of the carbon–metal bond to give various products resulting from hydrogen transfer and radical combination. Reaction of ethyl (trimethyl-silylthio)acetate, $Me_3SiSCH_2CO_2Et$, with aromatic aldehydes in the presence of NaH gives[95], after acidification, the hitherto unknown α-mercaptoacrylic esters $ArCH{=}C(SH)CO_2Et$ in good yield. No evidence for the existence of the tautomeric thione could be obtained by n.m.r. spectroscopy. β-Mercapto-crotonates are obtained[96] from the keto-ester and H_2S in acidic acetonitrile at $-60\,°C$; at higher temperatures the *gem*-dithiols are formed. With O-ethyl α-chlorothionoacetate under basic conditions, β-mercaptocrotonates give[97] the thiophen derivatives (23). β-Aryl-α-mercaptoacrylic acids add to Schiff's bases to give[98] 4-thiazolidinone derivatives (24); similar reactions occur with ω-nitrostyrenes and phenyl isothiocyanate. Thioketones (X = alkyl), dithiocarboxylic acids (X = SR), thioamides (X = NH$_2$) and, in general, compounds containing a carbon–sulphur double bond, add[99] to benzonitrile oxide, prepared *in situ*, to give 1,4,2-oxathiazoles (25).

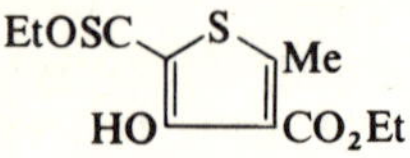

(23)   (24)   (25)

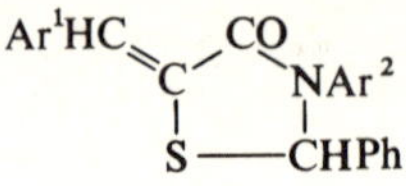

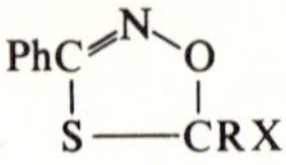

6.1.6.2 Thioureas

The preparation of *N*-substituted thioureas from isothiocyanates and amines, and their practical applications, have been reviewed[100]. Acyl isothiocyanates, RCON=C=S, and secondary amines similarly[101] give acylthioureas, which can be hydrolysed by acid to the 1,1-disubstituted thioureas. 1-Alkyl-1-cyanothioureas are obtained[102] by treating the recently obtained alkyldicyanamides, RN(CN)$_2$, with H$_2$S in ether. Unsymmetrically substituted thioureas have been found[103] to disproportionate by dissociation into the amine and the isothiocyanate. 1,3-Thiazinium salts (26; Z = NH$_2$ or Ar)

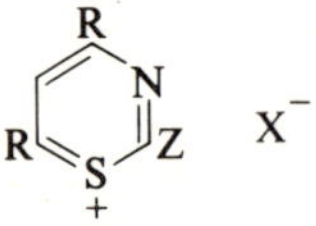

(26)

are obtained[104] by condensing thiourea with β-dicarbonyl compounds, or thioamides with αβ-unsaturated carbonyl compounds. Thioureas and thioamides are converted into the oxygen analogues with acidic DMSO at room temperature[105].

6.1.6.3 Thiocarbamates, thiocarbonates and xanthates

Contrary to earlier reports, *N*-monosubstituted dithiocarbamates, R^{1}NHCSSR^2, are obtained[106] by mixing equimolar amounts of the thiol and the isothiocyanate. The reaction is reversible above 100 °C and on electron impact. *S*-(2-Acylalkyl)dithiocarbamates, R^{1}COCHR^{2}SCSNR^{3}R^4, can be obtained[107] by the action of CS$_2$ on ketone Mannich bases, or by the displacement of the amino group from such bases with a dithiocarbamate anion. The readily accessible alkyl(alkoxythiocarbonyl)carbamates, e.g. (27), and alkyl 4,4-dialkyl-3-thioallophanates, R^{1}O$_2$CNHCSNR^2$_2$, show considerable potential as intermediates in heterocyclic synthesis[108] by virtue of the reactions of their methylated derivatives with difunctional nucleophiles, which can react at the 1,3- or the 3,3-position.

$$\text{ClC(OEt)}_3 \xrightarrow[\text{MeOH}]{\text{KCNS}} \underset{(27)}{\text{EtO}_2\text{CNHCSOMe}} \xrightarrow[\text{(ii) NH}_2\text{OH}]{\text{(i) Me}_2\text{SO}_4 \text{ OH}^-} \quad \overset{\text{CO}}{\underset{\text{MeOC}=\!\!=\text{N}}{\text{HN}\diagup\!\!\diagdown\text{O}}}$$

Reaction of alcohols with O-(4-methylphenyl)chlorothiocarbonate, prepared from p-cresol and thiophosgene, gives[109] the thiocarbonate O-ester, which on thermolysis smoothly gives the alkene. The elimination apparently occurs by an ionic mechanism when a cyclic *syn*-elimination is not possible.

Xanthates are most efficiently prepared[110] by using the methylsulphinyl carbanion in DMSO as the base. The thermolysis temperature is lower in this medium, or if an aryl ester is used. These factors make the reaction more efficient with primary and tertiary alcohols. The thermolysis of neopentyl-type xanthate S-methyl esters gives[111] an alkene and a dithiolcarbonate, $R_3\text{CCH}_2\text{SCOSMe}$, by two competing processes. Formation of the latter does not involve skeletal rearrangement, and a concerted process involving a four-membered cyclic transition state is postulated. t-Alkyl xanthates react with secondary amines to give[112] dithiourethanes, $R^1\text{SCSNR}^2\text{R}^3$, in good yield by displacing the bulky alkoxide ion. In contrast, primary and secondary alkyl xanthates give the thionourethane, $R^1\text{OCSNR}^2\text{R}^3$.

6.1.6.4 Dithiocarboxylic acids and their derivatives

Dithioesters are readily obtained[113] by treating alkylmagnesium chlorides with CS_2 in THF followed by alkylation. The anion of diethyl malonate reacts[114] with CS_2 to give the dithioacid anion and hence, by loss of a further proton, the corresponding dianion (28). Alkylation of the reaction mixture gives a mixture of the dithioester and the ketene dithioacetal.

$$\text{CH}_2\text{XY} \xrightarrow[\text{(ii) CS}_2]{\text{(i) Base}} \{\text{CHYXCS}\bar{\text{S}} \rightleftharpoons \text{CXY}=\text{C(SH)}\bar{\text{S}}\} \xrightarrow{-\text{H}^\cdot} \text{CXY}=\text{CS}_2^{2-}$$
$$(28)$$
$$(X = \text{CO}_2\text{Et}; \; Y = \text{CO}_2\text{Et, CN})$$

Except for their reactions with amines to give thioamides, little attention has been given to dithiocarboxylic acids. Dithioacetic acid reduces[115] sulphoxides, sulphilimines and related compounds to sulphides, but sulphones are unaffected. Thiolacetic acid reacts similarly but much more slowly, the difference being attributed[115] to the greater acidity of dithioacetic acid ($pK_a = 2.55$). The dithiocarboxylic acids add electrophilically to alkenes and nucleophilically to Michael acceptors, thus providing a useful synthetic route[115] to complex dithioesters. In contrast to previous methods, dithioesters can now be converted[116] into esters in essentially quantitative yield with equimolar amounts of copper(II) oxide and copper(II) chloride in ethanol at 25 °C.

Thioformamides are converted[117] into aldehydes in good yield by alkylation and treatment with Grignard reagents. Other thioamides give ketones in lower yields:

$$\text{HCSNMe}_2 \xrightarrow{\text{MeI}} \text{Me}_2\overset{+}{\text{N}}=\text{CHSMe} \xrightarrow[\text{(ii) H}^+]{\text{(i) RMgX–THF}} \text{RCHO}$$

N,N-Dimethylthioformamide and lithium di-isopropylamide at $-100\,^{\circ}\mathrm{C}$ give[118] dimethylthiocarbamoyl-lithium, Me_2NCSLi, instead of the expected derivative, $S{=}CHNMeCH_2Li$. The former reacts smoothly with aldehydes, ketones and esters. Thioformamides are chlorinated[119] in boiling CCl_4 to give thiocarbamoyl chlorides. The influence of substituents on their stability has been discussed[119]. Oxidation of C-tosyl-N,N-dimethylthioformamide with most reagents gives[120] S-tosyl-N,N-dimethylthiourethane, $TsSCONMe_2$, presumably by a $C \rightarrow S$ rearrangement of the tosyl group in the unstable S-oxide intermediate. Oxidation with ozone, however, gives C-tosyl-N,N-dimethylformamide and SO_2. It is suggested that the ozone adds to the carbon–sulphur double bond, forming an intermediate which then decomposes. Thioamides are converted[121] into nitriles by heating them with phenylpropiolamidines in methanol. The facile fission of the carbon–sulphur bond is attributed to a fragmentation reaction of the addition product (29). Thioamides react with aldehydes in the presence of acid to give thioamido-alkyl ions, $R^1CS\overset{+}{N}H{=}CHR^2$, which add[122] to alkenes by a stereospecific cis-1,4-cycloaddition, forming 5,6-dihydro-$4H$-1,3-thiazines (30). This method avoids difficult syntheses of intermediates which are usual for this ring system.

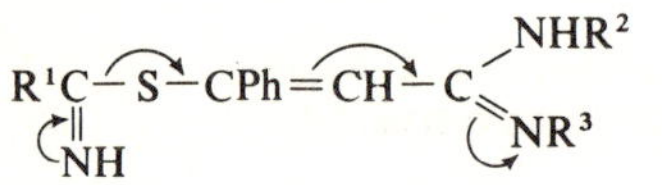

$$R^1C{-}S{-}CPh{=}CH{-}C\underset{NR^3}{\overset{NHR^2}{\diagup}}$$

(29)

$$\underset{S}{\overset{R^1C}{\diagdown}}\ \text{-1,3-thiazine ring with } CHR^2,\ CR^3R^4,\ CR^5R^6,\ N$$

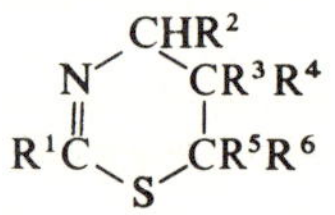

(30)

6.1.7 Thiolesters

Butyl dibutylthioboronite reacts with ketene and a carbonyl compound to give[123] β-hydroxythiolesters in excellent yield:

$$Bu^n_2BSBu^n + CH_2{=}CO + PhCHO \xrightarrow{\text{ether}} PhCH(OH)CH_2COSBu^n$$
$$(90\%)$$

The corresponding β-(alkylthio)thiolesters are obtained[124] by the insertion of ketene into a carbon–sulphur bond of the thioacetals $R^1CH(SR^2)_2$, or dithioacetals, $R^1R^2C(SR^3)_2$. The relatively inaccessible O-alkyl thioformates are formed[125] by treating alkyl dichloromethyl ethers with potassium O-ethyldithiocarbonate and heating the resultant bis-(O-ethyl dithiocarbonate) derivative, $ROCH(SCSOEt)_2$, at $200\,^{\circ}\mathrm{C}$. S-$(-)$-3-acetoxyundec-5-enyl thioacetate has been isolated[56] from a Hawaiian seaweed of the genus *Dictyopteris*.

Thermolysis of S-alkyl thiolacetates gives alkenes and thiolacetic acid by a mechanism similar to that for acetates[126], except that charge development on the α-carbon is less marked[127]. The rates of alkaline hydrolysis of thiolacetates in aqueous dioxan are correlated[128] by $\log k = 1.14 + 0.63\sigma^* + 0.23E_s$ and those for methyl thiolesters by $\log k = 1.14 + 0.95\sigma^* + 0.92E_s - 0.32 - (n - 3)$ where n is the number of α-hydrogen atoms. In contrast to the oxygen analogues, steric effects are of major importance. Interest in the mechanistic significance of the pH–rate profiles for the hydrolysis of thiol-esters[129] and in the intra-[130,131] and inter-molecular[131] catalysis of amino-lysis continues in view of their relevance to natural processes.

6.2 COMPOUNDS OF TETRAVALENT SULPHUR

6.2.1 Sulphonium salts

Sulphides react with 1-chlorobenzotriazole to give[132] 1-(dialkylsulphonium)-benzotriazole chlorides, or the corresponding covalent sulphuranes, which react with alcohols or amines in the presence of silver tetrafluoroborate to give alkoxysulphonium salts, $R^1_2\overset{+}{S}OR^2\ BF_4^-$, or aminosulphonium salts, $R^1_2\overset{+}{S}NR^2R^3\ BF_4^-$, respectively. Aminosulphonium salts, unlike alkoxy-sulphonium salts, are stable in aqueous media. N-Chlorosuccinimide[133] and chlorine[134] can also be used as the chlorinating agent. Alkoxy-sulphonium salts are also obtained[135] by treating DMSO with 1,2-epoxides in acid solution. Methoxy- and ethoxy-sulphonium salts, unlike the higher homo-logues, react[136] with dialkyl cadmium compounds at carbon to give the parent sulphoxide. Aminosulphonium salts derived from methyl sulphides and arylamines rearrange on treatment with a strong base to form[134,137] o-alkylthiomethylanilines by a [2,3]-sigmatropic rearrangement[137] of the intermediate ylide. β-Ketosulphides and α-alkoxycarbonyl sulphides simi-larly[134] give 3-methylthioindoles and 3-methylthio-oxindoles, respectively, in good to excellent yields. Thermolysis at 100 °C in DMF gives[134] a mixture of methyl sulphides. Buta-1,3-dienyldimethylsulphonium salts react in an alcohol with the alkoxide ion to give[138] substituted butenyl sulphides, $MeSCH_2CH(CH_2OR)CH=CH_2$, by a [2,3]sigmatropic rearrangement of the ylide formed after 1,4-addition of the alcohol.

6.2.2 Sulphoxides

6.2.2.1 Preparation

Sulphides react with sulphuryl chloride at −40 to −70 °C to give[139] dichloro-sulphuranes, which give sulphoxides on low-temperature hydrolysis without over-oxidation to sulphones. Either enantiomer of a chiral sulphoxide (31) can be obtained[140] by altering the order of the alkylation steps in the prepa-

ration from the chiral oxathiazolidine 2-oxide obtained from L-ephedrine. The alkylation proceeds with inversion. Optically active dialkyl sulphoxides can be prepared[141] from optically active alkyl aryl sulphoxides with alkyl-lithium compounds.

6.2.2.2 Halogenation

Reaction of sulphoxides with iodobenzene dichloride or with bromine and pyridine at $-40\,^{\circ}\text{C}$ gives[142] the α-halogenosulphoxide (32). The same product is formed from the sulphide by using two equivalents of halogenating

$$PhICl_2 \rightleftharpoons PhI + Cl_2$$

agent in aqueous pyridine. Iodobenzene dichloride acts only as a controlled source of chlorine and in some cases chlorine formation is rate limiting[143]. For substrates which are sterically hindered or which have electron withdrawing groups, e.g. 2-chlorophenyl methyl sulphoxide, the last step in the given mechanism is rate limiting. The lack of deuterium exchange, the isotope effect, $k_H/k_D = 5.5$, and the high negative ρ value (-2.55) are held[144] to indicate chlorination at sulphur, and a concerted chlorine cation migration and hydrogen abstraction involving retention at carbon and sulphur[145]. Alternatively, an intermediate ylide might be involved for which the rate of migration is greater than that of protonation. α-Chlorination occurs at the most highly substituted carbon atom of the sulphoxide, which is considered[146] to be in accord with a positive intermediate, such as an oxosulphenium ion, $R^1\overset{+}{S}(O){=}CHR^2$, rather than an ylide. In cyclic systems, however, chlorine is introduced *trans* to the lone pair on the sulphur[147], and an intermediate ylide, or *trans*-diaxial elimination of hydrogen chloride[148], is proposed.

Halogenation in the presence of silver nitrate involves inversion at sulphur. It is suggested[144] that the silver ion coordinates with the sulphur, thus promoting[145] migration of chlorine as an *anion*, and electron-pair transfer in a *trans*-periplanar conformation. In these chlorinations[144] the sulphoxide retains ^{18}O.

β, γ- or δ-hydroxyalkyl sulphoxides, and β-sulphinyl carboxylic acids or amides, give[146] the β-, γ- or δ-chloroalkyl sulphones, and the β-sulphonyl acid chlorides or nitriles, respectively, on chlorination. Similarly, the addition products of ketones and DMSO give[149] $\alpha\beta$-unsaturated methyl sulphones on chlorination. Oxygen transfer involving a cyclic intermediate is proposed. Reaction of alkyl t-butyl sulphoxides with *N*-halogenosuccinimides in chloroform–ethanol gives[150] the ethyl alkanesulphinate. The expulsion of a relatively stable carbocation from the chlorosulphoxonium ion was also noted[151] in the chlorination of benzhydryl benzyl sulphoxide, which gave benzhydryl chloride and toluene-α-sulphinic acid. t-Butyl β-, γ- and δ-hydroxy-alkyl sulphoxides give[152] excellent yields of β-, γ- and δ-sultines with sulphuryl chloride in dichloromethane. The suggested mechanism involves chloro-sulphoxonium ion formation, cyclisation and expulsion of the t-butyl cation.

6.2.2.3 Hydrogen exchange and reactions with acids

The relative acidities of the α-hydrogen atoms in sulphoxides are strongly solvent dependent[153] and therefore are not always in accord with those based on theoretical calculations. (RR/SS)-α-Monodeuteriobenzyl methyl sulphoxide (33) was treated with butyl-lithium in THF and then with water

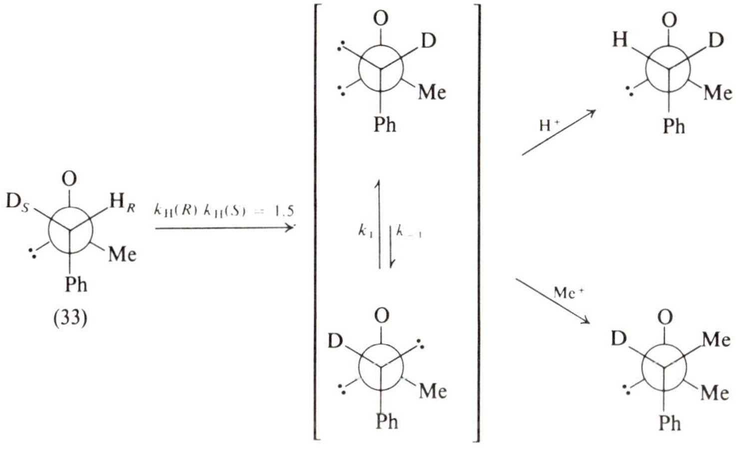

to give[154] the (RS/SR)-isomer (81%) with a high deuterium content (87%). The (RS/SR)-diastereoisomer, in contrast, on similar treatment also gave the (RS/SR)-isomer (71%), again with a high deuterium content (76%). These results and those involving quenching with methyl iodide agree with the following mechanism in which there is interconversion between the anions $(k_1/k_{-1} \approx 20)$, as is also found[155] for methyl 1-phenylethyl sulphoxide: preferential removal of the *pro-R* hydrogen in the (S)-sulphoxide $[k_H(R)/k_H(S) = 1.5]$ modified by an isotope effect; reprotonation with retention[156]; and methylation with inversion[156].

Equilibria and reactions of sulphoxides in moderately acidic media have been reviewed[157] and it has been confirmed[158] by ^{13}C n.m.r. that sulphoxides are protonated at oxygen rather than at sulphur.

6.2.2.4 Oxidation and reduction

Oxidation of sulphoxides to sulphones by organic hydroperoxide anions is retarded[159] by protic solvents and does not occur in aqueous media. The reduction in reactivity is attributed to solvation. It is suggested that reaction involves sulphur–oxygen bond formation followed by slow heterolysis of the oxygen–oxygen bond. Kinetic studies on the oxidation of DMSO by bromine[160] are consistent with a rapid reversible formation of the bromo-sulphoxonium ion, which is then slowly hydrolysed to the sulphone. This mechanism is similar to that for the oxidation of sulphides to sulphoxides by bromine.

Sulphoxides are rapidly and selectively reduced[161] to sulphides in excellent yield by dichloroborane, $HBCl_2$, in THF at 0 °C. Sulphones and common reducible groups are unaffected. This is one of the best of many methods. Other methods recently proposed include the use of sodium hydrogen sulphite[162], acetyl chloride[163], tin(II) chloride[164] and titanium(III) chloride[165].

6.2.2.5 *Other reactions*

The stereochemistry of substitution reactions at chiral sulphur involving reactions of sulphoxides, sulphilimines and sulphoximines has been reviewed[166]. Further details of the first monoligostatic (one common ligand) reaction cycle are reported[167] and additional di- and tri-ligostatic cycles described.

A further asymmetric synthesis of alcohols has been developed[168] in which optically active methyl *p*-tolyl sulphoxide is converted into the anion, added to an aldehyde, ketone or 1,2-epoxide and then desulphurised with Raney nickel. The addition products of lithio-sulphinyl carbanions and non-enolisable aldehydes or ketones are simultaneously dehydrated and desulphurised with *o*-phenylene phosphorochloridite to give alkenes[169]:

$$Ph_2CO + MeS(O)CH_2Li \xrightarrow{THF} Ph_2C(OLi)CH_2S(O)Me \xrightarrow{C_6H_4O_2PCl} Ph_2C=CH_2$$
$$(91\%)$$

Substituted allyl aryl sulphoxides (34) can be alkylated in the α-position and the sulphenate esters with which they are in equilibrium intercepted by trimethyl phosphite[170], or a thiol[171], to give the rearranged alcohols, thus providing a synthesis of trisubstituted ethylene derivatives. A similar synthesis

$$PhS(=O)CH_2CMe=CH_2 \xrightarrow[\text{(ii) RX}]{\text{(i) Bu}^n\text{Li}} PhS(=O)CHRCMe=CH_2$$

$$(34)$$
$$\swarrow (MeO)_3P$$

$$RCH=CMeCH_2OH$$

of trisubstituted ethylenes involving α-substituted methylallylsulphonium ylides has also been reported[172].

Optically active oxosulphonium salts, $EtMePh\overset{+}{S}(O)\,ClO_4{}^-$, have been prepared[173] by the methylation of optically active sulphoxides.

6.2.3 Sulphilimines

'Free' sulphilimines, $R_2S=NH$, are prepared[174] by treating *N-p*-tolylsulphilimines with concentrated H_2SO_4 and quenching the product with ether.

Acid-catalysed hydrolysis in 1–6 M acid gives the sulphoxide; kinetic studies[175] show that the rate-limiting step is nucleophilic attack by water at the sulphur atom of the protonated substrate. The free sulphilimines can also be prepared from the sulphide by reaction with O-mesitylenesulphonyl-hydroxylamine followed by anion exchange[176]. They readily decompose to the sulphide[174] but can be halogenated with N-halogenosuccinimides or calcium hypohalites to give[177] the N-halogenosulphilimines, which give the N-cyanosulphilimines with cyanide ion. Alkaline hydrolysis of N-chlorosulphilimines gives[177] the sulphoximine by nucleophilic reaction at sulphur. Sulphilimines are readily N-acylated[176]. They mediate the addition of the elements of ammonia to Michael acceptors and undergo substitution reactions with 1-fluoro-2,4-dinitrobenzene[176]. N-Carbamoylsulphilimines, $R_2S=NCONH_2$, are formed[178] by treating the sulphide with N-chlorourea in acetonitrile. Sulphilimines with β-hydrogen atoms, e.g. 1-phenylethyl phenyl N-tosylsulphilimine[179], give alkenes on being heated, by an E_i mechanism, although thermolysis in methanol gives the α-methoxysulphide. Optically active sulphilimines readily[180] racemise by pyramidal inversion on being heated. The slow rate of pyramidal inversion with sulphoxides and the greater rapidity with sulphonium salts is attributed to the decreasing strength of the 2p–3d π bond in the series, S=O, S=N and S—C. Sulphilimines are reduced[163] to sulphides with acetyl chloride.

6.2.4   Sulphuranes

The chemistry of π- and σ-sulphuranes has been reviewed[181]. σ-Sulphuranes contain a sulphur atom associated with four σ-bonds and a lone pair of electrons, whereas in π-sulphuranes it is associated with three σ-bonds, a π-bond and a lone-pair.

6.2.5   Sulphinic acids and their derivatives

Oxidation of aliphatic thiols with m-chloroperoxybenzoic acid (2 mole equivalents) in dichloromethane gives[182] sulphinic acids in high yields. Thiophthalimides, useful sulphenylating agents, are similarly oxidised by m-chloroperoxybenzoic acid to the N-alkylsulphinylphthalimides, which readily transfer[183] the alkylsulphinyl group to nucleophiles and which are more stable than sulphinyl chlorides. The effect of external factors and substitution on the ambient reactivity of sulphinate ions has been studied[184]. Hydroxymethanesulphinate ion adds to 1,2- and 1,4-benzoquinone to give[185] the symmetrical dihydroxyaryl sulphones and formaldehyde. Simple alkyl sulphinates have been resolved[186] by inclusion complex formation with cyclodextrins.

Chlorination of tricholoroalkyl sulphoxides in dichloromethane provides a potentially useful route to sulphinyl chlorides[187]:

$$ClCH_2S(O)CHCl_2 \xrightarrow[CH_2Cl_2]{Cl_2} ClCH_2SOCl$$

Reaction of alkanesulphinyl chlorides with equimolar quantities of diazomethane and sodium iodide in THF gives[188] iodomethyl sulphoxides, $RS(O)CH_2I$, which hitherto could only be obtained by halogen exchange. In the absence of iodide ion the chloride is obtained[189] directly.

Dichlorosulphine, $Cl_2C=S=O$, reacts as a dienophile with 1,3-dienes in [2 + 4]-type cycloadditions to give[190] cyclic α,α-dichlorosulphoxides, e.g. (35). It reacts stereospecifically with 2-diazopropane in ether at $-25\,^{\circ}C$ to give[191] the Δ^3-1,3,4-thiadiazoline 1-oxide (36), but with diazodiphenylmethane[192] the corresponding product apparently loses nitrogen to give the episulphoxide, which undergoes ring expansion and loss of chloride ion to give 2-chloro-3-phenylbenzo[b]thiophen 1-oxide (37).

(35)   (36)   (37)

The reaction between dialkyl sulphites and the anion of phenylacetonitrile, which gives[193] 3-hydroxy-4,5-diphenylisothiazoles, is considered to involve cyanophenylsulphine $PhC(CN)=S=O$ as an intermediate.

6.3 COMPOUNDS OF HEXAVALENT SULPHUR

6.3.1 Sulphones

Iodine is[194] a useful catalyst for the oxidation of sulphides to sulphones by potassium metaperiodate in aqueous ethanol.

Wurtz-type coupling of two halides can be effected by converting one into the sulphone with sodium benzenesulphinate and then into the anion with a strong base. The anion is utilised as a nucleophile for reaction with the other halide and the activating phenylsulphonyl group removed from the dimer with sodium amalgam and ethanol. The method was applied to the synthesis[195] of vitamin A from vinyl-β-ionol.

Aryl chloromethyl sulphones react with Grignard reagents to give[196] α-chloromethylmagnesium compounds, $ArSO_2CHClMgX$, which are thermostable and undergo the usual reactions of Grignard reagents. α-Lithio-(38) or α-magnesio-sulphones react with ketones to give β-hydroxysulphones, which give[197] trisubstituted ethylenes on dehydration and desulphuration. Tetrasubstituted ethylenes can be obtained by successive tosylation and desulphuration.

$$PhSO_2CR^1R^2Li \xrightarrow{R^3R^4CO} PhSO_2CR^1R^2CR^3R^4OH \xrightarrow[\text{(ii) Na Hg ROH}]{\text{(i) POCl}_3 \text{ pyr}} R^1CH=CR^3R^4$$

(38)   (i) TsCl $(R^2 = H)$
        (ii) Na Hg ROH

$$R^1R^2C=CR^3R^4$$

The α,α-dilithio derivatives (39) of alkyl or benzyl phenyl sulphones react with aldehydes and ketones to give the αβ-unsaturated sulphones directly[198]. Reduction with aluminium amalgam in aqueous THF gives the alkene.

$$\text{PhSO}_2\text{CR}^1\text{Li}_2 \xrightarrow[\text{(ii) R}^2\text{R}^3\text{CO}]{\text{(i) MgI}_2} \text{PhSO}_2\text{CR}^1\!=\!\text{CR}^2\text{R}^3 \xrightarrow[\text{THF/H}_2\text{O}]{\text{Al/Hg}} \text{R}^1\text{CH}\!=\!\text{CR}^2\text{R}^3$$
$$(39)$$

Lithium di-n-alkylcuprates in ether add an alkyl group to the β-position of αβ-alkenyl aryl sulphones. When this is followed by reduction with sodium amalgam in ethanol it provides a method[199] for the conversion of aldehydes and ketones into *gem*-dialkylalkanes. The monoanions of sulphones react with CS_2 in DMSO, and alkylation of the product gives[200] bis(alkylthio)-methylene derivatives, $(R^1S)_2C\!=\!CR^2SO_2R^3$, by a similar series of reactions to those described for the anion of diethyl malonate[114] (Section 6.1.6.4).

Primary dialkyl sulphones are surprisingly converted[201] in high yield into potassium *cis*-dialkylethylenesulphonates by KOH and CCl_4:

$$\text{R}^1\text{CH}_2\text{SO}_2\text{CH}_2\text{R}^2 \xrightarrow[80\,^{\circ}\text{C}]{\text{KOH–Bu}^t\text{OH–CCl}_4} \text{R}^1\text{CH}\!=\!\text{CR}^2\text{SO}_3\text{K}$$

Evidence is adduced[202] that the reaction involves *gem*-dichlorination, formation of a chlorothi-irane 1,1-dioxide, elimination and ring opening. Di-s-alkyl sulphones form mono-α-chlorosulphones, which undergo 1,3-elimination to give alkenes[202]. The *meso*- and the (±)-isomers of bis-(α-bromobenzyl) sulphone undergo[203] 1,3-elimination with triphenylphosphine to give *cis*- and *trans*-stilbene, respectively, with a much higher degree of stereospecificity than the analogous Ramberg–Bäcklund reaction.

Kinetic studies[204] of the reactions of 2-phenylsulphonylethyl halides with methanolic methoxide ion indicate that the bromide and chloride follow an $E2$ mechanism with C—H bond fission leading. The mechanism for the fluoride is much closer to the $E1$ cB borderline. With triethylamine in acetonitrile or benzene, the fluoride follows the $(E1\,\text{cB})_{ip}$ mechanism. The use[205] of phenoxide ion as a base is considered to promote the $E1$ cB mechanism.

The difficult reduction of sulphones to sulphides can be effected[206] in high yield with di-isobutylaluminium hydride.

6.3.2 Sulphoximines

Sulphoxides are aminated under mild conditions in high yield by *O*-mesitylenesulphonylhydroxylamine to give[207] *S*-aminosulphoxonium salts, $R^1R^2\overset{+}{S}(O)NH_2\,X^-$, which give sulphoximines with methanolic methoxide ion. Sulphoximines (40) are also formed by the addition[208] of *N*-tosylarenesulphonimidoyl chlorides to styrene.

$$\text{TsN}\!=\!\text{SAr}(\!=\!O)\text{Cl} + \text{PhCH}\!=\!\text{CH}_2 \xrightarrow[\text{Et}_3\text{N/MeCN}]{\text{Cu}^I\text{Cl}} \text{TsN}\!=\!\text{SAr}(\!=\!O)\text{CH}_2\text{CHClPh}$$
$$(40)$$

The uses of sulphoximines and their derivatives in synthesis have been reviewed[209]. A series of new chiral heterocyclic compounds (41) have been prepared[210] from sulphoximines by anion formation.

$$NH\!=\!SPh(\!=\!O)Me \xrightarrow[\text{(iii) } H_2O]{\substack{\text{(i) } 2Bu^nLi \\ \text{(ii) } CO_2}} \begin{array}{c} O \\ \parallel \\ PhS\!-\!CH_2 \\ \parallel \quad | \\ N\!-\!CO \end{array}$$

(41)

Deimination of sulphoximines by sulphur or disulphides occurs[211] with retention of configuration at sulphur. This may indicate that the nitrogen atom acts as a nucleophilic centre in reaction with the sulphur or disulphide, and then is attacked in turn by a sulphur anion which displaces the sulphoxide group.

6.3.3 Sulphonic acids and their derivatives

6.3.3.1 Trifluoromethanesulphonic acid and its derivatives

The interest in trifluoromethanesulphonic acid and its derivatives centres around its high acidity and the high leaving-group ability of its derivatives (triflates). Conductivity measurements[212] show that it is the strongest known acid in glacial acetic acid, $K_{CF_3SO_3H}/K_{FSO_3H} \approx 25$. Nitronium triflate is an excellent nitrating agent[213] in many solvents, giving high yields and an exceptional positional selectivity. Trifluoromethanesulphonic (triflic) acid acts[214] as a Friedel–Crafts catalyst. The mixed anhydrides of triflic acid and aliphatic acids can be prepared *in situ* from the acyl chloride and the sulphonic acid. They acylate benzene and even chlorobenzene on being heated[215], but are thermolabile and liberate CO, presumably from the decomposition of the acylium ion. N-Triflylamides, $RCONHSO_2CF_3$, are also acylating agents[216]. They can be prepared from the amide by successive treatment with NaH and triflic anhydride, or from N-phenyltriflamide, $PhNHSO_2CF_3$, and the acyl chloride. The replacement of two amino-hydrogens by triflyl groups converts the amino group into a good leaving group; thus N-benzyltriflimide, $PhCH_2N(SO_2CF_3)_2$, undergoes S_N2 reactions[217]. Benzyltriflamide anion can be alkylated and, after base-catalysed elimination and hydrolysis, gives[216] the primary amine as in the Gabriel synthesis:

$$PhCH_2\bar{N}SO_2CF_3 \xrightarrow{RX} PhCH_2NRSO_2CF_3 \xrightarrow[\text{(ii) } 10\% \text{ HCl}]{\text{(i) NaH–DMF}} RNH_2 + PhCHO$$

The reactions of triflamides have been reviewed[217].

The nonafluorobutanesulphonate anion is an even better leaving group than the triflate anion. Hydrolysis of cyclic vinyl nonafluorobutanesulphonates (nonaflates) to give ketones occurs[218] at almost twice the rate for the corresponding triflates. These nonaflates are prepared[219] by treating the sodium enolate in glyme with the sulphonyl fluoride; sodium aryloxides react similarly.

6.3.3.2 *Sulphonyl chlorides and sulphenes*

Chlorination of α-chloroalkyl sulphides in water gives[174] the sulphonyl chlorides in good yields. High substrate and positional selectivities are shown[220] by sulphonations with methane- and ethane-sulphonyl chloride catalysed by $AlCl_3$ and by $SbCl_5$. An early transition state (π-complex nature) is proposed. 1,2,4-Thiadiazine 1,1-dioxides (42) are obtained by the condensation of 2-chloroethanesulphonyl chloride with *gem*-diamines and imines, e.g. $HN=C(NH_2)Y$.

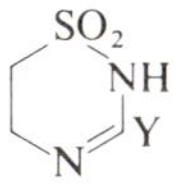

The rate of hydrolysis of chlorinated sulphonyl chlorides in aqueous dioxan has the following order[222]: $ClCH_2SO_2Cl < MeSO_2Cl \ll Cl_2CHSO_2Cl$. The first two hydrolyses are similar and are considered to have an S_N2-type mechanism, but the last reaction may involve rapid dichlorosulphene formation. This has been confirmed by deuterium incorporation.

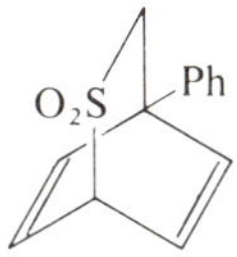

Sulphene has been generated[223] by the thermolysis *in vacuo* at 600 °C of *N*-methylsulphonylphthalimide, and trapped as the sulphonamide, thus confirming that sulphene can exist as a free entity in the gas phase. The *N*-benzyl derivative under similar conditions gave benzaldehyde because of the lower stability of phenylsulphene. The reverse Diels–Alder reaction of the bicyclic compound (43) also gives[224] sulphene, and the sulphonyl chloride in the original preparation can be replaced by the nitro-[225] or the dinitro-phenyl[226] sulphonates. In contrast to early work where monodeuteriation of sulphene was always observed, three deuterium atoms can be incorporated[227] into methanesulphonyl chloride by using less bulky bases, e.g. 1,4-diazobicyclo-[2,2,2]octane. The second and subsequent deuterium atoms are incorporated by a reversible zwitterion formation between the sulphene and the amine:

$$CH_2=SO_2 + R_3N \rightleftharpoons \overset{-}{C}H_2SO_2\overset{+}{N}R_3 \underset{\pm D^+/H^+}{\overset{}{\rightleftharpoons}} CH_2DSO_2\overset{+}{N}R_3$$

Vinylsulphene, derived from prop-2-ene-1-sulphonyl chloride, reacts[228] with enamines to give the 1,2-cycloadduct and the allyl aminovinyl sulphone, the proportions depending on the enamine; 1,4-cycloaddition does not occur directly. Similar results are obtained[229] with the corresponding acetylenic sulphonyl chloride. Sulphenes and phenylsulphenes substituted with electron withdrawing groups undergo cycloaddition[230] with tropone to give *cis*-γ-sultones, which extrude SO_2 stereospecifically on being heated to give substituted styrenes or stilbenes. With triphenylphosphine and sulphene a cyclic sulphone is considered to be formed which loses sulphur dioxide and forms[226] a methyltriphenylphosphonium salt on protonation.

6.3.3.3 Sulphonamides

Primary, secondary, and the relatively inaccessible tertiary alkanesulphon-amides are obtained[231] by reaction of the sulphinyl chloride with a hydroxyl-amine.

$$R^1SOCl + 2R^2R^3NOH \xrightarrow{\text{ether–CH}_2\text{Cl}_2} R^1SO_2NR^2R^3 + R^2R^3NOH \cdot HCl$$

Sodium bis-(2-methoxyethoxy)dihydridoaluminate in glyme or aromatic hydrocarbons under reflux reduces sulphonamides to give[232] the amines in good or acceptable yields. The thiol is presumably also formed.

6.3.3.4 Sulphates and sulphonates

The addition of fluorosulphonic acid to alkenes, halogenoalkenes or cyclopropanes provides a convenient general method[233] for the preparation of alkyl fluorosulphonates. s-Alkyl fluorosulphonates are more labile than the primary compounds and alkylate[233] benzene directly. The use of methyl fluorosulphonate as a methylating agent has been reviewed[234] and it is reported[235] that it can also act as a sulphonylating agent. Kinetic evidence[236] suggests a degree of bonding between the nucleophile and the carbon atom in the transition state for the hydrolysis of alkyl chlorosulphates, $ClSO_2OR$, and hence an S_N2-type reaction. Methyl p-nitrophenyl sulphate undergoes methanolysis[237] in alkaline solution by alkyl–oxygen fission to give dimethyl ether and p-nitrophenyl sulphate ion; reaction at sulphur or aryl carbon was not observed. The mechanism is held to be similar to that for the hydrolysis of dimethyl sulphate. The rapid solvolysis of the sulphate esters of steroids and carbohydrates in moist non-polar nucleophilic solvents has been shown[238] to be initiated by electrophilic impurities or the glass surface. Reaction does not occur in Teflon vessels.

6.3.3.5 Sulphamic acid and its derivatives

Monoalkyl sulphamyl chlorides are obtained[239] by the thermolysis of the N-chlorosulphonylurethanes derived from alcohols, which give relatively stable carbonium ions:

$$ROCONHSO_2Cl \xrightarrow{\text{heat}} RNHSO_2Cl + CO_2$$

Successive reaction[239] with t-butoxycarbonyl hydrazide, $Bu^tO_2CNHNH_2$, and lead tetra-acetate converts these sulphamyl chlorides into the amine. The overall reaction sequence parallels the Ritter reaction. Treatment of

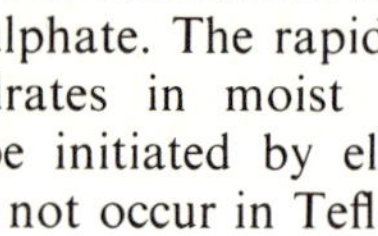

        (44) 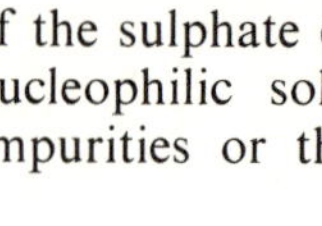   (45)

methoxycarbonylsulphamyl chloride with NaH in THF gives[240] a complex which adds to alkenes to form thiazetidines (44) and oxathiazines (45).

N-n-Butyl- and *N*-cyclohexyl-sulphamic acids sulphonate[241] arylamines and aryloxymethanes at 150–195 °C to give the *p*-substituted benzenesulphonic acids in high yield, and appear to be midway in reactivity between the unsubstituted and fully substituted sulphur trioxide complexes.

References

1. Klayman, D. L., Shine, R. J. and Bower, J. D. (1972). *J. Org. Chem.*, **37**, 1532
2. Yanagawa, H., Kato, T., Sagami, H. and Kitahara, Y. (1973). *Synthesis*, 607
3. Bradshaw, J. S., Chen, E. Y., Hales, R. H. and South, J. A. (1972). *J. Org. Chem.*, **37**, 2051, 2381
4. Feutrill, G. I. and Mirrington, R. N. (1972). *Aust. J. Chem.*, **25**, 1719, 1731
5. Aufauvre, Y., Verny, M. and Vessière, R. (1973). *Bull. Soc. Chim. Fr.*, 1373
6. Truce, W. E. and Tichenor, G. J. W. (1972). *J. Org. Chem.*, **37**, 2391
7. McIntosh, J. M. and Goodbrand, H. B. (1973). *Tetrahedron Lett.*, 3157
8. Makriyannis, A., Günther, W. H. H. and Mautner, H. G. (1973). *J. Amer. Chem. Soc.*, **95**, 8403
9. Tschesche, H. and Jering, H. (1973). *Angew. Chem. Int. Ed. Engl.*, **12**, 756
10. Gosden, A., Stevenson, D. and Young, G. T. (1972). *J. Chem. Soc. Chem. Commun.*, 1123
11. Foye, W. O. (1973). *Int. J. Sulfur Chem.*, **8**, 161
12. Zuika, I. V. and Bankovskii, Yu. A. (1973). *Russ. Chem. Rev.*, **42**, 23
13. Wilson, G. E. and Riley, J. G. (1972). *Tetrahedron Lett.*, 379
14. Fredga, A. (1972). *Chem. Scripta*, **2**, 47
15. Pryor, W. A. and Bickley, H. T. (1972). *J. Org. Chem.*, **37**, 2885
16. Burighel, A., Modena, G. and Tonellato, U. (1972). *J. Chem. Soc. Perkin Trans. II*, 2026
17. Denes, A. S., Csizmadia, I. G. and Modena, G. (1972). *J. Chem. Soc. Chem. Commun.*, 8; Csizmadia, I. G. (1974). Personal communication
18. Bird, R. and Stirling, C. J. M. (1973). *J. Chem. Soc. Perkin Trans. II*, 1221
19. Wilson, G. E. Jnr. and Martin, J. H. E. (1972). *J. Org. Chem.*, **37**, 2510
20. Sowerby, R. L. and Coates, R. M. (1972). *J. Amer. Chem. Soc.*, **94**, 4758
21. Shanklin, J. R., Johnson, C. R., Ollinger, J. and Coates, R. M. (1973). *J. Amer. Chem. Soc.*, **95**, 3429
22. Wright, A., Ling, D., Boudjouk, P. and West, R. (1972). *J. Amer. Chem. Soc.*, **94**, 4784
23. Corey, E. J. and Kim, C. U. (1972). *J. Amer. Chem. Soc.*, **94**, 7586; (1973). *J. Org. Chem.*, **38**, 1233
24. Corey, E. J. and Kim, C. U. (1973). *Tetrahedron Lett.*, 919
25. Furukawa, N., Inoue, T., Aida, T. and Oae, S. (1973). *J. Chem. Soc. Chem. Commun.*, 212
26. Saitkulova, F. G., Abashev, G. G. and Lapkin, I. I. (1973). *Zh. Org. Khim.*, **9**, 1405
27. Ho, T.-L. and Wong, C. M. (1973). *J. Chem. Soc. Chem. Commun.*, 224
28. Bohlmann, F. and Hopf, P.-D. (1973). *Chem. Ber.*, **106**, 3621
29. Meijer, J. and Brandsma, L. (1972). *Rec. Trav. Chim. Pays-Bas.*, **91**, 578; (1973). *ibid.*, **92**, 1331
30. Brandsma, L., Hoff, S. and Verkruijsse, H. D. (1973). *Rec. Trav. Chim. Pays-Bas*, **92**, 272
31. Meijer, J., Vermeer, P., Verkruijsse, H. D. and Brandsma, L. (1973). *Rec. Trav. Chim. Pays-Bas*, **92**, 1326
32. Oshima, K., Shimoji, K., Takahashi, H., Yamamoto, H. and Nozaki, H. (1973). *J. Amer. Chem. Soc.*, **95**, 2694
33. Mukaiyama, T., Kamio, K., Kobayashi, S. and Takei, H. (1972). *Bull. Chem. Soc. Jap.*, **45**, 3723
34. Baechler, R. D., Hummel, J. P. and Mislow, K. (1973). *J. Amer. Chem. Soc.*, **95**, 4442
35. Ellison, R. A. and Woessner, W. D. (1972). *J. Chem. Soc. Chem. Commun.*, 529

36. Ellison, R. A., Woessner, W. D. and Williams, C. C. (1972). *J. Org. Chem.*, **37**, 2757
37. Eliel, E. L. and Hartmann, A. A. (1972). *J. Org. Chem.*, **37**, 505
38. Corey, E. J. and Walinsky, S. W. (1972). *J. Amer. Chem. Soc.*, **94**, 8932
39. Carey, F. A. and Court, A. S. (1972). *J. Org. Chem.*, **37**, 1926; Jones, P. F., Lappert, M. F. and Szary, A. C. (1973). *J. Chem. Soc. Perkin Trans. I*, 2272; Seebach, D., Kolb, M. and Gröbel, B.-T. (1973). *Chem. Ber.*, **106**, 2277
40. Carey, F. A. and Court, A. S. (1972). *J. Org. Chem.*, **37**, 4474
41. Seebach, D., Kolb, M. and Gröbel, B.-T. (1973). *Angew. Chem. Int. Ed. Engl.*, **12**, 69
42. Corey, E. J. and Chen, R. H. K. (1973). *Tetrahedron Lett.*, 3817
43. Shahak, I. and Sasson, Y. (1973). *Tetrahedron Lett.*, 4207
44. Tamura, K., Sumoto, K., Fujii, S., Satoh, H. and Ikeda, M. (1973). *Synthesis*, 312; Ho, T.-K., Ho, H. C. and Wong, C. M. (1973). *Can. J. Chem.*, **51**, 153 and references therein
45. Corey, E. J. and Beames, D. J. (1973). *J. Amer. Chem. Soc.*, **95**, 5829
46. Ogura, K. and Tsuchihashi, G.-i. (1972). *Tetrahedron Lett.*, 2681
47. Ogura, K. and Tsuchihashi, G.-i. (1972). *Tetrahedron Lett.*, 1383
48. Richman, J. E., Herrmann, J. L. and Schlessinger, R. H. (1973). *Tetrahedron Lett.*, 3267, 3271, 3275
49. Herrmann, J. L., Richman, J. E., Wepplo, P. J. and Schlessinger, R. H. (1973). *Tetrahedron Lett.*, 4707
50. Herrmann, J. L., Kieczykowski, G. R., Romanet, R. F., Wepplo, P. J. and Schlessinger, R. H. (1973). *Tetrahedron Lett.*, 4711, 4715
51. Field, L. (1972). *Synthesis*, 101
52. Armitage, D. A., Clark, M. J. and Tso, C. C. (1972). *J. Chem. Soc. Perkin Trans. I*, 680
53. Cragg, R. H. and Weston, A. F. (1973). *Tetrahedron Lett.*, 655
54. Lalancette, J. M. and Laliberté, M. (1973). *Tetrahedron Lett.*, 1401
55. Block, E. and O'Connor, J. (1973). *J. Amer. Chem. Soc.*, **95**, 5048
56. Moore, R. E., Mistysyn, J. and Pettus, J. A., Jnr. (1972). *J. Chem. Soc. Chem. Commun.*, 326
57. Nelander, B. and Sunner, S. (1972). *J. Amer. Chem. Soc.*, **94**, 3576
58. Khim, Y. H. and Field, L. (1972). *J. Org. Chem.*, **37**, 2714
59. Field, L. and Khim, Y. H. (1972). *J. Med. Chem.*, **15**, 312
60. Danehy, J. P. and Elia, V. J. (1972). *J. Org. Chem.*, **37**, 369
61. Bentley, M. D., Douglass, I. B. and Lacadie, J. A. (1972). *J. Org. Chem.*, **37**, 333
62. Trost, B. M. and Salzmann, T. N. (1973). *J. Amer. Chem. Soc.*, **95**, 6840
63. Boelens, H. and Brandsma, L. (1972). *Recl. Trav. Chim. Pays-Bas*, **91**, 141
64. Block, E. (1972). *J. Amer. Chem. Soc.*, **94**, 642, 644
65. Block, E. and Weidman, S. W. (1973). *J. Amer. Chem. Soc.*, **95**, 5046
66. Kondo, K., Negishi, A. and Ojima, I. (1972). *J. Amer. Chem. Soc.*, **94**, 5786
67. Kice, J. L. and Cleveland, J. P. (1973). *J. Amer. Chem. Soc.*, **95**, 104, 109
68. Krimer, M. Z., Smit, V. A. and Shamshurin, A. A. (1973). *Dokl. Akad. Nauk SSSR*, **208**, 90
69. Kil'disheva, O. V., Lin'kova, M. G., Rasteikene, L. P., Zabelaite, V. A., Potsyute, N. K. and Knunyants, I. L. (1972). *Dokl. Akad. Nauk SSSR*, **203**, 331
70. Carbin, E., Helmkamp, G. K., Barnes, W. M. and Sundaralingam, M. (1972). *Int. J. Sulfur Chem. A*, **2**, 129
71. Oki, M., Nakamura, A. and Kobayashi, K. (1973). *Bull. Chem. Soc. Jap.*, **46**, 3610
72. Rieke, R. D., Bales, S. E. and Roberts, L. C. (1972). *J. Chem. Soc. Chem. Commun.*, 974
73. Krubsack, A. J. and Higa, T. (1973). *Tetrahedron Lett.*, 125, 4515
74. Senning, A. and Kelly, P. (1972). *Acta Chem. Scand.*, **26**, 2877
75. Kristinsson, H. (1973). *Tetrahedron Lett.*, 4489
76. Seebach, D. and Tescher, M. (1973). *Tetrahedron Lett.*, 5113
77. Leir, C. M. (1972). *J. Org. Chem.*, **37**, 887
78. Senning, A. and Schmidt, J. (1973). *Acta Chem. Scand.*, **27**, 2161
79. Clive, D. L. J. and Denzer, C. V. (1972). *J. Chem. Soc. Chem. Commun.*, 773
80. Kobayashi, M., Osawa, A. and Fujisawa, T. (1973). *J. Polymer Sci. Polymer Lett.*, **11**, 679

81. Braverman, S. and Reisman, D. (1973). *Tetrahedron Lett.*, 3563
82. Maricich, T. J. and Harrington, C. K. (1972). *J. Amer. Chem. Soc.*, **94**, 5116
83. Evans, D. A. and Andrews, G. C. (1972). *J. Amer. Chem. Soc.*, **94**, 3672
84. Numata, T. and Oae, S. (1972). *Bull. Chem. Soc. Jap.*, **45**, 2794
85. Hagberg, C.-E., Bohman, O. and Engdahl, C. (1972). *Tetrahedron Lett.*, 3689
86. Phadnis, S. P. (1972). *Ind. J. Chem.*, **10**, 699
87. Klayman, D. L. and Woods, T. S. (1973). *Int. J. Sulfur Chem.*, **8**, 5
88. Weidner, J. P. and Block, S. S. (1972). *J. Med. Chem.*, **15**, 564
89. Murray, R. W. and Jindal, S. L. (1972). *J. Org. Chem.*, **37**, 3516
90. Scheeren, J. W., Ooms, P. H. J. and Nivard, R. J. F. (1973). *Synthesis*, 149
91. Oae, S., Nakanishi, A. and Tsujimoto, N. (1972). *Chem. Ind. (London)*, 575
92. Paquer, D. and Vialle, J. (1972). *C.R. Acad. Sci. (Paris)*, **275**, 589
93. Metzner, P. and Vialle, J. (1973). *Bull. Soc. Chim. Fr.*, 1703
94. Dagonneau, M., Metzner, P. and Vialle, J. (1973). *Tetrahedron Lett.*, 3675
95. Hayashi, T. and Midorikawa, H. (1973). *Tetrahedron Lett.*, 2461
96. Duus, F. (1972). *Tetrahedron*, **28**, 5923
97. Hartke, K. and Gölz, G. (1973). *Annalen*, 1644
98. Harhash, A. H., Elnagdi, M. H. and Abdallah, S. O. (1973). *Ind. J. Chem.*, **11**, 128
99. Huisgen, R. and Mack, W. (1972). *Chem. Ber.*, **105**, 2815
100. Mozolis, V. V. and Iokubaitite, S. P. (1973). *Usp. Khim.*, **42**, 587
101. Hartmann, H. and Reuther, J. (1973). *J. Prakt. Chem.*, **315**, 144
102. Benders, P. H. (1973). *Tetrahedron Lett.*, 3653
103. Galstukhova, N. B., Berzina, I. M. and Shchukina, M. N. (1973). *Zh. Org. Khim.*, **9**, 1094
104. Hartmann, H. (1972). *Tetrahedron Lett.*, 3977
105. Mikolajczyk, M. and Luczak, J. (1972). *Chem. Ind. (London)*, 76
106. Larsen, C. and Jakobsen, P. (1973). *Acta Chem. Scand.*, **27**, 2001
107. Matolcsy, G. and Bordás, B. (1973). *Chem. Ber.*, **106**, 1483
108. Atkins, P. R., Glue, S. E. J. and Kay, I. T. (1973). *J. Chem. Soc. Perkin Trans. I*, 2644
109. Gerlach, H., Huong, T. T. and Müller, W. (1972). *J. Chem. Soc. Chem. Commun.*, 1215
110. Meurling, P., Sjöberg, K. and Sjöberg, B. (1972). *Acta Chem. Scand.*, **26**, 279
111. Rutherford, K. G., Tang, B. K., Lam, L. K. M. and Fung, D. P. C. (1972). *Can. J. Chem.*, **50**, 3288
112. Barrett, G. C. and Martins, C. M. O. A. (1972). *J. Chem. Soc. Chem. Commun.*, 638
113. Meijer, J., Vermeer, P. and Brandsma, L. (1973). *Rec. Trav. Chim. Pays-Bas.*, **92**, 601
114. Dalgaard, L., Kolind-Anderson, H. and Lawesson, S.-O. (1973). *Tetrahedron*, **29**, 2077
115. Oae, S., Yagihara, T. and Okabe, T. (1972). *Tetrahedron*, **28**, 3203
116. Takahashi, H., Oshima, K., Yamamoto, H. and Nozaki, H. (1973). *J. Amer. Chem. Soc.*, **95**, 5803
117. Yamaguchi, T., Shimizu, Y. and Suzuki, T. (1972). *Chem. Ind. (London)*, 380
118. Enders, D. and Seebach, D. (1973). *Angew. Chem. Int. Ed. Engl.*, **12**, 1014
119. Walter, W. and Becker, R. F. (1972). *Annalen*, **755**, 145
120. Nilsson, N. H. and Senning, A. (1972). *Angew. Chem. Int. Ed. Engl.*, **11**, 295
121. Fujita, H., Endo, R. and Murayama, K. (1972). *Bull. Chem. Soc. Jap.*, **45**, 1582
122. Abis, L. and Giordano, C. (1973). *J. Chem. Soc. Perkin Trans. I*, 771
123. Inomata, K., Muraki, M. and Mukaiyama, T. (1973). *Bull. Chem. Soc. Jap.*, **46**, 1807
124. Ecke, H. and Prigge, H. (1972). *Annalen*, **755**, 177
125. Holsboer, D. H. and Kloosterziel, H. (1972). *Rec. Trav. Chim. Pays-Bas*, **91**, 1371
126. Oele, P. C., Tinkelenberg, A. and Louw, R. (1972). *Tetrahedron Lett.*, 2375
127. Bigley, D. B. and Gabbott, R. E. (1973). *J. Chem. Soc. Perkin Trans. II*, 1293
128. Idoux, J. P., Hwang, P. T. R. and Hancock, C. K. (1973). *J. Org. Chem.*, **38**, 4239
129. Hershfield, R. and Schmir, G. L. (1972). *J. Amer. Chem. Soc.*, **94**, 1263, 6788; Zygmunt, R. J. and Barnett, R. E. (1972). *ibid.*, **94**, 1996
130. Kasperek, G. J. and Bruice, T. C. (1972). *J. Org. Chem.*, **37**, 1456
131. Anderson, P. E., Blackburn, G. M. and Murphy, S. (1972). *J. Chem. Soc. Chem. Commun.*, 171
132. Johnson, C. R., Bacon, C. C. and Kingsbury, W. D. (1972). *Tetrahedron Lett.*, 501

133. Vilsmaier, E. and Sprügel, W. (1972). *Tetrahedron Lett.*, 625
134. Gassman, P. G., van Bergen, T. J. and Gruetzmacher, G. (1973). *J. Amer. Chem. Soc.*, **95**, 6508
135. Khuddus, M. A. and Swern, D. (1973). *J. Amer. Chem. Soc.*, **95**, 8393
136. Anderson, K. K., Caret, R. L. and Ladd, D. L. (1972). *Int. J. Sulfur Chem. A*, **2**, 196
137. Claus, P. and Rieder, W. (1972). *Monatsh.*, **103**, 1163
138. Braun, H., Mayer, N., Strobl, G. and Kresze, G. (1973). *Annalen*, 1317
139. Traynelis, V. J., Yoshikawa, Y., Tarka, S. M. and Livingston, J. R. (1973). *J. Org. Chem.*, **38**, 3986
140. Wudl, F. and Lee, T. B. K. (1972). *J. Chem. Soc. Chem. Commun.*, 61
141. Lockard, J. P., Schroeck, C. W. and Johnson, C. R. (1973). *Synthesis*, 485
142. Cinquini, M. and Colonna, S. (1972). *J. Chem. Soc. Perkin Trans. I*, 1883
143. Cinquini, M., Colonna, S. and Landini, D. (1972). *J. Chem. Soc. Perkin Trans. II*, 296
144. Cinquini, M., Colonna, S., Fornasier, R. and Montanari, F. (1972). *J. Chem. Soc. Perkin Trans. I*, 1886
145. Calzavara, P., Cinquini, M., Colonna, S., Fornasier, R. and Montanari, F. (1973). *J. Amer. Chem. Soc.*, **95**, 7431
146. Durst, T., Tin, K.-C. and Marcil, M. J. V. (1973). *Can. J. Chem.*, **51**, 1704
147. Iriuchijima, S., Ishibashi, M. and Tsuchihashi, G. (1973). *Bull. Chem. Soc. Jap.*, **46**, 921
148. Klein, J. and Stollar, H. (1973). *J. Amer. Chem. Soc.*, **95**, 7437
149. Taguchi, H., Yamamoto, H. and Nozaki, H. (1973). *Tetrahedron Lett.*, 2463
150. Jung, F. and Durst, T. (1973). *J. Chem. Soc. Chem. Commun.*, 4
151. Meyers, C. Y. and McCollum, G. J. (1973). *Tetrahedron Lett.*, 289
152. Sharma, N. K., Jung, F. and Durst, T. (1973). *J. Amer. Chem. Soc.*, **95**, 3420
153. Fraser, R. R., Schuber, F. J. and Wigfield, Y. Y. (1972). *J. Amer. Chem. Soc.*, **94**, 8795; King, J. F. and du Manior, J. R. (1973). *Can. J. Chem.*, **51**, 4082
154. Nishihata, K. and Nishio, M. (1972). *J. Chem. Soc. Perkin Trans. II*, 1730; (1972). *Tetrahedron Lett.*, 4839
155. D'Amore, M. B. and Brauman, J. I. (1973). *J. Chem. Soc. Chem. Commun.*, 398
156. Durst, T., Viau, R. and McClorey, M. R. (1971). *J. Amer. Chem. Soc.*, **93**, 3077
157. Scorrano, G. (1973). *Accounts Chem. Res.*, **6**, 132; Modena, G. (1972). *Int. J. Sulfur Chem. C*, **7**, 95
158. Gatti, G., Levi, A., Lucchini, V., Modena, G. and Scorrano, G. (1973). *J. Chem. Soc. Chem. Commun.*, 251
159. Ogata, Y. and Suyama, S. (1973). *J. Chem. Soc. Perkin Trans. II*, 755
160. Cox, B. G. and Gibson, A. (1973). *J. Chem. Soc. Perkin Trans. II*, 1355
161. Brown, H. C. and Ravindran, N. (1973). *Synthesis*, 42
162. Johnson, C. R., Bacon, C. C. and Rigau, J. J. (1972). *J. Org. Chem.*, **37**, 919
163. Numata, T. and Oae, S. (1973). *Chem. Ind. (London)*, 277
164. Ho, T.-L. and Wong, C. M. (1973). *Synthesis*, 206
165. Ho, T.-L. and Wong, C. M (1973). *Synth. Commun.*, **3**, 37
166. Cram, D. J. and Cram, J. M. (1972). *Fortschr. Chem. Forsch.*, **31**, 1; Cram, D. J., Day, J., Garwood, D. C., Rayner, D. R., von Schriltz, D. M., Williams, T. R., Nudelman, A., Yamagishi, F. G., Booms, R. E. and Jones, M. R. (1972). *Int. J. Sulfur Chem. C*, **7**, 103; Garwood, D. C., Jones, M. R. and Cram, D. J. (1973). *J. Amer. Chem. Soc.*, **95**, 1925
167. Williams, T. R., Nudelman, A., Booms, R. E. and Cram, D. J. (1972). *J. Amer. Chem. Soc.*, **94**, 4684
168. Tsuchihashi, G., Iriuchijima, S. and Ishibashi, M. (1972). *Tetrahedron Lett.*, 4605
169. Kuwajimi, I. and Uchida, M. (1972). *Tetrahedron Lett.*, 649, 737
170. Evans, D. A., Andrews, G. C., Fujimoto, T. T. and Wells, D. (1973). *Tetrahedron Lett.*, 1385, 1389
171. Grieco, P. A. (1972). *J. Chem. Soc. Chem. Commun.*, 702
172. Grieco, P. A., Boxler, D. and Hiroi, K. (1973). *J. Org. Chem.*, **38**, 2573
173. Kamiyama, K., Minato, H. and Kobayashi, M. (1973). *Bull. Chem. Soc. Jap.*, **46**, 3895
174. Furukawa, N., Omata, T., Yoshimura, T., Aida, T. and Oae, S. (1972). *Tetrahedron Lett.*, 1619

175. Kapovits, I., Ruff, F. and Kucsman, A. (1972). *Tetrahedron*, **28**, 4405, 4413
176. Tamura, Y., Sumoto, K., Matsushima, H., Taniguchi, H. and Ikeda, M. (1973). *J. Org. Chem.*, **38**, 4324
177. Furukawa, N., Yoshimura, T. and Oae, S. (1973). *Tetrahedron Lett.*, 2113
178. Oae, S., Matsuda, T., Tsujihara, K. and Furukawa, N. (1972). *Bull. Chem. Soc. Jap.*, **45**, 3856
179. Oae, S., Harada, K., Tsujihara, K. and Furukawa, N. (1973). *Bull. Chem. Soc. Jap.*, **46**, 3482
180. Furukawa, N., Harada, K. and Oae, S. (1972). *Tetrahedron Lett.*, 1377
181. Trost, B. M. (1973). *Fortschr. Chem. Forsch.*, **41**, 1
182. Filby, W. G., Günther, K. and Penzhorn, R. D. (1973). *J. Org. Chem.*, **38**, 4070
183. Harpp, D. N. and Back, T. G. (1973). *J. Org. Chem.*, **38**, 4328
184. Schrank, K. and Weber, A. (1972). *Chem. Ber.*, **105**, 2188
185. Kerber, R. and Gestrich, W. (1973). *Chem. Ber.*, **106**, 798
186. Mikolajczyk, M. and Drabowicz, J. (1972). *Int. J. Sulfur Chem. A*, **2**, 200
187. Grossert, J. S., Hardstaff, W. R. and Langler, R. F. (1973). *Chem. Commun.*, 50
188. Venier, C. G. and Barager, H. J. (1973). *J. Chem. Soc. Chem. Commun.*, 319
189. Venier, C. G., Hsin-Hsiong, H. and Barager, H. J. (1973). *J. Org. Chem.*, **38**, 17
190. Zwanenburg, B., Thijs, L., Broens, J. B. and Strating, J. (1972). *Rec. Trav. Chim. Pays-Bas*, **91**, 443
191. Thijs, L., Wagenaar, A., van Rens, E. M. M. and Zwanenburg, B. (1973). *Tetrahedron Lett.*, 3589
192. Thijs, L., Strating, J. and Zwanenburg, B. (1972). *Rec. Trav. Chim. Pays-Bas*, **91**, 1345
193. Scott, M. D. (1972). *J. Chem. Soc. Perkin Trans. I*, 1432
194. Field, L. and Khim, Y. H. (1972). *J. Org. Chem.*, **37**, 2710
195. Julia, M. and Arnould, D. (1973). *Bull. Soc. Chim. Fr.*, 743, 746
196. Stetter, H. and Steinbeck, K. (1972). *Annalen*, **766**, 89
197. Julia, M. and Paris, J.-M. (1973). *Tetrahedron Lett.*, 4833
198. Pascali, V., Tangari, N. and Umani-Ronchi, A. (1973). *J. Chem. Soc. Perkin Trans. I*, 1166; Pascali, V. and Umani-Ronchi, A. (1973). *J. Chem. Soc. Chem. Commun.*, 351
199. Posner, G. H. and Brunelle, D. J. (1973). *J. Org. Chem.*, **38**, 2747
200. Ladurée, D., Rioult, P. and Vialle, J. (1973). *Bull. Soc. Chim. Fr.*, 637
201. Meyers, C. Y. and Ho, L. L. (1972). *Tetrahedron Lett.*, 4319
202. Meyers, C. Y., Ho, L. L., McCollum, G. J. and Branca, J. (1973). *Tetrahedron Lett.*, 1843
203. Bordwell, F. G. and Jarvis, B. B. (1973). *J. Amer. Chem. Soc.*, **95**, 3585
204. Fiandanese, V., Marchese, G. and Naso, F. (1973). *J. Chem. Soc. Perkin Trans. II*, 1538
205. Oae, S., Kadoma, Y. and Yano, Y. (1972). *Int. J. Sulfur Chem. A*, **2**, 29
206. Gardner, J. N., Kaiser, S., Krubiner, A. and Lucas, H. (1973). *Can. J. Chem.*, **51**, 1419
207. Tamura, Y., Sumoto, K., Minamikawa, J. and Ikeda, M. (1972). *Tetrahedron Lett.*, 4137
208. Derkach, N. Ya., Pasmurtseva, N. A., Markovskii, L. N. and Lerchenko, E. S. (1973). *Zh. Org. Khim.*, **9**, 1411
209. Johnson, C. R. (1973). *Accounts Chem. Res.*, **6**, 341
210. Williams, T. R. and Cram, D. J. (1973). *J. Org. Chem.*, **38**, 20
211. Oae, S., Tsuchida, Y. and Furukawa, N. (1973). *Bull. Chem. Soc. Jap.*, **46**, 648
212. Engelbrecht, A. and Rode, B. M. (1972). *Monatsh.*, **103**, 1315
213. Coon, C. L., Blucher, W. G. and Hill, M. E. (1973). *J. Org. Chem.*, **38**, 4243
214. Nyberg, K. (1973). *Chem. Scripta*, **4**, 143
215. Effenberger, F. and Epple, G. (1972). *Angew. Chem. Int. Ed. Engl.*, **11**, 299, 300
216. Hendrickson, J. B. and Bergeron, R. (1973). *Tetrahedron Lett.*, 3839, 4607
217. Hendrickson, J. B., Bergeron, R., Giga, A. and Sternbach, D. (1973). *J. Amer. Chem. Soc.*, **95**, 3412
218. Subramanian, L. R. and Hanack, M. (1972). *Chem. Ber.*, **105**, 1465
219. Subramanian, L. R., Bentz, H. and Hanack, M. (1973). *Synthesis*, 293
220. Olah, G. A., Kobayashi, S. and Nishimura, J. (1973). *J. Amer. Chem. Soc.*, **95**, 564
221. Étienne, A., Le Berre, A. and Giorgetti, J. P. (1973). *Bull. Soc. Chim. Fr.*, 985

222. Seifert, R., Zbirovsky, M. and Sauer, M. (1973). *Coll. Czech. Chem. Commun.*, **38**, 2477
223. Mijs, W. T., Reesink, J. B. and Wiersum, U. E. (1972). *J. Chem. Soc. Chem. Commun.*, 412
224. King, J. F. and Lewars, E. G. (1973). *Can. J. Chem.*, **51**, 3044
225. Christensen, L. W. (1973). *Synthesis*, 534
226. King, J. F., Lewars, E. G. and Danks, L. J. (1972). *Can. J. Chem.*, **50**, 866
227. King, J. F., Luinstra, E. A. and Harding, D. R. K. (1973). *J. Chem. Soc. Chem.. Commun.*, 1313
228. Bradamante, S., Maiorana, S. and Pagani, G. (1972). *J. Chem. Soc. Perkin Trans. I*, 282
229. Bradamante, S., Del Buttero, P. and Maiorana, S. (1973). *J. Chem. Soc. Perkin Trans. I*, 612
230. Truce, W. E. and Lin, C.-I. M. (1973). *J. Amer. Chem. Soc.*, **95**, 4426
231. Hovius, K. and Engberts, J. B. F. N. (1972). *Tetrahedron Lett.*, 181
232. Gold, E. H. and Babad, E. (1972). *J. Org. Chem.*, **37**, 2208
233. Olah, G. A., Nishimura, J. and Mo, Y. K. (1973). *Synthesis*, 661
234. Alder, R. W. (1973). *Chem. Ind. (London)*, 983
235. Kametani, T., Takahashi, K. and Ogasawara, K. (1972). *Synthesis*, 473
236. Ko, E. C. F. and Robertson, R. E. (1972). *Can. J. Chem.*, **50**, 434
237. Buncel, E., Raoult, A. and Wiltshire, J. F. (1973). *J. Amer. Chem. Soc.*, **95**, 799
238. Goren, M. B. and Kochansky, M. E. (1973). *J. Org. Chem.*, **38**, 3510
239. Hendrickson, J. B. and Joffee, I. (1973). *J. Amer. Chem. Soc.*, **95**, 4083
240. Burgess, E. M. and Williams, W. M. (1972). *J. Amer. Chem. Soc.*, **94**, 4386
241. Scott, F. L., Barry, J. A. and Spillane, W. J. (1972). *J. Chem. Soc. Perkin Trans. I*, 2663

7
Carbonyl Compounds

R. BRETTLE
University of Sheffield

7.1 MONOCARBONYL COMPOUNDS: ALDEHYDES AND KETONES

7.1.1 Synthesis

7.1.1.1 Syntheses based on dithianes and related compounds

Interest in syntheses involving the production of carbonyl groups via thio-acetals and related sulphur-containing compounds has continued. At least ten new methods for the generation of aldehydes and ketones from their thio-acetals have been studied. Amongst those which give high yields with aliphatic compounds are hydrolysis with copper(II) chloride[1], oxidation[2] with chloramine-T or with cerium(IV) ammonium nitrate[3], reaction with O-mesitylenesulphonylhydroxylamine[4] (a potentially dangerous reagent[5]), which was only satisfactory with the derivative of an α,β-alkenic aldehyde, and concentrated sulphuric acid[6], which was used in conjunction with a Wichterle ketone synthesis from a substituted vinyl chloride to produce simultaneously both carbonyl groups of a γ-diketone.

One route to the versatile alkylidenedithianes was described by three groups[7]; routes therefrom to aldehydes had been described earlier. The method involves the condensation of 2-lithio-2-trimethylsilyl-1,3-dithiane (1) with aldehydes or ketones. The reagent is prepared by lithiation of 1,3-dithiane, reaction with trimethylsilyl chloride and a second lithiation. The complete sequence can be conducted in a single flask. α,β-Alkenic aldehydes and ketones condense with (1) to give alkadienyldithianes, which react with alkyl-lithium compounds followed by iodomethane to give the thio-acetals of $trans$-α,β-alkenic methyl ketones[8]. An example is shown in (1) $\rightarrow$ (2). Another route to alkylidenedithianes involves 2-formyl-1,3-dithiane (from 2-lithio-1,3-dithiane and DMF in a Wittig reaction[9]); base-catalysed isomerisation of the alkenic compound may be necessary.

The related ketene thioacetal monoxides (3) can be obtained[10] in several ways, e.g. by the reaction of the anion from (4; R = Et) with aldehydes, followed by acetylation of the intermediate alkoxide and elimination of acetic acid[11]. The ketene thioacetal S-monoxides (3) act as Michael acceptors, thus providing routes to γ-dicarbonyl compounds, which are discussed on

p. 314. Conditions have been found for both the mono- and di-alkylation of (4; R = Et) leading to protected forms of aldehydes or ketones, which can be efficiently regenerated; a synthesis restricted to aldehydes based on (4; R = Me) was reported earlier*.

$$\begin{array}{cc}
\overset{R^2S(O)}{\underset{R^2S}{\diagdown}}C=C\overset{H}{\underset{R^1}{\diagup}} & R^2S(O)CH_2SR^2 \\
(3) & (4)
\end{array}$$

(5)

(6)

A novel route[13] to 2-allyl-1,3-dithianes (6) and thence to β,γ-alkenic aldehydes involves the [2,3]-sigmatropic rearrangement of sulphur ylides (5), which are readily available from primary allylic bromides.

7.1.1.2 Syntheses based on dihydro-oxazines

Full details have appeared[14] of the synthesis of ketones by the reaction of Grignard reagents with the N-methyl quaternary salts of 2-alkyl-5,6-dihydro-4,4,6-trimethyl-4H-1,3-oxazines, e.g. (7), followed by hydrolysis. The reagent in the aldehyde synthesis[15] shown is the 2-methyl substituted salt.

(7)

RCH_2CHO

This synthesis has important advantages over earlier methods based on dihydro-oxazines; full details of the one based on the unquaternised tetra-methyldihydro-oxazine have appeared[16]. A route to α-branched ketones (9) from 2-isoalkyldihydro-oxazines (8) via ketenimines has been described in full[17] and has been made more versatile by the variation shown[17]. 2-Alkyl-

---

* Series One, Vol. 2, p. 290.

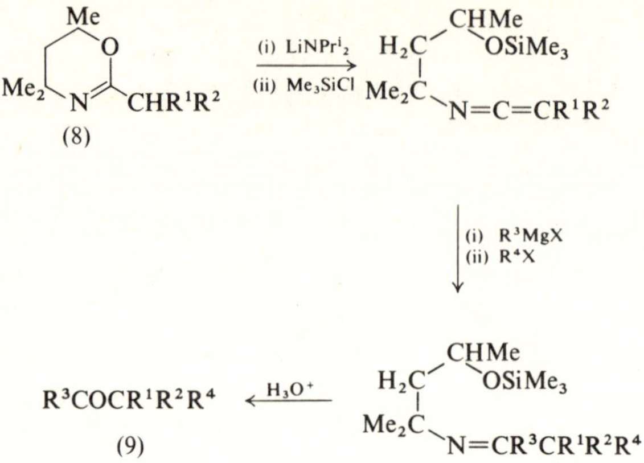

(8)

(i) LiNPr$^i_2$; (ii) Me$_3$SiCl →

(9)

idenedihydro-oxazines, except for the vinyl derivative, react by 1,4-addition with alkyl-lithium compounds to give ketenimines which can then be converted into ketones by known methods[18]. A method closely related to the original procedure in the oxazine series involves the reaction of an alkyl-lithium with a 2-alkyl-4,4-dimethyloxazoline (10) to generate a ketimine

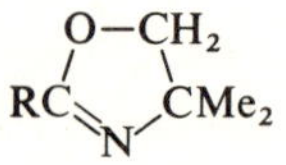

(10)

precursor[19]. Alkylmagnesium halides cannot be used here, but allylmagnesium halides can, so a route to β,γ-alkenic ketones[20] is provided. The corresponding *N*-methyl quaternary salts react with both alkyl-lithium compounds and unhindered alkylmagnesium halides to give products which on hydrolysis yield ketones[19].

The substitution of 2-methyl-2-thiazoline for 2,4-dimethylthiazole, and other modifications, have improved the aldehyde synthesis via 2-alkylthiazolidines, and the method can now be used to add a two-carbon unit to a ketone to give the corresponding β-hydroxyaldehyde[21].

7.1.1.3　Syntheses based on organoboranes

A major new synthesis of ketones is based on the alkylation with rearrangement of lithium trialkylalkynylborates (11), followed by oxidative hydrolysis[22]. Three alkyl groups, each of which comes from a separate starting material, can be incorporated into the resultant ketone. The lithium trialkylalkynylborates can also be acylated with acid chlorides to give 2-oxoboronenes (12) which on Jones oxidation give α,β-alkenic ketones[23]. Alternatively, they react with α-bromoketones to give a rearranged intermediate which on protonolysis gives a β,γ-alkenic ketone[24]; oxidative hydrolysis of the same intermediate gives a γ-diketone. Lithium trialkylalkynylborates also react regioselectively with oxiranes to give intermediates (13) which are converted by oxidative hydrolysis into γ-hydroxyketones[25].

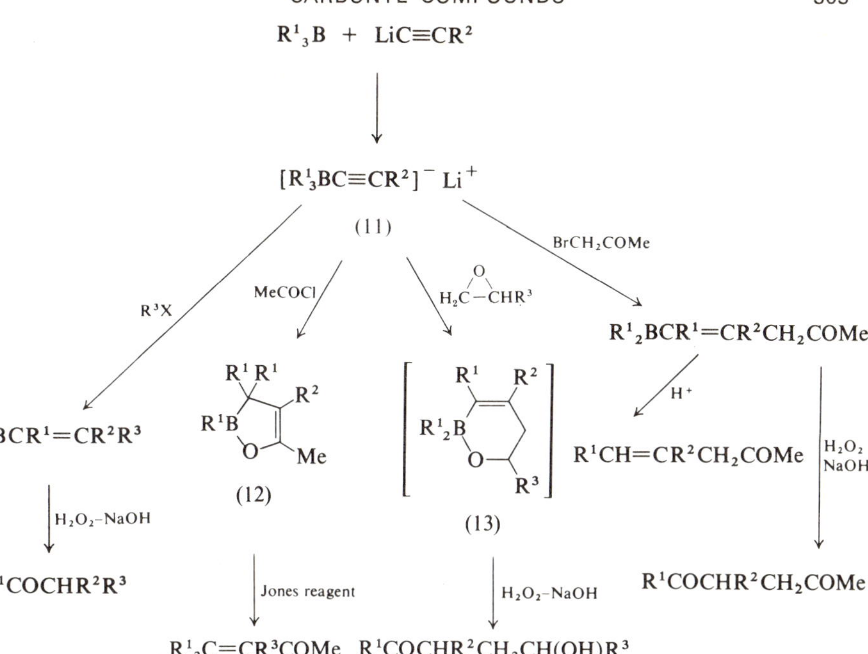

A full account has appeared of work on the cyclic hydroboration of 1,4-
and 1,5-dienes; the resultant *B*-alkylboracyclanes react with methyl vinyl
ketone to transfer the alkyl group to the β-carbon atom[26]. Alkenes react with
monochloroborane etherate to give dialkylchloroboranes, which with alkanols
give alkyl dialkylborinates[27]. Treatment of these esters with dichloromethyl
methyl ether in the presence of base gives a rearranged intermediate which
on oxidative hydrolysis gives a ketone; nonan-5-one is formed in 96% yield
from methyl dibutylborinate[28]. Hydroboration of a 1-chloroalkyne with a
monoalkylborane gives an alkyl-1-chloroalk-1-enylborane, which can then
be used for the hydroboration of an alk-1-yne. Rearrangement of the trisubstituted boron compound with NaOMe gives an intermediate which H_2O_2–
NaAOc converts into a *trans*-α,β-alkenic ketone. In this way, *trans*-undec-4-
en-6-one is built up from 1-chlorohex-1-yne and pent-1-yne[29].

7.1.1.4 Syntheses from acids and acid derivatives

Some useful extensions of the use of organocopper reagents have appeared.
The most important of these is the introduction of the lithium t-butoxy-s-
and -t-alkylcuprates, which now seem to be the best reagents for converting
acyl chlorides into the corresponding s- or t-alkyl ketones[30]. The reagents
are compatible with the presence of bromo, acyl and alkoxycarbonyl substituents in the acyl group. The analogous primary alkyl ketones can be
prepared by using the appropriate lithium dialkylcuprates[31]. Copper(I)
alkynides react with acyl chlorides in the presence of lithium iodide to give
conjugated acetylenic ketones[32]. The method works well for saturated and

α,β-alkenoyl chlorides and was also applied successfully to the unstable copper(I) 2-ethoxyethynide.

The alkylrhodium(I) complex (14), which does not react with aldehydes, reacts with acyl chlorides efficiently to give ketones[33]. Acyl chlorides can also be converted into ketones via acylcarbonylferrates (15) by successive treatment with disodium tetracarbonylferrate and an alkyl halide[34].

$$RhR(CO)(PPh_3)_2 \qquad\qquad [RCOFe(CO)_4]^-$$

$$(14) \qquad\qquad\qquad\qquad (15)$$

$$(16)$$

Ketones can be prepared by the reaction of alkylmagnesium halides in HMPT with esters[35]; propylmagnesium chloride and methyl 3,3-dimethyl-butanoate give propyl t-butyl ketone in 98% yield. S-(2-Pyridyl)thioates react with Grignard reagents to give ketones via complexes of the type (16)[36].

7.1.1.5 *Syntheses from alkenes*

Eschenmoser has introduced an important group of syntheses based on the silver-ion catalysed reaction of alkenes and alkynes with α-chloroaldo-nitrones (17) (for preparation see Ref. 37). In liquid SO_2 alkenes form an initial substitution product which upon acidic hydrolysis gives a β,γ-alkenic ketone, the position and configuration of the original double bond being retained[38]. With an alkyne a cycloaddition product is formed, which is readily cleaved by alkaline hydrolysis using a moist alumina column to give an α,β-alkenic ketone[39]. In non-polar solvents alkenes also undergo an initial cycloaddition to give products which can be converted into α,β-alkenic aldehydes by a multistep process[40]. This reaction has been successfully used with methyl (*E*)-geranate[38]. By starting with a cycloalkene the method can be adapted to give very high yields of dicarbonyl compounds[40]. For the details of the novel chemistry involved in these transformations the original papers must be consulted. The overall routes to carbonyl compounds that are now available are summarised above.

An ingenious synthesis of ketones by what is formally the insertion of a carbonyl group between the two doubly bonded atoms in an alkene has been reported[41]. It is based on the transformation of the dibromocarbene adduct into a cyclopropanone thioacetal, which, in the presence of an equimolar amount of water, is converted into the ketone by treatment with trifluoro-acetic acid.

The stereochemistry of the rhodium-catalysed 'oxo' reaction has been investigated[42]. The hydroformylation involves the *syn* addition of H and CHO, e.g. *E*-3-methylpent-2-ene gives 94% of *threo*-2,3-dimethylpentan-1-ol.

An interesting reaction from the biomimetic point of view is the conversion of alk-1-enes in modest yield into the corresponding methyl ketones by direct oxidation with molecular oxygen in the presence of chlorotris(triphenyl-phosphine)rhodium[43].

7.1.1.6 *Syntheses from alcohols*

The search for improved mild reagents for the oxidation of primary and secondary alcohols continues. Corey has investigated several new systems. The adduct from NCS and dimethyl sulphide[44] oxidises both primary and secondary alcohols to the corresponding carbonyl compounds in very high yields, but cannot be used with allylic alcohols. The complexes from DMSO and chlorine[45] and methyl phenyl sulphide and chlorine[46] also oxidise satur-ated alcohols very efficiently. Chromium(VI) oxide dissolves in dichloro-methane containing 3,5-dimethylpyrazole and the resultant solution is a convenient alternative to the Collins reagent (in which the base is pyridine)[47]. After oxidation with the Collins reagent the pyridine is conveniently and quantitatively removed by washing the reaction mixture with dilute aqueous nickel(II) chloride[48]. Chromium(VI) oxide intercalated in graphite can be made by heating the two components together under reduced pressure, and is available commercially as 'Seloxcette'. It is a useful reagent for the hetero-geneous oxidation of saturated and allylic primary alcohols[49]. Another insoluble reagent is a polymer-bound carbodi-imide, which has been used in the Moffatt oxidation with DMSO of the same types of alcohol[50]. The simplicity of the apparatus needed for the efficient small scale oxidation of alcohols, including deuteriated alcohols, in the vapour phase by copper(II) oxide has been emphasised[51].

7.1.1.7 Syntheses based on the alkylation of aldehydes and ketones

As well as the use of organocopper reagents discussed in Section 7.1.1.4, many examples of the synthesis of ketones by the conjugate addition of organocopper reagents to α,β-alkenic ketones have been reported, although many of the reports do not deal with applications in the aliphatic series. Syntheses with organocopper reagents have been reviewed by Normant[52], with emphasis on the influence of the conditions on the course of the reactions, and the conjugate addition of organocopper reagents to α,β-alkenic ketones is covered by the definitive article by Posner in *Organic Reactions*[53]. Lithium dialkylcuprates react with β-acetoxy-α,β-alkenic ketones (i.e. enol acetates of 1,3-diketones) to give β-alkyl-α,β-alkenic ketones, with displacement of the hetero-atom substituent; excess of the reagent leads to subsequent conjugate addition to give β,β-dialkyl-substituted ketones[54]. The hetero-atom substituent is similarly replaced in β-alkoxy- and β-alkylthio-α,β-alkenic ketones; sometimes the β-alkyl-α,β-alkenic ketone can be isolated, sometimes the conjugate addition cannot be prevented[55]. New reagents for the transfer of organic groups to the β-position of simple α,β-alkenic ketones include lithium di-[3-(1-ethoxyethoxy)]propyl-cuprate, containing a protected alcohol function[56], several related oxygen-functionalised lithium dialk-1-enylcuprates[57] and the lithium bis-[1,1-di-(phenylthio)]alkylcuprates, which contain the protected form of a ketone function and hence provide a route to 1,4-diketones[58]. An important new development is the discovery of reagents in which one of the two groups attached to the copper atom can be selectively transferred to the alkenic ketone, thus allowing all of a difficultly accessible group to be transferred. These reagents include lithium t-butoxy- and phenylthio-alkylcuprate[30], which transfer the alkyl group; lithium pent-1-ynyl-t-butylcuprate and a related mixed alkynyl reagent, in which the groups transferred are respectively t-butyl and an O-protected 3-hydroxy-*trans*-oct-1-enyl group[59]; and lithium alkylcyanocuprates, which transfer the alkyl group[60]. The stereospecific transfer of an alk-1-enyl group to an α,β-alkenic ketone, to give a specific configuration in the resultant γ,δ-alkenic ketone, can also be accomplished by using the appropriate dialkylalk-1-enylalane[61]; stereospecific routes to such organoaluminium reagents are available.

Organocopper reagents are also involved in a method for the α-alkylation of ketones based on their reaction with α,α'-dibromoketones (18)[30,62]. The reaction can be used to introduce even a t-butyl group by using, e.g. lithium t-butoxy-t-butylcuprate, and with secondary and tertiary alkyl groups alkylation of unsymmetrical ketones takes place predominantly at the less hindered position. The course of the reaction is shown above. Cyclopropane intermediates are involved in two other new alkylation procedures: methylenation of aldehyde enamines or trimethylsilyl enol ethers gives products cleaved by aqueous alkanols to give α-methyl substituted aldehydes[63]. α-Dialkylthiomethylene ketones are prepared by the action of a sterically hindered base on a ketone in the presence of CS_2 and an alkyl halide, and these ketones react, e.g. with lithium dimethylcuprate to give α-isopropylidene ketones, which with an excess of the reagent can be converted into the α-t-butyl substituted ketones[64]. This is the method of choice for the introduction of s-alkylidene or t-alkyl groups into the α-position of a ketone.

The C-alkylation of α-monoalkyl-aldehydes by alkyl halides can be carried out in a water–benzene system in the presence of NaOH and a catalytic amount of a quaternary ammonium salt as a phase-transfer agent[65]. Semi-empirical calculations have been made on the orientation in the base-catalysed alkylation of unsymmetrical ketones and of α,β-alkenic ketones[66]. The agreement with the experimental findings, including the prediction of the α-alkylation of α,β-alkenic ketones, is excellent.

The reactions of alkenes with aldehydes, initiated by manganese(III) acetate dihydrate, have been studied in detail. In the absence of solvent the main products from alk-1-enes are ketones of the type formed in the peroxide-catalysed addition reactions, but in acetic acid with high concentrations of the manganese salt the main products are the saturated and γ-acetoxy α-branched aldehydes, which are of interest in perfumery[67]. With branched alkenes, mixtures of saturated and alkenic aldehydes are formed in acetic acid[68]. The addition of copper(II) acetate leads mainly to a mixture of β,γ- and γ,δ-alkenic α-branched aldehydes[68,69]. Acetone can be alkylated by alk-1-enes under similar sets of conditions; in the presence of copper(II) acetate it gives predominantly γ,δ-alkenic and γ-acetoxy-α,β-alkenic methyl ketones[70].

7.1.1.8 Syntheses based on pericyclic reactions

Further syntheses based on a full understanding of the Claisen and the Cope rearrangements have been reported, but lack of space prevents most of this work being discussed in detail here. The use of 3-methoxyisoprene (19) as a chain extender is shown below; reduction of the aldehyde, formed with complete stereospecificity, produces an analogue of the starting allylic alcohol, which is then capable of further chain extension by the same method[71]. One of several examples[72] of the use of a Claisen rearrangement followed by two consecutive Cope rearrangements is shown below. The condensation of aldehyde acetals with 3,3-dialkylallyl alcohols leads to α-1,1-dialkylprop-2-enyl substituted aldehydes[73]. Propargyl acetates can be converted thermally into allenic acetates. The simple unsubstituted acetate rearranges and is decarbonylated at 630°C to give butenone[74], but appropriate alkenyl-substituted allenic acetates undergo the Cope rearrangement

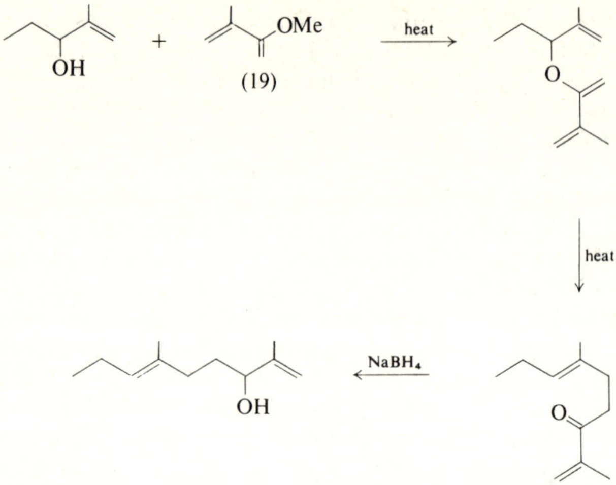

(19)

heat

NaBH₄

at lower temperatures to give enol acetates, hydrolysable to dialkenic aldehydes[75]. Condensation of appropriate alkenyl-substituted propargyl alcohols
with isopropenyl methyl ether leads similarly through allenic intermediates to
trialkenic ketones[75]. This brief account scarcely brings out the versatility of
these routes, which must always be considered when highly unsaturated
aldehydes or ketones are required.

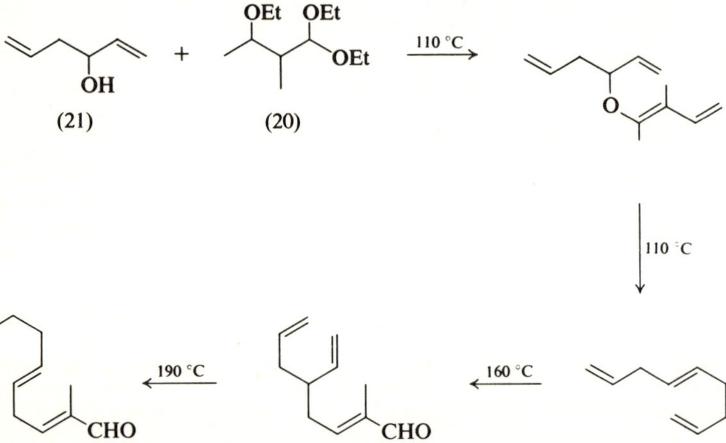

(21)   (20) $110\ ^\circ$C

$110\ ^\circ$C

$190\ ^\circ$C $160\ ^\circ$C

Alkylation of the anion from allyl vinyl sulphide gives a sulphide (22)
which undergoes the thio-Claisen rearrangement in refluxing aqueous DMF,
which also hydrolyses the initial alkenic thioaldehyde to a *trans*-γ,δ-alkenic
aldehyde[76]. Use of 2-ethoxyallyl vinyl sulphide leads via an enol ether to a

$$CH_2\text{=}CHCHRSCH\text{=}CH_2$$

(22)

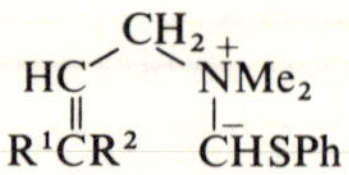

$$HC\underset{\underset{R^1CR^2}{\parallel}}{\overset{CH_2}{\diagup}}\overset{+}{N}Me_2$$
$$CHSPh$$

(23)

$$CH_2\text{=}CHCR^1R^2CHO$$

(24)

γ-keto-aldehyde[77]. The [2,3]-sigmatropic rearrangement of the ylides (23) derived from the reaction of allylic chlorides with dimethylaminomethyl phenyl sulphide leads to products which can be hydrolysed to the β,γ-olefinic aldehydes (24)[78].

7.1.1.9 Miscellaneous syntheses

The metallation of propargyl and allenyl ethers, e.g. (25), gives organometallic intermediates which can be transformed into a variety of α,β-alkenic aldehydes and ketones, as shown below[79]. 1-(Ethylthio)vinyl-lithium (26) reacts

$$R^1C{\equiv}CCH_2OR^2 \xrightarrow{\text{Bu}^n\text{Li}} R^1\bar{C}{=}C{=}CHOR^2 \xrightarrow[\text{K}_2\text{CO}_3]{\text{MeOH-H}_2\text{O-}} R^1CH{=}C{=}CHOR^2$$

$$\text{(25)}$$

$$\downarrow R^3X \qquad\qquad \downarrow H_3O^+$$

$$R^1R^3C{=}C{=}CR^4OR^2 \xleftarrow[\text{(ii) } R^4X]{\text{(i) Bu}^n\text{Li}} R^1R^3C{=}C{=}CHOR^2 \qquad R^1CH{=}CHCHO$$

$$\downarrow H_3O^+$$

$$R^1R^3C{=}CHCOR^4$$

with alkyl halides and aldehydes to give products, from which methyl ketones can be obtained by mercury(II) ion catalysed hydrolysis[80]. Lithio-triphenylphosphonioacetonide (27) reacts with alkyl and allyl halides to give ylides[81a] which can be hydrolysed to methyl ketones containing three additional carbon atoms[81b]. Dichloromethyl-lithium reacts with aldehydes and

$$CH_2{=}C(SEt)Li \qquad\qquad LiCH_2CO\bar{C}H\overset{+}{P}Ph_3$$

$$\text{(26)} \qquad\qquad\qquad \text{(27)}$$

ketones to give adducts which on lithiation undergo α-elimination with rearrangement to give α-chloroenolates; the latter on hydrolysis give α-chloro-aldehydes, which readily undergo dehydrohalogenation to give α,β-alkenic aldehydes; octan-2-one gives E-2-methyloct-2-enol by this method[83].

S-(+)-Oct-2-yl tosylate is converted stereospecifically into R-(−)-3-methylnonan-2-one by successive treatment with disodium tetracarbonyl-ferrate, carbon monoxide [which promotes the formation of the ion (15; R = 2-s-octyl)] and iodomethane[34].

Hydridoiron complexes, obtainable from iron pentacarbonyl and alkali metal hydroxides, selectively reduce the alkenic double bond in α,β-alkenic aldehydes and ketones[84]; α-methylene ketones prepared in situ can thus be converted into α-methyl-substituted ketones[85]. The conjugated alkenic linkage in citral can be selectively reduced by treatment with triethylsilane and chlorotris(triphenylphosphine)rhodium, which gives citronellal as its readily hydrolysable triethylsilyl enol ether[86].

Alkynic aldehydes and ketones can be prepared by the thermolysis of hydrazones derived from α,β-epoxyketones and trans-2,3-diphenyl-1-amino-aziridines. For example, hex-5-ynal can be obtained in this way from α,β-epoxycyclohexanone[87].

2-Methylpent-2-enal (29) has been synthesised from ethyl isopropenyl ketone (28) by the route shown, which involves a novel interchange of functionality[88].

7.1.2 Protective groups

The first book devoted entirely to protective groups includes a long chapter[89] on the protection of aldehydes and ketones. Recent work on the regeneration of carbonyl compounds from thioacetals is discussed in Section 7.1.1.1. Cyclic acetals prepared from 3-bromopropane-1,2-diol revert to the carbonyl compound on treatment under neutral conditions with zinc in methanol[90]. Similarly acetals prepared by exchange between 2,2,2-trichloroethanol and a simple acetal are cleaved by zinc in aprotic solvents such as ethyl acetate or THF[91]. Ketones can be protected as their oxime O-(phenylthio)methyl ethers and as such are unaffected by mild oxidative and reductive procedures; the ketone can be regenerated by a mild three-step procedure[92]. Several groups[93] have investigated cyanohydrin trimethylsilyl ethers as protected forms of carbonyl compounds. They can be prepared by direct addition of cyanotrimethylsilane; α,β-alkenic carbonyl compounds undergo exclusively 1,2-addition. They are stable in aprotic media and are inert to organometallic reagents, but regenerate the carbonyl compounds with aqueous acids or bases. Alkylation of the aldehyde derivatives, thus providing a route to ketones, is a likely development.

7.1.3 Reactions

The α-selenation of aldehydes and ketones can be accomplished in several ways and in the presence of various functional groups. A simple reaction with benzeneselenyl chloride may be sufficient[94]. Alternatively, the enolates generated by means of a lithium N,N-dialkylamide can be treated with diphenyl diselenide[95] or benzeneselenyl bromide[96], or the enol acetates can be treated with benzeneselenyl bromide in the presence of silver trifluoro-acetate[97]. The importance of these α-phenylselenyl derivatives lies in the fact that the corresponding selenoxides, prepared by oxidation with periodate, decompose at room temperature to give the α,β-alkenic aldehydes or ketones[94-97].

The recent challenge to the long-accepted mechanism for the halogenation of ketones in which the rate-determining step is the formation of the enol or enolate ion has been firmly rebutted by several papers on the bromination of butanone. A major problem in obtaining experimental support for the accepted mechanism is that the initial monobrominated products undergo further bromination and other reactions; for example, 1-bromo- and 3-bromo-butan-2-one react with hydroxide ion to give the hydroxy-ketones faster than they undergo further bromination[98]. The initial product ratios thus have to be inferred from the proportions of the stable end-products, but in fact this can be done with considerable precision. There is excellent agreement between the rates of bromination and of deuterium exchange determined for butanone in an aqueous acetate buffer[99], in aqueous sodium hydroxide[100] and in aqueous sulphuric acid[101]. The textbooks do not have to be rewritten! The relative proportions of 3-bromo- and 1-bromo-butan-2-one do, of course, depend on the particular conditions employed[99-101], and this aspect of the reaction has also been further explored[101,102]. These papers incidentally give alternative routes to many of the possible bromine-substituted butan-2-ones. A very versatile new method for the regiospecific preparation of monobromoketones, e.g. (30), is by the action of NBS on enol

$$R^1COCBrR^2R^3$$

(30)

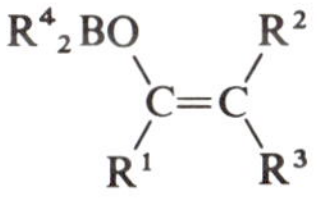

$$R^4{}_2BO \diagdown \quad \diagup R^2$$
$$C=C$$
$$R^1 \diagup \quad \diagdown R^3$$

$$R^1COCR^2R^3CH_2NMe_2$$

(32)

(31)

borinates (31), which can be prepared from trialkylboranes and either diazoketones or α,β-alkenic aldehydes or ketones[103].

The Mannich reaction, when applied to unsymmetrical ketones, leads to problems of regiospecificity akin to those encountered in bromination. This difficulty too can be overcome by preparation of the appropriate enol borinate (31), which on reaction with methylenedimethylammonium iodide gives a specific Mannich base (32)[104].

A new reducing agent, tetrabutylammonium cyanotrihydidoborate, will reduce an aldehyde selectively in the presence of a ketone[105]. The methods

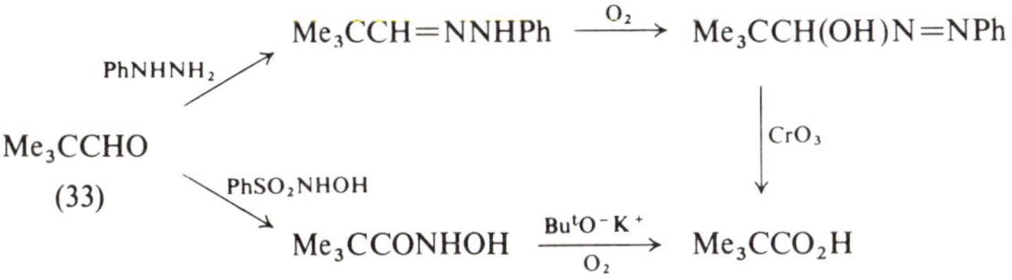

$$Me_3CCH{=}NNHPh \xrightarrow{O_2} Me_3CCH(OH)N{=}NPh$$

$$\xleftarrow{PhNHNH_2}$$

$$Me_3CCHO$$

(33)

$$\xrightarrow{PhSO_2NHOH}$$

$$\Bigg\downarrow CrO_3$$

$$Me_3CCONHOH \xrightarrow[O_2]{Bu^tO^-K^+} Me_3CCO_2H$$

available for the oxidation of aldehydes to acids still need improvement. Several ingenious methods for the oxidation of highly hindered aldehydes have been investigated by using pivalaldehyde (33) as a model[160]. The best of these are illustrated above. Unsymmetrical ketones react with chlorosulphonyl isocyanate to give β-ketonitriles by substitution at the more highly alkylated α-position[107]. The reactions of conjugated alkynyl ketones have been comprehensively reviewed[108].

7.1.4 Enols

An exciting development during the period under review has been the discovery of ways of generating the enol forms of aldehydes and ketones in high concentration. Hoffmann has used the reactions of 2-dimethylamino-4-methylenedioxolans with nucleophiles for this purpose. These heterocyclic precursors can be obtained by the debromination of α,α′-dibromoketones with a zinc–copper couple in DMF[109]. The tetramethyl derivative (34) with a trace of benzoic acid in DMF or CCl_4 generates 2,4-dimethylpenta-1,3-dien-3-ol (35), characterised by its ^1H n.m.r. spectrum, which tautomerises

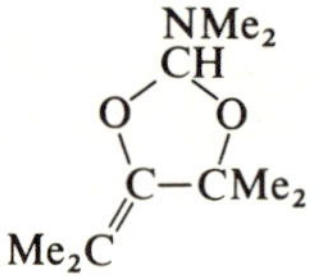

$$Me_2C=C(OH)CMe=CH_2$$

(35)

(34)

$$Me_2C=C(OH)CMe_2(OR)$$

(36)

to isopropenyl isopropyl ketone only during many hours[110]. The dienol cannot be trapped by diazomethane, but the hydrogen bound to oxygen undergoes virtually instantaneous deuterium exchange; the deuteriodienol tautomerises less rapidly than the protiodienol (35). Treatment of the dioxolan (34) with methanol in DMF gives the non-conjugated methoxyenol (36; R = Me) as well as the dienol (35). The non-conjugated enol is formed more rapidly but tautomerises less rapidly than the dienol. A solution containing the methoxyenol (36; R = Me) at a concentration of 140 g l^{-1} can be obtained! Water reacts with the dioxolan (34) to give the enediol (36; R = H) and although the acetate (36; R = Ac) tautomerises too rapidly to be detected, the deuterio analogue can be detected after treatment of the dioxolan (34) with deuterioacetic acid[110].

A simple dialkylated enol analogous to the methoxyenol (36; R = Me) has also been prepared, but this tautomerises to the methoxyketone in minutes rather than hours[111]. It has even been possible to generate the enol forms of acetaldehyde and acetone photochemically and to record their ^1H n.m.r. spectra. Vinyl alcohol is formed by the disproportionation of photochemically generated α-hydroxyethyl radicals formed, for example, by a Type I cleavage from acetoin, and the ^1H n.m.r. spectrum shows the expected CIDNP effects[112]. 2-Hydroxypropene can be prepared similarly from 3-hydroxy-3-methylbutan-2-one[113].

7.2 DI- AND TRI-CARBONYL COMPOUNDS

7.2.1 β-Diketones

7.2.1.1 *Synthesis*

An attempt to extend the monoacylation of alkenes to alkynes unexpectedly led to a new β-diketone synthesis. Reaction of a mono- or di-alkyl substituted acetylene with an acylium salt, which can be preformed or prepared *in situ* from an acyl chloride, in a *primary* nitroalkane as solvent at low temperatures, followed by treatment with a nucleophilic reagent such as aqueous sodium hydrogen carbonate, leads to a β-diketone. For example 3-n-butyl-octane-2,4-dione was prepared in 73% yield from dec-5-yne and acetyl hexafluoroantimonate[114]. Another new route to α-alkylated β-diketones is based on the reaction of zinc with a mixture of an α-bromoketone and an acyl chloride[115]. The reaction of dimethyloxosulphonium methylide with non-enolisable cyclic β-diketones gives acyclic β-keto-oxosulphonium ylides, which can be cleaved by zinc and acetic acid to give diketones. 3,3,5-Trimethylhexane-2,4-dione therefore becomes readily available from 2,2,4,4-tetramethylcyclobutane-1,3-dione[116]. Some chiral β-diketones of interest for the preparation of lanthanide shift reagents for n.m.r. studies have been prepared starting from *S*-s-butyl methyl ketone. The less highly substituted lithium enolate is prepared by using lithium di-isopropylamide to effect kinetically controlled deprotonation, without racemisation of the chiral centre, and reaction with an acid chloride then gives the lithium complex of the chiral β-diketone from which the free ligand can be obtained[117]. No *O*- or α'-acylation is observed.

7.2.1.2 *Alkylation*

Contrary to an earlier report, the use of the thallium(I) salts of β-diketones offers no special synthetic advantages for the preparation of the *C*-mono-alkylated derivatives[118]. The thallium salt of pentane-2,4-dione gives only *C*-alkylation with iodomethane, but a few per cent of the dimethylated product is now also reported; with iodoethane not only is some of the *C*,*C*-dialkylated product formed, but this is accompanied by the *O*-alkylated product. The reaction of pentane-2,4-dione with 3,3-dimethylallyl bromide in the presence of potassium carbonate in aprotic solvents gives as the major product the *C*-3,3-dimethylallylated compound (37) and none of the 1,1-dimethylallyl isomer (38). However, the copper chelate of pentane-2,4-dione under similar conditions gives some 20% of (38), having a tertiary carbon atom attached

$$Me_2C{=}CHCH_2CH(COCH_3)_2 \qquad\qquad CH_2{=}CHCMe_2CH(COCH_3)_2$$

$$(37) \qquad\qquad\qquad\qquad\qquad (38)$$

to the central carbon atom. Compound (38) seems to be formed by direct *C*-alkylation[119]. The anion formed by the action of butyl-lithium on the pyrrolidine enamine from pentane-2,4-dione undergoes alkylation exclusively at the γ-position[120].

7.2.2 γ-Diketones

A full account has appeared of the synthesis of γ-diketones from γ-nitroketones by reduction with titanium(III) chloride to the iminoketone followed by hydrolysis[121]. On a larger scale the Nef reaction, involving treatment with NaOH followed by HCl, may be more suitable, though for solubility reasons ethanol must be used as the solvent in the first step[122]. The reaction of a secondary nitroalkane with an alkyl nitrite and sodium nitrite may also be used to generate the second ketone function[123]. A more involved, but versatile, synthesis involves the cyanoethylation of a nitroalkane, followed by the Nef reaction, to give a γ-keto-nitrile. This is then protected as its thioacetal and converted into a protected γ-diketone by a Grignard reagent. Dethioacetalisation (*cf.* Section 7.1.1.1) finally gives the γ-diketone[124].

The anion (39; $R^1 = R^2 = Et$), which can be obtained in three simple steps from the compound (4; $R^2 = Et$), reacts with methyl vinyl ketone to give the 1,4-addition product; the latter can be converted into heptane-2,5-dione by the methods described above[125] (see Section 7.1.1.1). The unsubstituted anion (39; $R^1 = Et$; $R^2 = H$) gives only the 1,2-addition product. The method has been extended to the alkynic anion (39; $R^1 = Et$; $R^2 = $ hex-3-ynyl)[125].

$$\begin{array}{c} R^1S(O) \\ \diagdown \\ C^- \\ \diagup \diagdown \\ R^1S \quad R^2 \end{array}$$

(39)

$$EtCOCMe_2CHMeCOPr^i$$

(40)

$$EtCOCMe_2CMe_2COEt$$

(41)

$$Pr^iCO(CHMe)_2COPr^i$$

(42)

A new synthesis of γ-diketones by the action of zinc on α,α′-dibromoketones in *N*-methylformamide has been mentioned[126]. The α,α′-dibromo derivative of ethyl isopropyl ketone gives the diketones (40), (41) and (42) in the ratios 3.1 : 1.4 : 1.

Collins oxidation of 2,5-dialkylfurans gives α,β-alkenic γ-diketones[127]. The reaction can be applied to 2-methyl-5-(pent-2-enyl)furan, giving deca-3,7-diene-2,5-dione, in which the conjugated double bond can be selectively reduced by sodium dithionite.

7.2.3 Other diketones, keto-aldehydes and dialdehydes

The anions (39; $R^1 = Et$, $R^2 = H$ or alkyl) react with acyl chlorides or esters (for $R^2 = H$) to give the protected forms of α-keto-aldehydes or α-diketones, which can be converted into the dicarbonyl compounds through the dimethyl acetals by the action of trimethyl orthoformate followed by treatment with aqueous acid (*cf.* Section 7.1.1.1)[11]. The catalytic oxidation of 1,2-diols to α-diketones can be accomplished under mild conditions by the rhodium-catalysed transfer of hydrogen to benzylideneacetone[128]. Alkyl- and alkynyl-magnesium bromides react selectively at the aldehyde group of δ- and ε-keto-aldehydes to give ketols which can then be oxidised to δ- and ε-diketones with chromium(VI) oxide[129]. ε-Diketones are also formed in the reactions of carboxylic acids and esters with vinylmagnesium chloride in the presence of copper(I) chloride; methyl pentanoate gives tetradecane-5,10-dione and an equal amount of 5-oxonon-1-ene, but pentanoic acid gives only the diketone, which is then formed in 53% yield[130].

The oxidation of cycloalkene epoxides with aqueous periodic acid gives dialdehydes in good yield[131]. Vicinal cycloalkanediols can be cleaved to dicarbonyl compounds by manganese(IV) oxide in dichloromethane[132]. Butanone is oxidised to butanedione by 2,2,6,6-tetramethyl-1-oxopiperidinium chloride (43)[133].

7.2.4 Triketones and diketo-aldehydes

Reagent (43) also oxidises pentane-2,4-dione, giving pentane-2,3,4-trione[133]. The alkylation of the dianion of pentane-2,4-dione with bromo- and chloro-acetals of the type (44) is of preparative use only when $n \geqslant 3$; hydrolysis of the products leads to triketones[134]. 3-Hydroxyhex-3-ene-2,5-dione, the enol

$$
\begin{array}{cc}
\text{(43)} & \text{(44)}
\end{array}
$$

$$\text{MeCOCH=CHCOCMe=CH(CH}_2)_2\text{CMe=CHCHO}$$

(45)

form of hexane-2,3,5-trione, has been prepared by the dehydration of *threo*-3,4-dihydroxyhexane-2,5-dione, which was itself prepared by the bimolecular reduction of methylglyoxal[135].

The defence glands of the whirlygig beetle secrete gyrinal, 3,7-dimethyldo-deca-2,6,9-triene-8,11-dional (45)[136].

7.2.5 α-Keto-acids and their derivatives

A detailed account of the synthesis of α-keto-acids based on the Ritter reaction of cyanohydrins has been published[137]. α-Keto-esters can be prepared from their thioacetals, and a new route to these based on the alkylation of the anion of ethyl 1,3-dithiane-2-carboxylate has been developed[138]. The anion can conveniently be generated with sodium hydride; an earlier route to the thioacetals involved the use of butyl-lithium. Monohydroboration of alkyl alk-2-ynoates gives intermediates which on oxidative hydrolysis give α-keto-esters[139]. 1-Bromoalk-1-ynes react with ozone in methanol to give peroxy intermediates which are converted into α-keto-esters by the action of potassium iodide[140].

7.2.6 β-Keto-acids and their derivatives

7.2.6.1 Synthesis

A new synthesis[141] permits the isolation of several aliphatic β-keto-acids as well characterised crystalline solids; the pure acids are clearly more stable than is generally realised. The synthesis, which involves the acylation of the dianion of an acid by an ester, followed by careful liberation of the free β-keto-acid from its salt, is shown below. Thermal decarboxylation of the β-keto-acids leads quantitatively to ketones.

$$R^1R^2CHCO_2H \xrightarrow[\text{naphthalenide}]{\text{lithium}} R^1R^2\bar{C}CO_2^- \xrightarrow{R^3CO_2R^4} R^3COCR^1R^2CO_2^-$$

$$\Big\downarrow H_3O^+$$

$$R^3COCHR^1R^2 \xleftarrow{\text{heat}} R^3COCR^1R^2CO_2H$$

1-Thiomethyl-3,3-diethoxyprop-1-yne (46) gives an allenic anion with lithium diethylamide which can be alkylated to give an intermediate allene; on acid-catalysed hydrolysis this gives the diastereoisomeric enethiol ethers of a β-keto-ester, from which the β-keto-ester itself can be obtained[142]. O-Silyl ketene acetals of type (47), available by the reaction of ester enolates with the appropriate trialkylchlorosilane, can be acylated by various acid

$$MeSC\equiv CCH(OEt)_2$$
(46)

$$RCH=C\begin{smallmatrix} OSiMe_2Bu^t \\ \\ OEt \end{smallmatrix}$$
(47)

$$H_2C\begin{smallmatrix} CO_2SiMe_3 \\ \\ CO_2Et \end{smallmatrix}$$
(48)

chlorides to give enol silyl ethers which on hydrolysis give β-keto-esters[143]. Acid chlorides also react with the lithium salt of the malonic acid derivative (48) to give β-keto-esters[144]. The reaction fails for acrylyl chloride, though it works for some other α,β-alkenic acyl chlorides; ethyl acryloacetate can be made by using 3-ethoxypropanoyl chloride followed by acid-catalysed elimination of ethanol from the initial δ-ethoxy-β-keto-ester[144]. Ethyl acryloacetate can also be made in three steps from methyl vinyl ketone via alicyclic intermediates[145].

7.2.6.2 Alkylation

Several of the studies[118-120] of the alkylation of β-diketones discussed in Section 7.2.1.2 also dealt with the alkylation of β-keto-esters. The supposed advantages from using the thallium derivative of a β-keto-ester when *C*-monoalkylation is required are not supported by recent work[118]. The copper derivative of methyl acetoacetate behaves like the copper derivative of pentane-2,4-dione in the reaction with 3,3-dimethylallyl bromide[119]. The enamines of β-keto-esters can be alkylated at the γ-position through their anions[120].

The alkylation of ethyl acetoacetate by the bromide (49) to give a stereochemically pure product was achieved by the use of potassium t-amyloxide in t-amyl alcohol[146]. The very low solubility of KBr in this solvent prevented the inversion of the configuration at C-2 in the bromide (49) by bromide ion during the reaction. This process was observed when sodium ethoxide in

$$
\begin{array}{c}
\text{Me} \quad \text{Br} \\
\diagdown\diagup \\
| \text{—H} \\
| \text{—H} \\
\diagup\diagdown \\
\text{Me} \quad \text{D}
\end{array}
$$

(49)

ethanol was used to generate the anion owing to the greater solubility of KBr in ethanol.

A new palladium catalyst brings about the condensation of 2,3-dimethylbuta-1,3-diene with methyl acetoacetate to give selectively the 1 : 1 adduct ethyl α-(2,3-dimethylbut-2-enyl) acetoacetate[147].

7.2.7 γ-Aldehydo- and γ-keto-acids and their derivatives

Acylsuccinic esters, which can be obtained *inter alia* by Kharasch (radical) addition of aldehydes to dialkyl maleates, are converted directly into the corresponding γ-keto-esters by treatment with aqueous boric acid[148]. This is a much more convenient process than the sequence of hydrolysis, decarboxylation and re-esterification previously necessary.

Trialk-1-ynylboranes react with ethyl diazoacetate to give ethyl alk-3-ynoates; the latter undergo regiospecific mercury(II) ion catalysed hydration to give γ-keto-esters[149]. The enol trimethylsilyl ether of pinacolone reacts with ethoxycarbonylcarbene, generated by the copper-catalysed decomposition of ethyl diazoacetate, to give a high yield of ethyl 5,5-dimethyl-4-oxohexanoate[150].

The anion from compound (4) reacts with an α,β-alkenic ester to give the

protected form of a γ-aldehydo-ester and the analogous type of anion (39)
reacts similarly to give the corresponding protected form of a γ-keto-ester[125].
The free dicarbonyl compounds can be obtained by acidic hydrolysis in the
presence of mercury(II) chloride[12,125] (*cf.* Sections 7.1.1.1 and 7.2.2). The
ketene thioacetal monoxide (50), which can be prepared in an efficient six-
step synthesis from glyoxylic acid[151], acts as a Michael acceptor. The con-

$$
\begin{array}{c}
\text{O} \\
\parallel \\
\text{SMe} \\
/ \\
\text{H}_2\text{C}{=}\text{C} \\
\backslash \\
\text{SMe}
\end{array}
$$

(50)

jugate addition of ester enolates leads to the protected forms of γ-aldehydo-
esters from which the parent dicarbonyl compounds can be regenerated by
the action of aqueous acetonitrile containing perchloric acid[151]; methyl
butanoate gives methyl 2-ethyl-4-oxobutanoate in 90% yield. Alternatively,
the adduct anion may be trapped by low-temperature alkylation. Here the
product is determined by the stability of the initially formed sulphur-stabilised anion and that of the oxygen-stabilised enolate ion which can be formed
by an exchange reaction, so that the protected forms of either γ-keto-esters
or α-alkylated γ-aldehydo-esters may result[152]. Simple enolates give γ-keto-
esters but an α,β-alkenic enolate (which condenses at the α-position) leads
to aldehydo-esters.

β-Formylpropionic acid can be obtained in high yield by the action of
disodium tetracarbonylferrate on succinic anhydride[153].

7.2.8   Other aldehydo-acids and their derivatives

A significant improvement in the preparation of alkyl glyoxylates by the
cleavage of the corresponding tartrates is to carry out the oxidation with
paraperiodic acid in ether[154]. The addition of allyl alcohol to methyl propiolate gives methyl 3-allyloxyacrylate, which undergoes a thermal [3,3]-sigma-
tropic rearrangement to give methyl 2-formylpent-4-enoate[155]; further
extension of this type of reaction is possible. Ozonolysis of cyclohexene in
the presence of pyridine leads to some 5-formylpentanoic acid[156].

7.2.9   Diketo-esters, aldehydo-keto-esters, aldehydo-diesters and keto-diesters

Many of the new sulphur-containing synthons developed by Schlessinger,
and discussed in Sections 7.1.1.1, 7.2.2 and 7.2.7, can be used to build up
tricarbonyl systems. For example, the addition of the malonate anion to

$$[\text{R}^2\text{O}_2\text{C}]_2\text{CR}^1\text{CH}_2\text{CHO}$$

(51)

reagent (50) provides a route to aldehydo-diesters of the type (51)[152]. Further reagents developed by Schlessinger permit direct syntheses of various other tricarbonyl systems. For instance, the anions from ethyl 1,3-dithiolane-2-carboxylate and from methyl 2,2-di(ethylthio)acetate, which are synthons for an α-keto-ester grouping (*cf.* the use of ethyl 1,3-dithiane-2-carboxylate discussed in Section 7.2.5[138]) undergo Michael addition to α,β-alkenic esters and ketones to give the protected forms of keto-diesters and diketo-esters, respectively[157]. The cyclic anion seems to be the more useful reagent. It will also, for example, displace the chlorine from ethyl β-chlorocrotonate to give the cyclic thioacetal of diethyl 3-methyl-2-oxo-pent-3-enedioate[157]. Conversely, methyl α-(methylthio)acrylate acts as a Michael acceptor towards enamines and the enolates from simple and α,β-alkenic esters (which condense at the α-position). The products contain an α-(methylthio) substituted ester group, but this can be converted into an α,α-di(methylthio) substituted ester by treatment with base and thiomethylation of the anion with methyl toluene-*p*-thiolsulphonate; aldehyde or ketone groups must be protected as cyclic acetals during this sequence. Regeneration then gives aldehydo-keto-esters, diketo-esters or keto-diesters, depending on the starting materials[158]. Reaction of pentan-3-one with sodium hydride, carbon disulphide and iodomethane leads to the α,α'-bis(dithiomethylene) ketone (*cf.* Ref. 64), which can be converted into diethyl 2,4-dimethyl-3-oxoglutarate by the action of ethanol containing toluene-*p*-sulphonic acid[159]. Problems due to proton transfer reactions are circumvented in a new procedure for the preparation of β,δ-diketo-esters by the acylation of the dianions from β-keto-esters[160]. β,δ-Diketo-esters can also be prepared from γ,δ-alkynic β-keto-esters via the corresponding enamino-β-keto-esters obtained by the addition of a secondary amine to the conjugated acetylenic system; alkylation of the enamine at the γ-position before hydrolysis is possible[161]. The alkynic β-keto-esters can be prepared by the reaction between an alkynyl-magnesium bromide and the half-acid chloride half-ester of a malonic acid (ethoxycarbonylacetyl chloride).

7.3 PHOTOCHEMISTRY

The photochemistry of α-keto-aldehydes, α-diketones and α-keto-acids and their esters has been reviewed[162]. A shorter account of the photochemistry of alkanones in solution has also appeared[163], and it is clear that all of the main features such as α-cleavage (the Norrish type I cleavage), α,β-cleavage with γ-hydrogen abstraction (the Norrish type II cleavage), cycloaddition reactions, and the sensitisation of these processes, are now very well understood. Stereochemical evidence has been presented in support of a short-lived singlet radical intermediate in the Norrish type II cleavage of 4-methyl-hexan-2-one from the singlet excited state, and the differences between the singlet and triplet state processes have been discussed in terms of spin effects[146]. The photochemistry of the ketone (52) has been investigated. The usual products are carbon monoxide and the three possible C_{10} hydrocarbons from the recombination of dimethylallyl radicals, but in a glass vessel at −196 °C and at low conversions the only product is the rearranged

$$CH_2=CHCMe_2COCMe_2CH=CH_2$$

(52)

$$CH_2=CHCMe_2COCH_2CH=CMe_2$$

(53)

ketone (53). The solution photochemistry of hept-5-en-2-one, which contains both ketonic and alkenic chromophores, has been studied in detail. Ketone sensitised *cis–trans* isomerisation of the alkenic double bond is observed.

References

1. Narasaka, K., Sakashita, T. and Mukaiyama, T. (1972). *Bull. Chem. Soc. Jap.*, **45**, 3724
2. Huurdeman, W., Wynberg, H. and Emerson, D. W. (1972). *Synth. Commun.*, **2**, 7
3. Ho, T.-L, Ho, H. C. and Wong, C. M. (1972). *J. Chem. Soc. Chem. Commun.*, 791
4. Tamura, Y., Sumoto, K., Fujü, S., Satoh, H. and Ikeda, M. (1973). *Synthesis*, 312
5. *Chem. Eng. News.* (1973). Dec. 17th, p. 36
6. Ho, T.-L., Ho, H. C. and Wong, C. M. (1973). *Can. J. Chem.*, **51**, 153
7. Jones, P. F., Lappert, M. F. and Szary, A. C. (1973). *J. Chem. Soc. Perkin Trans. I*, 2272; Carey, F. A. and Court, A. S. (1972). *J. Org. Chem.*, **37**, 1926; Seebach, D., Gröbel, B.-Th., Beck, A. K., Braun, M. and Geiss, K.-H. (1972). *Angew. Chem. Int. Ed. Engl.*, **11**, 443
8. Seebach, D., Kolb, M. and Gröbel, B.-Th. (1973). *Angew. Chem. Int. Ed. Engl.*, **12**, 69
9. Meyers, A. I. and Strickland, R. C. (1972). *J. Org. Chem.*, **37**, 2579
10. Herrmann, J. L., Kieczykowski, G. R., Romanet, R. F., Wepplo, P. J. and Schlessinger, R. H. (1973). *Tetrahedron Lett.*, 4711
11. Herrmann, J. L., Richman, J. E., Wepplo, P. J. and Schlessinger, R. H. (1973). *Tetrahedron Lett.*, 4707
12. Richmann, J. E., Herrmann, J. L. and Schlessinger, R. H. (1973). *Tetrahedron Lett.*, 3267
13. Hunt, E. and Lythgoe, B. (1972). *J. Chem. Soc. Chem. Commun.*, 757
14. Meyers, A. I. and Smith, E. M. (1972). *J. Org. Chem.*, **37**, 4289
15. Meyers, A. I. and Nazarenko, N. (1972). *J. Amer. Chem. Soc.*, **94**, 3243
16. Meyers, A. I., Nabeya, A., Adickes, H. W., Politzer, I. R., Malone, G. R., Kovelesky, A. C., Nolen, R. L. and Portnoy, R. C. (1973). *J. Org. Chem.*, **38**, 36
17. Meyers, A. I., Smith, E. M. and Ao, M. S. (1973). *J. Org. Chem.*, **38**, 2129
18. Meyers, A. I., Kovelesky, A. C. and Jurjevich, A. F. (1973). *J. Org. Chem.*, **38**, 2136
19. Lion, C. and Dubois, J. E. (1973). *Tetrahedron*, **29**, 3417
20. Lion, C. and Dubois, J. E. (1973). *Bull. Soc. Chim. Fr.*, 2673
21. Meyers, A. I., Munavu, R. and Dwandetta, J. (1972). *Tetrahedron Lett.*, 3929
22. Pelter, A., Harrison, C. R. and Kirkpatrick, D. (1973). *J. Chem. Soc. Chem. Commun.*, 544
23. Naruse, M., Tomita, T., Utimoto, K. and Nozaki, H. (1973). *Tetrahedron Lett.*, 795
24. Pelter, A., Harrison, C. R. and Kirkpatrick, D. (1973). *Tetrahedron Lett.*, 4491
25. Naruse, M., Utimoto, K. and Nozaki, H. (1973). *Tetrahedron Lett.*, 2741
26. Negishi, E. and Brown, H. C. (1973). *J. Amer. Chem. Soc.*, **95**, 6757
27. Brown, H. C. and Ravindran, N. (1972). *J. Amer. Chem. Soc.*, **94**, 2112
28. Carlson, B. A. and Brown, H. C. (1973). *J. Amer. Chem. Soc.*, **95**, 6876
29. Negishi, E. and Yoshida, T. (1973). *J. Chem. Soc. Chem. Commun.*, 606
30. Posner, G. H., Whitten, C. E. and Sterling, J. J. (1973). *J. Amer. Chem. Soc.*, **95**, 7788
31. Posner, G. H., Whitten, C. E. and McFarland, P. E. (1972). *J. Amer. Chem. Soc.*, **94**, 5106; Luong, Thi, N. T., Riviere, H. and Spassky, A. (1973). *Bull. Soc. Chim. Fr.*, 2102
32. Bourgain, M. and Normant, J. F. (1973). *Bull. Soc. Chim. Fr.*, 2137

33. Hegedus, L. S., Lo, S. M. and Bloss, D. E. (1973). *J. Amer. Chem. Soc.*, **95**, 3040
34. Collman, J. P., Winter, S. R. and Clark, D. R. (1972). *J. Amer. Chem. Soc.*, **94**, 1788
35. Huet, F., Emptoz, G. and Jubier, A. (1973). *Tetrahedron*, **29**, 479
36. Mukaiyama, T., Araki, M. and Takei, H. (1973). *J. Amer. Chem. Soc.*, **95**, 4763
37. Kempe, U. M., DasGupta, T. K., Blatt, K., Cygax, P., Felix, D. and Eschenmoser, A. (1972). *Helv. Chim. Acta*, **55**, 2187
38. Shatzmiller, S., Gygax, P., Hall, D. and Eschenmoser, A. (1973). *Helv. Chim. Acta*, **56**, 2961
39. Shatzmiller, S. and Eschenmoser, A. (1973). *Helv. Chim. Acta*, **56**, 2975
40. Gygax, P., DasGupta, T. K. and Eschenmoser, A. (1972). *Helv. Chim. Acta*, **55**, 2205
41. Seebach, D., Braun, M. and Du Preez, N. (1973). *Tetrahedron Lett.*, 3509
42. Stefani, A., Consiglio, G., Botteghi, C. and Pino, P. (1973). *J. Amer. Chem. Soc.*, **95**, 6504
43. Dudley, C. and Read, G. (1972). *Tetrahedron Lett.*, 5273
44. Corey, E. J. and Kim, C. U. (1972). *J. Amer. Chem. Soc.*, **94**, 7586
45. Corey, E. J. and Kim, C. U. (1973). *Tetrahedron Lett.*, 919
46. Corey, E. J. and Kim, C. U. (1973). *J. Org. Chem.*, **38**, 1233
47. Corey, E. J. and Fleet, G. W. J. (1973). *Tetrahedron Lett.*, 4499
48. Brown, J. M., Golding, B. T. and Stofko, J. J., Jr. (1973). *J. Chem. Soc. Chem. Commun.*, 319
49. Lalancette, J. M., Rollin, G. and Dumas, A. P. (1972). *Can. J. Chem.*, **50**, 3058
50. Weinshenker, N. M. and Shen, C.-M. (1972). *Tetrahedron Lett.*, 3281 and 3285
51. Sheikh, M. Y. and Eadon, G. (1972). *Tetrahedron Lett.*, 257
52. Normant, J. F. (1972). *Synthesis*, 63
53. Posner, G. H. (1972). *Organic Reactions*, **19**, 1
54. Casey, C. P., Marten, D. F. and Boggs, R. A. (1973). *Tetrahedron Lett.*, 2071
55. Posner, G. H. and Brunelle, D. J. (1973). *J. Chem. Soc. Chem. Commun.*, 907
56. Eaton, P. and Mueller, R. H. (1972). *J. Amer. Chem. Soc.*, **94**, 1014
57. Kluge, A. F., Untch, K. G. and Fried, J. H. (1972). *J. Amer. Chem. Soc.*, **94**, 7827; Sih, C. J., Price, P., Sood, R., Salomon, R. G., Peruzzotti, G. and Casey, M. (1972). *J. Amer. Chem. Soc.*, **94**, 3643
58. Mukaiyama, T., Narasaka, K. and Furusato, M. (1972). *J. Amer. Chem. Soc.*, **94**, 8641
59. Corey, E. J. and Beames, D. J. (1972). *J. Amer. Chem. Soc.*, **94**, 7210
60. Gorlier, J.-P., Hamon, L., Levisalles, J. and Wagnon, J. (1973). *J. Chem. Soc. Chem. Commun.*, 88
61. Hooz, J. and Layton, R. B. (1973). *Can. J. Chem.*, **51**, 2098
62. Posner, G. H. and Sterling, J. J. (1973). *J. Amer. Chem. Soc.*, **95**, 3076
63. Kuehne, M. E. and King, J. C. (1973). *J. Org. Chem.*, **38**, 304; Conia, J. M. and Girard, C. (1973). *Tetrahedron Lett.*, 2767
64. Corey, E. J. and Chen, R. H. K. (1973). *Tetrahedron Lett.*, 3817
65. Dietl, H. K. and Brannock, K. C. (1973). *Tetrahedron Lett.*, 1273
66. Tran Huu Dan, M.-E., Fetizon, M. and Nguyen Trong Anh (1973). *Tetrahedron Lett.*, 851 and 855
67. Nikishin, G. I., Vinogradov, M. G., Verenchikov, S. P., Kostyukov, I. N and Kereselidze, R. V. (1972). *J. Org. Chem. USSR*, **8**, 544; Nikishin, G. I., Vinogradov, M. G. and Il'ina, G. P. (1972). *ibid.*, **8**, 1422
68. Vinogradov, M. G., Il'ina, G. P., Ignatenko, A. V. and Nikishin, G. A. (1972). *J. Org. Chem.*, *USSR*, **8**, 1425
69. Nikishin, G. I., Vinogradov, M. G. and Il'ina, G. P. (1972). *Synthesis*, 377
70. Vinogradov, M. G., Verenchikov, S. P. and Nikishin, G. I. (1972). *J. Org. Chem. USSR*, **8**, 2515
71. Faulkner, D. J. and Petersen, M. R. (1973). *J. Amer. Chem. Soc.*, **95**, 553
72. Cookson, R. C. and Rogers, N. R. (1973). *J. Chem. Soc. Perkin Trans. I*, 2741
73. Cookson, R. C. and Hughes, N. W. (1973). *J. Chem. Soc. Perkin Trans. I*, 2738
74. Trahanovsky, W. S. and Mullen, P. W. (1972). *J. Amer. Chem. Soc.*, **94**, 5086
75. Bowden, B., Cookson, R. C. and Davis, H. A. (1973). *J. Chem. Soc. Perkin Trans. I*, 2634
76. Oshima, K., Takahashi, H., Yamamoto, H. and Nozaki, H. (1973). *J. Amer. Chem. Soc.*, **95**, 2693

77. Oshima, K., Yamamoto, H. and Nozaki, H. (1973). *J. Amer. Chem. Soc.*, **95**, 4446
78. Huynh, C., Julia, S., Lorne, R. and Michelot, D. (1972). *Bull. Soc. Chim. Fr.*, 4057
79. Corey, E. J. and Terashima, S. (1972). *Tetrahedron Lett.*, 1815; Leroux, Y. and Roman, C. (1973). *ibid.*, 2585
80. Oshima, K., Shimoji, K., Takahashi, H., Yamamoto, H. and Nozaki, H. (1973). *J. Amer. Chem. Soc.*, **95**, 2694
81. (a) Taylor, J. D. and Wolf, J. F. (1972). *J. Chem. Soc. Chem. Commun.*, 876; (b) Cooke, M. P., Jr. (1973). *J. Org. Chem.*, **38**, 4083
82. Villieras, J., Bacquet, C. and Normant, J. F. (1972). *J. Organometal. Chem.*, **40**, C1; Taguchi, H., Yamamoto, H. and Nozaki, H. (1972). *Tetrahedron Lett.*, 4661
83. Taguchi, H., Tanaka, S., Yamamoto, H. and Nozaki, H. (1973). *Tetrahedron Lett.*, 2465
84. Noyori, R., Umeda, I. and Ishigami, T. (1972). *J. Org. Chem.*, **37**, 1542
85. Carnelli, G., Panunzio, M. and Umani-Ronchi, A. (1973). *Tetrahedron Lett.*, 2491
86. Ojima, I. and Kogure, T. (1972). *Tetrahedron Lett.*, 5035
87. Felix, D., Muller, R. K., Horn, U., Joos, R., Schreiber, J. and Eschenmoser, A. (1972). *Helv. Chim. Acta*, **55**, 1276
88. Büchi, G. and Vederas, J. C. (1972). *J. Amer. Chem. Soc.*, **94**, 9128
89. Loewenthal, H. J. E. (1973). *Protective Groups in Organic Chemistry* (J. F. W. McOmie, editor) (London and New York: Plenum Press)
90. Corey, E. J. and Ruden, R. A. (1973). *J. Org. Chem.*, **38**, 834
91. Isidor, J. L. and Carlson, R. M. (1973). *J. Org. Chem.*, **38**, 554
92. Vlattas, I., Della Vecchia, L. and Fitt, J. J. (1973). *J. Org. Chem.*, **38**, 3749
93. Evans, D. A., Truesdale, L. K. and Carroll, G. L. (1973). *J. Chem. Soc. Chem. Commun.*, 55; Evans, D. A., Hoffman, J. M. and Truesdale, L. K. (1973). *J. Amer. Chem. Soc.*, **95**, 5822; Lidy, W. and Sundmeyer, W. (1973). *Chem. Ber.*, **106**, 587; Deuchert, K., Hertenstein, U. and Hünig, S. (1973). *Synthesis*, 777
94. Sharples, K. B., Lauer, R. F. and Teranishi, A. Y. (1973). *J. Amer. Chem. Soc.*, **95**, 6137
95. Trost, B. M. and Salzmann, T. N. (1973). *J. Amer. Chem. Soc.*, **95**, 6840
96. Reich, H. J., Reich, I. L. and Renga, J. M. (1973). *J. Amer. Chem. Soc.*, **95**, 5813
97. Clive, D. L. J. (1973). *J. Chem. Soc. Chem. Commun.*, 695
98. Knipe, A. C. and Cox, B. G. (1973). *J. Org. Chem.*, **38**, 3429
99. Thorpe, J. W. and Warkentin, J. (1972). *Can. J. Chem.*, **50**, 3229; Cox, R. A. and Warkentin, J. (1972). *Can. J. Chem.*, **50**, 3239
100. Swain, C. G. and Dunlop, R. P. (1972). *J. Amer. Chem. Soc.*, **94**, 7204
101. Deno, N. C. and Fishbein, R. (1973). *J. Amer. Chem. Soc.*, **95**, 7445
102. Boyer, G. and de Aguirre, I. (1972). *Bull. Soc. Chim. Fr.*, 4007 and 4016
103. Hooz, J. and Bridson, J. N. (1972). *Can. J. Chem.*, **50**, 2387
104. Hooz, J. and Bridson, J. N. (1973). *J. Amer. Chem. Soc.*, **95**, 602
105. Hutchins, R. O. and Kandasamy, D. (1973). *J. Amer. Chem. Soc.*, **95**, 6131
106. Batten, P. L., Bentley, T. J., Boar, R. B., Draper, R. W., McGhie, J. F. and Barton, D. H. R. (1972). *J. Chem. Soc. Perkin Trans. I*, 739
107. Rasmussen, J. K. and Hassner, A. (1973). *Synthesis*, 682
108. Bol'shedvorskaya, R. L. and Vereshchagin, L. I. (1973). *Russ. Chem. Rev.*, **42**, 225
109. Hoffmann, H. M. R., Clemens, K. E., Schmidt, E. A. and Smithers, R. H. (1972). *J. Amer. Chem. Soc.*, **94**, 3201
110. Schmidt, E. A. and Hoffmann, H. M. R. (1972). *J. Amer. Chem. Soc.*, **94**, 7832
111. Hoffmann, H. M. R. and Schmidt, E. A. (1973). *Angew. Chem. Int. Ed. Engl.*, **12**, 238
112. Blank, B. and Fischer, H. (1973). *Helv. Chim. Acta*, **56**, 506
113. Laroff, G. P. and Fischer, H. (1973). *Helv. Chim. Acta*, **56**, 2011
114. Roitburd, G. V., Smith, W. A., Semenovsky, A. V., Schegolev, A. A., Kucherov, V. F., Chizhov, O. S. and Kadentsev, V. I. (1972). *Tetrahedron Lett.*, 4935; Roitburd, G. V., Smith, W. A., Semenovsky, A. V., Schegolev, A. A. and Kucherov, V. F. (1972). *Doklady*, **203**, 345
115. Lapkin, I. I. and Saitkulova, F. G. (1971). *J. Org. Chem. USSR*, **7**, 2586
116. Lund, E., Palmer, G. E. and Welland, R. P. (1972). *J. Chem. Soc. Chem. Commun.*, 136
117. Seebach, D. and Ehrig, V. (1972). *Angew. Chem. Int. Ed. Engl.*, **11**, 127
118. Hooz, J. and Smith, J. (1972). *J. Org. Chem.*, **37**, 4200

119. Carnduff, J., Larkin, J., Miller, J. A., Nonhebel, D. C., Stockdale, B. R. and Wood, H. C. S. (1972). *J. Chem. Soc. Perkin Trans. I*, 692; Miller, J. A., Scrimgeour, C. M., Black, R., Larkin, J., Nonhebel, D. C. and Wood, H. C. S. (1973). *ibid.*, 603
120. Yashimoto, M., Ishida, N. and Hiraoka, T. (1973). *Tetrahedron Lett.*, 39
121. McMurry, J. E. and Melton, J. (1973). *J. Org. Chem.*, **38**, 4367
122. Black, D. St. C. (1972). *Tetrahedron Lett.*, 1331 ·
123. Kornblum, N. and Wade, P. A. (1973). *J. Org. Chem.*, **38**, 1418
124. Ho, H. C., Ho, T.-L. and Wong, C. M. (1972). *Can. J. Chem.*, **50**, 2718
125. Herrmann, J. L., Richman, J. E. and Schlessinger, R. H. (1973). *Tetrahedron Lett.*, 3271 and 3275
126. Unpublished work by E. A. Schmidt quoted by Hoffmann, H. M. R. (1973). *Angew. Chem. Int. Ed. Engl.*, **12**, 819
127. Birch, A. J., Keogh, K. S. and Mamdapur, V. R. (1973). *Aust. J. Chem.*, **26**, 2670
128. Regen, S. L. and Whitesides, G. M. (1972). *J. Org. Chem.*, **37**, 1832
129. Vaskan, R. N. and Kovalev, B. G. (1973). *J. Org. Chem. USSR*, **9**, 501
130. Watanabe, S., Suga, K., Fujita, T. and Takahashi, Y. (1972). *Can. J. Chem.*, **50**, 2786
131. Nagarkatti, J. P. and Ashley, K. R. (1973). *Tetrahedron Lett.*, 4599
132. Ohloff, G. and Giersch, W. (1973). *Angew. Chem. Int. Ed. Engl.*, **12**, 401
133. Golubev, V. A. and Miklyush, R. V. (1972). *J. Org. Chem. USSR*, **8**, 1376
134. Flannery, R. E. and Hampton, K. G. (1972). *J. Org. Chem.*, **37**, 2806
135. Büchi, G., Demole, E. and Thomas, A. F. (1973). *J. Org. Chem.*, **38**, 123
136. Schildknecht, H., Neumaier, H. and Tauscher, B. (1972). *Annalen*, **756**, 155
137. Anatol, J. M. and Medete, A. (1972). *Bull. Soc. Chim. Fr.*, 189
138. Eliel, E. L. and Hartmann, A. A. (1972). *J. Org. Chem.*, **37**, 505
139. Plamondon, J., Snow, J. T. and Zweifel, G. (1971). *Organometal. Chem. Synth.*, **1**, 249
140. Cacchi, S., Caglioti, L. and Zappelli, P. (1973). *J. Org. Chem.*, **38**, 3653
141. Angelo, B. (1973). *C. R. Acad. Sci. (Paris), Ser. C*, **276**, 293
142. Carlson, R. M. and Isidor, J. L. (1973). *Tetrahedron Lett.*, 4819
143. Rathke, M. W. and Sullivan, D. F. (1973). *Tetrahedron Lett.*, 1297
144. Pichat, L. and Beaucourt, J.-P. (1973). *Synthesis*, 537
145. Stork, G. and Guthikonda, R. N. (1972). *Tetrahedron Lett.*, 2755
146. Casey, C. P. and Boggs, R. A. (1972). *J. Amer. Chem. Soc.*, **94**, 6457
147. Watanabe, S., Suga, K. and Fujita, T. (1973). *Can. J. Chem.*, **51**, 848
148. Wehrli, P. A. and Chu, V. (1973). *J. Org. Chem.*, **38**, 3436
149. Hooz, J. and Layton, R. B. (1972). *Can. J. Chem.*, **50**, 1105
150. Le Goaller, R. and Pierre, J.-L. (1973). *C. R. Acad. Sci. (Paris), Ser. C*, **276**, 193
151. Herrmann, J. L., Kieczykowski, G. R., Romanet, R. F., Wepplo, P. J. and Schlessinger, R. H. (1973). *Tetrahedron Lett.*, 4711
152. Herrmann, J. L., Kieczykowski, C .R., Romanet, R. F. and Schlessinger, R. H. (1973). *Tetrahedron Lett.*, 4715
153. Watanabe, Y., Yamashita, M., Mitsudo, T.-A., Tanaka, M. and Takegami, Y. (1973). *Tetrahedron Lett.*, 3535
154. Kelly, T. R., Schmidt, T. E. and Haggerty, J. G. (1972). *Synthesis*, 544
155. Cresson, P. (1973). *C. R. Acad. Sci. (Paris), Ser. C*, **276**, 1473
156. Odinokov, V. N., Odinokova, A. I. and Tolstikov, G. A. (1973). *J. Org. Chem. USSR*, **9**, 691
157. Cregge, R. J., Herrmann, J. L., Richman, J. E., Romanet, R. F. and Schlessinger, R. H. (1973). *Tetrahedron Lett.*, 2595; Herrmann, J. L., Richman, J. E. and Schlessinger, R. H. (1973). *ibid.*, 2599
158. Cregge, R. J., Herrmann, J. L. and Schlessinger, R. H. (1973). *Tetrahedron Lett.*, 2603
159. Shahak, I. and Sasson, Y. (1973). *Tetrahedron Lett.*, 4207
160. Huckin, S. N. and Weiler, L. (1972). *Tetrahedron Lett.*, 2405
161. Suzuki, E., Sekizaki, H. and Inoue, S. (1973). *J. Chem. Soc. Chem. Commun.*, 568
162. Monroe, B. M. (1971). *Advan. Photochem.*, **8**, 77
163. Turro, N. J., Dalton, J. C., Davies, K., Farrington, G., Hautala, R., Morton, D., Niemczyk, M. and Schore, N. (1972). *Accounts Chem. Res.*, 92
164. Engel, P. S. and Schexnayder, M. A. (1972). *J. Amer. Chem. Soc.*, **94**, 4357
165. Kurowsky, S. R. and Morrison, H. (1972). *J. Amer. Chem. Soc.*, **94**, 507

8
Carboxylic Acids

N. POLGAR
University of Oxford

8.1 GENERAL TECHNIQUES

8.1.1 Synthesis

A new method[1] for the synthesis of carboxylic acids and their derivatives is based upon the use of sodium tetracarbonylferrate (1). This reacts with

$$RX + Na_2Fe(CO)_4 \longrightarrow (2) \xrightarrow{CO} (3) \longrightarrow RCO_2H \ (4)$$

alkyl halides or tosylates to yield anionic alkyltetracarbonyliron complexes (2) which, in the presence of CO, give the anionic acyl complexes (3). Reaction with O_2 or aqueous NaOCl affords the carboxylic acids (4) or their derivatives. These syntheses have the advantage of selectivity in respect of the halogen substituents. Thus 1-bromo-6-chlorohexane, for example, is converted into 7-chloroheptanoic acid in high yield. Moreover, various functional groups, e.g. ester and ketone groups, are not affected under the conditions employed.

8.1.2 Degradation

A convenient procedure[2] for the stepwise degradation of aliphatic chains, as exemplified for ethyl decanoate (5), allowing the removal of one-, two- or three-carbon fragments from the chain, involves conversion of the ester via the intermediates (6) and (7) into 1,2-diphenyldecan-1-one (8), by utilising a rearrangement described earlier[3]. The further steps consist of the α-bromination of the ketone (8), followed by dehydrobromination of the resulting bromo derivative (9) to yield the unsaturated ketone (10). Reduction of this ketone to the allylic alcohol (11), followed by dehydration, affords the diene (12). Ozonolysis of (12) would lead to the loss of a three-carbon fragment from the original ester (5) while ozonolysis of the intermediates (7) and (10) would result in the loss of one and two carbon atoms, respectively.

$$Me(CH_2)_8CO_2Et \rightarrow Me(CH_2)_8C(OH)Ph_2 \rightarrow Me(CH_2)_7CH-CPh_2 \rightarrow$$
$$(5) \qquad\qquad (6) \qquad\qquad (7)$$

$$Me(CH_2)_7CHPhCOPh \rightarrow Me(CH_2)_7CPhBrCOPh \rightarrow Me(CH_2)_6CH=CPhCOPh \rightarrow$$
$$(8) \qquad\qquad (9) \qquad\qquad (10)$$

$$Me(CH_2)_6CH=CPhCH(OH)Ph \rightarrow Me(CH_2)_5CH=CHCPh=CHPh$$
$$(11) \qquad\qquad (12)$$

8.1.3 Esterification

Esterifications or transesterifications under mild conditions can be carried out by using a polymer-protected aluminium chloride as a catalyst[4]; the latter is a complex between aluminium chloride and polystyrene–divinyl-benzene copolymer beads.

Another mild esterification procedure for non-hydroxy aliphatic acids consists in keeping them at the room temperature with chloroform–methan-olic HCl–copper(II) acetate[5].

There is continued interest in devising suitable methods for the esterifi-cation of sterically hindered carboxylic acids. Attention has been drawn[6] to the advantages of dimethyl sulphate for the preparation of methyl esters, particularly when an acidic medium is to be avoided. A new mild procedure[7] involves esterification of the acids by means of alkyl halides by using HMPT as cosolvent under alkaline conditions.

Esters of long-chain alcohols with long-chain acids are obtainable in high yields by the reaction of alkyl methanesulphonates with the sodium or potassium salts of aliphatic acids[8].

8.1.4 Reactions

Reaction of an alkenyl orthoester (13) with one mole of a carboxylic acid (14) results essentially in the substitution of one alkenyloxy group by an acyloxy residue to give the dialkenyl derivative (15)[9]. Renewed treatment of such compounds with another mole of carboxylic acid is, however, shown to yield, depending on the structural features of the dialkenyl derivative (15), either predominantly alkenyl esters and acylals, or alkenyl esters, carboxylic anhydrides and carbonyl compounds.

$$R^3C(OCR^1{=}CHR^2)_3 + R^4CO_2H \rightarrow R^3C(OCR^1{=}CHR^2)_2O_2CR^4$$
$$\quad\;\; (13) \qquad\qquad (14) \qquad\qquad\qquad (15)$$

Esters and lactones can be directly cleaved to acyl halides by reaction with dihalogenotriphenylphosphoranes (16)[10,11].

$$R^1CO_2R^2 + Ph_3PX_2 \rightarrow R^1COX + R^2X + Ph_3P(O)$$
$$(16)$$

On treating an alkanoic acid with boiling $SOCl_2$ in the presence of intense visible or u.v. light, in the absence of base, a high percentage of an α-chloro-alkanoic acid is shown[12] to be obtainable, along with a small amount of the α,ω-dichloro derivative; the extent of the formation of the latter can be minimised by limiting the time of the reaction.

A new method[13] for the preparation of hydroxamic acids from secondary amides is based upon the oxidation of the N-trimethylsilylamides with diperoxo-oxohexamethylphosphoramidomolybdenum.

8.1.5 Protecting groups

A procedure for the preparation of methylthiomethyl esters has been reported[14]. These esters can be hydrolysed under neutral and acidic conditions; they are useful protecting groups for carboxylic acids.

A method[15] for the protection of lactones and esters against various nucleophilic reagents (e.g. RLi, AlH_4^-, OH^-) consists in converting them, by means of bis(dimethylaluminium) ethane-1,2-dithiolate (17) into 1,3-dithiolane derivatives. The reagent (17) can be generated from trimethylaluminium and ethane-1,2-dithiol. Thus γ-butyrolactone (18) is converted into the derivative (19).

In a similar fashion, esters can be protected by conversion into the corresponding ketene thioacetals, e.g. conversion of methyl stearate (20) into the ketene thioacetal (21).

$$Me_2AlS(CH_2)_2SAlMe_2$$

(17)

$$CH_2(CH_2)_2C{=}O \longrightarrow$$

(18)

(19)

$$Me(CH_2)_{15}CH_2CO_2Me \longrightarrow$$

(20)

(21)

A useful procedure for the protection of carboxylic acids against $LiAlH_4$ involves conversion of the carboxylic acids via acyl aziridines into 2-oxazolines[159].

8.2 MONOCARBOXYLIC ACIDS

8.2.1 Normal-chain acids

8.2.1.1 Saturated acids

The cleavage of saturated long-chain alkanoamides by anhydrous HF–BF_3 has been studied[16]. For the various *N*-n-butylamides investigated, branched-chain products having chain-lengths in the range C_7–C_{10} predominate, regardless of the alkyl chain-length of the starting amide.

8.2.1.2 Monoenoic acids

(a) *Syntheses*—A procedure[17] which could be a model for the flavin-promoted enzymic dehydrogenation of aliphatic acids involves dehydrogenation of the α-anions of linear alkanoic acids by the action of 2,3-dichloro-5,6-dicyanobenzoquinone. This dehydrogenation is stereoselective and leads to the (*E*)-αβ-unsaturated derivatives.

Acrylic esters can be obtained directly from acraldehyde and alcohols by reactions involving rhodium, ruthenium or iridium complexes[18].

A simple procedure[19] for the conversion of $\alpha\beta$-unsaturated carboxylic esters into the corresponding $\beta\gamma$-unsaturated esters involves conversion of the former, by means of lithium N-isopropylcyclohexylamide, into its enolate anion, followed by treating this with dilute HCl.

$\gamma\delta$-Unsaturated esters are obtainable from allylic halides by the action of ethoxycarbonylmethylcopper, which is prepared by adding lithium di-isopropylamide to ethyl acetate in the presence of copper(I) iodide at low temperatures[20]. Thus 2,3-dibromopropene (22) reacts with the above reagent to give ethyl 4-bromopent-4-enoate (23).

$$CH_2=CBrCH_2Br \xrightarrow{[CuCH_2CO_2Et]} CH_2=CBrCH_2CH_2CO_2Et$$
$$(22) \qquad\qquad (23)$$

Another route[21] to $\gamma\delta$-unsaturated acids involves the reaction of the esters (24) (diethyl malonate or diethyl alkylmalonates) with alkenes and manganese(III) or cobalt(III) acetate in the presence of copper(II) acetate according to the scheme shown (M = Mn^{III} or Co^{III}). In this procedure the oxidation

$$R^1CH(CO_2Et)_2 \xrightarrow{M(OAc)_3} \dot{C}R^1(CO_2Et)_2 \xrightarrow{R^2CH_2CR^3=CH_2}$$
$$(24)$$

$$R^2CH_2\dot{C}R^3CH_2CR^1(CO_2Et)_2 \xrightarrow{Cu(OAc)_2} R^2CH=CR^3CH_2CR^1(CO_2Et)_2$$
$$(25) \qquad\qquad (26) \quad\downarrow$$

$$R^2CH=CR^3CH_2CHR^1CO_2H$$
$$(27)$$

of the intermediate radical (25) proceeds selectively to give only the compound (26), which on hydrolysis and subsequent decarboxylation affords the acid (27) with the $\gamma\delta$-double bond.

(b) *Reactions*—$\beta\gamma$-Unsaturated acids were, until recently, believed to fail to undergo the Kolbe anodic coupling. The previous reports were based upon experiments in aqueous solutions, and it is now shown[22] that in methanolic solution $\beta\gamma$-alkenoic acids undergo the Kolbe reaction normally. The products resulting from a $\beta\gamma$-unsaturated acid anion (28) are the three dienes (30), (31) and (32), which would be expected to arise on dimerisation of the mesomeric radical (29a) $\leftrightarrow$ (29b). The non-terminal double bond in the products largely retains the stereochemistry of the starting material.

$$RCH=CHCH_2CO_2{}^-$$
$$(28)$$
$$\downarrow$$
$$RCH=CH\dot{C}H_2 \longleftrightarrow R\dot{C}HCH=CH_2$$
$$(29a) \qquad\qquad (29b)$$

$$(RCH=CHCH_2)_2 \qquad RCH=CHCH_2CHRCH=CH_2 \qquad (CH_2=CHCHR)_2$$
$$(30) \qquad\qquad (31) \qquad\qquad (32)$$

(c) *Miscellaneous*—The behaviour of all the methyl octadecenoates (*cis* and *trans*) in gas–liquid chromatography has been reported[23].

8.2.1.3 *Di- and poly-enoic acids*

(a) *Natural acids*—A series of trehalose esters derived from polyunsaturated acids have been isolated from certain mycobacterial strains[24]. The acids, designated phleic acids, have the structures (33) where $m = 14$ or 16 and $n = 4$, 5 or 6.

$$Me(CH_2)_m(CH{=}CHCH_2CH_2)_nCO_2H$$
(33)

(b) *Syntheses*—(i) *2,4-Dienoic acids.* Stereospecific syntheses of such acids by reaction of vinylic copper reagents with propargylic esters have been reported. Thus, addition of lithium divinylcuprate to methyl propynoate and methyl but-2-ynoate stereospecifically gave methyl *trans*-penta-2,4-dienoate (34) and methyl 3-methyl-*trans*-penta-2,4-dienoate (35), respectively[25].

$$CH_2{=}CHCH{\overset{t}{=}}CHCO_2Me \qquad CH_2{=}CHCMe{\overset{t}{=}}CHCO_2Me$$
(34)   (35)

Addition of the cuprolithium complex derived from hept-*cis*-1-enyl bromide to ethyl propiolate (36) afforded ethyl deca-*trans*-2,*cis*-4-dienoate (37), a constituent of Bartlett pear[26].

$$HC{\equiv}CCO_2Et \xrightarrow[\text{Li, Cu}^{I}]{\text{hept-\textit{cis}-1-enyl bromide}} Me(CH_2)_4CH{\overset{c}{=}}CHCH{\overset{t}{=}}CHCO_2Et$$
(36)   (37)

Another approach[27] to the synthesis of 'pear ester' and related compounds involves reaction of 4,5-epoxypent-*trans*-2-enal (38) with the ylide prepared from n-hexyltriphenylphosphonium bromide to give a mixture of the 1,2-epoxyundeca-3,5-diene isomers (39) and (40) in the ratio 85 : 15, thus resulting in a highly stereospecific introduction of the *cis* double bond at C-5. Cleavage at the single bond between C-1 and C-2 by means of periodic acid leads to the corresponding deca-2,4-dienals which, by the $NaCN{-}MnO_2$ procedure in the presence of ethanol, are converted into the corresponding ethyl esters.

$$\overset{\displaystyle O}{\overset{\diagdown}{CH_2{-}CHCH}}{\overset{t}{=}}CHCHO$$
(38)

$$\downarrow$$

$$\overset{\displaystyle O}{\overset{\diagdown}{CH_2{-}CHCH}}{\overset{t}{=}}CHCH{\overset{c}{=}}CH(CH_2)_4Me \quad + \quad \overset{\displaystyle O}{\overset{\diagdown}{CH_2{-}CHCH}}{\overset{t}{=}}CHCH{\overset{t}{=}}CH(CH_2)_4Me$$
(39)   (40)

(ii) *2,5-Dienoic acids.* The procedure[25] discussed above, but starting with allylcopper reagents, yields 1,4-diene derivatives. For example, allylcopper with methyl propynoate affords methyl *trans*-2,5-hexadienoate (41).

$$CH_2{=}CHCH_2CH{\overset{t}{=}}CHCO_2Me$$
(41)

A synthesis[28] of dodeca-*trans*-2,*cis*-5-dienoic acid (45) involves coupling of oct-1-ynylmagnesium bromide (42) with the bromocrotonic acid (43) to give the alkenynoic acid (44), which on catalytic hydrogenation over Lindlar's catalyst is converted into the dienoic acid (45) with a 2-*trans* double bond.

$$\text{Me(CH}_2)_5\text{C}\equiv\text{CMgBr} + \text{BrCH}_2\text{CH}\overset{t}{=}\text{CHCO}_2\text{H} \longrightarrow$$
$$\qquad\quad (42) \qquad\qquad\qquad\qquad (43)$$

$$\text{Me(CH}_2)_5\text{C}\equiv\text{CCH}_2\text{CH}\overset{t}{=}\text{CHCO}_2\text{H} \rightarrow \text{Me(CH}_2)_5\text{C}\overset{c}{=}\text{CCH}_2\text{CH}\overset{t}{=}\text{CHCO}_2\text{H}$$
$$\qquad\qquad (44) \qquad\qquad\qquad\qquad\qquad\qquad (45)$$

By using the appropriate Grignard reagents the above procedure has been applied to the synthesis of various *cis*-polyenoic acids with a 2-*trans* double bond; the latter itself can be removed selectively by continued hydrogenation. Thus arachidonic acid (47) was obtained via the tetraynoic acid (46) having a 2-*trans* double bond.

$$\text{Me(CH}_2)_4(\text{C}\equiv\text{CCH}_2)_4\text{CH}\overset{t}{=}\text{CHCO}_2\text{H}$$
$$(46)$$

$$\text{Me(CH}_2)_4(\text{CH}\overset{c}{=}\text{CHCH}_2)_4(\text{CH}_2)_2\text{CO}_2\text{H}$$
$$(47)$$

(iii) *2,6-Dienoic acids*. Ethyl and methyl dodeca-*trans*-2,*cis*-6-dienoate, constituents of the odoriferous principle of Bartlett pear, have been synthesised[29] by 1,6-addition of lithium hept-*cis*-1-enylcuprate (48) to ethyl or methyl *trans*-penta-2,4-dienoate (34; R = Et or Me). In this reaction a mixture of the *cis*-3- and *trans*-3-forms of the dodeca-3,*cis*-6-dienoates (49) results. Their isomerisation to the dodeca-*trans*-2,*cis*-6-dienoates (50) is readily accomplished by the action of KOBu^t in Bu^{t}OH.

$$\text{CH}_2=\text{CHCH}\overset{t}{=}\text{CHCO}_2\text{R} + [\text{Me(CH}_2)_4\text{CH}\overset{c}{=}\text{CH}]_2\text{Cu}^-\,\text{Li}^+$$
$$(34) \qquad\qquad\qquad\qquad (48)$$

$$\downarrow$$

$$\text{Me(CH}_2)_4\text{CH}\overset{c}{=}\text{CHCH}_2\text{CH}=\text{CHCH}_2\text{CO}_2\text{R}$$
$$(49) \quad \Big|\, \text{KOBu}^t\text{–Bu}^t\text{OH}$$
$$\downarrow$$

$$\text{Me(CH}_2)_4\text{CH}\overset{c}{=}\text{CH(CH}_2)_2\text{CH}\overset{t}{=}\text{CHCO}_2\text{R}$$
$$(50)$$

(iv) *3,6-Dienoic acids*. By reaction of the dienoate (34) with vinylcopper the *trans*-3-enoic ester (51) can also be obtained[25]. This reaction can be adapted as a method for the stereoselective synthesis of tri- and tetra-substituted ethylenes[30].

$$(34) \xrightarrow{\text{CH}_2=\text{CHCu}} \text{CH}_2=\text{CHCH}_2\text{CH}\overset{t}{=}\text{CHCH}_2\text{CO}_2\text{R}$$
$$(51)$$

(v) *Allenic acids*. Methyl but-2-ynoate (52) by the procedure[19] already discussed (p. 330) gives the allenic ester methyl buta-2,3-dienoate (53).

$$\text{MeC}\equiv\text{CCO}_2\text{Me} \xrightarrow[\text{(ii) H}_3\text{O}^+]{\text{(i) C}_6\text{H}_{11}\text{NPr}^i\text{Li}} \text{CH}_2=\text{C}=\text{CHCO}_2\text{Me}$$
$$(52) \qquad\qquad\qquad\qquad\qquad (53)$$

The racemic form of methyl tetradeca-*trans*-2,4,5-trienoate (57) [the (−)-form occurs in the male Dried Bean Beetle, and is presumed to be a sex attractant] has been synthesised[31] by a procedure involving the reductive elimination of the tetrahydropyran-2'-yloxy group from 4-tetrahydropyran-2'-yloxydodec-2-yn-1-ol (54) to give the allenic alcohol (55). Oxidation of (55) to the aldehyde (56), followed by treatment of (56) with the anion of trimethyl phosphonoacetate, resulted in the ester (57).

$$Me(CH_2)_7CHC{\equiv}CCH_2OH \xrightarrow{\ LiAlH_4\ } Me(CH_2)_7CH{=}C{=}CHCH_2OH$$

$$|$$
$$Othp \qquad\qquad\qquad\qquad\qquad (55)$$

(54) thp = tetrahydropyran-2-yl

$$Me(CH_2)_7CH{=}C{=}CHCH{\overset{t}{=}}CHCO_2Me \longleftarrow Me(CH_2)_7CH{=}C{=}CHCHO$$
$$(57) \qquad\qquad\qquad\qquad\qquad (56)$$

Another synthesis[32] is based upon the reaction of lithium dioctylcuprate with the dimethyl acetal (58) to give the allene (59), careful hydrolysis of which gave the unstable aldehyde (60); the latter on oxidation with MnO_2–NaCN in methanol afforded the methyl ester (57).

$$HC{\equiv}CCH(OAc)CH{=}CHCH(OMe)_2 \xrightarrow{\ (C_8H_{17})_2CuLi\ } RCH{=}C{=}CHCH{=}CHCH(OMe)_2$$
$$(58) \qquad\qquad\qquad\qquad\qquad\qquad\qquad (59)$$

$$(57)\ R\ =\ Me(CH_2)_7 \qquad\qquad \longleftarrow \qquad RCH{=}C{=}CHCH{=}CHCHO$$
$$(60)$$

The naturally occurring allenic acids (−)-octadeca-5,6-dienoic acid (laballenic acid) and (−)-octadeca-5,6-*trans*-16-trienoic acid (lamenallenic acid) were synthesised[33] according to the following scheme, where R = $Me(CH_2)_{10}$ and $MeCH{=}CH(CH_2)_8$ for laballenic acid and lamenallenic

$$HC{\equiv}CCH{=}CHCH_2Othp \to RC{\equiv}CCH{=}CHCH_2X \qquad X = Othp \qquad X = OH$$
$$(61) \qquad\qquad\qquad\qquad\quad (62) \qquad\qquad (63)$$

$$RCH{=}C{=}CH(CH_2)_3CO_2H \leftarrow RCH{=}C{=}CHCH_2CH_2Z \qquad Z = OH \qquad Z = Br$$
$$(66) \qquad\qquad\qquad\qquad\qquad (64) \qquad\qquad (65)$$

acid, respectively. The key intermediates were the (−)-forms of the allenic alcohols (64), obtained from the compound (61) by reaction with the respective alkyl bromide, followed by reduction of the alcohols (63) with the $LiAlH_4$–3-*O*-benzyl-1,2-*O*-cyclohexylidene-α-D-glucofuranose complex; they were shown to have the *R*-configuration. The corresponding bromides (65) gave, by the malonic ester method, the acids (66).

(vi) *Other di- and poly-enoic acids.* Octadeca-9c,12t-dienoic acid and its 9t,12c-isomer have been prepared[34] from natural sources, viz. vernolic (12,13-epoxyoleic) acid and crepenynic acid, respectively.

A review of the synthetic routes leading to polyunsaturated acids containing methylene-interrupted ('skipped') double bonds has been published[35]. This review includes an account of syntheses of labelled polyunsaturated acids.

(c) *Reactions*—Although lipoxygenase normally reacts with polyunsaturated acids having a *cis,cis*-penta-1,4-diene system (e.g. linoleic or linolenic acid) to produce hydroperoxides with a *cis,trans*-conjugated system, in an aqueous wheat-flour suspension this enzyme is shown[36] to convert linolenic acid (67) immediately into secondary oxidation products. Three of these products were identified as 9-hydroxyoctadeca-10*t*,12*c*,15*c*-trienoic acid (68), 9-hydroxyoctadeca-10-oxo-12*c*,15*c*-dienoic acid (69) and 9,12,13-trihydroxyoctadeca-10*t*,15*c*-dienoic acid (70).

$$EtCH \overset{c}{=} CHCH_2CH \overset{c}{=} CHCH_2CH \overset{c}{=} CH(CH_2)_7CO_2H$$
$$(67)$$

$$EtCH \overset{c}{=} CHCH_2CH \overset{c}{=} CHCH \overset{t}{=} CHCH(OH)(CH_2)_7CO_2H$$
$$(68)$$

$$EtCH \overset{c}{=} CHCH_2CH \overset{c}{=} CHCH_2COCH(OH)(CH_2)_7CO_2H$$
$$(69)$$

$$EtCH \overset{c}{=} CHCH_2CH(OH)CH(OH)CH \overset{t}{=} CHCH(OH)(CH_2)_7CO_2H$$
$$(70)$$

Another study[37] concerns the thermochemistry of aliphatic acid hydroperoxides. 13-L-Hydroperoxyoctadeca-9*c*,11*t*-dienoic acid (71), obtainable by lipoxygenase-catalysed oxygenation of linoleic acid, was found to form *threo*-12,13-epoxy-11-hydroxyoctadec-9-enoic acid (72) when heated anaerobically at 100 °C in aqueous ethanol. The major part of the compound (72)

$$Me(CH_2)_4CH(OOH)CH \overset{t}{=} CHCH \overset{c}{=} CH(CH_2)_7CO_2H$$

$$(71)$$

$$\downarrow$$

$$\overset{\displaystyle O}{\overset{\displaystyle /\backslash}{Me(CH_2)_4CHCHCH(OH)CH=CH(CH_2)_7CO_2H}}$$

$$(72)$$

is formed from the hydroperoxide (71) in a reaction involving *cis* addition to the C-11 double bond of the proximally linked hydroperoxide oxygen and a hydroxide ion or hydroxyl radical from the solvent.

In the above investigation 12,13-epoxy-11-hydroxyoctadec-9-enoic acid and its isomer, regarded as 9,10-epoxy-11-hydroxyoctadec-12-enoic acid, were also isolated from a sample of autoxidised linoleic acid.

8.2.1.4 Alkynoic acids

(a) *Natural acids*—The acids (73), (74) and (75) have been isolated[38] from fungal cultures.

$$Me(C\equiv C)_3CO_2H \qquad\qquad H(C\equiv C)_3CH_2CH_2CO_2H$$
$$(73) \qquad\qquad\qquad (74)$$

$$HO_2CCH \overset{c}{=} CH(C\equiv C)_2CH_2CH_2CO_2H$$
$$(75)$$

The new acetylenic hydroxy acids (76) and (77) have been found[39] in fungal sporophores.

$$HOCH_2CH \overset{t}{=} CH(C\equiv C)_2CH \overset{t}{=} CHCO_2H$$
$$(76)$$

$$HOCH_2CH \overset{c}{=} CH(C\equiv C)_2CH_2CH_2CO_2H$$
$$(77)$$

Studies[40] on the biogenesis of matricaria ester (78) involving feeding labelled compounds to plants indicate that it arises via the C_{18} ester (79) and the keto-ester (80).

$$MeCH=CH(C\equiv C)_2CH=CHCO_2Me$$
$$(78)$$

$$Pr^n(C\equiv C)_2CH_2CH=CH(CH_2)_7CO_2Me$$
$$(79)$$

$$Pr^n(C\equiv C)_2CH \overset{t}{=} CHCO(CH_2)_7CO_2Me$$
$$(80)$$

Acetylenedicarboxamide (cellocidin) has been shown[41] to arise in *Streptomyces* cultures directly from carbon atoms 2–5 of 2-oxoglutarate, as an offshoot of the Krebs cycle.

(b) *Syntheses*—Syntheses of various alkynoic acids as intermediates for the synthesis of di- and poly-enoic acids have already been discussed in Section 8.2.1.3.

A convenient one-step synthesis[42] of alk-2-ynoic esters (85) involves

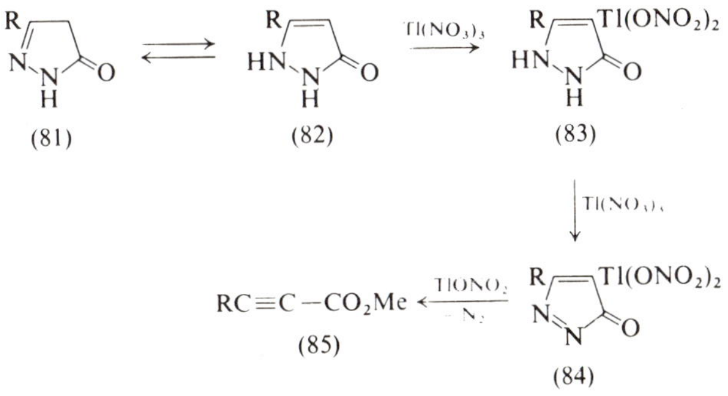

(81) ⇌ (82) → (83)

(84)

$$RC\equiv C-CO_2Me$$
$$(85)$$

treatment of pyrazol-5-ones (81) with two equivalents of thallium(III) nitrate in methanol. It is suggested that this procedure involves initial electrophilic thalliation of the 3-pyrazolin-5-one tautomer (82) with formation of the intermediate (83), the latter undergoing oxidation by the second equivalent of thallium(III) nitrate to give the oxopyrazole derivative (84). Solvolysis by methanol, with elimination of nitrogen and thallium(I) nitrate, results in the alkynoic ester (85).

Since pyrazol-5-ones are readily obtainable by reaction of β-keto-esters with hydrazine, the above synthesis represents the conversion of β-keto-esters into alk-2-ynoic esters. Direct conversion of a β-keto-ester into the corresponding alk-2-ynoic ester can also be achieved by its treatment in methanol solution first with hydrazine and then with thallium(III) nitrate.

Undeca-2t,8c-diene-4,6-diynoic acid (86) and the corresponding 8t-isomer have been synthesised[43] by standard procedures from the *cis*- and *trans*-forms, respectively, of 1-bromohex-3-en-1-yne.

$$EtCH \overset{c}{=} CH(C{\equiv}C)_2CH \overset{t}{=} CHCO_2H$$
$$(86)$$

Specifically labelled methyl crepenynate and linoleate were synthesised[44] via the Wittig reaction involving the phosphorane (87) and the aldehydo ester (88); labelled methyl oleate is prepared by using the phosphorane (89) and the aldehydo-ester (88).

$$Me(CH_2)_4C{\equiv}CCH_2CH{=}PPh_3 \qquad OCH(CH_2)_7CO_2Me$$
$$(87) \qquad\qquad\qquad (88)$$

$$Me(CH_2)_7CH{=}PPh_3$$
$$(89)$$

Various polyacetylenic esters with 9-ene-12,14-diyne unsaturation, including several specifically labelled samples, have been synthesised[45] from the Wittig salt (90) via the esters (91) and (92). The resulting esters have the general structure (93) in which R is, for example, MeC$\equiv$C, MeCH$=$CH or MeO$_2$CCH$=$CH.

$$Me_3SiC{\equiv}CCH_2CH_2\overset{+}{P}Ph_3\ I^- \longrightarrow Me_3SiC{\equiv}CCH_2CH\overset{c}{=}CH(CH_2)_7CO_2Me$$
$$(90) \qquad\qquad\qquad\qquad (91) \Big\downarrow$$

$$R(C{\equiv}C)_2CH_2CH\overset{c}{=}CH(CH_2)_7CO_2Me \longleftarrow IC{\equiv}CCH_2CH\overset{c}{=}CH(CH_2)_7CO_2Me$$
$$(93) \qquad\qquad\qquad\qquad (92)$$

Syntheses of a series of di- and tri-ynoic acids containing alkynic groups separated by a tetramethylene chain have also been reported[46]. The methods employed included alkylation of the *NN*-dimethylamide of a terminal alkynoic acid, or of an alkynol. Eicosa-5,11,14-triynoic acid is prepared by alkylation of the Grignard complex of dodeca-5,11-diynoic acid with oct-2-ynyl bromide.

The behaviour of all the methyl octadecynoates in argentation chromatography and gas–liquid chromatography has been reported[23].

8.2.2 Cycloalkanoic and cycloalkenoic long-chain acids

8.2.2.1 Natural acids

In studies of the cyclopropenoid acids of *Pavonia sepium* seed oil the homologous cyclopropenoid esters methyl sterculate (94; $n = 7$) and malvalate (94; $n = 6$), differing in chain length by only one methylene group, were separated by countercurrent distribution in a hexane–acetonitrile system[47].

$$Me(CH_2)_7C=C(CH_2)_nCO_2Me$$
$$\overset{CH_2}{\overset{/\;\backslash}{}}$$

(94)

11-Cyclohexylundecanoic and 13-cyclohexyltridecanoic acid, identified as components of the lipids of a thermophilic bacterium[48], are now shown[49] to be formed by a biosynthetic route involving chain elongation of a starter unit containing the intact C_7 carbon skeleton of shikimate.

8.2.2.2 Structural studies

The position of the ring in aliphatic acids derived from cyclopropane has been determined[50] by BF_3 catalysed methoxylation, followed by mass spectrometry. The mass spectra of the methoxylated esters are characterised by intense peaks due to cleavage adjacent to methoxy groups, thus indicating the position of the ring in the original ester.

A method[51] for the determination of the absolute configuration of *cis*-1,2-dialkylcyclopropane derivatives, e.g. lactobacillic acid (95), is based upon a gentle chromic acid oxidation of the cyclopropane compounds to give α-cyclopropyl ketones. The determination of the configuration of these ketones is achieved by reference to analogous synthetic[52] ketones of known configuration. Lactobacillic acid (95) is thus shown to be $11R,12S$-methyleneoctadecanoic acid.

$$Me(CH_2)_5CH-CH(CH_2)_9CO_2H$$
$$\underset{CH_2}{\overset{\backslash\;/}{}}$$

(95)

8.2.2.3 Syntheses

The whole series of methyl *trans*-methyleneoctadecanoates has been prepared from the corresponding *trans*-octadecenoates by the Simmons–Smith reaction; their g.l.c. behaviour, as well as their spectra, are reported[53]. An improved Simmons–Smith synthesis has also been described[54].

A number of lower homologues of methyl sterculate (94; $n = 7$) can be synthesised[55] by a procedure recently reported[56]. The intermediate di-esters (96), on decarbonylation by means of fluorosulphonic acid give the cyclopropenium cation esters (97) which are converted into the esters (94) by the action of $NaBH_4$.

$$\underset{(96)}{Me(CH_2)_7\overset{\displaystyle \overset{CHCO_2Et}{\diagup\diagdown}}{C}=C(CH_2)_nCO_2Me} \longrightarrow \underset{(97)}{Me(CH_2)_7\overset{\displaystyle \overset{\overset{+}{C}H}{\diagup\diagdown}}{C}=C(CH_2)_nCO_2Me} \longrightarrow (94)$$

In a similar procedure[57] leading to syntheses of methyl malvalate and methyl sterculate, the intermediate esters (96) are hydrolysed, and the acids are decarbonylated with perchloric acid.

8.2.3 Alkyl-branched acids

8.2.3.1 *General syntheses*

In studies[58] of the Koch synthesis of carboxylic acids (carbonylation of alkenes by the action of formic acid and sulphuric acid) the preferred modes of rearrangement of an n-alkyl chain have been investigated by the application of mass spectrometry and ^1H–^2H exchange. It is shown that the main constituents of the product obtained from oct-1-ene (98) by this procedure are carboxylic acids (99) and (100), each having a t-alkyl group.

$$Me(CH_2)_5CH=CH_2$$
$$(98)$$

$$Me(CH_2)_3CMeEtCO_2H \qquad\qquad Me(CH_2)_4CMe_2CO_2H$$
$$(99) \qquad\qquad\qquad\qquad (100)$$

A carbonylation reaction recently proposed[59] uses as catalyst Cu^I carbonyl. It is shown that, in H_2SO_4 containing Cu^I compounds, alkenes react at room temperature and atmospheric pressure with the formation, in high yield, of carboxylic acids having t-alkyl groups.

8.2.3.2 *Monoalkyl-branched acids*

(a) *Natural acids*—An acid occurring in the liver oil of the ocean sunfish *Mola mola* has been shown[60] to be 7-methylhexadec-7-enoic acid (101).

$$Me(CH_2)_7CH=CMe(CH_2)_5CO_2H$$
$$(101)$$

(b) *Syntheses*—The formation of α-metallated carboxylates (dianions) presents a convenient route to the preparation of α-substituted long-chain carboxylic acids. The extent of the metallation of normal-chain and α-branched carboxylic acids by reaction with lithium di-isopropylamide has now been investigated[61] comparatively by using solutions in THF and in THF containing HMPT. The degree to which α-anions of the acids are formed is unaffected by HMPT, but it assists the alkylation of α-metallated acids by solubilising insoluble dianions of normal-chain acids and by accelerating their alkylation.

The α-metallated carboxylates (102) also react[62] readily with formaldehyde to give α-alkylhydracrylic acids (103); the latter are readily dehydrated by acid catalysts to α-alkylacrylic acids (104).

$$\text{R}\bar{\text{C}}\text{HCO}_2{}^- \xrightarrow[\text{(ii) H}^+]{\text{(i) HCHO}} \text{HOCH}_2\text{CHRCO}_2\text{H} \xrightarrow[\text{distillation}]{\text{H}^+} \text{CH}_2{=}\text{CRCO}_2\text{H}$$

$$(102) \qquad\qquad\qquad (103) \qquad\qquad\qquad (104)$$

In another procedure[63], methyl 3-hydroxypropanoate (105) is converted into its lithium enolate which can be alkylated in high yields with various alkylating agents to give the α-alkyl derivatives (103).

$$\text{HOCH}_2\text{CH}_2\text{CO}_2\text{Me} \rightarrow \text{methyl ester of (103)}$$

$$(105)$$

A deconjugative mono- and di-alkylation of the enolate anion derived from ethyl crotonate by using lithium di-isopropylamide–HMPT has also been described[64].

A convenient one-flask procedure[65] for ester alkylation consists essentially in adding a given ester to a solution of lithium di-isopropylamide in THF, followed by alkylation of the resulting ester enolate.

8.2.3.3   Isoprenoid acids

(a) *Natural acids*—The chemistry and biochemistry of phytanic, pristanic and related acids has been reviewed[66].

(b) *Syntheses*—The DDD-(R,R,R)-stereoisomers of 4,8,12,16-tetramethyl-heptadecanoic acid (106; $n = 2$) and 5,9,13,17-tetramethyloctadecanoic acid (106; $n = 3$), both higher homologues of phytanic acid, have been synthesised[67] from $3R,7R,11R,15$-tetramethylhexadecyl iodide (derived from the lipids of *Halobacterium cutirubrum*) by standard homologation procedures.

$$\text{Me}_2\text{CH(CH}_2\text{CH}_2\text{CH}_2\text{CHMe)}_3\text{(CH}_2\text{)}_n\text{CO}_2\text{H}$$

$$(106)$$

8.2.3.4   *The* Cecropia *juvenile hormones*

(a) *Natural compounds*—Two juvenile hormones isolated from the tobacco hornworm moth, *Manduca sexta*, have been identified as methyl (2*E*,6*E*)-(10*R*)-10,11-epoxy-3,7,11-trimethyldodeca-2,6-dienoate (107) and (2*E*,6*E*)-

$$\text{R}^1\text{MeC}{-}\text{CH(CH}_2\text{)}_2\text{CR}^2{=}\text{CH(CH}_2\text{)}_2\text{CMe}{=}\text{CHCO}_2\text{Me}$$
$$\backslash\!\diagup$$
$$\text{O}$$

$$(107) \quad \text{R}^1 = \text{R}^2 = \text{Me}$$
$$(108) \quad \text{R}^1 = \text{Et}; \ \text{R}^2 = \text{Me}$$

(10*R*,11*S*)-10,11-epoxy-3,7,11-trimethyltrideca-2,6-dienoate (108)[68]; the compound (107) is a new natural hormone.

(b) *Syntheses*—A new approach[69] to the stereospecific total synthesis of a racemic *Cecropia* juvenile hormone is based upon the use of the dimeric dihydrothiopyran (109) as a structural unit.

(109)

The racemic imino analogue (113) of the C_{18} juvenile hormone and its C-11 stereoisomer have been synthesised[70] from the chloro derivative (110) as starting point via the azido derivative (111) and the diastereoisomeric azido-alcohols (112) obtained from the ketone (111) by reduction with $NaBH_4$. The mesylates derived from these alcohols are converted into the aziridines (113) by reduction with hydrazine hydrate and Raney nickel in ethanol.

$$EtMeCXCO(CH_2)_2CEt{=}CH(CH_2)_2CMe{=}CHCO_2Me$$

(110) X = Cl
(111) X = N_3

$$EtMeC(N_3)CH(OH)(CH_2)_2CEt{=}CH(CH_2)_2CMe{=}CHCO_2Me$$

(112)

$$EtMeC{-}CH(CH_2)_2CEt{=}CH(CH_2)_2CMe{=}CHCO_2Me$$
$$\diagdown\diagup$$
$$NH$$

(113)

A series of new *Cecropia* juvenile hormone analogues has been prepared[71] for comparative biological studies; the compounds (114) and (115) showed the highest activity.

$$Bu^nMeC{-}CH(CH_2)_2CR{=}CH(CH_2)_2CMe{=}CHCO_2Me$$
$$\diagdown\diagup$$
$$O$$

(114) R = Me
(115) R = Et

8.3 DI- AND POLY-CARBOXYLIC ACIDS

8.3.1 Natural acids

Radiclonic acid, isolated from the mycelial extracts of an unidentified fungus, has been assigned[72] the structure (116).

$$EtMeCHCH_2CHMeCH_2CH(CH_2OH)CH_2CHMe{-}$$
$$CH{=}CMeCH{=}(CO_2H)CH_2CHMeCO_2H$$

(116)

From n.m.r. and mass spectrometric studies a new structure has been proposed[73] for the toxic antibiotic bongkrekic acid. The latter is now formulated as 3-carboxymethyl-17-methoxy-6,18,21-trimethyldocosa-2,4,8,12,14,-18,20-heptaenedioic acid (117). The chiralities of the two asymmetric centres

$$HO_2CCH = C(CH_2CO_2H)CH = CH\overset{6}{C}HMeCH_2CH = CH(CH_2)_2(CH = CH)_2CH_2-$$
$$\overset{17}{C}H(OMe)CMe = CHCH = CMeCO_2H$$

(117)

of this acid have now also been determined[74,75] by degradation resulting in the isolation of the fragments C-5 to C-8 and C-15 to C-18. These fragments containing the asymmetric centres have been synthesised from optically active precursors of known configuration, thus indicating that the C-6 centre of bongkrekic acid has the S-configuration and the C-17 centre the R-configuration.

8.3.2 Syntheses

8.3.2.1 General procedures

A shortened procedure[76] for lengthening the chain of certain dicarboxylic acids by one carbon with the aid of the Arndt–Eistert synthesis is based on the preparation of the intermediate diazo-ketone ester directly from the cyclic anhydride derived from the acid.

A method[77] for the synthesis of long-chain dicarboxylic acids involves Claisen condensation of methyl N,N-dimethylsebacamate (118) to give the bis-N,N-dimethylamide (119). Hydrolysis of the latter with 48% hydrobromic acid yields 10-oxononadecanedioic acid (120) which is converted into nonadecanedioic acid (121) by a Clemmensen reduction.

$$Me_2NCO(CH_2)_8CO_2Me \longrightarrow Me_2NCO(CH_2)_7CH(CO_2Me)CO(CH_2)_8CONMe_2$$
(118)   (119)

$$HO_2C(CH_2)_{17}CO_2H \longleftarrow HO_2C(CH_2)_8CO(CH_2)_8CO_2H$$
(121)   (120)

In studies[78] of the metathesis of the methyl esters of unsaturated fatty acids by a WCl_6–$SnMe_4$ catalyst, methyl oleate and methyl elaidate have been shown to give octadec-9-ene and dimethyl octadec-9-enedioate.

A C_{100} dicarboxylic acid has been obtained[79] by degradation of polyethylene crystals with fuming HNO_3, followed by treatment of the product with concentrated H_2SO_4. The resulting dicarboxylic acid is then subjected to chain extension by means of several difunctional coupling agents.

8.3.2.2 Syntheses of alkyl-branched acids

A catalytic one-step synthesis converting oleic acid into a mixture of 9- and 10-carboxystearic acids in high yields involves carboxylation of the unsaturated acid with the aid of palladium catalysts in the presence of triphenylphosphine[80]. Another procedure[81] for the preparation of carboxystearic acids is based upon the catalytic autoxidation of hydroformylated oleic acid.

A detailed study[82] on the structure of the carboxystearic acids resulting from the Koch reaction (*cf.* Section 8.2.3.1) has also been published. All the carboxy groups formed are α-methyl branched, and distributed between C-7 and C-16 of the chain.

A procedure[83] for the selective alkylation of monoesters derived from unsymmetrical alkylsuccinic acids consists in converting these monoesters into dianions (in which the proton α to the ester group has been lost); the dianion then yields on reaction with alkyl halides or with carbonyl compounds a single product of predictable structure. Thus the dianion (123), obtained from the alkyl-substituted methyl hydrogen succinate (122) by the action of two moles of lithium amide in liquid ammonia, is alkylated with R^2I to form the ester (124).

$$HO_2CCH_2CHR^1CO_2Me \xrightarrow[NH_3]{2\,NH_2^-} \ ^-O_2CCH_2\bar{C}R^1CO_2Me$$

$$(122) \qquad\qquad (123)$$

$$\downarrow R^2I$$

$$R^1R^2C(CO_2Me)CH_2CO_2^-$$

$$(124)$$

2,2,9,9-Tetramethylsebacic acid (125; $n = 6$) and 2,2,8,8-tetramethylazelaic acid (125; $n = 5$) have been synthesised[84] by the free-radical addition of isobutanoic acid to hexa-1,5-diene and penta-1,4-diene, respectively; di-t-butyl peroxide was used as initiator.

$$HO_2CCMe_2(CH_2)_nCMe_2CO_2H \qquad\qquad HO_2CCH_2CMePr^iCO_2Me$$

$$(125) \qquad\qquad\qquad\qquad (126)$$

The absolute configurations of 2-isopropyl-2-methyl-succinic and -glutaric acids have been studied[85] in view of various inconsistencies in the literature. The (+)-form of the half ester (126) derived from 2-isopropyl-2-methyl-succinic acid was resolved into the (+)- and the (−)-half esters, and the (+)-form was converted into (+)-2-isopropyl-2-methylglutaric acid. By an x-ray crystallographic examination of (+)-rubidium 1-methyl-2-isopropyl-2-methylsuccinate, the absolute configuration of the (+)-acids has been defined as *S*.

8.3.3 Reactions

Decarboxylation of geminal diesters, β-keto-esters and α-cyano-esters[86] to the corresponding esters, ketones and nitriles can be effected in excellent yields by NaCl in wet DMSO at 140–186 °C.

8.4 HYDROXY ACIDS

8.4.1 Normal-chain acids

8.4.1.1 *Natural acids*

(+)-*erythro*-2,3-Dihydroxyhexacosanoic acid and homologous acids with 24–27 carbon atoms have been isolated from yeast cerebrins[87,88].

8.4.1.2 *Syntheses*

The *erythro-* and *threo*-forms of 2,3-dihydroxyhexacosanoic acid have been synthesised[88] via *trans*-hexacos-2-enoic acid; the latter is obtained, together with *trans*-hexacos-3-enoic acid, by dehydrobromination of 2-bromohexacosanoic acid, and a procedure for separating these isomeric acids is also described.

2-Hydroxyundec-10-enoic acid, synthesised from undec-10-enoic acid, has been resolved[89]; the optically pure enantiomers are obtained by fractional crystallisation of their salts with optically active α-phenylethylamine.

The hydroxy-*cis*-enynoic acids (127) and (128), as well as the *trans*-isomer of the acid (127), have been prepared in the course of syntheses of acids with smooth muscle stimulant activity[90, 91].

$$Me(CH_2)_4CH(OH)CH \overset{c}{=} CHC \equiv C(CH_2)_6CO_2H$$
(127)

$$Me(CH_2)_4CH(OH)C \equiv CCH \overset{c}{=} CH(CH_2)_6CO_2H$$
(128)

8.4.1.3 *Reactions*

In preparing the above *cis*-enynoic acid (127) via the corresponding nitrile, the related acid (129) is also formed, and *cis*-enynoic acids of the type (127) cyclise to furanoid acids on treatment with alkali or on chromatography on argentous silica gel[90]. In contrast, the corresponding *trans*-acid and the isomeric *cis*-acid (128) are unaffected by prolonged contact with argentous silica gel. The *cis*-acids (127) and (128) show activity in stimulating smooth muscle, but the furanoid acid (129) is inactive.

$$Me(CH_2)_4 \overset{\displaystyle\bigsqcup}{\underset{O}{}}(CH_2)_7CO_2H$$
(129)

8.4.2 The mycolic acids

A study[92] of the mycolic acids from *Mycobacterium phlei* indicates the complexity of the mixture of mycolic acids present in this organism. In this study the acids were separated by chromatography of their methyl esters to give esters derived (i) from mycolic acids containing the β-hydroxy-acid system as sole oxygen function, (ii) from ketomycolic acids, and (iii) from dicarboxylic acids. Each of these esters, represented by the general formula (130) for the main homologues, was converted by thermolysis into a meromycolic aldehyde (131). Reduction of the latter, followed by acetylation of the resulting alcohols, gave the meromycolyl acetates (132) (the meromycolic aldehydes derived from ketomycolic acids yielded diacetates by these procedures). The acetates (132) were subjected to argentation chromatography, and the resulting fractions studied by mass spectrometry.

$$R^1CH(OH)CHR^2CO_2Me \rightarrow R^1CHO \rightarrow R^1CH_2OAc$$
$$(130)\ R^2 = n\text{-}C_{22}H_{45} \qquad (131) \qquad (132)$$

8.4.3 Other alkyl-branched acids

In connection with studies of the CH(OMe)CHMe system of the phthiocerols, the diastereoisomeric 3-hydroxy-2-methylbutanoic acids (135) were prepared[93] by oxymercuration of tiglic acid (133) by reaction with aqueous mercuric acetate, followed by reduction of the resulting acetoxymercury compound (134) with $NaBH_4$.

$$MeCH(OH)CMe(HgOAc)CO_2H \qquad MeCH(OH)CHMeCO_2H$$

$$(133) \qquad\qquad (134) \qquad\qquad\qquad (135)$$

The stereochemistry of the products formed by the demercuration of the above acetoxymercury compound varied depending on the dermercuration procedure. Thus, on reducing the compound (134) with $NaBH_4$ in alkaline solution the diastereoisomers of the acid (135) were formed in approximately equal amounts, whereas by the action of H_2S in aqueous NaOH a product containing mainly one (about 95%) of the diastereoisomers is formed.

The stereochemistry of the above diastereoisomers has been studied[93] by comparing their i.r., n.m.r. and mass spectrometric properties with those of the diastereoisomeric ethyl 2-allyl-3-hydroxybutanoates (136) and 3-hydroxy-2-propylbutanoates (137). Evidence is presented that the predominant diastereoisomer arising on demercuration of the acetoxymercuration adduct (134) with H_2S has the configuration shown in the Fischer projection (138).

The corresponding 3-methoxy-2-methylbutanoic acids are discussed in Section 8.5.

$$MeCH(OH)CH(CO_2Et)CH_2CH{=}CH_2 \qquad MeCH(OH)CH(CO_2Et)CH_2Et$$
$$(136) \qquad\qquad\qquad (137)$$

$$(138)$$

A new hydroxy-acid isolated from lichens, bourgeanic acid, has the structure (139)[94]; it is derived by esterification of the hydroxy group of the acid (140) with the carboxy group of another molecule of this acid. The acid (140) is structurally related to the polymethyl-substituted acids with methyl-

$$EtMeCHCH_2CHMeCH(OH)CHMeCO_2CH(CHMeCO_2H)CHMeCH_2CHMeEt$$
$$(139)$$

$$EtMeCHCH_2CHMeCH(OH)CHMeCO_2H$$
$$(140)$$

ene-interrupted branching, occurring as constituents of mycobacterial lipids, as well as component acids of the preen-gland wax of various birds[95].

2,3-Dihydroxy-2-methylpropanoic acid (141) and 2,3-dihydroxy-2-methyl-butanoic acid (142) have been synthesised via acetol cyanhydrin and acetoin cyanhydrin, respectively, with the aid of the techniques of ion-exclusion and ion-exchange[96,97]. In the preparation of the acid (142) a high proportion of one of the diastereoisomers is obtained[97], thus indicating that the reaction of acetoin with anhydrous HCN is stereospecific.

$$HOCH_2C(OH)MeCO_2H \qquad\qquad MeCH(OH)C(OH)MeCO_2H$$
$$(141) \qquad\qquad\qquad\qquad (142)$$

In connection with another synthesis, the asymmetric synthesis of the $(+)$-form of diethyl 2-methylmalate (citramalate) (143) via ethyl $(-)$-menthyl citramalate has been investigated[98].

$$EtO_2CC(OH)MeCH_2CO_2Et \qquad\qquad HO_2CCPr^i(OH)CH_2CO_2H$$
$$(143) \qquad\qquad\qquad\qquad (144)$$

In the course of studies on the absolute configuration of 2-hydroxy-2-isopropylsuccinic acid (α-isopropylmalic acid) (144), an intermediate in the biosynthesis of leucine, x-ray crystallographic examination of the mono-potassium salt showed that the natural form has the 2S-configuration[99].

A precursor of isoleucine is the $(-)$-form of 2,3-dihydroxy-3-methyl-pentanoic acid (145). This has been shown[100] to have the 2R,3R-configuration.

$$EtMeC(OH)CH(OH)CO_2H$$
$$(145)$$

Studies[101] involving optical rotatory dispersion and circular dichroism indicate that $(+)$-2,3-dimethyltartaric acid has the S,S-configuration.

8.4.4 The prostaglandins

The extensive literature dealing with this field has been the subject of numerous reviews (e.g. Refs. 102–106).

The more recent syntheses of prostaglandins include a new approach[107,108] involving partial syntheses from material available from a natural source, the soft coral *Plexaura homomalla*; the latter contains esterified derivatives of prostaglandin A$_2$ and its epimer.

8.5 ALKOXY ACIDS

8.5.1 Natural acids

Two novel unsaturated fatty acids isolated from extracts of *Solanum tuber-*

osum have been shown to have the structures (146) and (147), respectively, both containing a divinyl ether function[109,110]. The acid (146), colneleic acid, is formed by enzymic oxidation of linoleic acid, while the acid (147) is derived enzymically from linolenic acid. Their formation thus involves insertion of an oxygen function into the carbon chain of an aliphatic acid.

$$Me(CH_2)_4CH \overset{c}{=} CHCH \overset{t}{=} CHOCH \overset{t}{=} CH(CH_2)_6CO_2H$$
$$(146)$$

$$MeCH_2CH \overset{c}{=} CHCH_2CH \overset{c}{=} CHCH \overset{t}{=} CHOCH \overset{t}{=} CH(CH_2)_6CO_2H$$
$$(147)$$

The ether structure of the above acids and their carbon skeletons have been confirmed by the synthesis[111] of methyl 9-nonyloxynonanoate (148), a saturated derivative of these acids. This synthesis is described in Section 8.5.2.

$$Me(CH_2)_7CH_2OCH_2(CH_2)_7CO_2Me$$
$$(148)$$

8.5.2 Syntheses

Methyl 9-nonyloxynonanoate (148) was synthesised[111] via the oxazoline ester (149), obtained from methyl hydrogen azelate by the action of 2,2-dimethylaziridine and dicyclohexylcarbodi-imide, thus using the 5,5-di-methyloxazolinyl residue as a protecting group during the subsequent stages. Reduction of this ester with $LiAlH_4$ gave the hydroxy derivative (150), which was converted into the bromide (151). Reaction of the latter with one equivalent of the sodium non-1-oxide gave the ether derivative (152), which on being boiled with aqueous KOH, followed by esterification, yielded the ester (148).

$$MeO_2C(CH_2)_7C\underset{O-CMe_2}{\overset{N-CH_2}{<}} \longrightarrow XCH_2(CH_2)_7C\underset{O-CMe_2}{\overset{N-CH_2}{<}}$$

$$(149)$$

$$(150)\ X = OH$$
$$(151)\ X = Br$$
$$(152)\ X = Me(CH_2)_7CH_2O \longrightarrow (148)$$

A series of papers deal with the alkoxyhalogenation of alkenoic acids. Methyl octadec-9-enoate is converted by the action of t-butyl hypobromite in the presence of an alcohol into a bromo-ether which, on reduction, yields the respective ether derived from the hydroxy-octadecanoic acid[112]. Methyl undec-10-enoate by the above procedure yields preferentially the 10-alkoxy-11-bromo isomer; with the αβ-unsaturated ester methyl heptadec-2-enoate, the 3-alkoxy-2-bromo-derivative is the main product[113]. Methyl 12-hydroxy-octadec-*cis*-9-enoate and its acetate are also readily methoxybrominated. Both possible methoxybromides are formed, with the 10-bromo-9-methoxy isomer as the most abundant product[114].

The four possible stereoisomers of methyl 2,4-dimethoxytridec-12-enoate (158) have been synthesised[115] from the optically active 2-methoxyundec-10-enoic acids. For the preparation of the 4D-isomers, methyl 2D-methoxyundec-10-enoate (153) is reduced with $LiAlH_4$ to the alcohol (154) which is converted, via the tosyl derivative (155), into the iodide (156). The next step is the alkylation of diethyl methoxymalonate with the optically active iodide (156), in the presence of $KOBu^t$, to give the substituted malonic ester (157), which yields a mixture of diastereoisomers derived from the 4D-form of the ester (158) is the usual way. These diastereoisomers were separated by

$$CH_2{=}CH(CH_2)_7\overset{D}{C}H(OMe)CO_2Me$$
(153)

$$CH_2{=}CH(CH_2)_7\overset{D}{C}H(OMe)CH_2X$$

(154) X = OH   (155) X = OTs   (156) X = I

$$CH_2{=}CH(CH_2)_7\overset{D}{C}H(OMe)CH_2C(OMe)(CO_2Et)_2$$
(157)

(i) hydrolysis
(ii) $-CO_2$
(iii) esterification

$$CH_2{=}CH(CH_2)_7CH(OMe)CH_2CH(OMe)CO_2Me$$
(158)

preparative g.l.c. or chromatography on silicic acid. Starting with the 2L-isomer of the ester (153), the enantiomers of the above compounds were obtained.

In further studies[115], the 2D,4D- and 2L,4D-isomers of the ester (158) were oxidised to the respective isomers of the half-ester (159), which by chain-elongation gave the 2D,4D- and 2L,4D-isomers of methyl 2,4-dimethoxy-

$$HO_2C(CH_2)_7CH(OMe)CH_2CH(OMe)CO_2Me$$
(159)

$$Me(CH_2)_{22}CH(OMe)CH_2CH(OMe)CO_2Me$$
(160)

heptacosanoate (160). This procedure provides a method of stereospecific synthesis of methylene-interrupted di- and poly-methoxy-substituted aliphatic acids.

The diastereoisomeric 3-methoxy-2-methylbutanoic acids (162), the methoxy derivatives of the hydroxy-acids already discussed (Section 8.4.3), were synthesised[93] during studies of the stereochemistry of the CH(OMe)CHMe system in the phthiocerols. Methoxymercuriation of tiglic acid (133) with mercuric acetate in methanol gave the adduct (161). Reduction of (161) with $NaBH_4$ in alkaline solution resulted in about equal amounts of the diastereoisomers of the acid (162), whereas on reduction without added alkali one of the diastereoisomers predominated by about 4:1. The predominant diastereoisomer was formed almost entirely by demercuriation of the adduct (161) with H_2S in aqueous alkali. Evidence is presented which indicates that this diastereoisomer has the configuration (163).

$$H\diagdown C=C \diagup CO_2H$$
$$Me \diagup \qquad \diagdown Me$$

(133)

MeCH(OMe)CMe(HgOAc)CO$_2$H

(161)

MeCH(OMe)CHMeCO$_2$H

(162)

CO$_2$H
Me—|—H
H—|—OMe
Me

(163)

Me — H
CO$_2$H
Me — H
OMe

(163a)

By the Fischer convention the diastereoisomer having the configuration (163) would be named the *threo*-acid. For the studies described above different nomenclature has been used. The isomer (163), which by rotation about the central carbon–carbon bond has in one configuration (163a) two groups (Me and H) at one asymmetric centre eclipsed by the same groups at the adjacent centre, is designated *erythro*[93]. Accordingly, the acid of configuration (163) is named *erythro*-3-methoxy-2-methylbutyric acid.

The above isomer (163) has been resolved[93] via the quinine and cinchonidine salts. The (−)-acid was used for the synthesis[116] of a diastereoisomer of methoxyphthiocerane B by chain-elongation to give the optically active methoxy-acids (164) and (165) as intermediates.

The (+)- and (−)-enantiomers of the acid (163) have also been shown to have the *S*- and *R*-configuration, respectively, at the asymmetric centre bearing the methyl group[117].

$$MeCH(OMe)CHMe(CH_2)_nCO_2H$$

(164) $n = 2$ (165) $n = 10$

8.6 EPOXY ACIDS

8.6.1 Natural acids

The *Cecropia* juvenile hormones have already been discussed in Section 8.2.3.4.

An acid from the cutin of young apple fruits is shown to be 18-oxo-9,10-epoxystearic acid; it was identified by procedures involving hydrogenolysis with LiAlH$_4$, deuteriolysis with LiAlD$_4$, and chromic acid oxidation of the deuteriated material, followed by reduction with LiAlH$_4$ and mass spectrometry of the products[118].

8.6.2 Syntheses

The complete series of the methyl epoxyoctadecanoates has been prepared by epoxidation of the respective octadecenoates, and their g.l.c., n.m.r. and mass spectrometric properties have been reported[119].

In studies on the halogenation of various unsaturated hydroxy-esters it is found[120] that the expected dihalogeno derivative is sometimes accompanied by a halogen-containing cyclic ether resulting from neighbouring group participation of the hydroxy group in the reaction occurring at the unsaturated centre. Thus, the products formed from methyl 9-hydroxyoctadec-*cis*-12-enoate (166) by the action of bromine include methyl 13-bromo-9,12-epoxystearate (167).

$$Me(CH_2)_4CH\overset{c}{=}CHCH_2CH(OH)(CH_2)_7CO_2Me$$

(166)

$$Me(CH_2)_4CHBrCHCH_2CH_2CH(CH_2)_7CO_2Me$$

(167)

8.7 LACTONES

8.7.1 α-Lactones

α-Lactones, which occur as intermediates in various reactions, have been further studied. A series of α-lactones (169; R = alkyl) have been synthesised[121] by irradiation of the malonyl peroxide (168) under conditions which

(168) → (169)

permit spectroscopic studies of their chemistry and photochemistry. In another study[122] the major products of vapour-phase thermolysis of the dimeric dimethylmalonyl peroxide (170) and monomeric trimethylene-malonyl peroxide (171) have been shown to be carbon dioxide and ketones (acetone and cyclobutanone, respectively), expected to result from the formation and subsequent decarboxylation of an α-lactonic intermediate.

(170)   (171)

In connection with suggestions that α-peroxylactones are involved in bioluminescence and chemiluminescence, the α-peroxylactone 4-t-butyl-1,2-dioxetan-3-one (175) has been synthesised[123]. The key intermediate, the α-hydroperoxy-acid (174), was obtained by photo-oxidation of the bis-

(trimethylsilyl)acetal (172) to give the ester (173), which on hydrolysis afforded the acid (174). The cyclisation of (174) to the lactone (175) was readily achieved by means of dicyclohexylcarbodi-imide.

$$Bu^t\ CH{=}C(OSiMe_3)_2 \qquad\qquad Bu^tCH(CO_2SiMe_3)OOSiMe_3$$

$$(172) \qquad\qquad\qquad\qquad\qquad (173)$$

$$Bu^tCH(OOH)CO_2H$$

$$(174)$$

$$\begin{array}{c} Bu^tCH{-}CO \\ |\qquad\ | \\ O{-}{-}O \end{array}$$

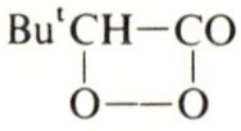

$$(175)$$

8.7.2 β-Lactones

βγ-Unsaturated acids, when treated with I_2 in the standard iodolactonisation procedure, yield γ-lactones. However, under certain conditions (short reaction time, absence of KI), some βγ-unsaturated acids form β-lactones[124]. Thus 2,2-dimethylbut-3-enoic acid (176) yields the β-lactone (177). The latter is readily isomerised[125] to the more stable γ-lactone (178).

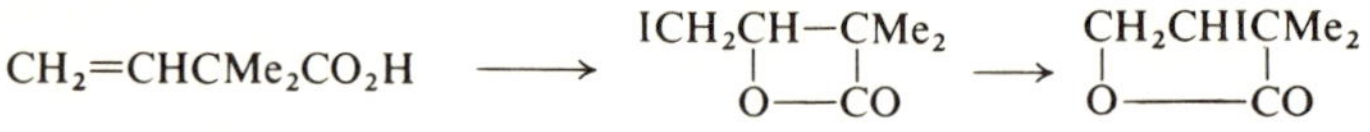

$$CH_2{=}CHCMe_2CO_2H \longrightarrow \begin{array}{c} ICH_2CH{-}CMe_2 \\ |\qquad\ | \\ O{-}CO \end{array} \longrightarrow \begin{array}{c} CH_2CHICMe_2 \\ |\qquad\ | \\ O{-}{-}{-}CO \end{array}$$

$$(176) \qquad\qquad\qquad\qquad (177) \qquad\qquad (178)$$

The oligomerisation of β-propiolactone (179) by the action of a Cu_2O–isocyanide complex in the presence of an alkyl halide has been studied[126]. Oligomers (180) having an acrylic carbon–carbon double bond as the terminal group are found to be selectively produced.

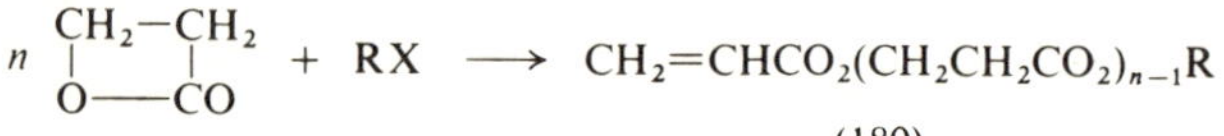

$$n\ \begin{array}{c} CH_2{-}CH_2 \\ |\qquad\ | \\ O{-}{-}CO \end{array} + RX \longrightarrow CH_2{=}CHCO_2(CH_2CH_2CO_2)_{n-1}R$$

$$(179) \qquad\qquad\qquad\qquad\qquad (180)$$

8.7.3 γ-Lactones

8.7.3.1 Structural studies

The hydroxy-lactone (181) is obtained by ozonolysis of an alkaloid from *Skimmia japonica*[127]; its structure is established by asymmetric synthesis of its monoformate, and its absolute configuration has been determined by correlation with that of (−)-4-methylpentane-1,3-diol.

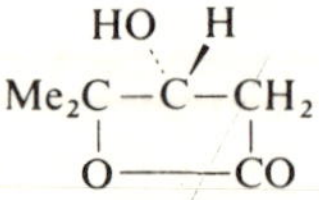

$$\begin{array}{c} HO\quad H \\ \diagup \\ Me_2C{-}C{-}CH_2 \\ |\qquad\quad | \\ O{-}{-}{-}CO \end{array}$$

$$(181)$$

8.7.3.2 Synthesis of α-methyl- and α-methylene-γ-butyrolactones

Interest in the synthesis of α-methylated and α-methylenated γ-butyrolactones continues, owing to their presence as structural units in many naturally occurring compounds. The direct conversion of γ-butyrolactone (182) into its α-methyl derivative (183) has been accomplished by using a lithium dialkylamide as a non-nucleophilic base to form the lactone enolate, which reacts under mild conditions with methyl iodide[128]. A convenient procedure

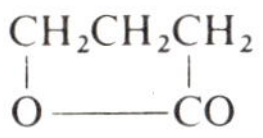

$$
\begin{array}{cc}
\mathrm{CH_2CH_2CH_2} & \mathrm{CH_2CH_2CHMe} \\
\mathrm{O\!-\!\!-\!\!-\!CO} & \mathrm{O\!-\!\!-\!\!-\!CO} \\
(182) & (183)
\end{array}
$$

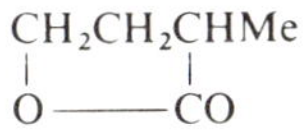

for the controlled alkylation of lactone enolates has also been described[129].

The α-methyl derivatives of γ-butyrolactones have been converted into the corresponding α-methylene-lactones by bromination and subsequent dehydrobromination[130,131]. A lithium dialkylamide has also been employed[132] for the direct conversion of a γ-butyrolactone into an α-hydroxymethyl-γ-butyrolactone. Thus, by reaction of the lactone enolate (formed by means of lithium di-isopropylamide) with formaldehyde, the γ-lactone (184) affords the γ-hydroxymethyl derivative (185) which, via the mesyl derivative (186), readily gives the α-methylene-γ-butyrolactone (187). This procedure is also

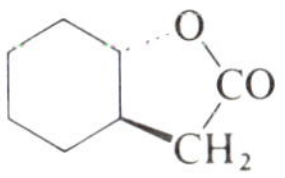

$$
\begin{array}{ccc}
(184) & \begin{array}{l}(185)\ \ R = H \\ (186)\ \ R = MeSO_2\end{array} & (187)
\end{array}
$$

applicable to the α-hydroxymethylation of δ-valerolactone.

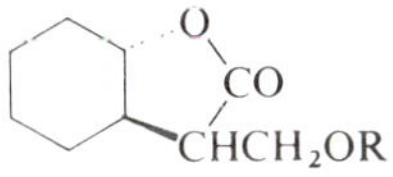

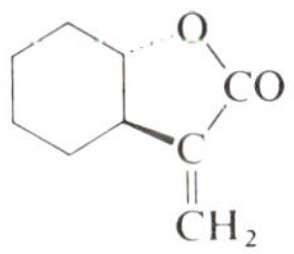

A general method for the preparation of α-methylene-lactones is based upon the reductive amination of α-formyl-lactones[133]. According to this procedure, γ-butyrolactone (182) is converted, by reaction with ethyl formate in the presence of NaH, into the sodio derivative of α-hydroxymethylene-γ-butyrolactone (188). The latter is reductively aminated with dimethylamine–dimethylamine hydrochloride and sodium cyanotrihydridoborate to give α-dimethylaminomethyl-γ-butyrolactone (189), which is converted, via the methiodide, into the α-methylene-lactone (190). According to a modifica-

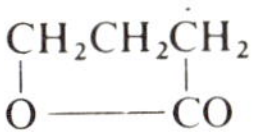

$$
\begin{array}{ccc}
\mathrm{CH_2CH_2CH_2} & \mathrm{CH_2CH_2C\!=\!CHOH} & \mathrm{CH_2CH_2CHCH_2NMe_2} \\
\mathrm{O\!-\!\!-\!\!-\!CO} & \mathrm{O\!-\!\!-\!\!-\!CO} & \mathrm{O\!-\!\!-\!\!-\!CO} \\
(182) & (188) & (189)
\end{array}
$$

$$
\begin{array}{c}
\mathrm{CH_2CH_2C\!=\!CH_2} \\
\mathrm{O\!-\!\!-\!\!-\!CO} \\
(190)
\end{array}
$$

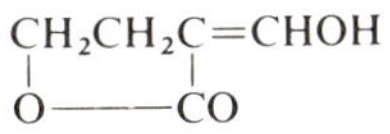

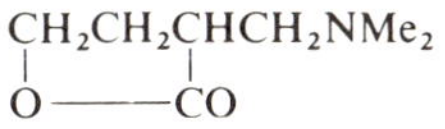

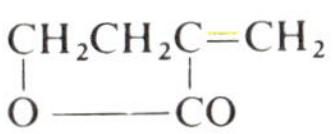

tion of this procedure the hydroxymethylene intermediate is converted directly with Et_2NH into an aminomethylene derivative; the latter, by catalytic hydrogenation followed by heating the product with NaOAc–AcOH, yields the α-methylene-lactone[134].

Another general route to α-methylene-lactones is provided by the α-carboxylation of lactones[135]. The α-carboxy-lactone (191) is converted into the α-methylene derivative (192) by heating it with diethylamine and formaldehyde. Evidence is presented[136] which indicates that the formation of the lactone (192) from lactone (191) involves a concerted decarboxylation–elimination mechanism.

$$CMe_2CHMeCHCO_2H \qquad\qquad CMe_2CHMeC{=}CH_2$$
$$O\text{———}{-}CO \qquad\qquad\qquad O\text{———}CO$$

$$(191) \qquad\qquad\qquad\qquad (192)$$

The recently reported methods[134] for the α-methylenation of γ-butyrolactones include the carboxylation of the γ-butyrolactone (193) with methyl methoxymagnesium carbonate to give the acid (194), which on reaction first with N,N'-carbonyldi-imidazole, and then with $Zn(BH_4)_2$, yields the hydroxymethyl derivative (195). Dehydration of (195) with $(PhO)_3\overset{+}{P}Me\ I^-$ gives the methylene-lactone (196).

$$ArCHCH_2CHR \qquad\qquad ArCHCH_2C{=}CH_2$$
$$O\text{——}CO \qquad\qquad\qquad O\text{——}CO$$

(193) R = H
(194) R = CO_2H
(195) R = CH_2OH

$$Ar = MeO\text{—}C_6H_4\text{—} \qquad (196)$$

The α-carboxylation of lactones has also been utilised in a procedure[137] which aims at constructing the α-methylene moiety in a protected form. The α-carboxylated lactone (197) on reaction with two moles of lithium di-iso-propylamide gives a dianion which with iodomethyl methyl sulphide, followed by acidification, yields the acid (198). When heated in xylene, the acid (198) lost CO_2 and the lactone (199) was obtained. Removal of the methyl-thio group in a two-step procedure liberated the α-methylene-lactone.

(197) → (198) → (199)

A different approach[138] to the synthesis of α-methylene-γ-butyrolactones is based upon a rearrangement of cyclopropanes involving the interconversion of cyclopropylcarbinylic and homoallylic systems. For example, ethyl 1-hydroxymethylcyclopropanecarboxylate (200) with hydrobromic acid and zinc bromide affords α-methylene-γ-butyrolactone (190).

H₂C—CH₂OH / C / H₂C—CO₂Et (200) ⟶ [CH₂CH₂C=CH₂ | OH CO₂Et] ⟶ CH₂CH₂C=CH₂ | O----CO (190)

8.7.3.3 Avenaciolide and 4-isoavenaciolide

The bislactone avenaciolide (211) is an antifungal metabolite of *Aspergillus avenaceus*; a minor metabolite is shown to be 4-isoavenaciolide (210)[139]. The racemic form of the latter has been synthesised[140] as follows. The unsaturated γ-lactone acid (201), prepared from 2-nonylidenesuccinic acid, gives on reduction with zinc in acetic acid the *trans*-γ-lactonic acid (202), which, via the corresponding diazoketone (203), affords the α-acetoxyketone (204). Hydrolysis to the α-ketol (205), followed by oxidation with copper(II) acetate in methanol, gives the mixed diastereoisomeric hydroxy-esters (206) which are oxidised with chromic acid in pyridine to the α-keto-ester (207). The latter, under acidic conditions, was transformed into the α-hydroxy-αβ-unsaturated γ-lactone (208). Catalytic hydrogenation, followed by treatment with acid, gave the bis-γ-lactone (209); this lactone is also obtainable, in lower yield, directly from the hydroxy-esters (206) by alkaline hydrolysis, followed by treatment with acid. Carboxylation at the 3-position and treat-

(201) (202)

(203) X = CHN₂
(204) X = CH₂OAc
(205) X = CH₂OH

(206) (207)

(208)

(209)

(210) R¹ = H, R² = n-C₈H₁₇
(211) R¹ = n-C₈H₁₇, R² = H

ment of the resulting acid with aqueous formaldehyde and diethylamine gives ($\pm$)-4-isoavenaciolide (210). From ^{13}C-labelling studies of avenaciolide (211) it has been shown[141] that it is biosynthesised from 3-oxododecanoic acid and succinic acid.

8.7.3.4 Miscellaneous syntheses

The lactone (214) was synthesised[25] via the intermediate methyl 4-trimethylsiloxynon-2-ynoate (212). Reaction with lithium divinylcuprate gave the cis-adduct (213), which on exposure to methanolic HCl (resulting in cleavage of the trimethylsilyl ether and lactonisation) yielded the lactone (214).

$$Me(CH_2)_4CH(OSiMe_3)C{\equiv}CCO_2Me$$

(212)

$$Me(CH_2)_4CH(OSiMe_3)C(CH{=}CH_2){=}CHCO_2Me$$

(213) H and vinyl cis

$$\begin{array}{c} CH{=}CH_2 \\ | \\ Me(CH_2)_4CHC{=}CH \\ |\qquad\quad| \\ O{-\!-}CO \end{array}$$

(214)

A procedure[21] already discussed for the preparation of $\gamma\delta$-unsaturated acids (see Section 8.2.1.2) affords the γ-lactone (216) by decomposition of the alkenyl-substituted malonic ester (215) in acidic solution.

$$CH_2{=}CMeCH_2CH(CO_2Et)_2 \longrightarrow \begin{array}{c} Me_2CCH_2CH_2 \\ |\qquad\quad| \\ O{-\!-}CO \end{array}$$

(215)   (216)

A recent synthesis of butyrolactones is based upon the catalytic cyclisation of diazomalonates with copper[142]. Thus isobutyl methyl diazomalonate (217) gives the lactone (218).

$$N_2C(CO_2Me)CO_2Bu^i$$

(217)

$$\begin{array}{c} Me_2 \\ C \\ {/}\ \ {\diagdown}CHCO_2Me \\ H_2C\qquad\ | \\ {\diagdown}O{-}CO \end{array}$$

(218)

Tetramethylcyclobutane-1,4-dione (219) is readily converted into the keto-butyrolactone (220) by Baeyer–Villiger oxidation with hydrogen peroxide[143].

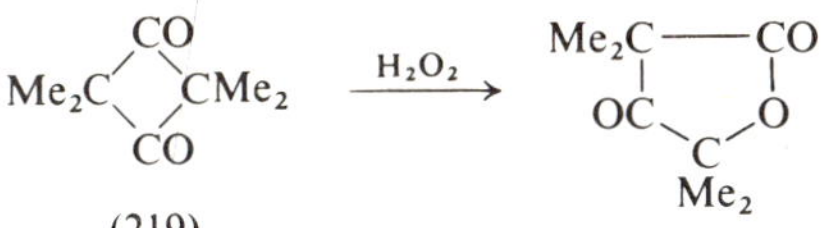

$$\text{Me}_2\text{C}\underset{\text{CO}}{\overset{\text{CO}}{\diagup\diagdown}}\text{CMe}_2 \quad \xrightarrow{\text{H}_2\text{O}_2} \quad \text{Me}_2\text{C}\text{---}\text{CO}$$

(219)   (220)

A synthesis of 3-(hydroxymethyl)-4,4-dimethylpentanoic acid γ-lactone (226) in four steps from ethyl 3-hydroxy-3,4,4-trimethylpentanoate (221) has been described[144]. Dehydration of the ester (221) results in a mixture of the ethyl alkenoates (222) and (223); this mixture on bromination with NBS gives only the allylic bromide (224). Thermal cyclisation of the bromide results in the unsaturated lactone (225), which on catalytic hydrogenation affords the lactone (226).

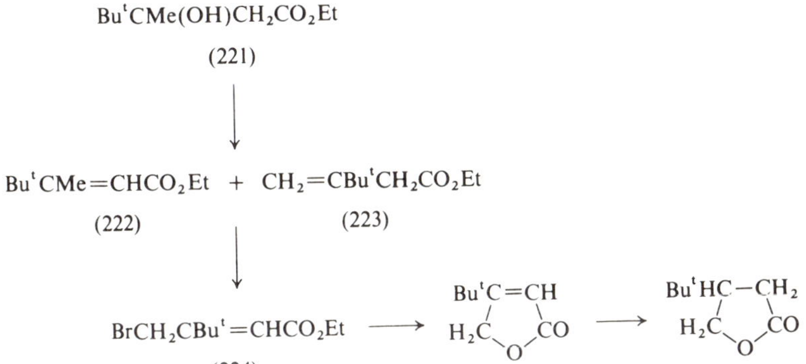

$$\text{Bu}^t\text{CMe(OH)CH}_2\text{CO}_2\text{Et}$$

(221)

$$\text{Bu}^t\text{CMe}{=}\text{CHCO}_2\text{Et} \quad + \quad \text{CH}_2{=}\text{CBu}^t\text{CH}_2\text{CO}_2\text{Et}$$

(222)   (223)

$$\text{BrCH}_2\text{CBu}^t{=}\text{CHCO}_2\text{Et} \longrightarrow \quad \longrightarrow$$

(224)   (225)   (226)

8.7.4 δ-Lactones

Several syntheses of α-methyl- and α-methylene-γ-butyrolactones discussed in Section 8.7.3.2 are also applicable to δ-lactones (see Refs. 129 and 132–135).

A general synthesis of δ-lactones[145] starts from glutaraldehyde (227). A Grignard reaction between this dialdehyde and an alkyl or substituted-alkyl Grignard reagent results in a δ-hydroxy-aldehyde; the latter exist preferentially in the cyclic hemiacetal form (228). Oxidation of this hydroxy-aldehyde affords the δ-lactone (229); for this oxidation, Ag_2O in NaOH–MeOH is suitable.

$$(\text{CH}_2)_3(\text{CHO})_2$$

(227)   (228)   (229)

8.7.5 Higher lactones

8.7.5.1 *Natural compounds*

Vermiculine[146], a lactone from *Penicillium vermiculatum*, has the structure
(230).

$$\begin{array}{l} CO(CH_2)_3CHCOMe \\ \;|\qquad\qquad\quad\;\;{>}O \\ CH{=}CH{-}CO \end{array}$$

(230)

8.7.5.2 *Syntheses*

Various 10- to 16-membered ring ketolactones have been synthesised by a
procedure involving oxidation of bicyclic enol ethers[147,148]. For example,
9-ketotridecanolide (234) has been obtained by the following sequence of
reactions. The *C*-alkylation of the sodium enolate derived from ethoxy-
carbonylcyclononanone (231) with 1-bromo-4-acetoxybutane gives 2-(4'-
hydroxybutyl)cyclononanone (232). The latter was converted by hemiacetal
formation and dehydration into the enol ether (233), which was oxidised
with *m*-chloroperoxybenzoic acid to give 9-ketotridecanolide (234).

$$(CH_2)_7 \begin{array}{l} {\diagup} C{=}O \\ \;\;| \\ {\diagdown} CHCO_2Et \end{array} \longrightarrow (CH_2)_7 \begin{array}{l} {\diagup} C{=}O \\ \;\;| \\ {\diagdown} CH(CH_2)_4OH \end{array}$$

(231)   (232)

$$(CH_2)_7 \begin{array}{l} \quad\; O \\ \quad\;\; \| \\ {\diagup} C{-}O \\ \;\;| \\ {\diagdown} C{-}(CH_2)_4 \\ \;\; \| \\ \quad\; O \end{array} \longleftarrow (CH_2)_7 \begin{array}{l} {\diagup} C{-}O \\ \;\;\| \quad\;| \\ {\diagdown} C{-}{-}(CH_2)_4 \end{array}$$

(234)   (233)

8.8 LACTAMS

8.8.1 β-Lactams

The chemistry of the β-lactams has been reviewed[149]. A method for the
synthesis of β-lactams by ring enlargement of cyclopropanones consists in the
addition of cyclopropanone in the form of its acetic acid addition product
(235) to the α-cyano-hydroxylamine (236) to give the *N*-hydroxycarbinol-

amine (237); the latter on treatment with toluene-*p*-sulphonyl chloride is rearranged to the β-lactam (238)[150].

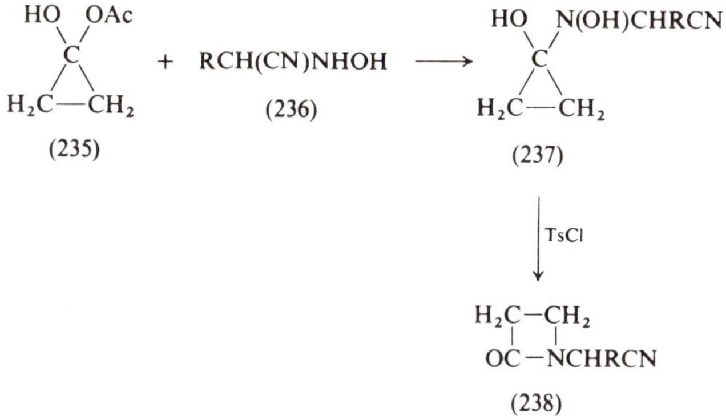

$$\begin{array}{ccc} \overset{HO}{\underset{}{\diagdown}}\overset{OAc}{\underset{}{\diagup}} & & \overset{HO}{\underset{}{\diagdown}}\overset{N(OH)CHRCN}{\underset{}{\diagup}} \\ C & + \ RCH(CN)NHOH \longrightarrow & C \\ H_2C\!-\!CH_2 & (236) & H_2C\!-\!CH_2 \\ (235) & & (237) \end{array}$$

$$\downarrow \ TsCl$$

$$\begin{array}{c} H_2C\!-\!CH_2 \\ | \quad \ | \\ OC\!-\!NCHRCN \\ (238) \end{array}$$

A new method for the synthesis of β-lactams involves a photolytic ring-contraction of 3-diazopyrrolidine-2,4-diones[151-153]. Thus 3-diazo-5-methyl-pyrrolidine-2,4-dione (239) gives, on photolysis in benzene in the presence of t-butyl carbazate, the *cis*-β-lactam (240) and the *trans*-β-lactam (241).

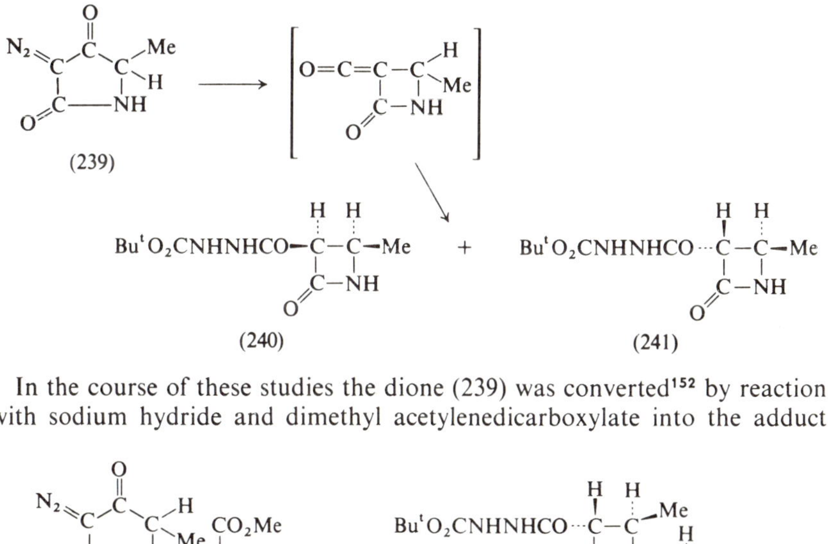

(239)

$$\longrightarrow \left[O\!=\!C\!=\!C\!-\!C \begin{array}{c} H \\ Me \end{array} \right]$$

$$Bu^tO_2CNHNHCO\!-\!C\!-\!C\!-\!Me \quad (240) \qquad + \qquad Bu^tO_2CNHNHCO\cdots C\!-\!C\!-\!Me \quad (241)$$

In the course of these studies the dione (239) was converted[152] by reaction with sodium hydride and dimethyl acetylenedicarboxylate into the adduct

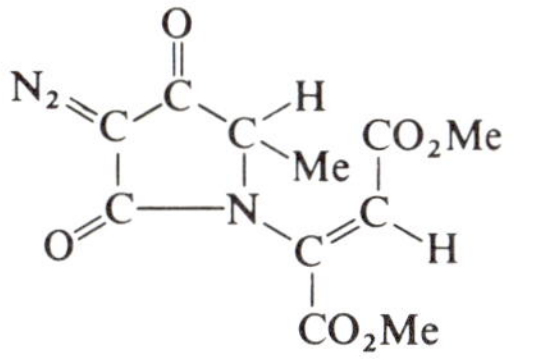

(242)

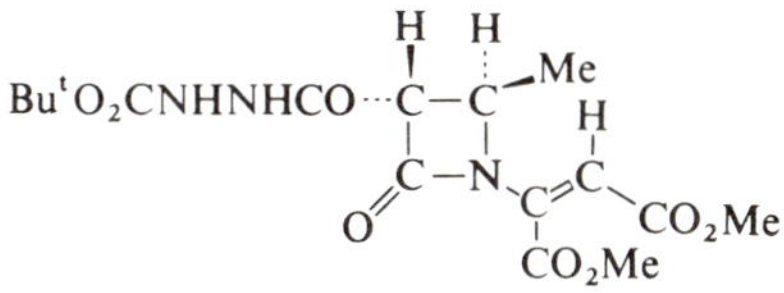

(243)

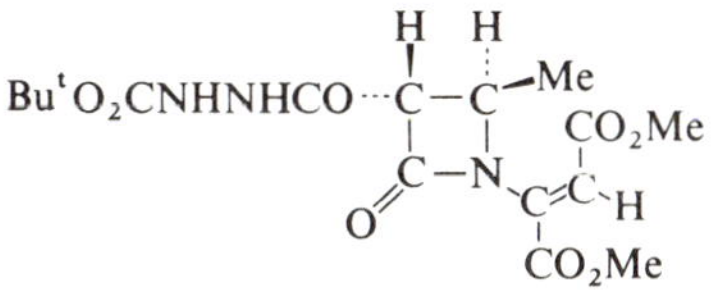

(244)

(242). The adduct, when irradiated for a short time in the presence of t-butyl carbazate, gives both the (*E*)-*trans*-β-lactam (243) and the (*Z*)-*trans*-β-lactam (244); on prolonged irradiation the (*E*)-*trans*-β-lactam (243) is found to be generated exclusively.

Another procedure[154] is based upon an oxidative ring contraction of α-keto-γ-lactams by means of periodate. Thus, 1-methyl-2,3-pyrrolidinedione (245) rearranges by the action of periodate to 3-carboxy-1-methyl-2-azetidinone (246).

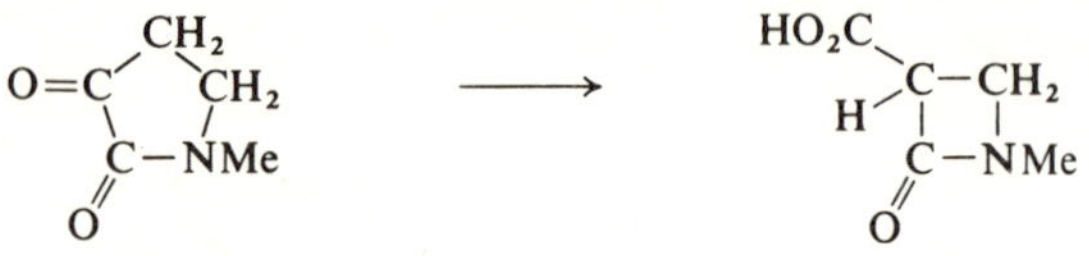

(245) ⟶ (246)

A stereospecific synthesis of *cis*-β-lactams (248; R = MeO, PhO, etc.) has been described[155]. This synthesis depends upon the desulphurisation of a thioether (247) with Raney nickel.

(247) ⟶ (248)

α-Azido-β-lactams have been prepared by using mixed anhydrides in place of acid chlorides in reactions with imines[156].

In studies on the synthesis of substances structurally related to penicillins and cephalosporins, a series of β-lactams (249) having a nitrogen atom in common with valine methyl ester and carrying a benzylthio substituent in the 4-position, as well as various substituents at the 3-position, have been prepared as model compounds[157].

$$XHC-CHSCH_2Ph$$
$$OC-NCHPr^iCO_2H$$

(249)

8.8.2 Higher lactams

The reaction of α-bromo-ε-caprolactam (250) with NaOMe gives[158] the β-methoxy-ε-caprolactam (251), and not the α-methoxy derivative. This is explained by a mechanism which involves elimination of HBr, followed by addition of the methoxide ion to the resulting unsaturated lactam.

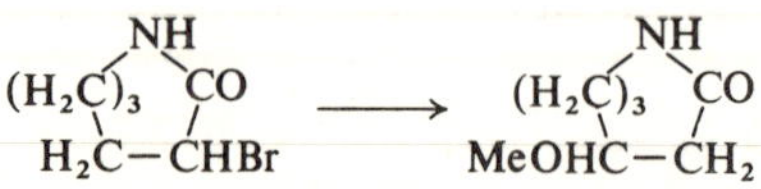

(250) ⟶ (251)

References

1. Collman, J. P., Winter, S. R. and Komoto, R. G. (1973). *J. Amer. Chem. Soc.*, **95**, 249
2. Fétizon, M., Kakis, F. J. and Ignatiadou-Ragoussis, V. (1973). *J. Org. Chem.*, **38**, 1732
3. Kakis, F. J., Brase, D. and Oshima, A. (1971). *J. Org. Chem.*, **36**, 4117
4. Blossey, E. C., Turner, L. M. and Neckers, D. C. (1973). *Tetrahedron Lett.*, 1823
5. Hoshi, M., Williams, M. and Kishimoto, Y. (1973). *J. Lipid Res.*, **14**, 599
6. Grundy, J., James, B. G. and Pattenden, G. (1972). *Tetrahedron Lett.*, 757
7. Pfeffer, P. E., Foglia, T. A., Barr, P. A., Schmeltz, I. and Silbert, L. S. (1972). *Tetrahedron Lett.*, 4063
8. Spener, F. and Mangold, H. K. (1972). *Chem. Phys. Lipids*, **8**, 180
9. van Melick, J. E. W., Scheeren, J. W. and Nivard, R. J. F. (1973). *Rec. Trav. Chim.*, **92**, 775
10. Burton, D. J. and Koppes, W. M. (1973). *J. Chem. Soc. Chem. Commun.*, 425
11. Anderson, A. G., Jr. and Kono, D. H. (1973). *Tetrahedron Lett.*, 5121
12. Rodin, R. L. and Gershon, H. (1973). *J. Org. Chem.*, **38**, 3919
13. Matlin, S. A. and Sammes, P. G. (1972). *J. Chem. Soc. Chem. Commun.*, 1222
14. Ho, T. L. and Wong, C. M. (1973). *J. Chem. Soc. Chem. Commun.*, 224
15. Corey, E. J. and Beames, D. J. (1973). *J. Amer. Chem. Soc.*, **95**, 5829
16. Hecht, S. S. and Rothman, E. S. (1973). *J. Org. Chem.*, **38**, 3733
17. Cainelli, G., Cardillo, G. and Umani Ronchi, A. (1973). *J. Chem. Soc. Chem. Commun.*, 94
18. Hidai, M., Ishimi, K., Iwase, M., Tanaka, E. and Uchida, J. (1973). *Tetrahedron Lett.*, 1189
19. Rathke, M. W. and Sullivan, D. (1972). *Tetrahedron Lett.*, 4249
20. Kuwajima, I. and Doi, Y. (1972). *Tetrahedron Lett.*, 1163
21. Nikishin, G. I., Vinogradov, M. G. and Fedorova, T. M. (1973). *J. Chem. Soc. Chem. Commun.*, 693
22. Garwood, R. F., Naser-ud-Din, Scott, C. J. and Weedon, B. C. L. (1973). *J. Chem. Soc. Perkin Trans. I*, 2714
23. Barve, J. A., Gunstone, F. D., Jacobsberg, F.-R. and Winlow, P. (1972). *Chem. Phys. Lipids*, **8**, 117
24. Asselineau, C. P., Montrozier, H. L., Promé, J.-C., Savagnac, A. M. and Welby, M. (1972). *Eur. J. Biochem.*, **28**, 102
25. Corey, E. J., Kim, C. U., Chen, R. H. K. and Takeda, M. (1972). *J. Amer. Chem. Soc.*, **94**, 4395
26. Näf, F. and Degen, P. (1971). *Helv. Chim. Acta*, **54**, 1939
27. Ohloff, G. and Pawlak, M. (1973). *Helv. Chim. Acta*, **56**, 1176
28. Heslinga, L., Pabon, H. J. J. and van Dorp, D. A. (1973). *Rec. Trav. Chim.*, **92**, 287
29. Näf, F., Degen, P. and Ohloff, G. (1972). *Helv. Chim. Acta*, **55**, 82
30. Corey, E. J. and Chen, R. H. K. (1973). *Tetrahedron Lett.*, 1611
31. Landor, P. D., Landor, S. R. and Mukasa, S. (1971). *Chem. Commun.*, 1638
32. Descoins, C., Henrick, C. A. and Siddall, J. B. (1972). *Tetrahedron Lett.*, 3777
33. Cowie, J. S., Landor, P. D., Landor, S. R. and Punja, N. (1972). *J. Chem. Soc. Perkin Trans. I*, 2197
34. Gunstone, F. D. and Jacobsberg, F. R. (1972). *Chem. Phys. Lipids*, **9**, 112
35. Kunau, W.-H. (1973). *Chem. Phys. Lipids*, **11**, 254
36. Graveland, A. (1973). *Lipids*, **8**, 606
37. Hamberg, M. and Gotthammar, B. (1973). *Lipids*, **8**, 737
38. Hearn, M. T. W., Jones, Sir E. R. H., Pellatt, M. G., Thaller, V. and Turner, J. L. (1973). *J. Chem. Soc. Perkin Trans. I*, 2785
39. Farrell, I. W., Keeping, J. W., Pellatt, M. G. and Thaller, V. (1973). *J. Chem. Soc. Perkin Trans. I*, 2642
40. Bohlmann, F. and Burkhardt, T. (1972). *Chem. Ber.*, **105**, 521
41. Jones, Sir E. R. H., Keeping, J. W., Pellatt, M. G. and Thaller, V. (1973). *J. Chem. Soc. Perkin Trans. I*, 148
42. Taylor, E. C., Robey, R. L. and McKillop, A. (1972). *Angew. Chem. Int. Ed. Engl.*, **11**, 48

43. Thaller, V. and Turner, J. L. (1972). *J. Chem. Soc. Perkin Trans. I*, 552
44. Barley, G. C., Jones, Sir E. R. H., Thaller, V. and Vere Hodge, R. A. (1973). *J. Chem. Soc. Perkin Trans. I*, 151
45. Fallis, A. G., Hearn, M. T. W., Jones, Sir E. R. H., Thaller, V. and Turner, J. L. (1973). *J. Chem. Soc. Perkin Trans. I*, 743
46. Ames, D. E. and Binns, S. H. (1972). *J. Chem. Soc. Perkin Trans. I*, 255
47. Madrigal, R. V. and Smith, C. R., Jr. (1973). *Lipids*, **8**, 407
48. *Cf.* Polgar, N. (1973). In *MTP International Review of Science, Organic Chemistry, Series One*, Vol. 2, *Aliphatic Compounds* (N. B. Chapman, editor), 327 (London: Butterworths)
49. De Rosa, M., Gambacorta, A., Minale, L. and Bu'lock, J. D. (1972). *Biochem. J.*, **128**, 751
50. Minnikin, D. E. (1972). *Lipids*, **7**, 398
51. Tocanne, J. F. (1972). *Tetrahedron*, **28**, 363
52. Tocanne, J. F. and Bergmann, R. G. (1972). *Tetrahedron*, **28**, 373
53. Gunstone, F. D. and Perera, B. S. (1973). *Chem. Phys. Lipids*, **10**, 303
54. Denis, J. M., Girard, C. and Conia, J. M. (1972). *Synthesis*, 549
55. Williams, J. L. and Sgoutas, D. S. (1972). *Chem. Phys. Lipids*, **9**, 295
56. *Cf.* Ref. 48, p. 328
57. Pawlowski, N. E., Lee, D. J. and Sinnhuber, R. O. (1972). *J. Org. Chem.*, **37**, 3245
58. Kell, D. R. and McQuillin, F. J. (1972). *J. Chem. Soc. Perkin Trans. I*, 2096
59. Souma, Y., Sano, H. and Iyoda, J. (1973). *J. Org. Chem.*, **38**, 2016
60. Ackman, R. G., Safe, L., Hooper, S. N., Paradis, M. and Safe, S. (1973). *Lipids*, **8**, 21
61. Pfeffer, P. E., Silbert, L. S. and Chirinko, J. M., Jr. (1972). *J. Org. Chem.*, **37**, 451
62. Pfeffer, P. E., Kinsel, E. and Silbert, L. S. (1972). *J. Org. Chem.*, **37**, 1256
63. Herrmann, J. L. and Schlessinger, R. H. (1973). *Tetrahedron Lett.*, 2429
64. Herrmann, J. L., Kieczykowski, G. R. and Schlessinger, R. H. (1973). *Tetrahedron Lett.*, 2433
65. Cregge, R. J., Herrmann, J. L., Lee, C. S., Richman, J. E. and Schlessinger, R. H. (1973). *Tetrahedron Lett.*, 2425
66. Lough, A. K. (1973). *Progr. Chem. Fats Lipids*, **14**, 1
67. Kates, M., Hancock, A. J., Ackman, R. G. and Hooper, S. N. (1972). *Chem. Phys. Lipids*, **8**, 32
68. Judy, K. J., Schooley, D. A., Dunham, L. L., Hall, M. S., Bergot, B. J. and Siddall, J. B. (1973). *Proc. Nat. Acad. Sci. USA*, **70**, 1509
69. Kondo, K., Negishi, A., Matsui, K. and Tunemoto, D. (1972). *J. Chem. Soc. Chem. Commun.*, 1311
70. Anderson, R. J., Henrick, C. A. and Siddall, J. B. (1972). *J. Org. Chem.*, **37**, 1266
71. Ozawa, Y., Mori, K. and Matsui, M. (1973). *Agr. Biol. Chem.*, **37**, 2373 (*Chem. Abstr.*, **80**, 37 321)
72. Sassa, T., Takemura, T., Ikeda, M. and Miura, Y. (1973). *Tetrahedron Lett.*, 2333
73. de Bruijn, J., Frost, D. J., Nugteren, D. H., Gaudemer, A., Lumbach, G. W. M., Cox, H. C. and Berends, W. (1973). *Tetrahedron*, **29**, 1541
74. Zylber, J., Gaudemer, F. and Gaudemer, A. (1973). *Experientia*, **29**, 387
75. Zylber, J., Gaudemer, F. and Gaudemer, A. (1973). *Experientia*, **29**, 648
76. Della, E. W. and Kendall, M. (1973). *J. Chem. Soc. Perkin Trans. I*, 2729
77. Cohen, H. and Shubart, R. (1973). *J. Org. Chem.*, **38**, 1424
78. van Dam, P. B., Mittelmeijer, M. C. and Boelhouwer, C. (1972). *J. Chem. Soc. Chem. Commun.*, 1221
79. Ballard, D. G. H. and Dawkins, J. W. (1973). *Eur. Polym. J.*, **9**, 211
80. Frankel, E. N. and Thomas, F. L. (1973). *J. Amer. Oil Chem. Soc.*, **50**, 39
81. Schwab, A. W. (1973). *J. Amer. Oil Chem. Soc.*, **50**, 74
82. Lai, R., Comeau, D., Ucciani, E. and Naudet, M. (1973). *Chem. Phys. Lipids*, **10**, 291
83. Kofron, W. G. and Wideman, L. G. (1972). *J. Org. Chem.*, **37**, 555
84. Bradney, M. A. M., Forbes, A. D. and Wood, J. (1973). *J. Chem. Soc. Perkin Trans. II*, 1655
85. Cox, M. R., Koch, H. P. and Whalley, W. B. (1973). *J. Chem. Soc. Perkin Trans. I*, 174
86. Krapcho, A. P. and Lovey, A. J. (1973). *Tetrahedron Lett.*, 957
87. Proštenik, M., Kljaic, K. and Weinert, M. (1973). *Lipids*, **8**, 325
88. Hoshi, M., Kishimoto, Y. and Hignite, C. (1973). *J. Lipid Res.*, **14**, 406

 89. Eliasson, B., Odham, G. and Pettersson, B. (1971). *Acta Chem. Scand.*, **25**, 2217
 90. Crundwell, E. and Cripps, A. L. (1973). *Chem. Phys. Lipids*, **11**, 39
 91. Crundwell, E. and Cripps, A. L. (1972). *J. Med. Chem.*, **15**, 754
 92. Kusamran, K., Polgar, N. and Minnikin, D. E. (1972). *J. Chem. Soc. Chem. Commun.*, 111
 93. Maskens, K. and Polgar, N. (1973). *J. Chem. Soc. Perkin Trans. I*, 109
 94. Bodo, B., Hebrard, P., Molho, L. and Molho, D. (1973). *Tetrahedron Lett.*, 1631
 95. *Cf.* Ref. 48, p. 331
 96. Powell, J. E., Burkholder, H. R. and Farrell, J. L. (1971). *J. Chromatogr.*, **57**, 309
 97. Powell, J. E., Kulprathipanja, S., Johnson, D. A. and Burkholder, H. (1972). *J. Chromatogr.*, **74**, 265
 98. Brandänge, S., Josephson, S. and Vallén, S. (1973). *Acta Chem. Scand.*, **27**, 1084
 99. Cole, F. E., Kalyanpur, M. G. and Stevens, C. M. (1973). *Biochemistry*, **12**, 3346
100. Crout, D. H. G. and Whitehouse, D. (1972). *J. Chem. Soc. Chem. Commun.*, 398
101. Muroi, M., Oda, J. and Inouye, Y. (1973). *Bull. Inst. Chem. Res., Kyoto Univ.*, **51**, 182 (*Chem. Abstr.*, **79**, 145 929)
102. (1972). *The Prostaglandins* (S. M. M. Karim, editor) (Oxford and Lancaster: MTP Medical and Technical Publishing Co.)
103. Axen, U., Pike, J. E. and Schneider, W. P. (1973). *The Total Synthesis of Natural Products*, Vol. 1, 81 (J. ApSimon, editor) (New York: Wiley-Interscience)
104. Bentley, P. H. (1973). *Chem. Soc. Rev.*, **2**, 29
105. Bindra, J. S. and Bindra, R. (1973). *Progr. Drug Res.*, **17**, 410
106. Weinshenker, N. M. and Andersen, N. H. (1973). In *The Prostaglandins*, Vol. 1, 5 (P. W. Ramwell, editor) (New York: Plenum Press)
107. Schneider, W. P., Hamilton, R. D. and Rhuland, L. E. (1972). *J. Amer. Chem. Soc.*, **94**, 2122
108. Bundy, G. L., Schneider, W. P., Lincoln, F. H. and Pike, J. E. (1972). *J. Amer. Chem. Soc.*, **94**, 2123
109. Galliard, T. and Phillips, D. R. (1972). *Biochem. J.*, **129**, 743
110. Galliard, T., Phillips, D. R. and Frost, D. J. (1973). *Chem. Phys. Lipids*, **11**, 173
111. Curtis, R. F. and Fenwick, R. G. (1973). *Chem. Phys. Lipids*, **11**, 11
112. Siouffi, A. M., Truphémus, Y. and Naudet, M. (1972). *Chem. Phys. Lipids*, **8**, 91
113. Siouffi, A. M. and Naudet, M. (1973). *Chem. Phys. Lipids*, **11**, 66
114. Siouffi, A. M. and Naudet, M. (1973). *Chem. Phys. Lipids*, **11**, 103
115. Eliasson, B., Odham, G. and Pettersson, B. (1971). *Acta Chem. Scand.*, **25**, 3404
116. Maskens, K. and Polgar, N. (1973). *J. Chem. Soc. Perkin Trans. I*, 1117
117. Maskens, K. and Polgar, N. (1973). *J. Chem. Soc. Perkin Trans. I*, 1909
118. Kolattukudy, P. E. (1973). *Lipids*, **8**, 90
119. Gunstone, F. D. and Jacobsberg, F. R. (1972). *Chem. Phys. Lipids*, **9**, 26
120. Gunstone, F. D. and Perera, B. S. (1973). *Chem. Phys. Lipids*, **11**, 43
121. Chapman, O. L., Wojtkowski, P. W., Adam, W., Rodriguez, O. and Rucktäschel, R. (1972). *J. Amer. Chem. Soc.*, **94**, 1365
122. Martin, M. M., Hammer, F. T. and Zador, E. (1973). *J. Org. Chem.*, **38**, 3422
123. Adam, W. and Liu, J.-C. (1972). *J. Amer. Chem. Soc.*, **94**, 2894
124. Barnett, W. E. and Sohn, W. H. (1972). *Tetrahedron Lett.*, 1777
125. Barnett, W. E. and Sohn, W. H. (1972). *J. Chem. Soc. Chem. Commun.*, 472
126. Saegusa, T., Murase, I., Nakai, M. and Ito, Y. (1972). *Bull. Chem. Soc. Jap.*, **45**, 3604
127. Collins, J. F. and Grundon, M. F. (1973). *J. Chem. Soc. Perkin Trans. I*, 161
128. Posner, G. H. and Loomis, G. L. (1972). *J. Chem. Soc. Chem. Commun.*, 892
129. Herrmann, J. L. and Schlessinger, R. H. (1973). *J. Chem. Soc. Chem. Commun.*, 711
130. Greene, A. E., Muller, J.-C. and Ourisson, G. (1972). *Tetrahedron Lett.*, 2489
131. Greene, A. E., Muller, J.-C. and Ourisson, G. (1972). *Tetrahedron Lett.*, 3375
132. Grieco, P. A. and Hiroi, K. (1972). *J. Chem. Soc. Chem. Commun.*, 1317
133. Harmon, A. D. and Hutchinson, C. R. (1973). *Tetrahedron Lett.*, 1293
134. Yamada, K., Kato, M. and Hirata, Y. (1973). *Tetrahedron Lett.*, 2745
135. Grieco, P. A. and Hiroi, K. (1973). *J. Chem. Soc. Chem. Commun.*, 500
136. Dalton, L. K. and Elmes, B. C. (1972). *Aust. J. Chem.*, **25**, 625
137. Ronald, R. C. (1973). *Tetrahedron Lett.*, 3831
138. Hudrlik, P. F., Rudnick, L. R. and Korzeniowski, S. H. (1973). *J. Amer. Chem. Soc.*, **95**, 6848

139. Aldridge, D. C. and Turner, W. B. (1971). *J. Chem. Soc. C*, 2431
140. Yamada, K., Kato, M., Iyoda, M. and Hirata, Y. (1973). *J. Chem. Soc. Chem. Commun.*, 499
141. Tanabe, M., Hamasaki, T. and Suzuki, Y. (1973). *J. Chem. Soc. Chem. Commun.*, 212
142. Ledon, H., Linstrumelle, G. and Julia, S. (1973). *Bull. Soc. Chim. Fr.*, 2071
143. Bischoff, C. (1973). *Z. Chem.*, **13**, 11
144. DeBoer, A. and Hunter, J. A. (1973). *J. Org. Chem.*, **38**, 144
145. Rosenberger, M., Andrews, D., DiMaria, F., Duggan, A. J. and Saucy, G. (1972). *Helv. Chim. Acta*, **55**, 249
146. Sedmera, P., Vokoun, J., Podojil, M., Vanek, Z., Fuska, J., Nemec, P. and Kuhr, I. (1973). *Tetrahedron Lett.*, 1347
147. Borowitz, I. J., Williams, G. J., Gross, L., Beller, H., Kurland, D., Suciu, N., Bandurco, V. and Rigby, R. D. G. (1972). *J. Org. Chem.*, **37**, 581
148. Borowitz, I. J., Bandurco, V., Heyman, M., Rigby, R. D. G. and Ueng, S.-N. (1973). *J. Org. Chem.*, **38**, 1234
149. Manhas, M. S. and Bose, A. K. (1971). *Beta-Lactams*, Part I (New York: Wiley-Interscience)
150. Wasserman, H. H., Glazer, E. A. and Hearn, M. J. (1973). *Tetrahedron Lett.*, 4855
151. Lowe, G. and Ridley, D. D. (1973). *J. Chem. Soc. Chem. Commun.*, 328
152. Lowe, G. and Ridley, D. D. (1973). *J. Chem. Soc. Perkin Trans. I*, 2024
153. Lowe, G. and Wing Yeung, H. (1973). *J. Chem. Soc. Perkin Trans. I*, 2907
154. Bender, D. R., Bjeldanes, L. F., Knapp, D. R., McKean, D. R. and Rapoport, H. (1973). *J. Org. Chem.*, **38**, 3439
155. Bose, A. K., Dayal, B., Chawla, H. P. S. and Manhas, M. S. (1972). *Tetrahedron Lett.*, 2823
156. Bose, A. K., Kapur, J. C., Sharma, S. D. and Manhas, M. S. (1973). *Tetrahedron Lett.*, 2319
157. Bachi, M. D. and Goldberg, O. (1972). *J. Chem. Soc. Chem. Commun.*, 319
158. Reimschuessel, H. K. (1973). *J. Org. Chem.*, **38**, 169
159. Haidukewych, D. and Meyers, A. I. (1972). *Tetrahedron Lett.*, 3031

9
Boron Compounds

K. J. TOYNE
University of Hull

9.1 INTRODUCTION AND REVIEWS

As in the previous volume[1], this chapter is restricted principally to the synthetically useful reactions of aliphatic organoboron compounds; such topics as carboranes, boron hydrides and the inorganic aspects of boron chemistry are excluded.

The productivity of 'boron chemists' has been maintained during 1972–73 (approximately 600 publications have been considered for this review), and the increasing utility of organoboron compounds in synthesis is well illustrated by the frequency with which the reactions reported warrant the use of terms such as 'simple', 'facile', 'convenient', 'stereospecific', 'regiospecific', 'in excellent yield', 'without isomerisation', 'synthesised from predetermined units', 'suitable for bulky groups', etc. The synthesis and use of organochloroboranes and the use of alkynyl- and alkenyl-organoborates have become particularly important areas recently, and there are also indications that photochemical and electrochemical reactions of boron compounds and the reactions of carbanions stabilised by boron atoms will prove increasingly useful in synthesis.

A single reaction sequence may, for example, involve hydroboration, photochemical and radical reactions, nucleophilically or electrophilically induced rearrangements, and boronic acid derivatives, etc., so it cannot be classified uniquely. The titles of sections of this review are therefore, to a large extent, chosen arbitrarily, with the intention of reviewing the selected work as concisely as possible.

Many books and reviews on boron compounds have been published in 1972–73 and the following list gives an indication of the variety of topics covered: an important book on 'Boranes in Organic Chemistry' by H. C. Brown[2] and one on 'Organoboranes in Organic Synthesis' by G. M. L. Cragg[3]; annual surveys of carboranes and hydroboration for 1971[4] and 1972[5], and of other aspects of boron chemistry for 1971[6] and 1972[7]; chapters in various volumes of *Inorganic Chemistry Series One* of the *MTP International Reviews* which are complementary to the present review[8–10]; a chapter on the organometallic chemistry of Group III elements (mainly boron) in a Chemical Society Special Publication[11]; two chapters, in the first of a four-volume series on free radicals, entitled 'Rate constants for free radical reactions in solution' (S_H2 reactions at multivalent atoms. The autoxidation of organoboranes)[12] and 'Bimolecular homolytic substitution at metal centres'[13]; reviews of (*a*) organic synthesis by using radical reactions of organoboranes[14,15], (*b*) radical reactions of organometallic compounds (reactions at a boron centre figure prominently)[16–18], (*c*) the synthetic utility of organoboranes[19], (*d*) organoboranes as alkylating and arylating agents[20], (*e*) the hydroboration of dienes and the use of the products of hydroboration in organic synthesis[21,22], (*f*) the chemistry of allylboron compounds[23], (*g*) boron–nitrogen betaines[24], (*h*) iminoboranes[25], (*i*) boron derivatives as selective reagents for organic synthesis[26], and (*j*) rearrangements of organo-boron compounds involving the cleavage of a boron σ-bond[27].

In addition, several lengthy papers on the following topics, which cannot be summarised adequately in this limited chapter, have been published: hydroboration with diborane in THF of acyclic and cyclic dienes[28], of penta-1,4-diene (including a synthesis of bisborinane and *B*-alkylborinanes)[29], of hexa-1,5-diene[30], and of 2,4-dimethylpenta-1,4-diene and 2,5-dimethyl-hexa-1,5-diene[31]; hydroboration of dienes with thexylborane*[32]; the reduction of organic functional groups with thexylborane in THF and a comparison of the reducing characteristics of diborane, thexylborane and disiamyl-borane† and their advantages and disadvantages for specific reductions[33];

---

* Thexyl = Me_2CHCMe_2
† Siamyl = $Me_2CHCHMe$

and the reduction of aliphatic and aromatic carboxylic acids with diborane in THF[34].

9.2 HYDROBORATION. SYNTHESIS OF ORGANOBORANES
(See also Sections 9.1, 9.3, 9.6, 9.10.4)

A solution of diborane in dichloromethane, suitable for all the usual reductions and hydroborations, can be prepared by the reaction of a tetra-alkylammonium borohydride (from $NaBH_4$ and a tetra-alkylammonium sulphate) with an alkyl halide[35].

Hydroborations are usually regarded as involving a four-centre transition state, but an alternative mechanism has been proposed[36] which involves the formation of a three-centre two-electron π-complex intermediate, followed by a rate-limiting concerted formation of products from the intermediate [equation (9.1)].

$$\text{>C=C<} + \text{H-B<} \rightleftharpoons \text{>C-C<} \rightarrow \text{-C-C-} \quad (9.1)$$

Hydroboration–oxidation has been shown to be a *cis*-hydration process[37] for the reaction of *cis*- and *trans*-but-2-ene with diborane. Anti-Markovnikov hydration of unsymmetrical alkenes with diborane sometimes gives significant amounts of the minor isomer, but reactions of monochloroborane diethyl etherate[38] (see p. 366) with terminal alkenes give primary alcohols of greater purity yet show little regioselectivity with internal alkenes[38,39]. 9-Borabicyclo-[3.3.1]nonane in refluxing THF[39], however, reacts rapidly with terminal, internal or even highly branched alkenes and eventually yields the anti-Markovnikov alcohols with high regioselectivity (often $> 99.5\%$ isomeric purity).

Monoalkylboranes coordinated to triethylamine can be prepared from the reaction of thexylmonoalkylboranes (1) with triethylamine[40] [equation (9.2)]

$$R^2BH_2 + \text{>C=C<} \longrightarrow R^1R^2BH \xrightarrow{NEt_3} R^1BH_2 \cdot NEt_3 + Me_2C=CMe_2 \quad (9.2)$$
$$(1)$$
$$(R^2 = \text{thexyl} = Me_2CHCMe_2)$$

and the product hydroborates alkenes at a convenient rate to give mixed trialkylboranes, $R^1R^3_2B$. Thexylmonoalkylboranes (1) can also be used to hydroborate ω-acetoxyalk-1-enes, $CH_2=CH(CH_2)_nOAc$, and subsequent carbonylation of the product gives the ketones $RCO(CH_2)_{n+2}OAc$, of predetermined chain length[41]. The 'dialkylboranes', e.g. (1), mentioned previously really exist as dimers (tetra-alkyldiboranes), but a monomeric dialkylborane (dithexylborane), which shows i.r. absorption at 2470 cm^{-1} (B—H bond) and the absence of the B—H bridge absorptions, has now been detected[42].

Although methods of preparation of t-alkyldialkylboranes were previously available, an alternative route[43] involving the reduction of 2-t-alkylbenzo-[1.3.2]dioxaboroles with aluminium hydride in pyridine gives a mono-t-alkylborane–pyridine complex initially, from which the free borane can be generated and used to hydroborate alkenes. The trialkylboranes so obtained can be successively carbonylated and oxidised to give highly branched trialkylcarbinols in which the t-alkyl group has migrated without isomerisation [equation (9.3)].

$$R^1R^2R^3CBR^4R^5 \longrightarrow \longrightarrow R^1R^2R^3CC(OH)R^4R^5 \qquad (9.3)$$

A very convenient new route to organoboranes has been reported[44]. In it a Grignard reagent formed from an alkyl or aryl halide in the presence of diborane is converted almost quantitatively into the organoborane by heating it under reflux in THF. This simple method provides a route to organoboranes not accessible by direct hydroboration and the replacement of magnesium by boron takes place with retention of configuration[45]. Alkyl- and arylboranes prepared by this new method have been oxidised to alcohols or phenols in good yield[44] and react with 'alkaline silver nitrate' to give good yields of bialkyls (secondary and primary alkyl groups) and biaryls[46].

Trialkylboranes of the type R_3^1B or $R_2^1R^2B$ can be prepared by reaction of 2-alkylbenzo[1.3.2]dioxaboroles with a Grignard reagent (R^1MgX); both oxygen–boron bonds are cleaved and boranes are obtained in good yield[47].

Two other syntheses of mixed trialkylboranes have been reported. First[48], a dialkyl thioboronite prepared from a trialkylborane is reduced by lithium aluminium hydride in the presence of an alkene as shown in equation (9.4).

$$R^1_3B \longrightarrow R^1_2BSR^2 \xrightarrow[\text{LiAlH}_4]{\text{alkene}} R^1_2R^3B \qquad (9.4)$$

Secondly, cycloalkylthexylboranes react with hindered alkenes by elimination of 2,3-dimethylbut-2-ene to give dialkylcycloalkylboranes[49]; for example, cyclopentylthexylborane reacts with 2,4,4-trimethylpent-1-ene by elimination of 2,3-dimethylbut-2-ene to form cyclopentyldi-(2,4,4-trimethylpent-1-yl)-borane, but hydroborates 2-methylbut-1-ene normally.

9.3 ORGANOCHLOROBORANES: SYNTHESIS AND REACTIONS

Monochloroborane diethyl etherate or dichloroborane diethyl etherate offer many advantages over diborane itself in hydroboration, and the development of their use has been a major advance during the past two years.

Monochloroborane in THF hydroborates alkenes slowly and incompletely because of the strongly bound complex it forms with the solvent, but monochloroborane diethyl etherate, formed from lithium borohydride and boron trichloride diethyl etherate in diethyl ether, readily hydroborates alkenes to give dialkylchloroboranes [equation (9.5)] which are easily isolated in excellent yield by distillation, or may be converted into borinic acids or esters by reaction with water or alcohols[50].

$$LiBH_4 + BCl_3 + 2Et_2O \longrightarrow LiCl + 2BH_2Cl\cdot OEt_2 \xrightarrow{\underset{}{C=C}} 2R_2BCl \qquad (9.5)$$

Alcohols formed by anti-Markovnikov addition are obtained in much greater isomeric purity ($> 99.5\%$) by hydroboration of alkenes with monochloroborane diethyl etherate followed by oxidation than by hydroboration–oxidation with diborane in THF[38]; thus the synthesis of regiospecifically and stereochemically pure derivatives is markedly improved by using monochloroborane (see p. 366; Refs. 38 and 39). Monochloroborane diethyl etherate also reacts readily with terminal and internal alkynes to give, for the first time, a convenient procedure for the synthesis of dialkenylchloroboranes, which are easily isolated, or protonolysed to alkenes, or oxidised to aldehydes or ketones or converted into *cis,trans*-dienes on treatment with sodium hydroxide and iodine[51] [e.g. as in equation (9.6); X = R^2 or H].

$$R^1C{\equiv}CX \xrightarrow{\text{BH}_2\text{Cl}} \begin{array}{c} R^1 \quad X \\ \diagdown C{=}C\diagup \\ H \diagup \qquad \diagdown \\ \quad\quad B{-}Cl \\ H \diagdown \qquad \diagup \\ \diagup C{=}C\diagdown \\ R^1 \quad X \end{array} \xrightarrow{\text{NaOH I}_2} \begin{array}{c} R^1 \quad X \\ \diagdown C{=}C\diagup \quad R^1 \\ H\diagup \qquad \diagdown C{=}C\diagup \\ \quad\quad X\diagup \quad\quad \diagdown H \end{array} \qquad (9.6)$$

In contrast to the facile reactions of monochloroborane diethyl etherate with alkenes[50], dichloroborane in diethyl ether (or THF) fails to react at a significant rate[52], presumably because of the formation of a more stable complex with the solvent. However, the solvent molecules coordinated to boron can be displaced by boron trichloride, which is an even stronger Lewis acid and so the addition of dichloroborane diethyl etherate to a mixture of an alkene, boron trichloride and an inert solvent (e.g. pentane) gives a precipitate of boron trichloride diethyl etherate and a solution of an alkyl-

$$\text{LiBH}_4 + 3\text{BCl}_3 + 4\text{Et}_2\text{O}$$
$$\downarrow$$
$$4\text{BHCl}_2{\cdot}\text{OEt}_2 + \text{LiCl}{\downarrow} \xrightarrow[\text{BCl}_3,\ \text{pentane}]{\diagdown C{=}C\diagup} \text{RBCl}_2 + \text{BCl}_3{\cdot}\text{Et}_2\text{O} \quad (9.7)$$

dichloroborane[52] [equation (9.7)]. The alkyldichloroborane can be isolated by distillation or can be converted into a boronic acid or ester. In a similar fashion the hydroboration of alkynes with dichloroborane gives alkenyl-dichloroboranes[53], $R^1CH{=}CR^2BCl_2$ or $R^1CH{=}CHBCl_2$, and so provides the first simple general synthesis of these compounds. The alkenyldichloroboranes can be protonolysed to stereoisomerically pure *cis*-dienes, oxidised to ketones or converted into vinylboronic acids or esters and hence into vinyl iodides and bromides (see p. 372).

Alkyldichloroboranes can also be prepared[54] from trialkylboranes and boron trichloride (in the presence of a small amount of diborane) at 110 °C

$$\text{R}_3\text{B} + 2\text{BCl}_3 \longrightarrow 3\text{RBCl}_2 \qquad (9.8)$$

[equation (9.8)]. This synthesis is less general than that given above[52] and trialkylboranes derived from internal alkenes undergo the redistribution reaction sluggishly and suffer from migration of the boron atom within the alkyl group.

Dialkylchloroboranes (R^1_2BCl)[55] and alkyldichloroboranes (R^1BCl$_2$)[56]

each react stereospecifically with alkyl or aryl azides (R^2N_3) to produce secondary amines (R^1R^2NH) in excellent yields. Both these reactions are faster than the corresponding reactions with trialkylboranes because the increased Lewis acidity of the chloroboranes gives stronger coordination with the azide, and even chloroboranes and azides with secondary alkyl groups react readily; the additional advantage of reactions involving alkyl-(or aryl)dichloroboranes is that complete utilisation of the alkyl group in the chloroborane is achieved [equation (9.9)].

$$R^1BCl_2 + R^2N_3 \longrightarrow \left[\begin{array}{c} R^1 - \bar{B}Cl_2 \\ \mid \quad \quad + \\ R^2 - N - N_2 \end{array} \right] \longrightarrow R^1R^2NBCl_2 \longrightarrow R^1R^2NH \quad (9.9)$$

Dialkylchloroboranes[57] and alkyldichloroboranes[58] have also been used to alkylate ethyl diazoacetate to give ethyl alkanoates, RCH_2CO_2Et, in excellent yields. The reaction probably proceeds by coordination of, for example, an alkyldichloroborane to give the intermediate $R\bar{B}Cl_2CH(\overset{+}{N_2})$-$CO_2Et$, which loses nitrogen with subsequent or concurrent migration of the alkyl group. Once again use of both types of chloroboranes allows milder reaction conditions to be employed and permits the use of bulkier alkyl groups than in the reaction with trialkylboranes; in addition, alkyldichloroboranes give complete utilisation of the alkyl group.

The radical-chain autoxidation of trialkylboranes to yield alcohols and alkyl hydroperoxides was described in the previous review[1], but only two alkyl compounds had been converted into the hydroperoxide. The autoxidation of alkyldichloroboranes, however, gives the alkylperoxydichloroborane

$$RBCl_2 + O_2 \longrightarrow R\cdot + \dot{O}_2BCl_2 \quad (9.10)$$
$$\dot{R} + O_2 \longrightarrow R\dot{O}_2 \quad (9.11)$$
$$R\dot{O}_2 + RBCl_2 \longrightarrow RO_2BCl_2 + \dot{R} \quad (9.12)$$
$$RO_2BCl_2 + RBCl_2 \longrightarrow 2ROBCl_2 \quad (9.13)$$

[equations (9.10)–(9.12)][59], which is lost in a rapid subsequent reaction [equation (9.13)]. In ether, reaction (9.13) is strongly suppressed because of coordination of the alkyldichloroborane with the solvent and the products in equation (9.12) can be hydrolysed to yield alkyl hydroperoxides efficiently.

The reaction of alkali-metal dibutylboron compounds (Bu^n_2BM; formed from dibutylchloroborane and a sodium–potassium alloy) with alkyl iodides or bromides gives very little alkyldibutylborane, but mainly alkanes and bialkyls by electron transfer reactions [equation (9.14)][60].

$$Bu^n_2BM + RI \longrightarrow Bu^n_2B\cdot + R\bar{I}^{\cdot} + M^+$$

$$\downarrow$$

$$R\cdot + I^- \quad (9.14)$$

$$\swarrow \quad \searrow$$

$$RH \quad \quad RR$$

9.4 ALKYNYL- AND ALKENYL-ORGANOBORATES
(See also Section 9.5)

Lithium trialkylalk-1-ynylborates, prepared as shown in equations (9.15) and (9.16), react with iodine in THF under mild conditions to give almost

$$HC{\equiv}CR^1 + Bu^nLi \longrightarrow LiC{\equiv}CR^1 + Bu^nH \qquad (9.15)$$

$$R^2{}_3B + LiC{\equiv}CR^1 \longrightarrow Li[R^2{}_3\overset{-}{B}C{\equiv}CR^1] \qquad (9.16)$$

quantitative yields of internal alkynes[61] [equation (9.17)]. Unlike alkylations of terminal alkynes, which involve the corresponding anion in nucleophilic

$$\overset{+}{Li}[R^2{}_3\overset{-}{B}C{\equiv}CR^1] + I_2 \longrightarrow R^1C{\equiv}CR^2 + R^2{}_2BI + LiI \qquad (9.17)$$

substitutions, this reaction sequence permits the introduction of aryl groups, as well as primary and secondary alkyl groups. Electrophilic attack by iodine on the triple bond of the trialkylalkynylborate probably generates a partial positive charge at C-1 and so initiates migration of an alkyl group from boron to the electron-deficient centre, with subsequent dehalogeno-boronation of the intermediate $R^1CI{=}CR^2BR^2_2$ to yield the internal alkyne. In an alternative method[62] of alkylating terminal alkynes, the vigorous exo-thermic reaction of lithium trialkylalk-1-ynylborates with methanesulphinyl chloride at 0 °C is used. By electrophilic attack on the triple bond, the methanesulphinyl chloride probably induces alkyl group migration from boron to give the vinyl intermediate and hence the product [equation (9.18)].

$$\begin{array}{c} R^1 \\ | \\ R^1{-}B{-}C{\equiv}C{-}R^2 \\ | \\ R^1 \quad Me{-}S{-}Cl \\ \quad\quad \| \\ \quad\quad O \end{array} \longrightarrow \left[\begin{array}{c} R^1 \quad\quad R^2 \\ \diagdown\;\;\diagup \\ C{=}C \\ \diagup\quad\quad\diagdown \\ R^1{}_2B \quad\quad SMe \\ \quad\quad \| \\ \quad\quad O \end{array} \right] \longrightarrow R^1C{\equiv}CR^2 \qquad (9.18)$$

Mixed trialkylboranes $R^1R^2R^3B$ have been used in this reaction to determine an order of selectivity in the transfer of the *B*-alkyl group, but mixed products were often obtained[62].

Lithium trialkylalk-1-ynylborates can be alkylated or protonated and the intermediate vinylboranes then oxidatively hydrolysed to give specifically substituted ketones[63] [equation (9.19)]. The alkylation has also been achieved

$$R^1{}_3\overset{-}{B}C{\equiv}CR^2 \xrightarrow[\text{or } H^+]{R^3X} \begin{array}{c} R^1{}_2BCR^1{=}CR^2R_3 \\ \text{or} \\ R^1{}_2BCR^1{=}CHR^2 \end{array}$$

$$\Big\downarrow \text{oxidation}$$

$$R^1COCHR^2R^3 \text{ or } R^1COCH_2R^2 \qquad (9.19)$$

with α-halogenocarbonyl compounds[64] ($BrCH_2COX$) to give the inter-mediate (2), which can either be oxidised to γ-diketones (X = alkyl or Ph) or γ-ketoesters (X = OR), or can be protonolysed to β,γ-unsaturated ketones (X = alkyl or Ph) or esters (X = OR) [equation (9.20)]; the migration from

$$R^1_3\bar{B}C\equiv CR^2 \xrightarrow{BrCH_2COX} \underset{R^1_2B}{\overset{R^1}{}}C=C\underset{CH_2COX}{\overset{R^2}{}} \xrightarrow{oxidation} R^1COCHR^2CH_2COX$$

$$(2)$$

$$\downarrow H\cdot$$

$$(9.20)$$

$$\underset{H}{\overset{R^1}{}}C=C\underset{CH_2COX}{\overset{R^2}{}}$$

boron takes place with secondary as well as primary groups. Alkylations with triethyloxonium tetrafluoroborate, dimethyl sulphate, allyl bromide, benzyl bromide and methyl toluene-p-sulphonate, and protonation with methanesulphonic acid, have all been carried out to give ketones in good to excellent yield. This sequence of reactions represents a major extension of alkyne chemistry and allows a ketone to be synthesised from independently variable units, i.e. R^1 from R^1_3B, R^2 from $R^2C\equiv CH$, R^3 (or H) from R^3X (or H^+).

Acylation of trialkylalk-1-ynylborates is possible and provides another example of their use as 'building-blocks' in synthesis[65]; α,β-unsaturated ketones ($R^1_2C=CR^2COR^3$) can be constructed from trialkylboranes (R^1_3B), alkynes ($HC\equiv CR^2$) and acyl chlorides (R^3COCl) [equation (9.21)]. The

$$(9.21)$$

$$\text{(Cyclic borinate)} \qquad (3)$$

cyclic borinate (enol borinate, vinyloxyborane) resists the usual oxidation with alkaline H_2O_2, possibly because of a major contribution from structure (3), but Jones oxidation gives the α,β-unsaturated ketone ($R^1_2C=CR^2COR^3$) in moderate yield.

Oxiranes have been used[66] as alkylating agents and probably react as

$$(9.22)$$

shown in equation (9.22). The reaction mixture has been worked up in three ways: treatment with (a) alkaline H_2O_2 to give γ-hydroxyketones, (b) acetic' acid to give trisubstituted ethylenes of known stereochemistry, (c) sodium hydroxide and iodine to give a tetrasubstituted ethylene.

Vinyl-lithium coordinates with trialkylboranes to give lithium alk-1-enyl-trialkylborates; the latter react[67] with methyloxirane in a similar fashion to the reaction of lithium trialkylalk-1-ynylborates shown in equation (9.22). The cyclic borate intermediate [dihydro derivative of (4; $R^2 = H$)] on treatment with alkaline H_2O_2 gives 1,4-diols.

The product (5) from the reaction of propargyl chloride and a dialkyl-borane has been treated with methyl-lithium to give a lithium alk-1-enyl-trialkylborate which undergoes an alkyl group migration with displacement

$$(9.23)$$

$$R\,MeBCHRCH\!=\!CH_2 \longrightarrow RCH\!=\!CHCH_2BMeR$$

$$(6) \qquad\qquad (7)$$

of chloride ion [equation (9.23)][68]. The initial product (6) rearranges to the more stable isomer (see Ref. 69 for a study of the rearrangement of allyl-boranes) and so provides a new route to allylboranes (7), from which the terminal alkene $RCH_2CH\!=\!CH_2$ can be obtained by protonolysis.

A similar rearrangement to that shown in equation (9.23), which represents an intramolecular Michael reaction, is induced when the alkenylboranes obtained from propiolic esters and a monoalkylthexylborane [equation

$$(9.24)$$

$$(R^1 = \text{thexyl})$$

$$R^3CH(OH)CH_2CO_2H$$

(9.24)] are treated with base. The product can be oxidised to provide a route from alkenes to β-hydroxyalkanoic acids with groups at the β-position stereochemically defined[70].

9.5 ALKENYLBORON COMPOUNDS. VINYLOXYBORANES
(See also Sections 9.3, 9.4, 9.6, 9.10.1, 9.10.2)

Many studies have been reported of the reactions of alkynes with organo-boranes and of the subsequent behaviour of the alkenylboranes.

Benzo[1.3.2]dioxaborole (8) reacts readily with an alkyne to give, stereospecifically and almost regioselectively, the monohydroboration product

$$(9.25)$$

(8)

(9) (X = R² or H)

(9) [equation (9.25)] from which the alkeneboronic acid can be easily obtained[71]. Protonolysis of (9) yields the *cis*-alkene (X = R^2), oxidation of (9) gives an aldehyde (R^1CH_2CHO) or ketone ($R^1CH_2COR^2$)[71] and reaction of (9) with mercuric acetate gives alkenylmercuric acetates[72]. An alternative route to alkenylmercuric acetates is via the monohydroboration product of alkynes with dicyclohexylborane[73].

trans-Alken-1-ylboronic acids react rapidly with I_2 and NaOH to form the *trans*-alken-1-yl iodides in excellent yield and of high stereochemical purity[74] [equation (9.26)], but the boronic acids or their catechol esters react successively with bromine and base to give the *cis*-alken-1-yl bromides, also

$$(9.26)$$

in excellent yield and of high stereochemical purity[75]. Therefore alkynes can be converted into vinyl bromides and iodides of opposite configuration and then into vinylmagnesium halides or vinyl-lithium reagents of known configuration.

Monohydroboration of a 1-bromoalk-1-yne and the subsequent reactions of the alkenylborane produced provide excellent routes to *trans*-alkenes[76] and trialkylsubstituted ethylenes[77], and the advantages of these routes over alternative methods have been discussed. The reaction of 1-bromoalk-1-ynes with monoalkylthexylboranes gives an intermediate (10; R^1 = thexyl) which on rearrangement with NaOMe followed by protonolysis gives *trans*-

$$(9.27)$$

(10)   (11)   (12)

alkenes[76,78] [equation (9.27)]. The intermediate (10; $R^1 = R^2$), formed from the reaction of 1-bromoalk-1-ynes with dialkylboranes, reacts with aqueous NaOH to give (11), and iodine causes a migration of the second alkyl group on boron to give, after deboroniodination, the trisubstituted ethylene (12)[77]; boracycloalkanes (e.g. bisborinane) can be used as the 'dialkylborane' to prepare alkylidenecycloalkanes.

The NaOMe induced rearrangement of vinylboranes of the type shown in equation (9.27) has been extended to provide routes to *trans,trans*-dienes or to *trans*-alkenyl ketones[79]. Thexylborane reacts successively with 1-chloroalk-1-ynes and alk-1-ynes to give (13) and the rearrangement product (14)

$$\text{(9.28)}$$

(13) (14) (R^1 = thexyl)

[equation (9.28)] can be protonolysed to conjugated *trans,trans*-dienes or oxidised to conjugated *trans*-alkenyl ketones[79].

The conversion of dialkylvinylboranes (15; X = H or R), obtained from the reaction of alkynes ($R^1C{\equiv}CX$) with dialkylboranes (R^2_2BH), into *cis*-disubstituted or trisubstituted ethylenes has been reported previously. A stereospecific synthesis of *trans*-disubstituted alkenes (16; X = H) or tri-substituted alkenes (16; X = R) of predictable stereochemistry has been reported[80] from the same intermediate (15) by treatment with cyanogen bromide or cyanogen iodide in dichloromethane [equation (9.29)].

$$\text{(9.29)}$$

(15) (16)

Vinyloxyboranes formed, for example, by the reaction of thioboronites ($R^3_2BSR^3$) with ketene or by the reaction of trialkylboranes with butenone or diazoketones, react with compounds containing carbonyl groups to give β-hydroxythiolesters or β-hydroxyketones [equation (9.30); X = SPh or

$$R^1R^2C(OH)CH_2COX \quad \text{(9.30)}$$

(X = SPh or R^3)

R^3][81,82]. Vinyloxyboranes also react with NBS or NCS to give the α-halo-genocarbonyl compound[83] [equation (9.31); X = R^1 or OR^1] in a regio-specific manner. By this method an α-halogenoketone can be assembled with specific alkyl groups (R and X) from the trialkylborane and from the diazo-

$$R_3B + N_2CHCOX \longrightarrow [R_2BCHRCOX]$$

$$\downarrow$$

$$RCH{=}CX(OBR_2) \longrightarrow RCHBrCOX \qquad (9.31)$$

ketone, respectively, whereas direct bromination of a ketone may give α- and/or α'-bromination.

9.6 NUCLEOPHILICALLY AND ELECTROPHILICALLY INDUCED MIGRATION OF ALKYL GROUPS FROM BORON
(See also Sections 9.3, 9.4, 9.5)

Several new methods for preparing tertiary carbinols from trialkylboranes have been reported. The lithium triethylcarboxide $(LiOCEt_3)$ induced reaction of dichloromethyl methyl ether with trialkylboranes at moderate temperature provides a route to tertiary carbinols[84,85] in excellent yields

$$R_3B + CHCl_2OMe + LiOCEt_3 \longrightarrow R_3CBCl(OMe) \longrightarrow R_3COH \qquad (9.32)$$

[equation (9.32)]. This method has advantages over, or is a valuable alternative, to other methods (e.g. carbonylation or the cyanoborate process) mentioned in the reports[84,85]. Trialkylboranes containing t-alkyl groups react rapidly[85] and the migration of the t-alkyl group occurs without isomerisation. For the cyanoborate process mentioned above, the migration of the third alkyl group was initially reported as being convenient only for primary alkyl groups. It has since been found that the use of pyridine or DMF as solvent allows the migration of three bulky groups and all three migrations proceed with retention of configuration, the ease of each migration of an alkyl group being in the order primary $>$ secondary $>$ tertiary[86]. A method reported in 1971 for preparing tertiary carbinols by the base-induced reaction of chlorodifluoromethane with trialkylboranes failed for secondary and hindered trialkylboranes because, as has now been shown[87], the highly hindered intermediate $[R_3CBF(OCEt_3)]$ is remarkably stable to oxidation. The triethylcarbinoxy group can, however, be cleaved by using a strong acid and subsequent oxidation of the resulting unhindered intermediate gives the trialkylcarbinols in excellent yields[87].

The lithium triethylcarboxide-induced reaction of borinic esters and dichloromethyl methyl ether under mild conditions has been used to yield an intermediate which can be oxidised to the corresponding ketone in high yield[88] {equation (9.33); see Ref. 89 for the preparation of bicyclo[3.3.1]-nonan-9-one by a similar method}. Dialkylborinates are now readily avail-

$$R_2BOMe + CHCl_2OMe + LiOCEt_3 \longrightarrow R_2CClB(OMe)_2 \longrightarrow R_2CO \quad (9.33)$$

able from the reaction of alkenes with monochloroborane diethyl etherate followed by alcoholysis (see p. 366, Ref. 50) so that a facile, versatile route from alkenes to ketones which accommodates bulky alkyl groups is available.

Trialk-1-ynylboranes have been prepared by the reaction of a lithium alk-1-ynide with boron trifluoride etherate in THF[90]. Subsequent reaction of the trialkynylborane with ethyl diazoacetate at $-20\,^\circ$C followed by hydrolysis gives the propargylic ester; mercuric ion-catalysed hydration of the β,γ-triple bond gives only the γ-keto-esters[90] [equation (9.34)].

$$(RC\equiv C)_3B + N_2CHCO_2Et \longrightarrow \longrightarrow RC\equiv CCH_2CO_2Et$$

$$\downarrow$$

$$RCOCH_2CH_2CO_2Et \qquad (9.34)$$

The lithium derivative of benzyl phenyl sulphide ($Li^+\ Ph\bar{C}HSPh$) reacts with trialkylboranes to give alkylbenzenes ($PhCH_2R$) after hydrolysis[91]. Further reaction of the intermediate, $R_2BCHRPh$, with the lithium derivative was prevented by the addition of copper(I) iodide at $-75\,^\circ$C or dimethyl sulphate at room temperature.

The bromination of trialkylboranes in the presence of water and the subsequent double nucleophilically induced migration were reported in the last review[1], but now a single migration can be achieved by the reaction of the α-bromo-organoborane with an electrophilic reagent, e.g. tin(IV) bromide[92]. The reagent (M) coordinates with the α-bromine atom to induce the migration of the alkyl group and so yields dialkylboron bromides by a simple new route [equation (9.35)].

$$\underset{\underset{M}{\overset{\mid}{\underset{Br}{\mid}}}{\overset{R^1}{\underset{\mid}{\overset{\mid}{C}}}-BR^2}} \longrightarrow \underset{\overset{\mid}{Br}}{\overset{R^1}{\underset{\mid}{C}-BR^2}} \qquad (9.35)$$

9.7 HOMOLYTIC REACTIONS

Dialkylborinic acids, which are the products of the first stage of the light-induced bromination reactions of trialkylboranes in the presence of water[1], have been prepared separately[50] and used in the bromination–oxidation sequence to give an alternative route to highly substituted alcohols[93].

The bromination–oxidation of trialkylboranes has been extended[94] to dialkylthexylboranes, $R^1R^2BCMe_2CHMe_2$, and boracyclanes[95]: α-bromination is conveniently carried out with NBS or sodium bromate and hydrobromic acid, which generate bromine slowly and automatically[96]. For the former compounds the migration of the thexyl group is very slow and tertiary alcohols derived from the different alkyl groups (R^1 and R^2, including groups with functional substituents) are formed[94]; the latter six-membered cyclic boron compounds give ring contraction [equation (9.36)] on migration of the 'alkyl' group in the intermediate to form alcohols [from (17; X = R^1)] on oxidation, or ketones [from (17; X = H)] on further α-bromination and oxidation[95].

$$\underset{(17)}{\overset{\sim\!\sim\!\sim}{\underset{\underset{R}{\overset{|}{B}}}{\underset{Br}{XC}}\!\!-\!\!CH_2}} \xrightarrow{\;H_2O\;} \underset{RBOH}{\overset{\sim\!\sim\!\sim}{XC\!-\!CH_2}} \qquad (9.36)$$

Trialkylboranes react with formaldehyde either by a radical process to give direct addition to the carbonyl group [equation (9.37)] or by reductive dealkylation [equation (9.38); R = Bu^n], probably involving the transition

$$R_2BOCH_2R \qquad\qquad (9.37)$$

$$R_3B\ +\ CH_2O$$

$$R_2BOMe\ +\ EtCH{=}CH_2 \qquad\qquad (9.38)$$

$$\underset{(18)}{\overset{\displaystyle CH_2{-}\overset{R}{CH}}{\underset{O{=}CH_2}{B}}\!\!\Big(H} \longrightarrow \overset{\displaystyle CH_2{=}\overset{R}{CH}}{\underset{O{=}CH_2}{B}\ \ H} \longrightarrow \overset{\displaystyle CH_2{=}\overset{R}{CH}}{\underset{O{-}CH_3}{B}} \qquad (9.39)$$

state (18) [equation (9.39)][97]. The non-radical process mentioned above is further exemplified by the reaction of trialkylboranes with *cis*-azobenzene to give an alkene[98].

An extension of the radical reactions of organoboranes with 3,4-epoxybut-1-ene ($CH_2{=}CHCHCH_2O$) to their reactions with substituted 3-4-epoxy-but-1-ynes ($HC{\equiv}CCR^2CH_2O$) provides a route to allene derivatives via 1,5-addition[99] [equation (9.40)]. Radicals derived from functionally sub-

$$R^1_3B\ +\ HC{\equiv}CCR^2{-}\overset{\overset{\displaystyle O}{\frown}}{CH_2}$$

$$\downarrow$$

$$R^1CH{=}C{=}CR^2CH_2OBR^1_2 \longrightarrow R^1CH{=}C{=}CR^2CH_2OH \qquad (9.40)$$

stituted organoboranes react with quinones to give functionally substituted hydroquinones[100], and tris(cycloalkyl)boranes react with crotonaldehyde in the presence of pyridine, possibly by a radical process, to give 3-cycloalkyl-butanals[101].

9.8 PHOTOCHEMICAL REACTIONS

The reactions of organoboron compounds with photochemically produced radicals have been studied for several years[1] (see also p. 375). In addition, there have been several further studies of the photochemical reactions of aminoboron compounds which are isoelectronic, polar π-bond analogues of alkenes [see equation (9.41)]. 1,2-Bis(dimethylamino)-1,2-diphenyldiborane,

$$\overset{\cdot\cdot}{\underset{/}{>}}N-B< \quad \longleftrightarrow \quad \overset{+}{\underset{/}{>}}N=\bar{B}< \qquad (9.41)$$

PhB(NMe$_2$)BPh(NMe$_2$), has previously been shown to undergo photolytic boron–boron homolysis, in contrast to the photochemical electrocyclic ring closure of 2,3-diphenylbuta-1,3-diene; a theoretical investigation[102] of the electronic structure of bis(amino)diborane (H$_2$NBHBHNH$_2$) and (amino)-vinylborane (CH$_2$=CHBHNH$_2$) has now been made to elucidate the reasons for differences between the excited state properties of these chromophores and those of butadiene systems.

A photochemical *cis–trans* isomerisation of aminoboranes has been detected[103] [equation (9.42)] which is mechanistically different from the

$$\underset{Me}{\overset{PhCH_2}{>}}N \rightleftharpoons B\underset{CH_2Ph}{\overset{Ph}{<}} \quad \underset{\longleftarrow}{\overset{h\nu}{\longrightarrow}} \quad \underset{Me}{\overset{PhCH_2}{>}}N \rightleftharpoons B\underset{Ph}{\overset{CH_2Ph}{<}} \qquad (9.42)$$

$$(19)$$

superficially similar isomerisation in alkenes. Irradiation of (19) in CCl$_4$ illustrates[104] a second primary process from aminoborane excited states, which gives homolysis of the boron–benzyl bond and then yields several products by alkyl and boron radical recombination and by reaction of the radicals with the solvent.

Photolysis of but-1-en-3-yl(dimethylamino)phenylborane in cyclohexane gives but-2-en-1-yl(dimethylamino)phenylborane by a novel 1,3-aminoboryl

$$\begin{array}{c} Ph \\ \backslash \\ B \lesseqgtr NMe_2 \\ / \\ CHMe \\ \backslash \\ HC=CH_2 \end{array} \quad \underset{\longleftarrow}{\overset{h\nu}{\longrightarrow}} \quad \begin{array}{c} Ph \\ \backslash \\ B \lesseqgtr NMe_2 \\ / \\ CH_2 \\ \backslash \\ HC=CHMe \end{array} \qquad (9.43)$$

shift [equation (9.43)]. This reaction is analogous to the reaction of β,γ-unsaturated ketones, for example, rather than of 1,4-dienes, which are isoelectronic with allyl(amino)boranes[105], and so demonstrates that similar photochemical reactions can be observed in systems of quite different electronic configuration.

The product of irradiation of cyclohexene and a trialkylborane in the presence of a sensitiser has been oxidised with alkaline H$_2$O$_2$ to give a *cis*-2-alkylcyclohexanol, which is probably produced by the reaction of the trialkyl-borane with the highly strained *trans*-cyclohexene to give the intermediate

$$(20) \qquad \longrightarrow \qquad \qquad (9.44)$$

(20)[106] [equation (9.44)]. Migration of an alkyl group from boron followed by oxidation leads to a *cis*-2-alkylcyclohexanol; similar reactions have been noted for 1-ethylcyclohexene and cycloheptene but cyclo-pentene, -octene and -dodecene did not react similarly.

Trialkylboranes also react[107] with the photoexcited state of enolic β-dicarbonyl compounds such as ethyl acetoacetate and acetylacetone and yield either β-alkylated-β-hydroxycarbonyl compounds via the intermediate (21), arising from 1,4-addition of a trialkylborane to the enolic system, or β-alkylated-α,β-unsaturated carbonyl compounds via the elimination of dialkylborinic acid from (22) [equation (9.45)].

$$\text{(9.45)}$$

9.9 ALLYLBORANES

Monohydroboration of 3-phenylbuta-1,2-diene with disiamylborane generates mainly the allyldisiamylborane (23), which reacts with various types of carbonyl group, possibly as shown in equation (9.46; X = Ph), to give

$$\text{MePhC}{=}\text{CHCH}_2\text{BR}_2 \qquad\qquad \text{Me}_2\text{C}{=}\text{CHCH}_2\text{BEt}_2$$

$$\text{(23)} \qquad\qquad\qquad\qquad \text{(24)}$$

$$\text{(R = Me}_2\text{CHCHMe)}$$

alk-3-en-1-ols and hence a route to substituted 1,4-dienes[108]. The allyldiethylborane (24) has also been prepared from 3-methylbuta-1,2-diene and diethylborane and it reacts similarly with ketones, as shown in equation

$$\text{(9.46)}$$

(9.46; X = Me), so that the isoprene unit (Me$_2$CCH=CH$_2$) can be introduced into a compound[109].

Triallylborane has been shown previously to react with alkynes through three successive stages[23]. In the first stage, *cis* addition to the triple bond in, for example, propargyl chloride (3-chloropropyne)[110-112] gives (25), which cyclises in the second stage with the formation of (26), and finally thermal cyclisation of (26) gives (27) [see scheme (9.47)]. Compound (28), produced by the allyl rearrangement of compound (26), is also formed in the second stage[110,112] and is cyclised to (29) on being rapidly heated[111,112].

$$HC \equiv CCH_2Cl + (C_3H_5)_3B$$

(9.47)

(25)   (26)   (27)

(28)   (29)

The reaction of triallylborane with 1-ethynylcyclohexene has also been studied and gives products analogous to (25) and (26) above at room temperature[113]; on distillation the second-stage product gives the product analogous to (27)[114].

Triallylborane also reacts with allene or 3-methylbuta-1,2-diene in three stages similar to those shown in equation (9.47). In the first stage the diallylboryl group adds to C-1 in allene but with 3-methylbuta-1,2-diene it adds to C-2 and the allyl group adds to C-3[115]. The reactions of tris-(2-methylallyl)-borane with acetylene and with hex-1-yne have also been reported[116].

Triallylborane[117] or tris(but-2-en-1-yl)borane[118] react with 1-methylcyclopropene either by *cis* addition of diallylboryl and allyl fragments (with allyl rearrangement) to the double bond of the cyclopropene to give (30) or by cleavage of the C-2—C-3 bond in the cyclopropene without rearrangement of the allyl group to give (31).

(30) R = H or Me   (31) R = H or Me

The reactions of trialkylboranes with 1-methylcyclopropene provides a synthesis of 2-methylalk-1-enes $CH_2 = CMeCH_2R$ [119].

9.10 MISCELLANEOUS TOPICS

9.10.1 Carbanions stabilised by a boron atom

An electron-deficient boron atom is similar to a carbonyl group in that it should be capable of stabilising an anion at the α-position, but the formation of such an anion from an organoborane by reaction with a base has never

$$>B-\bar{C}- \quad \longleftrightarrow \quad >\bar{B}=C<$$

(32)

been reported previously because the base coordinates with the boron atom. The anion (32) has now been formed[120] from *B*-methyl-9-borabicyclo[3.3.1]-nonane and a sterically hindered base, lithium 2,2,6,6-tetramethylpiperidide (LiTMP); the resultant anion has been alkylated with n-butyl bromide and the product oxidised to give pentan-1-ol.

Vinylboron compounds, prepared by monohydroboration of alkynes with disiamylborane, have also been treated with LiTMP to give a boron-stabil-

$$RCH_2CH=CHB< \quad \longrightarrow \quad R\bar{C}HCH=CHB< \quad \longleftrightarrow \quad RCH=CH-\bar{C}HB< \qquad (9.48)$$

$$R = Me(CH_2)_4 \qquad\qquad (33) \qquad\qquad\qquad (34)$$

ised anion[121] [equation (9.48)] which reacts with acetone or trimethylchloro-silane predominantly via (33) and with methyl iodide or water predominantly via (34).

$$RCH_2CH\genfrac{}{}{0pt}{}{B<}{B<} \quad \xrightarrow{MeLi} \quad RCH_2CH\genfrac{}{}{0pt}{}{B<}{\underset{Me}{\bar{B}<}} \quad \longrightarrow \quad RCH_2\bar{C}H-B< \qquad (9.49)$$

Another procedure[122] for producing (32) is shown in equation (9.49), and subsequent alkylation and oxidation of the carbanion gives a novel route to secondary alcohols. Clearly other applications of boron-stabilised carbanions in organic synthesis can be expected.

9.10.2 Metallations
(See also Section 9.5)

Organoboranes derived from terminal alkenes react with mercuric acetate in THF to give dialkylmercury compounds, whereas organoboranes derived from internal alkenes do not react under these conditions[123]; dialkylmercurials can therefore be prepared in excellent yield by the above reaction of alkyl-dicyclohexylboranes.

1,1-Bis(chloromercuri)alkanes, $RCH_2CH(HgCl)_2$, have been prepared from terminal alkynes by hydroboration with diborane, followed by methan-olysis to the tetramethyl diboronate, $RCH_2CH[B(OMe)_2]_2$, and treatment with mercuric chloride[124].

9.10.3 Electrolytic reactions

Anodic oxidation of trihexylborane in methanolic NaOMe (i.e. $R_3\bar{B}OMe$)

gives dodecane and, in the presence of buta-1,3-diene, products (35) and (36), which arise from mixed coupling of the buta-1,3-diene radical cation (BRC) and the alkyl radical according to equation (9.50), are also obtained[125].

$$R_3\bar{B}OMe$$

$$\downarrow$$

$$R\cdot \xrightarrow[\text{Me}\bar{O}]{(BRC)} McOCH_2CH=CHCH_2R \quad \text{and} \quad CH_2=CHCH(OMe)CH_2R \qquad (9.50)$$

$$(35) \qquad\qquad (36)$$

$$\downarrow$$

$$R_2$$

Another report[126] of the electrolysis of trialkylboranes in methanolic KOH to give bialkyls in good yield has also appeared.

9.10.4 Kinetic investigations

The kinetics of the hydroboration of 2,3-dimethylbut-2-ene with borane in THF have been studied[127] and the relative rate constants of the first and the second stages of the hydroboration of some alkenes[128] show that the rate constant for the formation of a dialkylborane from a monoalkylborane (second stage) is larger than that for the formation of the monoalkylborane (first stage).

Kinetic studies of the following radical processes have also been reported: the autoxidation of trialkylboranes (rate constants for the primary initiation stage)[129]; the autoxidation of benzylboranes and 1-phenylethylboranes[130]; the reactions of organoboranes with t-butyl hypochlorite (bimolecular t-butoxydealkylation)[131]; the reaction of t-butoxyl radicals with diborane[132]; and the photochemical iodination of triethylborane[133].

9.11 SUMMARY OF THE TYPES OF COMPOUND SYNTHESISED

The following types of compound have been prepared by procedures given in papers covered by this review, but the general references given in the introduction are excluded. β,γ-Acetylenic esters[90]; alcohols[38,39,120,122] (β,γ-unsaturated alcohols[108,109] and trialkylcarbinols[43,84–87,93,94]); aldehydes[51,71,101]; alkenes[53,66,68,77,80,119] (*trans*-alkenes[76,78,80] and *cis*-alkenes[51,71]); *cis*-alkenyl bromides[75]; *trans*-alkenyl iodides[74]; alkylbenzenes[91]; alkyl hydroperoxides[59]; alkynes[61,62]; allene derivatives[99]; amides[86], secondary amines[55,56]; bialkyls (or biaryls)[46,125,126]; *cis,trans*-dienes[51]; *trans,trans*-dienes[79]; 1,4-diketones[64]; 1,4-diols[67]; ethyl alkyl(or aryl)acetates[57,58]; α-halogenoesters[83]; α-halogenoketones[83]; hydroquinones[100]; β-hydroxycarboxylic acids[70]; β-hydroxyesters[81,82,107]; β-hydroxyketones[81,82,107];

γ-hydroxyketones[66]; γ-keto-esters[64,90]; ketones[41,51,53,63,71,86,88]; mercury derivatives[72,73,123,124]; β,γ-unsaturated esters[64]; α,β-unsaturated ketones[65,107] (*trans*-[79]); β,γ-unsaturated ketones[64].

References

1. Toyne, K. J. (1973). In *MTP International Review of Science, Organic Chemistry Series One*, Vol. 2, *Aliphatic Compounds*, 357 (N. B. Chapman, editor) (London: Butterworths)
2. Brown, H. C. (1972). *Boranes in Organic Chemistry* (New York: Cornell University Press)
3. Cragg, G. M. L. (1973). *Organoboranes in Organic Synthesis* (New York: M. Dekker)
4. Matteson, D. S. (1972). *J. Organometal. Chem.*, **41**, 13
5. Matteson, D. S. (1973). *J. Organometal. Chem.*, **58**, 1
6. Niedenzu, K. (1972). *J. Organometal. Chem.*, **41**, 79
7. Niedenzu, K. (1973). *J. Organometal. Chem.*, **53**, 337
8. Cragg, R. H. (1972). In *MTP International Review of Science, Inorganic Chemistry Series One*, Vol. 1, *Main Group Elements: Hydrogen and Groups I–IV*, 185 (M. F. Lappert, editor) (London: Butterworths)
9. Niedenzu, K. (1972). In *MTP International Review of Science, Inorganic Chemistry Series One*, Vol. 4, *Organometallic Derivatives of the Main Group Elements*, 73 (B. J. Aylett, editor) (London: Butterworths)
10. Lockhart, J. C. (1972). In *MTP International Review of Science, Inorganic Chemistry Series One*, Vol. 9, *Reaction Mechanisms in Inorganic Chemistry*, 87 (M. L. Tobe, editor) (London: Butterworths)
11. Maher, J. P. (1972). *Organometal. Chem.*, **1**, 40
12. Ingold, K. U. (1973). *Free Radicals*, **1**, 37 (J. K. Kochi, editor) (New York: Interscience)
13. Davies, A. G. and Roberts, B. P. (1973). *Free Radicals*, **1**, 547 (J. K. Kochi, editor) (New York: Interscience)
14. Brown, H. C. and Midland, M. M. (1972). *Angew. Chem. Int. Ed. Engl.*, **11**, 692
15. Suzuki, A. (1971). *Yuki Gosei Kagaku Kyokai Shi*, **29**, 995
16. Davies, A. G. and Roberts, B. P. (1972). *Accounts Chem. Res.*, **5**, 387
17. Davies, A. G. (1972). *Chem. Ind.*, 832
18. Milovskaya, E. B. (1973). *Usp. Khim.*, **42**, 881
19. Lane, C. F. (1973). *Aldrichim. Acta*, **6**, 21
20. Brown, H. C. and Rogić, M. M. (1972). *Organometal. Chem. Syn.*, **1**, 305
21. Brown, H. C. and Negishi, E. (1972). *Pure Appl. Chem.*, **29**, 527
22. Zaidlewicz, M., Uzarewicz, I. and Uzarewicz, A. (1972). *Wiad. Chem.*, **26**, 33
23. Mikhailov, B. M. (1972). *Organometal. Chem. Rev., Sect. A*, **8**, 1
24. Kliegel, W. (1972). *Organometal. Chem. Rev., Sect. A*, **8**, 153
25. Meller, A. (1972). *Fortschr. Chem. Forsch.*, **26**, 37
26. Pelter, A. (1973). *Chem. Ind.*, 206
27. Paetzold, P. I. and Grundke, H. (1973). *Synthesis*, 635
28. Brown, H. C., Negishi, E. and Burke, P. L. (1972). *J. Amer. Chem. Soc.*, **94**, 3561
29. Negishi, E., Burke, P. L. and Brown, H. C. (1972). *J. Amer. Chem. Soc.*, **94**, 7431
30. Burke, P. L., Negishi, E. and Brown, H. C. (1973). *J. Amer. Chem. Soc.*, **95**, 3654
31. Negishi, E. and Brown, H. C. (1973). *J. Amer. Chem. Soc.*, **95**, 6757
32. Brown, H. C. and Negishi, E. (1972). *J. Amer. Chem. Soc.*, **94**, 3567
33. Brown, H. C., Heim, P. and Yoon, N. M. (1972). *J. Org. Chem.*, **37**, 2942
34. Yoon, N. M., Pak, C. S., Brown, H. C., Krishnamurthy, S. and Stocky, T. P. (1973). *J. Org. Chem.*, **38**, 2786
35. Brändström, A., Junggren, U. and Lamm, B. (1972). *Tetrahedron Lett.*, 3173
36. Jones, P. R. (1972). *J. Org. Chem.*, **37**, 1886
37. Kabalka, G. W. and Bowman, N. S. (1973). *J. Org. Chem.*, **38**, 1607
38. Brown, H. C. and Ravindran, N. (1973). *J. Org. Chem.*, **38**, 182
39. Scouten, C. G. and Brown, H. C. (1973). *J. Org. Chem.*, **38**, 4092

40. Brown, H. C., Negishi, E. and Katz, J.-J. (1972). *J. Amer. Chem. Soc.*, **94**, 5893
41. Negishi, E. and Brown, H. C. (1972). *Synthesis*, 196
42. Negishi, E., Katz, J.-J. and Brown, H. C. (1972). *J. Amer. Chem. Soc.*, **94**, 4025
43. Negishi, E. and Brown, H. C. (1972). *Synthesis*, 197
44. Breuer, S. W. and Broster, F. A. (1972). *J. Organometal. Chem.*, **35**, C5
45. Breuer, S. W. (1972). *J. Chem. Soc. Chem. Commun.*, 671
46. Breuer, S. W. and Broster, F. A. (1972). *Tetrahedron Lett.*, 2193
47. Cabiddu, S., Maccioni, A. and Secci, M. (1972). *Gazz. Chim. Ital.*, **102**, 555
48. Pelter, A. and Sharrocks, D. N. (1972). *J. Chem. Soc. Chem. Commun.*, 566
49. Lane, C. F. and Brown, H. C. (1972). *J. Organometal. Chem.*, **34**, C29
50. Brown, H. C. and Ravindran, N. (1972). *J. Amer. Chem. Soc.*, **94**, 2112
51. Brown, H. C. and Ravindran, N. (1973). *J. Org. Chem.*, **38**, 1617
52. Brown, H. C. and Ravindran, N. (1973). *J. Amer. Chem. Soc.*, **95**, 2396
53. Brown, H. C. and Ravindran, N. (1973). *J. Organometal. Chem.*, **61**, C5
54. Brown, H. C. and Levy, A. B. (1972). *J. Organometal. Chem.*, **44**, 233
55. Brown, H. C., Midland, M. M. and Levy, A. B. (1972). *J. Amer. Chem. Soc.*, **94**, 2114
56. Brown, H. C., Midland, M. M. and Levy, A. B. (1973). *J. Amer. Chem. Soc.*, **95**, 2394
57. Brown, H. C., Midland, M. M. and Levy, A. B. (1972). *J. Amer. Chem. Soc.*, **94**, 3662
58. Hooz, J., Bridson, J. N., Calzada, J. G., Brown, H. C., Midland, M. M. and Levy, A. B. (1973). *J. Org. Chem.*, **38**, 2574
59. Midland, M. M. and Brown, H. C. (1973). *J. Amer. Chem. Soc.*, **95**, 4069
60. Pasto, D. J. and Wojtkowski, P. W. (1972). *J. Organometal. Chem.*, **34**, 251
61. Suzuki, A., Miyaura, N., Abiko, S., Itoh, M., Brown, H. C., Sinclair, J. A. and Midland, M. M. (1973). *J. Amer. Chem. Soc.*, **95**, 3080
62. Naruse, M., Utimoto, K. and Nozaki, H. (1973). *Tetrahedron Lett.*, 1847
63. Pelter, A., Harrison, C. R. and Kirkpatrick, D. (1973). *J. Chem. Soc. Chem. Commun.*, 544
64. Pelter, A., Harrison, C. R. and Kirkpatrick, D. (1973). *Tetrahedron Lett.*, 4491
65. Naruse, M., Tomita, T., Utimoto, K. and Nozaki, H. (1973). *Tetrahedron Lett.*, 795
66. Naruse, M., Utimoto, K. and Nozaki, H. (1973). *Tetrahedron Lett.*, 2741
67. Utimoto, K., Uchida, K. and Nozaki, H. (1973). *Tetrahedron Lett.*, 4527
68. Zweifel, G. and Horng, A. (1973). *Synthesis*, 672
69. Hancock, K. G. and Kramer, J. D. (1973). *J. Amer. Chem. Soc.*, **95**, 6463
70. Negishi, E. and Yoshida, T. (1973). *J. Amer. Chem. Soc.*, **95**, 6837
71. Brown, H. C. and Gupta, S. K. (1972). *J. Amer. Chem. Soc.*, **94**, 4370
72. Larock, R. C., Gupta, S. K. and Brown, H. C. (1972). *J. Amer. Chem. Soc.*, **94**, 4371
73. Larock, R. C. and Brown, H. C. (1972). *J. Organometal. Chem.*, **36**, 1
74. Brown, H. C., Hamaoka, T. and Ravindran, N. (1973). *J. Amer. Chem. Soc.*, **95**, 5786
75. Brown, H. C., Hamaoka, T. and Ravindran, N. (1973). *J. Amer. Chem. Soc.*, **95**, 6456
76. Negishi, E., Katz, J.-J. and Brown, H. C. (1972). *Synthesis*, 555
77. Zweifel, G. and Fisher, R. P. (1972). *Synthesis*, 557
78. Corey, E. J. and Ravindranathan, T. (1972). *J. Amer. Chem. Soc.*, **94**, 4013
79. Negishi, E. and Yoshida, T. (1973). *J. Chem. Soc. Chem. Commun.*, 606
80. Zweifel, G., Fisher, R. P., Snow, J. T. and Whitney, C. C. (1972). *J. Amer. Chem. Soc.*, **94**, 6560
81. Mukaiyama, T., Inomata, K. and Muraki, M. (1973). *J. Amer. Chem. Soc.*, **95**, 967
82. Inomata, K., Muraki, M. and Mukaiyama, T. (1973). *Bull. Chem. Soc. Jap.*, **46**, 1807
83. Hooz, J. and Bridson, J. N. (1972). *Can. J. Chem.*, **50**, 2387
84. Brown, H. C. and Carlson, B. A. (1973). *J. Org. Chem.*, **38**, 2422
85. Brown, H. C., Katz, J.-J. and Carlson, B. A. (1973). *J. Org. Chem.*, **38**, 3968
86. Pelter, A., Hutchings, M. G. and Smith, K. (1973). *J. Chem. Soc. Chem. Commun.*, 186
87. Brown, H. C. and Carlson, B. A. (1973). *J. Organometal. Chem.*, **54**, 61
88. Carlson, B. A. and Brown, H. C. (1973). *J. Amer. Chem. Soc.*, **95**, 6876
89. Carlson, B. A. and Brown, H. C. (1973). *Synthesis*, 776
90. Hooz, J. and Layton, R. B. (1972). *Can. J. Chem.*, **50**, 1105
91. Mukaiyama, T., Yamamoto, S. and Shiono, M. (1972). *Bull. Chem. Soc. Jap.*, **45**, 2244
92. Brown, H. C. and Yamamoto, Y. (1972). *J. Chem. Soc. Chem. Commun.*, 71
93. Brown, H. C. and Lane, C. F. (1972). *Synthesis*, 303

94. Brown, H. C., Yamamoto, Y. and Lane, C. F. (1972). *Synthesis*, 304
95. Yamamoto, Y. and Brown, H. C. (1973). *J. Chem. Soc. Chem. Commun.*, 801
96. Brown, H. C. and Yamamoto, Y. (1972). *Synthesis*, 699
97. Miyaura, N., Itoh, M., Suzuki, A., Brown, H. C., Midland, M. M. and Jacob, P. (1972). *J. Amer. Chem. Soc.*, **94**, 6549
98. Davies, A. G., Roberts, B. P. and Scaiano, J. C. (1972). *J. Chem. Soc. Perkin Trans. II*, 803
99. Suzuki, A., Miyaura, N., Itoh, M., Brown, H. C. and Jacob, P. (1973). *Synthesis*, 305
100. Kabalka, G. W. (1973). *Tetrahedron*, **29**, 1159
101. Utimoto, K., Tanaka, T., Furubayashi, T. and Nozaki, H. (1973). *Tetrahedron Lett.*, 787
102. Uriarte, A. K. and Hancock, K. G. (1973). *Inorg. Chem.*, **12**, 1428
103. Hancock, K. G. and Dickinson, D. A. (1972). *J. Amer. Chem. Soc.*, **94**, 4396
104. Hancock, K. G. and Dickinson, D. A. (1972). *J. Chem. Soc. Chem. Commun.*, 962
105. Hancock, K. G. and Kramer, J. D. (1973). *J. Amer. Chem. Soc.*, **95**, 3425
106. Miyamoto, N., Isiyama, S., Utimoto, K. and Nozaki, H. (1973). *Tetrahedron*, **29**, 2365
107. Utimoto, K., Tanaka, T. and Nozaki, H. (1972). *Tetrahedron Lett.*, 1167
108. Mehrotra, I. and Devaprabhakara, D. (1973). *J. Organometal. Chem.*, **51**, 93
109. Mikhailov, B. M., Smirnov, V. N. and Ryazonova, O. D. (1972). *Dokl. Akad. Nauk SSSR*, **204**, 612
110. Mikhailov, B. M. and Bryantsev, B. I. (1973). *Zh. Obshch. Khim.*, **43**, 1102
111. Mikhailov, B. M., Bryantsev, B. I. and Kozminskaya, T. K. (1973). *Zh. Obshch. Khim.*, **43**, 1108
112. Mikhailov, B. M., Bryantsev, B. I. and Kozminskaya, T. K. (1972). *Dokl. Akad. Nauk SSSR*, **203**, 837
113. Mikhailov, B. M. and Cherkasova, K. L. (1972). *Zh. Obshch. Khim.*, **42**, 138
114. Mikhailov, B. M. and Cherkasova, K. L. (1972). *Zh. Obshch. Khim.*, **42**, 1744
115. Mikhailov, B. M., Smirnov, V. N. and Prokof'ev, E. P. (1972). *Dokl. Akad. Nauk SSSR*, **206**, 125
116. Bubnov, Yu. N., Grigoryan, M. Sh. and Mikhailov, B. M. (1972). *Zh. Obshch. Khim.*, **42**, 1738
117. Bubnov, Yu. N., Nesmeyanova, O. A., Rudashevskaya, T. Yu., Mikhailov, B. M. and Kazanskii, B. A. (1973). *Zh. Obshch. Khim.*, **43**, 127
118. Bubnov, Yu. N., Nesmeyanova, O. A., Rudashevskaya, T. Yu., Mikhailov, B. M. and Kazanskii, B. A. (1973). *Zh. Obshch. Khim.*, **43**, 135
119. Mikhailov, B. M., Bubnov, Yu. N., Nesmeyanova, O. A., Kiselev, V. G., Rudashevskaya, T. Yu. and Kazanskii, B. A. (1972). *Tetrahedron Lett.*, 4627
120. Rathke, M. W. and Kow, R. (1972). *J. Amer. Chem. Soc.*, **94**, 6854
121. Kow, R. and Rathke, M. W. (1973). *J. Amer. Chem. Soc.*, **95**, 2715
122. Zweifel, G., Fisher, R. P. and Horng, A. (1973). *Synthesis*, 37
123. Buhler, J. D. and Brown, H. C. (1972). *J. Organometal. Chem.*, **40**, 265
124. Larock, R. C. (1973). *J. Organometal. Chem.*, **61**, 27
125. Schäfer, H. and Koch, D. (1972). *Angew. Chem. Int. Ed. Engl.*, **11**, 48
126. Taguchi, T., Itoh, M. and Suzuki, A. (1973). *Chem. Lett.*, 719
127. Pasto, D. J., Lepeska, B. and Cheng, T. C. (1972). *J. Amer. Chem. Soc.*, **94**, 6083
128. Pasto, D. J., Lepeska, B. and Balasubramaniyan, V. (1972). *J. Amer. Chem. Soc.*, **94**, 6090
129. Brindley, P. B. and Hodgson, J. C. (1972). *J. Chem. Soc. Chem. Commun.*, 202
130. Korcek, S., Watts, G. B. and Ingold, K. U. (1972). *J. Chem. Soc. Perkin Trans. II*, 242
131. Davies, A. G., Maki, T. and Roberts, B. P. (1972). *J. Chem. Soc. Perkin Trans. II*, 744
132. Ingold, K. U. (1973). *J. Chem. Soc. Perkin Trans. II*, 420
133. Abufhele, M., Andersen, C., Lissi, E. A. and Sanhueza, E. (1972). *J. Organometal. Chem.*, **42**, 19

Index

Alkenes *continued*
  electrophilic additions to, 84, 85
  epoxidation, metal ion-catalysed, 154, 155
  halogenation, 153
  cis-hydroxylation, 134
  ketone synthesis from, 304, 305
  nitro-, reactions with phosphenes, 224
  nucleophilic additions to, 85, 86
  organoboron compounds in preparation
    of, 381
  peroxymercuriation, 156
  preparation, 16–18
    from 1,2-diols, 136–138
    from epoxides, 148, 149
    from ketones, 270
  radical additions to, 86, 87
  reaction with dithiocarboxylic acids, 281
    with peroxy compounds, 143, 144
  structure, calculation, 5
  trans-, synthesis, 372
Alkoxyhalogenation
  alkenoic acids, 346
Alkylation
  aldehydes, 306, 307
  alkanes, 11
  amines, 197–199
  aromatic hydrocarbons with alkanes, 15
  β-diketones, 313, 314
  elemental phosphorus, 226
  β-keto acids, 317
  ketones, 306, 307
  N-, 177, 178
  nitro compounds, 170
Alkyl groups
  migration from boron, 374, 375
Alkyl halides (*see also* Bromides; Fluorides;
  Iodides)
  organophosphorus derivatives, 236–238
  preparation, 84–88
  reactions, 92–100
    with trialkyl phosphites, 225
Alkyl nitrates
  β-nitro-, reactions, 173
Alkynes
  acids, 335, 336
  carbonyl compound derivatives, prepa-
    ration, 309, 310
  co-oligomerisation with buta-1,3-diene,
    58
  electrophilic additions to, 88
  1-halogeno-, preparation, 89
    reactions, 102, 103
  nucleophilic addition to, 88, 89
  organoboron compounds in preparation
    of, 381
  organophosphorus derivatives, 243, 244
  preparation, 66
  radical addition to, 89
  reactions, 67–73
    with phosphenes, 223, 224

Allenes
  acids, synthesis, 332, 333
  light-induced cycloaddition to benzene,
    45
  organoboron compounds in preparation
    of, 381
  organophosphorus derivatives, 242–245
  oxidation, 47
  radical additions, 47
  reaction with ozone, 45
  transition-metal catalysed reactions, 48
Allene, tetramethyl-
  photo-oxygenation, 45
Allyl alcohol
  cis-diol preparation from, 134, 135
  2,3-epoxypropanal preparation from, 154
Allyloxy groups
  alcohol protection by, 133
Aluminates
  alkoxyhydrido-, monohydric alcohol
    preparation by, 113
  tetrahydro-, lithium salt, chiral reagents
    from, 121
Aluminium
  alkyls, asymmetric reduction of carbonyl
    compounds with, 121
  compounds, catalysts in diene reactions,
    62
  organophosphorus derivatives, reactions,
    227, 228
— —, trialkyl-
  reaction with epoxides, 152
— —, trimethyl-
  alcohol reaction with, 134
Aluminium halides
  reaction of alkyl halides with phosphorus
    halides and, 226
Aluminium hydride
  acyclic α-ketol reduction by, 135
Amides (*see also* specific amides)
  dehydration to nitriles, 174
  organoboron compounds in preparation
    of, 381
  organophosphorus derivatives, 230–232
  preparation from nitriles, 176
  reactions, 209, 210
  secondary, 189
  synthesis, 208
  thio-, dehydration to nitriles, 174
    preparation, 271
    reactions, 282
Amidines, 190
Amidoximes
  isocyanates, 184
Amidrazones
  alkylation, 208
Amines
  basicity, 201
  chiral, 201
  N-chloro, 205